RACE, CLASS, and GENDER

An Anthology
Second Edition

Margaret L. Andersen
University of Delaware

Patricia Hill Collins
University of Cincinnati

Wadsworth Publishing Company
I(T)P ™ An International Thomson Publishing Company

Belmont • Albany • Bonn • Boston • Cincinnati • Detroit • London • Madrid • Melbourne
Mexico City • New York • Paris • San Francisco • Singapore • Tokyo • Toronto • Washington

Editorial Assistant: Jason Moore
Production Editor: Karen Garrison
Managing Designer: Andrew Ogus
Print Buyer: Barbara Britton
Permissions Editor: Marion Hansen
Designer: Cynthia Schultz
Compositor: Omegatype Typography
Signing Representative: Ron Shelly
Printer: Arcata Graphics/Fairfield
Cover Art: Hans Hofmann, *Summer Night's Dream*, 1958, oil on canvas, 52 x 60 1/4.
Albright-Knox Art Gallery, Buffalo, New York. Gift of Seymour H. Knox, 1958.

This book is printed on
acid-free recycled paper

For more information, contact Wadsworth Publishing Company:

Wadsworth Publishing Company
10 Davis Drive
Belmont, California 94002, USA

International Thomson Publishing
Berkshire House 168-173
High Holborn
London, WC1V 7AA, England

Thomas Nelson Australia
102 Dodds Street
South Melbourne 3205
Victoria, Australia

Nelson Canada
1120 Birchmount Road
Scarborough, Ontario
Canada M1K 5G4

International Thomson Editores
Campos Eliseos 385, Piso 7
Col. Polanco
11560 México D.F. México

International Thomson Publishing GmbH
Königwinterer Strasse 418
53227 Bonn, Germany

International Thomson Publishing Asia
221 Henderson Road
#05-10 Henderson Building
Singapore 0315

International Thomson Publishing Japan
Hirakawacho Kyowa Building, 3F
2-2-1 Hirakawacho
Chiyoda-ku, Tokyo 102, Japan

3 4 5 6 7 8 9 10—01 00 99 98 97 96 95

Library of Congress Cataloging-in-Publication Data
Race, class, and gender : an anthology / [compiled by] Margaret L.
  Andersen, Patricia Hill Collins. — 2nd ed.
     p.   cm.
  Includes bibliographical references (p.   ).
  ISBN 0-534-24768-7 (acid-free paper)
     1. United States—Social conditions—1980– 2. United States—Race
relations. 3. Social classes—United States. 4. Sex role—United
States. 5. Homosexuality—United States. 6. Discrimination—United
States. I. Andersen, Margaret L. II. Collins, Patricia Hill.
HN59.2.R32   1994
305—dc20   94-19800

Contents

Preface

TOWARD INCLUSIVE THINKING THROUGH THE STUDY OF RACE, CLASS, AND GENDER

This book analyzes the interrelationship of race, class, and gender and how these structures have shaped the experiences of all people in the United States. Race, class, and gender are interlocking categories of experience that affect all aspects of human life; they simultaneously structure the experiences of all people in this society. Although at times race, class, or gender may feel more salient or meaningful in a given person's life, they are overlapping and cumulative in their effect on people's experience.

Because of this simultaneity of race, class, and gender in people's lives, we conceptualize them as different, but interrelated, axes of social structure. We use the approach of a matrix of domination to analyze them. In this matrix of domination there are multiple, interlocking levels of domination that stem from the societal configuration of race, class, and gender relations. This structural pattern affects individual consciousness, group interaction, and group access to institutional power and privileges (Collins 1990).

Approaching the study of race, class, and gender from this perspective is different from the usual additive model in which people study the independent effect of race, class, and gender on human experience. The additive model is reflected in terms like *double* and *triple jeopardy*—terms used to describe the oppression of women of color by race, as well as by gender and class. We do not think of race and gender oppression in simple additive terms, an implication of phrases like double and triple jeopardy. The effect of race, class, and gender does "add up"—both over time and in intensity of impact, but seeing race, class, and gender only in additive terms misses the social structural connections between them and the particular ways that different configurations of race, class, and gender affect group experience.

The additive model also tends only to look at the impact of race, class, and gender on the experience of their victims, such as women and people of color, rather than seeing them as an integral part of the social structures that shape the experience of all groups. The additive model also places people in either/or categories, as if one is either Black or White, oppressed or oppressor, powerful or powerless. We do not think that people can be dichotomized so easily. First, race is not a fixed category, as Michael Thornton clearly shows in his essay on mixed-race identity; rather, race is a socially constructed category whose meaning shifts over time. Second, the division between powerful and powerless, or oppressor and oppressed, is not clear-cut. One may be privileged by race but disempowered by virtue of gender; thus, White women may be disadvantaged because of gender but privileged by race and, perhaps, but not necessarily, by class. Third, race, class, and gender are manifested differently, depending on their configuration with the others. So, for example, while from an additive model, one might say Black men are privileged as men, this makes no sense when their race *and* gender *and* class are taken into account. Otherwise, how can we possibly explain the particular disadvantages African American men experience—in the criminal justice system, in education, and in the labor market? For that matter, how can we explain the experience of Native American women—disadvantaged by the particular and unique experiences that they have based on race, class, and gender—none of which are isolated from the effects of the other? Finally, White Americans are not in either/or categories—either oppressor or oppressed. Separating White Americans out as if their experience stands alone misses the way that White experience is intertwined with that of other groups.

Approaching the study of race, class, and gender from the perspective of the matrix of domination means that we also see our work somewhat differently than the language of "diversity" or "multiculturalism" suggests. Clearly, our work is part of these movements, but we find the framework of multiculturalism and diversity to be limited and therefore problematic. The concept of diversity has become the catchword for trying to understand the complexities of race, class, and gender in American society. The movement to promote diversity has made people more sensitive and aware of the intersections of race, class, and gender, and we are glad to be part of this. Thinking about diversity has also encouraged students and activists to see linkages to other forms of oppression, including sexuality, age, region, physical disability, national identity, and ethnicity, but "understanding diversity" is not the only point. The very term *diversity* implies that understanding race, class, and gender means only recognizing the plurality of views and experiences in society, as if race, class, and gender oppression were just the sum total of different experiences. We want readers to be more analytical. Analyzing race, class, and gender is more than "appreciating cultural diversity." It requires analysis and criticism of existing systems of power and privilege; otherwise, understanding diversity becomes just one more privilege for those with the most access to education—something that has always been a mark of the elite class. Analyzing race, class, and gender as they shape different group experiences is also about power, privilege, and equity. This means more than just knowing the culture of an array of human groups. It means recognizing and analyzing the hierarchies and systems of domination that permeate society and that systematically exploit and control people.

There has been a tendency in recent years to think of race, class, and gender in terms of a voices metaphor—a metaphor that appropriately comes from challenging the silence that has surrounded many group experiences. In this framework, people think about diversity as "listening to the voices" of a multitude of previously silenced groups. This is an important part of coming to understand race, class, and gender, but it is not enough. First, people may begin hearing the voices as if they were disembodied from particular historical and social conditions—a framework that has been exacerbated by postmodernist trends in feminist theory. This perspective sees experience as a matter of competing discourses, personifying "voice" as if the voice or discourse in itself constituted lived experience. Second, the voices framework makes it seem that any analysis is incomplete unless every voice is heard. In a sense, of course, this is true, since inclusion of silenced people is one of the goals of multicultural work. But, an analysis grounded in race, class, and gender can be complete even if centered in the experience of a single group—as long as that group's experience is situated within a framework that recognizes the influence of race, class, and gender on group experience.

Finally, we see limits to the multicultural movement because we think it is important to challenge the idea that diversity is important only at the level of culture—an implication of a term like *multiculturalism*. Culture is traditionally defined as the "total way of life" of a group of people. It includes material and symbolic culture and is an important dimension of understanding human life. Analysis of culture per se, however, looks at the group itself, not necessarily the broader conditions within which the group lives. Of course, as anthropologists know, a sound analysis of culture situates group experience within these external conditions. Still, the focus on culture frequently ignores social conditions of power, privilege, and prestige. The result is that multicultural studies often seem tangled up with cultural pluralism—as if knowing a culture other than one's own is the goal of multicultural education. We think it is important to see the diversity and plurality of different cultural forms, but this perspective in and of itself misses the broader point of understanding how racism, sexism, homophobia, and class relations have shaped the experience of groups. Imagine, for example, that we tried to study the oppression of gays and lesbians in terms of gay culture only. It becomes immediately obvious that doing so turns attention onto the group itself, instead of the dominant society. Likewise, studying race in terms of Latino culture, or Asian-American culture, or African-American culture, or studying women's oppression only by looking at women's culture, encourages thinking that blames the victims for their own oppression and ignores structural inequalities and systematic oppression.

For all of these reasons, the focus of this book is on the institutional, or structural, bases for race, class, and gender relations. We want readers to conceptualize race, class, and gender as interactive systems, not just as separate features of experience, variables in sociological equations, voices, or cultural experiences. We want readers to see race, class, and gender in an analytical and not just a descriptive way, although we recognize that description of group experience, both historically and now, is an important part of this process.

We have organized the book to follow this thinking. Part One, "Shifting the Center and Reconstructing Knowledge," contains personal reflections on the ways race, class, and gender shape individual and collective experiences. These articles provide a fresh beginning point for social thought because they put those who have traditionally been excluded at the center of our thought. These different accounts are not meant to represent everyone who has been excluded but to ground readers in thinking about the lives of people who are rarely put at the center of dominant group thinking. When excluded groups are put in the forefront of thought, they are typically stereotyped and distorted. When you look at society through the different perspective of those who are reflective about their location within systems of race, class, and gender oppression, you begin to see just how much race, class, and gender really matter.

In Part Two, "Conceptualizing Race, Class, and Gender," we provide a basic conceptual grounding for the three major concepts being examined—race, class, and gender. We separate them here only for analytical purposes. Although we do not see them as separate systems, we treat them individually so that students will understand how each operates and then be better able to see their interlocking nature. We also want students to see the continuing effects of race, class, and gender on people's experiences; our introduction to this section and the articles within document those effects and point to the interrelationships between them.

In Part Three, "Rethinking Institutions," we examine how race, class, and gender shape the organization of social institutions and, as a result, how these institutions affect group experience. Social scientists routinely document how different race, class, gender groups are affected by institutional structures. We know this is true, but we want to go beyond these analyses by also showing how institutions are constructed through race, class, and gender relations. It is not just that institutions affect people differently, but that the institutions themselves are constructed through race, class, and gender relations. As contemporary feminist scholars argue, institutions are "gendered" (Acker 1993), "racially formed" (Omi and Winant 1986), and organized as class systems.

Part Four, "Analyzing Social Issues," uses the inclusive perspective gained from understanding race, class, and gender to analyze contemporary social issues. We examine sexuality, violence, and American identity—social issues that are hotly contested in the public discourse and that we think can only be understood if they are located within a race, class, and gender analysis. Indeed, we think that these social issues are hotly contested *because* they evolve from contemporary race, class, and gender politics; thus, we think that understanding contemporary issues is misguided if we do not situate them within the context of race, class, and gender relations. While specific social issues may change from time to time, contemporary discussions about what constitutes American identity and culture, about sexual politics, and about the causes and consequences of violence are framed by implicit and explicit assumptions that stem from competing ideologies of race, class, and gender; therefore, understanding the social issues on the public agenda requires an analysis of race, class, and gender relations.

Finally, in Part Five, "Social Change and the Politics of Empowerment," we look at social change. When thinking about race, class, and gender, it is all too easy to think only in terms of how people are victimized and oppressed by these systems. We have avoided a social problems approach because we want readers to move away from thinking only in a problem-centered framework. Race, class, and gender are indeed the basis for many social problems, but a problems-based approach tends to portray people primarily as victims, while ignoring their independent views of themselves and their society. A problems-based approach also tends to see oppressed groups only through the perspective of the more privileged, relegating those who most suffer under race, class, and gender oppression to the status of "others." This reproduces the hierarchical viewpoints that have permeated traditional thinking. Limiting yourself to a problems-centered or victims framework also leaves you feeling that you have no power to make change in the society.

People are not just victims; they are creative and visionary. As a result, people organize to resist oppression and to make liberating social changes. In fact, oppression generates resistance. In Part Five we examine the meaning of activism and its connection to the conditions in which people live. This section also includes articles by writers whose ideas are both inclusive and visionary. We see these articles as providing the ideals for change that inform the spirit of this book. Change takes place through new actions, new social structures, and new meaning systems, not just through existing channels, but unless you have an inclusive framework from which to make change, your action is limited. Resistance to oppression on behalf of one's own group is not enough. Truly multicultural change takes place only when people and groups build coalitions with others. Change also takes place over the long run. Short-term actions are needed, but the long-term, institutional changes needed must be sustained by new visions; thus, we include in this final section the statements of several activists who provide a visionary approach to the future.

In selecting articles for the anthology, we tried to find those that would show the intersections of race, class, and gender. Not every piece does this, and we consider it equally important for students to learn to see connections between groups and social structural conditions even when they may not be obvious. We searched for articles that are conceptually and theoretically informed, and also accessible to undergraduate readers. Although it is important to think of race, class, and gender as analytical categories, we do not want to lose sight of how they affect human experiences and feelings; thus, we have included personal narratives that are reflective and analytical. We think that personal accounts generate empathy and also help us connect personal experiences to social structural conditions.

We have tried to be as inclusive of all groups as possible, representing the richness of difference and diversity within the United States. It is impossible to include all historically and presently marginalized groups in one book, so we encourage readers to explore the many other works available. We have selected materials that explain the relationships among race, class, and gender in ways that illuminate the experiences of many groups, not just those about whom an article is specifically written. As we begin to untangle the structure of race, class, and gender relations, we can better see both the commonalities and differences that various historical experiences have generated.

We also want students to learn that thinking about race, class, and gender is not just studying victims. We remind students that race, class, and gender have affected the experiences of all groups. As a result, we do not think we should talk only about women when talking about gender, or about people of color when talking about race. Because race, class, and gender affect the experience of all, it is important to study men when analyzing gender, to study Whites when analyzing race, and to study the experience of all classes when analyzing class. If we are thinking in an inclusive way, we will think about women and not just men when studying race, Latinos and people of color when thinking about class, and women and men of color when studying gender.

In addition, we should not relegate the study of racial-ethnic groups, the working class, and women only to subjects marked explicitly as race, class, or gender studies. As categories of social experience, race, class, and gender shape all social institutions and systems of meaning; thus, it is important to think about people of color, different class experiences, and women in analyses of all social institutions and belief systems.

Developing truly inclusive thinking and teaching is a long-term process, one involving both personal, intellectual, and political change. We do not claim to be models of perfection in this regard. We have been pleased by the strong response to the first edition of this book and we have found it fascinating to see how race, class, and gender studies have developed in the short period of time since the publication of the first edition. We were pleased to find more articles on African-American men for the second edition that were grounded in a race, class, and gender analysis. We were also pleased to see so much new work on Asian-American and Latino experience, but we know more work is needed. Our own teaching and thinking has been transformed by the process of developing this book; we imagine many changes still to come.

ABOUT THE SECOND EDITION

We have been pleased by the success of the first edition of *Race, Class, and Gender* because it represents the commitment of many people to become more inclusive in their teaching and thinking. We appreciate the reactions people have shared with us, particularly those who have reviewed the book for Wadsworth. We learned from these teachers that we needed more historical grounding in the second edition and that our analysis of class was not strong enough. We have addressed these by including more pieces with a historical perspective, and we have totally revised the section on class. We have also written longer analytical introductions to each part so that readers will get some of the basic concepts and theoretical orientation that the articles do not always directly provide; we think these essays give readers a stronger conceptual grounding and an understanding of what we are trying to accomplish. While this one book cannot do everything people want, we hope that the second edition furthers people's work in this area and assists teachers in the hard work of building a more inclusive curriculum.

The second edition differs from the first in the following ways. First, we have reorganized the book to include the new section on analyzing social issues. Second, we have included more articles that provide a historical perspective. We have also included longer analytical introductions by us; these are intended to enrich the analytical framework of the book. We have added a new section on sexuality, but because we do not want readers to conclude that heterosexual privilege is only significant when thinking about sexuality per se, we have included articles about gay and lesbian experience throughout the book. We have revised the section on class to include more discussion of different dimensions of the class system. Finally, we have tried to find a balance of different group experiences. Although not every group can be covered on every subject, we hope that the inclusion of different groups conveys a sense of the multiple, but related, ways that race, class, and gender shape American society and culture. We have maintained the theme of linking personal narratives with the analytical study of race, class, and gender in the second edition.

The second edition also includes an Instructor's Manual that shows the different ways that *Race, Class, and Gender* can be used in different courses. The Instructor's Manual includes suggestions for class exercises, discussion and examination questions, and course assignments. We appreciate the suggestion of earlier reviewers that we include such a manual and thank Martha Thompson for her inspired work in developing it. We also thank the different faculty members who contributed materials for the Instructor's Manual. We think it will be a valuable pedagogical aid for those teaching with this book.

A NOTE ON LANGUAGE

The reconstruction of existing ways of thinking to become more inclusive requires many transformations. One transformation needed is in the language we use to describe and define different groups. The language we use reflects many assumptions about race, class, and gender, and for that reason, language changes and evolves as knowledge changes. The term *minority*, for example, marginalizes groups, making them seem somehow outside the mainstream or majority culture. Even worse, the phrase *non-White*, routinely used by social scientists, defines groups in terms of what they are not and assumes Whites to have the universal experiences against which all other groups are measured. We have consciously avoided use of both terms throughout this book.

We have capitalized *Black* in our writing because of the specific historical experience, varied as it is, of African Americans in the United States. We also capitalize *White* when referring to a particular group experience; however, we recognize that "White American" is no more a uniform experience than is "African American." We realize these are arguable points, but we wanted to make our decision apparent and explicit. For those who are "purists" and like to follow the rules, we are also pleased to see that the fourteenth edition of the *Chicago Manual of Style* recognizes that these grammatical questions are also political ones. This most recent edition suggests that writers might want to capitalize Black and White to reflect the fact that one is speaking about properly named groups.

Language becomes especially problematic when we want to talk about features of experience that different groups share. Using shortcut terms, like *Hispanic, Latino, Native American,* and even *women of color,* homogenizes distinct historical experiences. Even the term *White* falsely unifies experiences across such factors as ethnicity, region, class, and gender, to name a few. At times, though, we want to talk of common experiences across different groups and so we have used labels like "Latino" and "women of color" to do so. Unfortunately, describing groups in this way reinforces basic categories of oppression. We do not know how to resolve this problem, but we want readers to be aware of the limitations and significance of language as they try to think more inclusively about diverse group experiences.

ACKNOWLEDGMENTS

An anthology rests on the efforts of more people than the editors alone. This book has been inspired by our work with scholars and teachers around the country who are working to make their teaching and writing more inclusive and sensitive to the experiences of all groups. Over the years of our own collaboration, we have each been enriched by the work of those trying to make higher education a more equitable and fair institution. This book grows from several of those projects, most notably the Memphis State Center for Research on Women Curriculum Workshops and the American Sociological Association Minority Opportunity Summer Training Program, now the Minority Opportunity Through School Transformation Program. Quite literally, these two programs provided the context for many of our discussions, as well as places to work together. We thank Maxine Baca Zinn, Chuck Bonjean, Marion Coleman, Bonnie Thornton Dill, Elizabeth Higginbotham, Clarence Lo, Lionel Maldonado, Carole Marks, Cora Marrett, Howard Taylor, and Lynn Weber for providing the space and time in which to work. More importantly, we thank them for the companionship, encouragement, and vision that inspires our work. In addition, we also want to thank the Wadsworth reviewers who provided such detailed and extensive commentary on the first edition of the book: J. Q. Adams, Western Illinois University; Tom Bidell, Boston College; Rose Brewer, University of Minnesota; Karen Hansen, Brandeis University; Lee Harrington, Miami University, Ohio; Barbara Heyl, Illinois State University, Normal; Dennis Kalob, Loyola University, New Orleans; Ligaya McGovern, Ferris State University; Ruth Ray, Wayne State University; Patricia Turner, University of California, Davis; Lynn Weber, Memphis State University; Karen Wolford, SUNY, Oswego; Shirley Yee, University of Washington. We have taken their suggestions for revision seriously and thank them for their commitment to more inclusive and progressive teaching.

Many other people contributed to the development of this book. We thank Heather Smith and Susan Freeman for assisting with library work and helping us locate articles. Tina Dunhour and Rachel Levy, now graduates of the University of Delaware, advised us from students' perspectives about the articles we selected. We appreciate the support provided by our two institutions, with special thanks to Richard B. Murray, Provost, University of Delaware; and Joseph Caruso, Dean of the College of Arts and Sciences, and Tony Perzigan, former Head of the Department of African-American Studies, and Norman Harris, Head of the Department of African-American Studies, University of Cincinnati, for their support. We also thank Linda Prusak, Anna Marie Brown, Nancy Benderoth, Sadie Wright Oliver, and Roderick W. Williams for their invaluable secretarial support. We hope they know how much we value them.

We are fortunate to have an editor who is unwavering in her enthusiasm for this book. Serina Beauparlant has given us extensive guidance, assistance, and support throughout the project; we thank her for everything she has done. Susan Shook coordinated the development of the Instructor's Manual and we thank her for her commitment to this project. We also thank Jason Moore, Karen Garrison, and the entire production staff at Wadsworth; they all contributed to the success of *Race, Class, and Gender* and we appreciate their help.

Developing this book has been an experience based on friendship, hard work, travel, and fun. We especially thank Valerie, Patrice, Roger, and Richard for giving us the support, love, and time we needed to do this work well. This book has deepened our friendship, as we have grown more committed to transformed ways of thinking and being. We are lucky to be working on a project that draws us closer together.

References:

Acker, Joan. 1992. "Gendered Institutions: From Sex Roles to Gendered Institutions." *Contemporary Sociology* 21 (September): 565–569.

Collins, Patricia Hill. 1990. *Black Feminist Thought: Knowledge, Consciousness, and Empowerment.* Cambridge: Unwin and Hyman.

Omi, Michael, and Howard Winant. 1986. *Racial Formation in the United States: From the 1960s to the 1980s.* New York: Routledge and Kegan Paul.

About the Editors

Margaret L. Andersen is Professor of Sociology and Women's Studies at the University of Delaware, where she also serves as Vice Provost for Academic Affairs. She is the author of *Thinking About Women* (Macmillan), *Social Problems* (with Frank R. Scarpitti), and a forthcoming book, *Sociology: Understanding a Diverse Society* (with Howard F. Taylor). She currently serves as Editor of *Gender & Society* and is on the Council of the American Sociological Association.

Patricia Hill Collins is Professor of African-American Studies and Sociology at the University of Cincinnati. She is the author of *Black Feminist Thought: Knowledge, Consciousness, and the Politics of Empowerment*, which won the C. Wright Mills Award in 1991. She serves on the Council of the American Sociological Association.

About the Contributors

Paula Gunn Allen is Professor of English at UCLA. She was awarded the Native American Prize for Literature in 1990. That same year her anthology of short stories, *Spider Woman's Granddaughters*, was awarded the American Book Award, sponsored by the Before Columbus Foundation, and the Susan Koppleman Award. A major Native American poet, writer, and scholar, she's published seven volumes of poetry, a novel, a collection of essays, and two anthologies. Her prose and poetry appear widely in anthologies, journals, and scholarly publications.

Teresa Amott is Associate Professor of Economics at Bucknell University. She is the author of *Race, Gender, and Work: A Multicultural Economic History of Women in the United States* (with Julie Matthaei, 1991) and *Caught in the Crisis: Women in the U.S. Economy Today* (1993), along with numerous articles. She is an Editorial Associate with *Dollars and Sense* magazine, and is committed to sharing economic analysis with unions, welfare rights organizations, women's organizations, and other progressive groups.

Elijah Anderson is the Charles and William L. Day Professor of the Social Sciences at the University of Pennsylvania. An expert on the sociology of black America, he is the author of the widely regarded sociological work, *A Place on the Corner: A Study of Black Street Corner Men* (1978) and numerous articles on the black experience. For his recently published ethnographic study, *Streetwise:*

Note: Not all contributors provided biographical information.

Race, Class and Change in an Urban Community (1990), he was honored with the Robert E. Park Award of the American Sociological Association. He has also won the Lindback Award for Distinguished Teaching at Penn. He is associate director of Penn's Center for Urban Ethnography and associate editor of *Qualitative Sociology*. Other topics with which he concerns himself are the social psychology of organizations, field methods of social research, social interaction, and social organization. He received a B.A. degree from Indiana University, an M.A. degree from the University of Chicago, and a Ph.D. degree from Northwestern University, where he was a Ford Foundation Fellow.

Evelyn Torton Beck is Professor and Director of the Women's Studies Program at the University of Maryland–College Park and a member of the Jewish Studies Program. She is the author of *Kafka and the Yiddish Theater* (1971), *The Prism of Sex: Essays in the Sociology of Knowledge* (with Julia Sherman, 1979), and *Nice Jewish Girls: A Lesbian Anthology* (1982, rev. ed. 1989). She has lectured and written widely on Jewish women's studies, lesbian studies, anti-Semitism in the women's movement, feminist transformations of knowledge, and feminist pedagogy.

Peter Blood is a marriage and family therapist for the Institute for Christian Counseling and Therapy. He writes about and leads retreats focusing on men's issues. He is the editor of *Rise Up Singing* and is editorial director for the *Sing Out* publication.

Olivia Castellano is Professor of English at California State University, Sacramento, where she teaches composition, Chicano literature, and creative writing. She was a Stanford Fellow in Modern Thought and Literature in 1976. She has published three collections of poetry: *Blue Mandolin–Yellow Field* (1980), *Blue Horse of Madness* (1983), and *Spaces That Time Missed* (1986). Her fourth book of poetry, *Thank God the Moon is Forgiving*, is ready for publication. She is working on a novel about growing up in Southwest Texas, tentatively titled *Borderghost* (or *Sotol: The Comstock Journals*).

Sucheng Chan is Professor of History and Chair of the Asian American Studies Program at the University of California, Santa Barbara. She is the author of *This Bittersweet Soil: The Chinese in California Agriculture, 1860–1910* (which won three awards), *Asian Californians*, and *Asian Americans: An Interpretive History*. She has also edited four books and written dozens of articles.

Sumi K. Cho is a Visiting Faculty Fellow at the University of Iowa College of Law, where she is pursuing her research on Asian Americans and Critical Race Theory. She is a graduate of the UC Berkeley Ethnic Studies doctoral program and Boalt Hall School of Law. As a campus/community organizer on issues of affirmative action, racial/gender discrimination, and sexual harassment, she hopes to bring a multidisciplinary approach and community-based orientation to the study of race, gender, and law.

Johnnetta B. Cole was named the first African-American woman president of Spelman College in 1987, the largest women's college in Georgia. Dr. Cole's scholarship centers on cultural anthropology, Afro-American studies, and women's studies.

Nancy Diao currently works in banking. She spent ten years as a community organizer in the San Francisco Bay Area Asian Community. She was born in Beijing, China.

Bonnie Thornton Dill earned her Ph.D. at New York University after working for a number of years in antipoverty and open admissions programs in New York. She is currently Professor of Women's Studies at the University of Maryland, College Park. She was the founding director of the Center for Research on Women at Memphis State University. She has contributed articles to such journals as *Signs, Journal of Family History*, and *Feminist Studies* and is co-editor with Maxine Baca Zinn of the book *Women of Color in American Society* for Temple University Press. She is also conducting research on single mothers, race, and poverty in the rural South with a grant from the Aspen Institute and the Ford Foundation.

Diana Dujon teaches welfare organizing and is an administrator at the College of Public and Community Service of the University of Massachusetts, Boston. She is a co-founder of Advocacy for Resources for Modern Survival, a Boston welfare-rights group.

Michael Eric Dyson, who has taught at Hartford Seminary, Chicago Theological Seminary, and Brown University, is presently a Professor of Communication Studies at the University of North Carolina, Chapel Hill. He is the author of the widely acclaimed *Reflecting Black: African-American Cultural Criticism* (1993). Dyson is presently writing a book on black males, *Boys to Men: Black Males in America*, to be published by Random House in 1997. He is co-editing an anthology of writings about Malcolm X's life with Wahneema Lubiano, entitled *Rethinking Malcolm X*, to be published by Basil Blackwell in

1995. And in the fall of 1994, his book, *Making Malcolm: The Myth & Meaning of Malcolm X*, will be published by Oxford University Press.

D. Stanley Eitzen is Professor of Sociology at Colorado State University. With Maxine Baca Zinn, he has written *Diversity in Families, In Conflict and Order*, and *Social Problems*. He has also written *Elite Deviance* (with David R. Simon), *Sociology of North American Sport* (with George H. Sage), *Criminology* (with Doug A. Timmer), and *Paths to Homelessness* (with Doug A. Timmer and Kathryn Talley).

Olivia M. Espín, Ph.D., is Professor of Women's Studies at San Diego State University and part-time Core Faculty at the California School of Professional Psychology. She has published on psychotherapy with Latinas, immigrant and refugee women, women's sexuality, and other topics. She has recently co-edited *Refugee Women and Their Mental Health: Shattered Societies, Shattered Lives*. Her book, *Power, Culture, and Tradition: The Lives of Latina Healers in Urban Centers in the United States* is forthcoming. Dr. Espín received a Distinguished Professional Contribution Award from the American Psychological Association in 1991.

Susan Faludi is a journalist and the author of *Backlash: The Undeclared War Against American Women* (1991). She has written for numerous magazines and newspapers, from the *San Jose Mercury News* to *Mother Jones* to *The Wall Street Journal*. She won the Pulitzer Prize in 1991 for her story in *The Wall Street Journal* about the human costs of a leveraged buyout of Safeway Stores. She is at work now on a book about the crisis of masculine identity facing American men.

Henry Louis Gates, Jr. is W.E.B. Du Bois Professor of the Humanities, Professor of English, and Chairman of the Department of Afro-American Studies at Harvard University. He writes for numerous publications, including *The New York Times Book Review, The New Yorker, The Nation, Harper's, The New Republic*, and *The Village Voice*, in addition to many scholarly journals. His books include *Colored People: A Memoir, Loose Canons, Figures in Black*, and *The Signifying Monkey*, for which he received the American Book Award; and he is the general editor of the forty-volume *Schomburg Library of Nineteenth Century Black Women Writers*, for which he received the Anisfield-Wolf Book Award.

Judy Gradford works with Transition House, a Cambridge, Massachusetts, battered-women's shelter. She is a co-founder of Advocacy for Resources for Modern Survival and a board member of DBHN.

Rayna Green is currently Director of the American Indian Program at the National Museum of American History, Smithsonian Institution. She has served on several faculties, public institutions, and nonprofit boards. She has written several books, including *That's What She Said: Contemporary Fiction and Poetry by Native American Women*, and *Native American Women: A Contextual Bibliography*, and many essays on American Indians and American culture. She is also known for her work in museum exhibitions, performance production, television, and radio.

Jacquelyn Dowd Hall earned her Ph.D. at Columbia University. She is Director of the Southern Oral History Program and Julia Cherry Spruill Professor of History at the University of North Carolina, Chapel Hill. She is currently at work on a study of class, race, and sexuality in the twentieth-century South.

Evelynn Hammonds earned her Ph.D. in the History of Science at Harvard University. She is Assistant Professor of the History of Science in the Program in Science, Technology, and Society at the Massachusetts Institute of Technology. She is currently working on a study of African-American women in the AIDS epidemic.

Elizabeth Higginbotham earned her doctorate in Sociology at Brandeis University. She is the Associate Director of the Center for Research on Women and Associate Professor in the Department Sociology and Social Work at The University of Memphis. She has published widely on issues of race, class, and gender. She is currently working with Lynn Weber on a research project, "Social Mobility, Race, and Women's Mental Health," which is a broad investigation of the role of upward mobility in the educational experiences, work life, family life, well-being, and mental health for 200 Black and White professional and managerial women in the Memphis area.

Robert A. Hummer is a doctoral candidate in Sociology at Florida State University. His areas of interest include social stratification and demography. Current projects include the specification of processes that lead to infant mortality differentials across race/ethnic groups and the understanding of health and mortality differentials between women and men.

June Jordan is Professor of African American Studies at University of California, Berkeley and the author of 21 books, to date. Her most recent titles include *Haruko/Love Poems* and *Technical Difficulties*. She is a regular political columnist for *The Progressive* and she has just completed a new opera in collaboration with the composer, John Adams, and the director, Peter Sellars.

Kathleen Kautzer wrote her Ph.D. dissertation on the Older Women's League for Brandeis University. She has been a union organizer of clerical workers and is the daughter of a displaced homemaker.

Tracy A. M. Lai teaches history/coordinated studies, with an emphasis on ethnic and women studies, at Seattle Central Community College. Her writing comes out of her involvement in the Asian community, student, labor and international support movements. She is active in Sisters of Color International, Unity Organizing Committee, and Asian Pacific American Labor Alliance.

George Lakey is a consultant and trainer, an activist and college teacher, and author of five books. He has conducted over 1,000 workshops on five continents and was already leading men's conferences in the mid-1970s. He has initiated social change projects in civil rights, neighborhood development, economic conversion, and peace, and was a founder of the Movement for a New Society. An openly gay grandfather of four, he frequently lectures at major colleges and universities.

Donna Langston has worked a variety of jobs: as secretary, waitress, factory worker, and on an oil refinery crew. She is a Jewish lesbian who was raised and spent the majority of her adult life in an urban, working class/poor background. She is co-editor of *Changing Our Power* and Associate Professor of Women's Studies at Mankato State University.

Audre Lorde, who passed away in 1992, grew up in the West Indian community of Harlem in the 1930s, the daughter of immigrants from Grenada. She attended Hunter College (later becoming Professor of English there), ventured to the American expatriate community in Mexico, and participated in the Greenwich Village scene of the early 1950s. She is a major figure in the lesbian and feminist movements. Among her works are *Sister Outsider, Zami: A New Spelling of My Name, Uses of the Erotic, Chosen Poems Old and New, The Black Unicorn,* and *From a Land Where Other People Live.*

Clarence Lusane is an activist, lecturer, and free-lance journalist in Washington, D.C. He is the author of *African Americans at the Crossroads: The Restructuring of Black Leadership and the 1992 Elections,* and *Pipe Dream Blues: Racism and the War on Drugs,* both from South End Press. His award-winning writings have appeared in many publications including *Z Magazine, Black*

Scholar, and *Covert Action Information Bulletin*. He has travelled widely and is Chairman of the National Alliance of Third World Journalists. Lusane holds a Masters in Political Science and is completing his Ph.D. at Howard University.

Arturo Madrid is a native of New Mexico but has lived elsewhere over the past thirty years, including California, New Hampshire, Minnesota, and Washington, D.C. Trained as an analyst of literary texts, he has spent most of his life reading and reacting to social texts. He has also worked as an administrator, government official, and institution builder. From 1984 to 1993 he served as founding President of the Tomàs Rivera Center, a national institute for policy studies on Latino issues. Currently he is Murchison Distinguished Professor of the Humanities at Trinity University, San Antonio, TX.

Patricia Yancey Martin is Daisy Parker Flory Alumni Professor of Sociology at Florida State University. She specializes in formal organizations, social stratification (primarily gender), and the social organization of work. She has recently published a theoretical paper on feminist organizations and is completing a book-length study of 130 organizations in Florida that deal with rape victims.

Peggy McIntosh is Associate Director of the Wellesley College Center for Research on Women. She is founder and codirector of the National SEED Project on Inclusive Curriculum. She consults widely throughout the country and the world with college and school faculty who are creating gender-fair and multicultural curricula. She has written many articles on women's studies, curriculum change, and systems of unearned privilege. She has also taught at the Bearley School, Harvard University, Trinity College (Washington, D.C.), the University of Denver, the University of Durham (England), and Wellesley College. She is contributing editor to *Women's Studies Quarterly* and a consulting editor to *SAGE: A Scholarly Journal on Black Women*.

Michael Messner is Professor of Sociology at the University of Southern California and is the author of *Sport, Men, and Gender Order: Critical Feminist Perspectives*, as well as *Men's Lives* (with Michael Kimmel).

Roslyn Arlin Mickelson, a former high school social studies teacher, is Associate Professor of Sociology and adjunct Associate Professor of Women's Studies at the University of North Carolina at Charlotte. She received her Ph.D. in the sociology of education (1984) from the University of California, Los Angeles. Prior to coming to the University of North Carolina in 1985,

V Social Change and the Politics of Empowerment   488

she completed a postdoctoral fellowship at the University of Michigan, Ann Arbor's Bush Program in Child Development and Social Policy. Her research examines the political economy of schooling; in particular, the ways that race, class, and gender shape educational processes and outcomes. Currently, she is researching the curricular and equity implications of business-led school reforms.

Joan Moore is Distinguished Professor of Sociology at the University of Wisconsin-Milwaukee. She has written extensively about urban poverty and about Latinos, with a particular concern for youth gangs.

Cherríe Moraga is a poet, playwright and essayist. She is the co-editor of *This Bridge Called My Back: Writing by Radical Women of Color* which won The Before Columbus American Book Award in 1986. She is the author of numerous plays including *Shadow of a Man*, winner of the 1990 Fund for New American Plays Award and *Heroes and Saints*, winner of the Will Glickman Prize and the Pen West Award in 1992. Her most recent book is a collection of poems and essays entitled *The Last Generation*. Ms. Moraga is also a recipient of the National Endowment for the Arts' Theatre Playwrights' Fellowship.

Raquel Pinderhughes is Assistant Professor of Urban Studies at San Francisco State University. She is the editor, with Joan Moore, of *In the Barrios: Latinos and the Underclass Debate*. Her research areas and interests include: urban restructuring, urban poverty, the relationship between race, poverty and environmental equity, the socioeconomic condition of Latino populations in the United States, and the legal rights of families in the United States. She is currently conducting research on the social and economic conditions of Latinos living in San Francisco.

Roberta Praeger is a long-time Cambridge, Massachusetts, activist who has worked on housing, welfare, and women's issues.

Bernice Johnson Reagon is currently employed by the Smithsonian Institution, American History Department, in Washington, D.C.

Holly Sklar is a writer and lecturer. Her latest book is *Streets of Hope: The Fall and Rise of an Urban Neighborhood* (with Peter Medoff). Her other books include *Trilateralism: The Trilateral Commission and Elite Planning for World Management, Washington's War on Nicaragua*, and *Poverty in the American Dream* (with Barbara Ehrenreich and Karin Stallard). She is a columnist for *Z Magazine* and has contributed to numerous other publications. She is presently co-authoring a book about democracy.

Stephen Samuel Smith is Assistant Professor of Political Science at Winthrop University in South Carolina where he teaches and does research on the politics of education, public policy, and urban politics. In addition, his scholarly interests include the effects of ideology and government violence on working class political participation in the United States. He also serves as an evaluator for a residential substance abuse treatment program for pregnant women and their children.

Gloria Steinem is among the country's most widely read and critically acclaimed feminist writers. Her contributions to the women's movement have made her one of the most influential women in America. She co-founded *Ms.* magazine, the national feminist monthly, in 1972. She is a best-selling author whose works include *Outrageous Acts and Everyday Rebellions* and *Marilyn: Norma Jean*. Her most recent book is *Revolution from Within: A Book of Self Esteem* (1992).

Dottie Stevens is a Boston-area welfare-rights and poor people's activist, a co-founder of Advocacy for Resources for Modern Survival, and a board member of DBHN.

Ronald Takaki is professor of Ethnic Studies at the University of California, Berkeley. He served as departmental chairperson and also the graduate advisor of the Ethnic Studies Ph.D. program. He is the author of several books, including *Iron Cages: Race and Culture in 19th Century America*, *Strangers from a Different Shore: A History of Asian Americans*, and *A Different Mirror: A History of Multicultural America*. He helped to establish Berkeley's multicultural requirement for graduation.

Ronald L. Taylor is Professor of Sociology and Director of the Institute for African-American Studies at the University of Connecticut, Storrs. His published works include *Minority Families in the United States*; *African American Youth: Perspectives on their Status in the United States*; *Black Adolescence* (with Patricia Bell-Scott, et al.), and *The Black Male in America* (with Doris Y. Wilkinson). He has published widely on psychosocial development among black children and youth, gender roles, and African American families.

Michael Charles Thornton is an associate professor of Afro-American Studies, and Acting Interim-Director of Asian American Studies at the University of Wisconsin-Madison. He was born in Kobe, Japan to a Japanese mother and an Afro-American father. This heritage underlies most of his life's and professional work. His primary area of interest revolves around race relations among groups of color, but especially black perceptions of Asian Americans and Latino Americans. He is particularly interested in factors related to cross-racial identification and how these factors relate to interracial coalitions among groups of color.

Lynn Weber earned her doctorate in Sociology at the University of Illinois-Urbana. She is the Director of the Center for Research on Women and Professor in the Department Sociology and Social Work at The University of Memphis. Co-author of *The American Perception of Class* (with Reeve Vanneman), she has published widely on social class, race, and gender issues. She is currently working with Elizabeth Higginbotham on a research project, "Social Mobility, Race, and Women's Mental Health," which is a broad investigation of the role of upward mobility in the educational experiences, work life, family life, well being, and mental health for 200 Black and White professional and managerial women in the Memphis area.

Gaye Williams graduated in 1983 from Radcliffe–Harvard College. She has served as Acting Archivist at the Bethune Museum Archives National Historic Site. She is also an activist who tries to combine scholarship, politics, and art to affect change.

Naomi Wolf was born in San Francisco in 1962. She was educated at Yale University and at New College, Oxford University where was a Rhodes Scholar. Her essays have appeared in various publications including: *The New Republic, Wall Street Journal, Glamour, Ms., Esquire, The Washington Post,* and *The New York Times.* She speaks widely on college campuses and is the author of *The Beauty Myth* and, most recently, *Fire With Fire. Promiscuities,* a history and memoir of female desire, is forthcoming. She lives in Washington D.C.

Deborah Woo is a sociologist in Community Studies at the University of California, Santa Cruz. She earned her Ph.D. from the University of California, Berkeley, in 1983. Her research focuses on issues of cultural diversity, ethnic stratification, and the ideological or theoretical framework that shapes social relations within a particular context. Her writings relate these issues to mental health, literature, law, and education.

Gloria Yamato is currently the Community Relations Associate for the Pacific Northwest Regional Office of the American Friends Service Committee. She is also the acting Executive Director for ACT/fra, an organization for community-based social/economic change. She has contributed essays to *Changing Our Power* and *Making Face, Making Soul/Haciendo Caras.*

Maxine Baca Zinn is Professor of Sociology at Michigan State University and Senior Research Associate in the Julian Samora Research Institute. She earned a Ph.D. in Sociology from the University of Oregon. She specializes in race relations, gender, and the sociology of the family. With D. Stanley Eitzen, she has written *Diversity in Families, In Conflict and Order,* and *Social Problems.* She is the co-editor (with Bonnie Thornton Dill) of *Women of Color in U.S. Society.*

I

Shifting the Center and Reconstructing Knowledge

We begin this book by asking, Who has been excluded from what is known and how might we see the world differently if we were to acknowledge and value the experiences and thoughts of those who have been excluded? Many groups whose experiences have been vital in the formation of American society and culture have been silenced in the construction of knowledge about this society. The result is that what we know—about the experiences of both these silenced groups and the dominant culture—is distorted and incomplete.

As examples, think about the large number of social science studies that routinely make general conclusions about *the* population, when they have been based on research done only about and by men. Or, think about the content of history courses (and, for that matter, music, art, and literature courses) and how little the work of women, Asian Americans, Latinos, African Americans, Native Americans, gays, and lesbians is included. By excluding or marginalizing these groups, do we not communicate that they have no history or that their history is less important and less central to the development of American culture than is the history of White-American men? What false conclusions does this exclusionary thinking generate? One example is the claim that democracy and egalitarianism were central cultural beliefs in the early history of the United States. If this is true, how do we explain the enslavement of millions of African Americans? the genocide of American Indians? the laws against intermarriage between Asian Americans and White Americans? Without more inclusive thinking, our knowledge of American society and history is limited or wrong, and we are left only with contemporary stereotypes and misleading information as the basis for our knowledge about each other.

"Shifting the center" means putting at the center of our thinking the experiences of groups who have formerly been excluded. Without doing so, many groups simply remain invisible. When they are seen, they are typically judged through the experiences of White people, rather than understood on their own terms; this establishes a false norm through which all groups are judged. This is well expressed by Scott Kurashige, who writes, "The public view of Asian Americans is a lot like that of Casper the Ghost: we're either white or we're invisible" (Kurashige, 1994).

Shifting the center is a shift in stance that illuminates the experiences of not only the oppressed groups but also of those in the dominant culture. For example, the development of women's studies has changed what we know and how we think about women; at the same time, it has changed what we know and how we think about men. This does not mean that women's studies is about "male-bashing." It means that we take the experiences of women and men seriously and analyze the way that gender, race, and class have shaped the experiences of both—although in different but interrelated ways. Likewise, the study of racial-ethnic groups begins by learning the diverse histories and experiences of these groups, but in doing so, we transform our understanding of White experiences, too. The exclusionary thinking we have relied upon in the past simply does not reveal the intricate interconnections that exist between the different groups comprising American society.

Exclusionary thinking is increasingly being challenged by scholars and teachers who want to include the diversity of human experience in the construction and transmission of knowledge. Thinking more inclusively opens up the way the world is viewed, making the experience of previously excluded groups more visible and central in the construction of knowledge. Inclusive thinking shifts our perspective from the white, male-centered forms of thinking that have characterized much of Western thought, helping us better understand the intersections of race, class, and gender in the experiences of all groups, including those with privilege and power.

Why would we want to do this? Does reconstructing knowledge matter? To begin with, knowledge is not just some abstract thing—good to have, but not all that important. There are real consequences to having partial or distorted knowledge. First, learning about other groups helps you realize the partiality of your own perspective; furthermore, this is true for both dominant and subordinate groups. Only knowing the history of African Americans, for example, or seeing that history only in single-minded terms will not reveal the historical linkages between the oppression of African Americans and the exclusionary and exploitative treatment of other groups. This is well pointed out by Ronald Takaki in his essay on the multicultural history of American society included here.

Second, having misleading and incorrect knowledge gives us a poor social analysis and leads to the formation of bad social policy—policy that then reproduces, rather than solves, social problems. Finally, knowledge is not just about content and information; it provides an orientation to the world. What you know frames how you behave and how you think about yourself and others. If what you know is wrong because it is based on exclusionary thought, you are likely to act in exclusionary ways, thereby reproducing the racism, anti-Semitism, sexism, class oppression, and homophobia of society. This may not be because you are overtly racist, anti-Semitic, sexist, elitist, or homophobic (although it may encourage these beliefs), but it is simply because you do not know any better. Challenging oppressive race, class, and gender relations in society requires a reconstruction of knowledge so that we have some basis from which to change these damaging and dehumanizing systems of oppression.

As we said in the preface, developing inclusive thinking is more than just "understanding diversity" or valuing cultural pluralism. Inclusive thinking begins with the recognition that the United States is a multicultural and diverse society. Population data reveals that simple truth, but inclusive thinking is more than recognizing the plurality of experiences in this society. Understanding race, class, and gender means coming to see the systematic exclusion and exploitation of different groups. This is more than just adding in different group experiences to already established frameworks of thought. It means constructing new analyses that are focused on the centrality, of race, class, and gender in the experiences of us all.

You can begin to develop a more inclusive perspective by asking, How does the world look different if we put the experiences of those who have been excluded at the center of our thinking? At first people might be tempted to simply assert the perspective and experience of their own group. Nationalist movements of different kinds have done this. They are valuable because they recognize the contributions and achievements of oppressed groups, but ultimately they are unidimensional frameworks because they are centered in a single group experience and they encourage, rather than discourage, exclusionary thought.

Seeing inclusively is more than just seeing the world through the perspective of any one group whose views have been distorted or ignored. Remember that group membership cuts across race, class, and gender categories: for example, one may be a working-class, Asian American woman or a Black, middle-class man or a gay, White, working-class woman. Inclusive thinking means seeing the interconnections between these experiences and not reducing a given person's or group's life to a single factor. In addition, developing inclusive thinking is more than just summing up the experiences of individual groups. Race, class, and gender are social structural categories. This means that they are embedded in the institutional structure of society. Understanding them requires a social structural analysis—by which we mean revealing the race, class, and gender patterns and processes that form the very framework of society.

We believe that shifting our perspective by thinking about the experiences of those who have been excluded from knowledge changes how we think about society, history, and culture. No longer do different groups seem just "different" or "deviant." Rather, specific patterns of the intersections of race, class, and gender are revealed, as are the connections that exist between groups. We then learn how our different experiences are linked, both historically and now.

Once we understand that race, class, and gender are simultaneous and intersecting systems of relationship and meaning, we come to see the different ways that other categories of experience intersect in society. Age, religion, sexual orientation, physical ability, region, and ethnicity also shape systems of privilege and inequality. We have tried to integrate these different experiences throughout the book, although we could not include as much as we would have liked.

Seeing these connections is what we think is most important about analyzing race, class, and gender. It is the reason for taking the perspective of people other than those like you. The purpose of the articles in "Shifting the Center" is not to appropriate others' experiences but to begin uncovering the structures of race, class, and gender that are embedded in the experiences of all people.

By providing personal accounts of what exclusion means and how it feels, the articles in Part One show the limits of existing knowledge. We have selected personal accounts that reflect the diverse experiences of race, class, gender, and/or sexual orientation. We intend for the personal nature of these accounts to build empathy between groups—an emotional stance that we think is critical to seeing linkages and connections. By describing how exclusion shapes individual and collective consciousness, these accounts also document the diversity of experience within the United States. As a result, we begin to see how diverse experiences can shape the concepts and theories we develop in the academy. Together, they suggest new possibilities for thinking inclusively and imagining a society that would be more enabling for everyone.

In "Missing People and Others," Arturo Madrid tells how exclusion and marginalization in the educational curriculum have affected us all. Madrid describes his schooling as a process of denial of the specific experience of Latinos, Asian Americans, Native Americans, and all other groups together considered to be "other." He asks us to consider what it feels like to be the missing person. As we read his account, we see how education has contributed to exclusion by marginalizing the history of Chicanos and groups considered to be "other."

In "La Güera" Cherríe Moraga writes about her developing awareness of her class, race, and sexual identity. Like the other authors here, Moraga tells us what exclusion has meant in her experience and how it feels. Her essay also reveals the intersections of race, class, and gender with the system of compulsory heterosexuality—that is, the institutionalized structures and beliefs that define and enforce heterosexual behavior as the only natural and permissible form of sexual expression. For Moraga, acknowledging her lesbian identity deepened her understanding of her mother's oppression as a poor Chicana. Moraga reminds us of the importance of breaking the silences that protect oppression.

Readers should keep the accounts of Madrid, Mura, and Moraga in mind as they read the remainder of this book. What new experiences, understandings, theories, histories, and analyses do these readings inspire? What does it take for a member of one group (say a White male) to be willing to learn from and value the experiences of another (for example, a Chicana lesbian)? June Jordan's essay, "Report from the Bahamas" is a model of this inclusive thinking. In it she begins from her own experience as a Black, middle-class, well-educated woman and reflects on her connections with other women in different locations in the race, class, gender system. In doing so, she finds connections in surprising places, while also understanding the differences that distinguish our experiences. She also shows her student, a White Catholic Irish woman, finding common ground with her South African classmate. There are no simple or singular answers to these questions, but recentering our frame of reference can reshape what we know, both formally and in everyday life. These essays show that, although we are caught in multiple systems, we can learn to see our connection to others. This is not just an intellectual exercise. As Paula Gunn Allen shows in her article on Native American women, "Angry Women Are Building," it is a matter of survival. Resisting systems of oppression means revising the ideas about ourselves and others that have been created as a part of a system of social control.

Reading these articles shows us the radical shift that thinking inclusively requires. Such a shift begins with valuing the experiences of those who have been excluded and questioning the assumptions made about all groups. When we begin to take a more inclusive perspective, divisions between privileged and underprivileged, oppressor and oppressed are challenged because we start to see race, class, and gender as intersecting systems of experience. For example, White women and women of color share some common experiences based on their gender, but their racial experiences are quite distinct; moreover, experiences within the race/gender system are further conditioned by one's class. (Johnnetta Cole points out in an article included in Part Two that gender is manifested differently, depending on one's race and class position.)

Engaging oneself at the personal level is critical to this process of thinking inclusively. Changing one's mind is not just a matter of assessing facts and data, though that is important; it also requires examining one's feelings. That is why we begin with several personal narratives. Unlike more conventional forms of sociological data (such as surveys, interviews, and even direct observations), personal accounts are more likely to elicit emotional responses. Traditionally, social science has defined emotional engagement as an impediment to objectivity. Sociology, for example, has emphasized rational thought as the basis for social action and has often discouraged more personalized reflection, but the capacity to reflect on one's experience makes us distinctly human. Personal documents tap the private, reflective dimension of life, enabling us to see the inner life of others and, in the process, revealing our own lives more completely. The idea that objectivity is best reached only through rational thought is a specifically Western and masculine way of thinking—one that we challenge throughout this book.

Including personal narratives is not meant to limit our level of understanding only to individuals. As sociologists, we study individuals in groups as a way of revealing the social structures shaping collective experiences. Through doing so, we discover our common experiences and see the impact of the social structures of race, class, and gender on our experiences. Marilyn Frye's article, "Oppression," introduces the structural perspective that lies at the heart of this book. She distinguishes oppression from suffering, pointing out that many individuals suffer in society, but that oppression is structured into the fabric of social institutions. Using the metaphor of a birdcage, Frye artfully explains the concept of social structure. Looking only at an individual wire in a cage does not reveal the network of wires that form a cage; likewise, social structure refers to the patterns of behavior, belief, resource distribution, and social control that constitute society.

Analysis of the multicultural basis of society is critical to understanding who we are as a society and a culture. The new focus on multiculturalism is, as Ronald Takaki says in "A Different Mirror," a debate over our national identity. Takaki makes a point of showing the common points of connection in the histories of African Americans, Chicanos, Irish Americans, Jews, and Indians. He argues that only when we understand this multidimensional history will we see ourselves in the full complexity of our humanity.

As all of the authors in this section show, the strength and richness of our society lies in its diversity, but the potential of this diversity will only be realized through social structures that value and protect all human experiences. To reconstruct what we know, we begin with the experiences of those who have been excluded. In so doing, we begin to build a system of knowledge that ensures the survival of us all.

Reference:

Kurashige, Scott. 1994. Cited in Karin Aguilar-San Juan, "Linking the Issues: From Identity to Activism." Pp. 1–15 in *The State of Asian America: Activism and Resistance in the 1990s.* ed. Karin Aguilar-San Juan. Boston: South End Press.

Shifting the Center

MISSING PEOPLE AND OTHERS:

Joining Together to Expand the Circle

Arturo Madrid

1

I am a citizen of the United States, as are my parents and as were their parents, grandparents, and great-grandparents. My ancestors' presence in what is now the United States antedates Plymouth Rock, even without taking into account any American Indian heritage I might have.

I do not, however, fit those mental sets that define America and Americans. My physical appearance, my speech patterns, my name, my profession (a professor of Spanish) create a text that confuses the reader. My normal experience is to be asked, "And where are *you* from?"

My response depends on my mood. Passive-aggressive, I answer, "From here." Aggressive-passive, I ask, "Do you mean where am I originally from?" But ultimately my answer to those follow-up questions that ask about origins will be that we have always been from here.

Overcoming my resentment I will try to educate, knowing that nine times out of ten my words fall on inattentive ears. I have spent most of my adult life explaining who I am not. I am exotic, but—as Richard Rodriguez of *Hunger of Memory* fame so painfully found out—not exotic enough . . . not Peruvian, or Pakistani, or Persian, or whatever.

I am, however, very clearly *the other*, if only your everyday, garden-variety, domestic *other*. I've always known that I was *the other*, even before I knew the vocabulary or understood the significance of being *the other*.

From: *Change* 20 (May/June 1988): 55–59. Reprinted by permission.

I grew up in an isolated and historically marginal part of the United States, a small mountain village in the state of New Mexico, the eldest child of parents native to that region and whose ancestors had always lived there. In those vast and empty spaces, people who look like me, speak as I do, and have names like mine predominate. But the *americanos* lived among us: the descendants of those nineteenth-century immigrants who dispossessed us of our lands; missionaries who came to convert us and stayed to live among us; artists who became enchanted with our land and humanscape and went native; refugees from unhealthy climes, crowded spaces, unpleasant circumstances; and, of course, the inhabitants of Los Alamos, whose socio-cultural distance from us was moreover accentuated by the fact that they occupied a space removed from and proscribed to us. More importantly, however, they—*los americanos*—were omnipresent (and almost exclusively so) in newspapers, newsmagazines, books, on radio, in movies and, ultimately, on television.

Despite the operating myth of the day, school did not erase my otherness. It did try to deny it, and in doing so only accentuated it. To this day, schooling is more socialization than education, but when I was in elementary school—and given where I was—socialization was everything. School was where one became an American. Because there was a pervasive and systematic denial by the society that surrounded us that we were Americans. That denial was both explicit and implicit. My earliest memory of the former was that there were two kinds of churches: theirs and ours. The more usual was the implicit denial, our absence from the larger cultural, economic, political and social spaces—the one that reminded us constantly that we were *the other*. And school was where we felt it most acutely.

Quite beyond saluting the flag and pledging allegiance to it (a very intense and meaningful action, given that the U.S. was involved in a war and our brothers, cousins, uncles, and fathers were on the front lines) becoming American was learning English and its corollary—not speaking Spanish. Until very recently ours was a proscribed language—either *de jure* (by rule, by policy, by law) or *de facto* (by practice, implicitly if not explicitly; through social and political and economic pressure). I do not argue that learning English was not appropriate. On the contrary. Like it or not, and we had no basis to make any judgments on that matter, we were Americans by virtue of having been born Americans, and English was the common language of Americans. And there was a myth, a pervasive myth, that said that if we only learned to speak English well—and particularly without an accent—we would be welcomed into the American fellowship.

Senator Sam Hayakawa notwithstanding, the true text was not our speech, but rather our names and our appearance, for we would always have an accent, however perfect our pronunciation, however excellent our enunciation, however divine our diction. That accent would be heard in our pigmentation, our physiognomy, our names. We were, in short, *the other*.

Being *the other* means feeling different; is awareness of being distinct; is consciousness of being dissimilar. It means being outside the game, outside the circle, outside the set. It means being on the edges, on the margins, on the periphery. Otherness means feeling excluded, closed out, precluded, even disdained and scorned. It produces a sense of isolation, of apartness, of disconnectedness, of alienation.

Being *the other* involves a contradictory phenomenon. On the one hand being *the other* frequently means being invisible. Ralph Ellison wrote eloquently about that experience in his magisterial novel *The Invisible Man*. On the other hand, being *the other* sometimes involves sticking out like a sore thumb. What is she/he doing here?

If one is *the other*, one will inevitably be perceived unidimensionally; will be seen stereotypically; will be defined and delimited by mental sets that may not bear much relation to existing realities. There is a darker side to otherness as well. *The other* disturbs, disquiets, discomforts. It provokes distrust and suspicion. *The other* makes people feel anxious, nervous, apprehensive, even fearful. *The other* frightens, scares.

For some of us being *the other* is only annoying; for others it is debilitating; for still others it is damning. Many try to flee otherness by taking on protective colorations that provide invisibility, whether of dress or speech or manner or name. Only a fortunate few succeed. For the majority, otherness is permanently sealed by physical appearance. For the rest, otherness is betrayed by ways of being, speaking or of doing.

I spent the first half of my life downplaying the significance and consequences of otherness. The second half has seen me wrestling to understand its complex and deeply ingrained realities; striving to fathom why otherness denies us a voice or visibility or validity in American society and its institutions; struggling to make otherness familiar, reasonable, even normal to my fellow Americans.

I am also a missing person. Growing up in northern New Mexico I had only a slight sense of our being missing persons. *Hispanos*, as we called (and call) ourselves in New Mexico, were very much a part of the fabric of the society and there were Hispano professionals everywhere about me: doctors, lawyers, school teachers, and administrators. My people owned businesses, ran organizations and were both appointed and elected public officials.

To be sure, we did not own the larger businesses, nor at the time were we permitted to be part of the banking world. Other than that, however, people who looked like me, spoke like me, and had names like mine, predominated. There was, to be sure, Los Alamos, but as I have said, it was removed from our realities.

My awareness of our absence from the larger institutional life of society became sharper when I went off to college, but even then it was attenuated by the circumstances of history and geography. The demography of Albuquerque still strongly reflected its historical and cultural origins, despite the influx of Midwesterners and Easterners. Moreover, many of my classmates at the University of New Mexico in Albuquerque were Hispanos, and even some of my professors were.

I thought that would obtain at UCLA, where I began graduate studies in 1960. Los Angeles already had a very large Mexican population, and that population was visible even in and around Westwood and on the campus. Many of the groundskeepers and food-service personnel at UCLA were Mexican. But Mexican-American students were few and mostly invisible, and I do not recall seeing or knowing a single Mexican-American (or, for that matter, black, Asian, or American Indian) professional on the staff or faculty of that institution during the five years I was there.

Needless to say, persons like me were not present in any capacity at Dartmouth College—the site of my first teaching appointment—and, of course, were not even part of the institutional or individual mind-set. I knew then that we—a "we" that had come to encompass American Indians, Asian-Americans, black Americans, Puerto Ricans, and women—were truly missing persons in American institutional life.

Over the past three decades, the *de jure* and *de facto* segregations that have historically characterized American institutions have been under assault. As a consequence, minorities and women have become part of American institutional life, and although there are still many areas where we are not to be found, the missing persons phenomenon is not as pervasive as it once was.

However, the presence of *the other*, particularly minorities, in institutions and in institutional life, is, as we say in Spanish, *a flor de tierra*; spare plants whose roots do not go deep, a surface phenomenon, vulnerable to inclemencies of an economic, political, or social nature.

Our entrance into and our status in institutional life is not unlike a scenario set forth by my grandmother's pastor when she informed him that she and her family were leaving their mountain village to relocate in the Rio Grande Valley. When he asked her to promise that she would remain true to the faith and continue to involve herself in the life of the church, she assured him that she would and asked him why he thought she would do otherwise.

"Doña Trinidad," he told her, "in the Valley there is no Spanish church. There is only an American church." "But," she protested, "I read and speak English and would be able to worship there." Her pastor's response was: "It is possible that they will not admit you, and even if they do, they might not accept you. And that is why I want you to promise me that you are going to go to church. Because if they don't let you in through the front door, I want you to go in through the back door. And if you can't get in through the back door, go in the side door. And if you are unable to enter through the side door I want you to go in through the window. What is important is that you enter and that you stay."

Some of us entered institutional life through the front door; others through the back door; and still others through side doors. Many, if not most of us, came in through windows and continue to come in through windows. Of those who entered through the front door, some never made it past the lobby; others were ushered into corners and niches. Those who entered through back and side doors inevitably have remained in back and side rooms. And those who entered through windows found enclosures built around them. For despite the lip service given to the goal of the integration of minorities into institutional life, what has occurred instead is ghettoization, marginalization, isolation.

Not only have the entry points been limited: in addition, the dynamics have been singularly conflictive. Gaining entry and its corollary—gaining space—have frequently come as a consequence of demands made on institutions and institutional officers. Rather than entering institutions more or less passively, minorities have, of necessity, entered them actively, even aggressively. Rather than taking, they have demanded. Institutional relations have thus been adversarial, infused with specific and generalized tensions.

The nature of the entrance and the nature of the space occupied have greatly influenced the view and attitudes of the majority population within those institutions. All of us are put into the same box; that is, no matter what the individual reality, the assessment of the individual is inevitably conditioned by a perception that is held of the class. Whatever our history, whatever our record, whatever our validations, whatever our accomplishments, by and large we are perceived unidimensionally and are dealt with accordingly.

My most recent experience in this regard is atypical only in its explicitness. A few years ago I allowed myself to be persuaded to seek the presidency of a large and prestigious state university. I was invited for an interview and presented myself before the selection committee, which included members of the board of trustees. The opening question of the brief but memorable interview was directed at me by a member of that august body. "Dr. Madrid," he asked, "why does a one-dimensional person like you think he can be the president of a multi-dimensional institution like ours?"

If, as I happen to believe, the well-being of a society is directly related to the degree and extent to which all of its citizens participate in its institutions, we have a challenge before us. One of the strengths of our society—perhaps its main strength—has been a tradition of struggle against clubbishness, exclusivity, and restriction.

Today, more than ever, given the extraordinary changes that are taking place in our society, we need to take up that struggle again—irritating, grating, troublesome, unfashionable, unpleasant as it is. As educated and educator members of this society, we have a special responsibility for leading the struggle against marginalization, exclusion, and alienation.

Let us work together to assure that all American institutions, not just its precollegiate educational and penal institutions, reflect the diversity of our society. Not to do so is to risk greater alienation on the part of a growing segment of our society. It is to risk increased social tension in an already conflictive world. And ultimately it is to risk the survival of a range of institutions that, for all their defects and deficiencies, permit us the space, the opportunity, and the freedom to improve our individual and collective lot; to guide the course of our government, and to redress whatever grievances we have. Let us join together to expand, not to close the circle.

LA GÜERA

2

Cherríe Moraga

It requires something more than personal experience to gain a philosophy or point of view from any specific event. It is the quality of our response to the event and our capacity to enter into the lives of others that help us to make their lives and experiences our own.

Emma Goldman[1]

I am the very well-educated daughter of a woman who, by the standards in this country, would be considered largely illiterate. My mother was born in Santa Paula, Southern California, at a time when much of the central valley there was still farm land. Nearly thirty-five years later, in 1948, she was the only daughter of six to marry an anglo, my father.

---

1. Alix Kates Shulman, "Was My Life Worth Living?" *Red Emma Speaks.* (New York: Random House, 1972), p. 388.

From: Cherríe Moraga and Gloria Anzaldúa (eds.), *This Bridge Called My Back: Radical Writings by Women of Color* (New York: Kitchen Table Press, 1983), pp. 27–34. Copyright © 1983 by Cherríe Moraga. Reprinted by permission of the author and the publisher.

I remember all of my mother's stories, probably much better than she realizes. She is a fine story-teller, recalling every event of her life with the vividness of the present, noting each detail right down to the cut and color of her dress. I remember stories of her being pulled out of school at the ages of five, seven, nine, and eleven to work in the fields, along with her brothers and sisters; stories of her father drinking away whatever small profit she was able to make for the family; of her going the long way home to avoid meeting him on the street, staggering toward the same destination. I remember stories of my mother lying about her age in order to get a job as a hat-check girl at Agua Caliente Racetrack in Tijuana. At fourteen, she was the main support of the family. I can still see her walking home alone at 3 a.m., only to turn all of her salary and tips over to her mother, who was pregnant again.

The stories continue through the war years and on: walnut-cracking factories, the Voit Rubber factory, and then the computer boom. I remember my mother doing piecework for the electronics plant in our neighborhood. In the late evening, she would sit in front of the T.V. set, wrapping copper wires into the backs of circuit boards, talking about "keeping up with the younger girls." By that time, she was already in her mid-fifties.

Meanwhile, I was college-prep in school. After classes, I would go with my mother to fill out job applications for her, or write checks for her at the supermarket. We would have the scenario all worked out ahead of time. My mother would sign the check before we'd get to the store. Then, as we'd approach the checkstand, she would say—within earshot of the cashier—"oh honey, you go 'head and make out the check," as if she couldn't be bothered with such an insignificant detail. No one asked any questions.

I was educated, and wore it with a keen sense of pride and satisfaction, my head propped up with the knowledge, from my mother, that my life would be easier than hers. I was educated; but more than this, I was "la güera": fair-skinned. Born with the features of my Chicana mother, but the skin of my Anglo father, I had it made.

No one ever quite told me this (that light was right), but I knew that being light was something valued in my family (who were all Chicano, with the exception of my father). In fact, everything about my upbringing (at least what occurred on a conscious level) attempted to bleach me of what color I did have. Although my mother was fluent in it, I was never taught much Spanish at home. I picked up what I did learn from school and from over-heard snatches of conversation among my relatives and mother. She often called other lower-income Mexicans "braceros," or "wet-backs," referring to herself and her family as "a different class of people." And yet, the real story was that my family, too, had been poor (some still are) and farmworkers. My mother can remember this in her blood as if it were yesterday. But this is something she would like to forget (and rightfully), for to her, on a basic economic level, being Chicana meant being "less." It was through my mother's desire to protect her children from poverty and illiteracy that we became "anglocized"; the more effectively we could pass in the white world, the better guaranteed our future.

From all of this, I experience, daily, a huge disparity between what I was born into and what I was to grow up to become. Because, (as Goldman suggests) these stories my mother told me crept under my "güera" skin. I had no choice but to enter into the life of my mother. *I had no choice.* I took her life into my heart, but managed to keep a lid on it as long as I feigned being the happy, upwardly mobile heterosexual.

When I finally lifted the lid to my lesbianism, a profound connection with my mother reawakened in me. It wasn't until I acknowledged and confronted my own lesbianism in the flesh, that my heartfelt identification with and empathy for my mother's oppression—due to being poor, uneducated, and Chicana—was realized. My lesbianism is the avenue through which I have learned the most about silence and oppression, and it continues to be the most tactile reminder to me that we are not free human beings.

You see, one follows the other. I had known for years that I was a lesbian, had felt it in my bones, had ached with the knowledge, gone crazed with the knowledge, wallowed in the silence of it. Silence *is* like starvation. Don't be fooled. It's nothing short of that, and felt most sharply when one has had a full belly most of her life. When we are not physically starving, we have the luxury to realize psychic and emotional starvation. It is from this starvation that other starvations can be recognized—if one is willing to take the risk of making the connection—if one is willing to be responsible to the result of the connection. For me, the connection is an inevitable one.

What I am saying is that the joys of looking like a white girl ain't so great since I realized I could be beaten on the street for being a dyke. If my sister's being beaten because she's Black, it's pretty much the same principle. We're both getting beaten any way you look at it. The connection is blatant; and in the case of my own family, the difference in the privileges attached to looking white instead of brown are merely a generation apart.

In this country, lesbianism is a poverty—as is being brown, as is being a woman, as is being just plain poor. The danger lies in ranking the oppressions. *The danger lies in failing to acknowledge the specificity of the oppression.* The danger lies in attempting to deal with oppression purely from a theoretical base. Without an emotional, heartfelt grappling with the source of our own oppression, without naming the enemy within ourselves and outside of us, no authentic, non-hierarchical connection among oppressed groups can take place.

When the going gets rough, will we abandon our so-called comrades in a flurry of racist/heterosexist/what-have-you panic? To whose camp, then, should the lesbian of color retreat? Her very presence violates the ranking and abstraction of oppression. Do we merely live hand to mouth? Do we merely struggle with the "ism" that's sitting on top of our own heads?

The answer is: yes, I think first we do; and we must do so thoroughly and deeply. But to fail to move out from there will only isolate us in our own oppression—will only insulate, rather than radicalize us.

To illustrate: a gay male friend of mine once confided to me that he continued to feel that, on some level, I didn't trust him because he was male; that he felt, really, if it ever came down to a "battle of the sexes," I might kill him. I admitted that I might very well. He wanted to understand the source of my distrust. I responded, "You're not a woman. Be a woman for a day. Imagine being a woman." He confessed that the thought terrified him because, to him, being a woman meant being raped by men. He *had* felt raped by men; he wanted to forget what that meant. What grew from that discussion was the realization that in order for him to create an authentic alliance with me, he must deal with the primary source of his own sense of oppression. He must, first, emotionally come to terms with what it feels like to be a victim. If he—or anyone—were to truly do this, it would be impossible to discount the oppression of others, except by again forgetting how we have been hurt.

And yet, oppressed groups are forgetting all the time. There are instances of this in the rising Black middle class, and certainly an obvious trend of such "unconsciousness" among white gay men. Because to remember may mean giving up whatever privileges we have managed to squeeze out of this society by virtue of our gender, race, class, or sexuality.

Within the women's movement, the connections among women of different backgrounds and sexual orientations have been fragile, at best. I think this phenomenon is indicative of our failure to seriously address ourselves to some very frightening questions: How have I internalized my own oppression? How have I oppressed? Instead, we have let rhetoric do the job of poetry. Even the word "oppression" has lost its power. We need a new language, better words that can more closely describe women's fear of and resistance to one another; words that will not always come out sounding like dogma.

What prompted me in the first place to work on an anthology by radical women of color was a deep sense that I had a valuable insight to contribute, by virtue of my birthright and background. And yet, I don't really understand first-hand what it feels like being shitted on for being brown. I understand much more about the joys of it—being Chicana and having family are synonymous for me. What I know about loving, singing, crying, telling stories, speaking with my heart and hands, even having a sense of my own soul comes from the love of my mother, aunts, cousins . . .

But at the age of twenty-seven, it is frightening to acknowledge that I have internalized a racism and classism, where the object of oppression is not only someone outside of my skin, but the someone inside my skin. In fact, to a large degree, the real battle with such oppression, for all of us, begins under the skin. I have had to confront the fact that much of what I value about being Chicana, about my family, has been subverted by anglo culture and my own cooperation with it. This realization did not occur to me overnight. For example, it wasn't until long after my graduation from the private college I'd attended in Los Angeles, that I realized the major reason for my total alienation from and fear of my classmates was rooted in class and culture. CLICK.

Three years after graduation, in an apple-orchard in Sonoma, a friend of mine (who comes from an Italian Irish working-class family) says to me, "Cherríe, no wonder you felt like such a nut in school. Most of the people there were white and rich." It was true. All along I had felt the difference, but not until I had put the words "class" and "color" to the experience, did my feelings make any sense. For years, I had berated myself for not being as "free" as my classmates. I completely bought that they simply had more guts than I did—to rebel against their parents and run around the country hitch-hiking, reading books and studying "art." They had enough privilege to be atheists, for chrissake. There was no one around filling in the disparity for me between their parents, who were Hollywood filmmakers, and my parents, who wouldn't know the name of a filmmaker if their lives depended on it (and precisely because their lives didn't depend on it, they couldn't be bothered). But I knew nothing about "privilege" then. White was right. Period. I could pass. If I got educated enough, there would never be any telling.

Three years after that, another CLICK. In a letter to Barbara Smith, I wrote:

> I went to a concert where Ntosake Shange was reading. There, everything exploded for me. She was speaking a language that I knew—in the deepest parts of me—existed, and that I had ignored in my own feminist studies and even in my own writing. What Ntosake caught in me is the realization that in my development as a poet, I have, in many ways, denied the voice of my brown mother—the brown in me. I have acclimated to the sound of a white language which, as my father represents it, does not speak to the emotions in my poems—emotions which stem from the love of my mother.
>
> The reading was agitating. Made me uncomfortable. Threw me into a week-long terror of how deeply I was affected. I felt that I had to start all over again. That I turned only to the perceptions of white middle-class women to speak for me and all women. I am shocked by my own ignorance.

Sitting in that auditorium chair was the first time I had realized to the core of me that for years I had disowned the language I knew best—ignored the words and rhythms that were the closest to me. The sounds of my mother and aunts gossiping—half in English, half in Spanish—while drinking cerveza in the kitchen. And the hands—I had cut off the hands in my poems. But not in conversation; still the hands could not be kept down. Still they insisted on moving.

The reading had forced me to remember that I knew things from my roots. But to remember puts me up against what I don't know. Shange's reading agitated me because she spoke with power about a world that is both alien and common to me: "the capacity to enter into the lives of others." But you can't just take the goods and run. I knew that then, sitting in the Oakland auditorium (as I know in my poetry), that the only thing worth writing about is what seems to be unknown and, therefore, fearful.

The "unknown" is often depicted in racist literature as the "darkness" within a person. Similarly, sexist writers will refer to fear in the form of the vagina, calling it "the orifice of death." In contrast, it is a pleasure to read works such as Maxine Hong Kingston's *Woman Warrior*, where fear and alienation are described as "the white ghosts." And yet, the bulk of literature in this country reinforces the myth that what is dark and female is evil. Consequently, each of us—whether dark, female, or both—has in some way *internalized* this oppressive imagery. What the oppressor often succeeds in doing is simply *externalizing* his fears, projecting them into the bodies of women, Asians, gays, disabled folks, whoever seems most "other."

> call me
> roach and presumptuous
> nightmare on your white pillow
> your itch to destroy
> the indestructible
> part of yourself
>
>   Audre Lorde[2]

But it is not really difference the oppressor fears so much as similarity. He fears he will discover in himself the same aches, the same longings as those of the people he has shitted on. He fears the immobilization threatened by his

---

2. From "The Brown Menace or Poem to the Survival of Roaches," *The New York Head Shop and Museum* (Detroit: Broadside, 1974), p. 48.

own incipient guilt. He fears he will have to change his life once he has seen himself in the bodies of the people he has called different. He fears the hatred, anger, and vengeance of those he has hurt.

This is the oppressor's nightmare, but it is not exclusive to him. We women have a similar nightmare, for each of us in some way has been both oppressed and the oppressor. We are afraid to look at how we have failed each other. We are afraid to see how we have taken the values of our oppressor into our hearts and turned them against ourselves and one another. We are afraid to admit how deeply "the man's" words have been ingrained in us.

To assess the damage is a dangerous act. I think of how, even as a feminist lesbian, I have so wanted to ignore my own homophobia, my own hatred of myself for being queer. I have not wanted to admit that my deepest personal sense of myself has not quite "caught up" with my "woman-identified" politics. I have been afraid to criticize lesbian writers who choose to "skip over" these issues in the name of feminism. In 1979, we talk of "old gay" and "butch and femme" roles as if they were ancient history. We toss them aside as merely patriarchal notions. And yet, the truth of the matter is that I have sometimes taken society's fear and hatred of lesbians to bed with me. I have sometimes hated my lover for loving me. I have sometimes felt "not woman enough" for her. I have sometimes felt "not man enough." For a lesbian trying to survive in a heterosexist society, there is no easy way around these emotions. Similarly, in a white-dominated world, there is little getting around racism and our own internalization of it. It's always there, embodied in some one we least expect to rub up against.

When we do rub up against this person, *there* then is the challenge. *There* then is the opportunity to look at the nightmare within us. But we usually shrink from such a challenge.

Time and time again, I have observed that the usual response among white women's groups when the "racism issue" comes up is to deny the difference. I have heard comments like, "Well, we're open to *all* women; why don't they (women of color) come? You can only do so much . . ." But there is seldom any analysis of how the very nature and structure of the group itself may be founded on racist or classist assumptions. More importantly, so often the women seem to feel no loss, no lack, no absence when women of color are not involved; therefore, there is little desire to change the situation. This has hurt me deeply. I have come to believe that the only reason women of a privileged class will dare to look at *how* it is that *they* oppress, is when they've come to know the meaning of their own oppression. And understand that the oppression of others hurts them personally.

The other side of the story is that women of color and working-class women often shrink from challenging white middle-class women. It is much easier to rank oppressions and set up a hierarchy, rather than take responsibility for changing our own lives. We have failed to demand that white women, particularly those who claim to be speaking for all women, be accountable for their racism.

The dialogue has simply not gone deep enough.

I have many times questioned my right to even work on an anthology which is to be written "exclusively by Third World women." I have had to look critically at my claim to color, at a time when, among white feminist ranks, it is a "politically correct" (and sometimes peripherally advantageous) assertion to make. I must acknowledge the fact that, physically, I have had a *choice* about making that claim, in contrast to women who have not had such a choice, and have been abused for their color. I must reckon with the fact that for most of my life, by virtue of the very fact that I am white-looking, I identified with and aspired toward white values, and that I rode the wave of that Southern California privilege as far as conscience would let me.

Well, now I feel both bleached and beached. I feel angry about this—the years when I refused to recognize privilege, both when it worked against me, and when I worked it, ignorantly, at the expense of others. These are not settled issues. That is why this work feels so risky to me. It continues to be discovery. It has brought me into contact with women who invariably know a hell of a lot more than I do about racism, as experienced in the flesh, as revealed in the flesh of their writing.

I think: what is my responsibility to my roots—both white and brown, Spanish-speaking and English? I am a woman with a foot in both worlds; and I refuse the split. I feel the necessity for dialogue. Sometimes I feel it urgently.

But one voice is not enough, nor two, although this is where dialogue begins. It is essential that radical feminists confront their fear of and resistance to each other, because without this, there *will* be no bread on the table. Simply, we will not survive. If we could make this connection in our heart of hearts, that if we are serious about a revolution—better—if we seriously believe there should be joy in our lives (real joy, not just "good times"), then we need one another. We women need each other. Because my/your solitary, self-asserting "go-for-the-throat-of-fear" power is not enough. The real power, as you and I well know, is collective. I can't afford to be afraid of you, nor you of me. If it takes head-on collisions, let's do it: this polite timidity is killing us.

As Lorde suggests in the passage I cited earlier, it is in looking to the nightmare that the dream is found. There, the survivor emerges to insist on a future, a vision, yes, born out of what is dark and female. The feminist movement must be a movement of such survivors, a movement with a future.

REPORT FROM THE BAHAMAS   3

June Jordan

I am staying in a hotel that calls itself The Sheraton British Colonial. One of the photographs advertising the place displays a middle-aged Black man in a waiter's tuxedo, smiling. What intrigues me most about the picture is just this: while the Black man bears a tray full of "colorful" drinks above his left shoulder, both of his feet, shoes and trouserlegs, up to ten inches above his ankles, stand in the also "colorful" Caribbean salt water. He is so delighted to serve you he will wade into the water to bring you Banana Daquiris while you float! More precisely, he will wade into the water, fully clothed, oblivious to the ruin of his shoes, his trousers, his health, and he will do it with a smile.

I am in the Bahamas. On the phone in my room, a spinning complement of plastic pages offers handy index clues such as CAR RENTAL and CASINOS. A message from the Ministry of Tourism appears among these travellers tips. Opening with a paragraph of "WELCOME," the message then proceeds to "A PAGE OF HISTORY," which reads as follows:

> New World History begins on the same day that modern Bahamian history begins—October 12, 1492. That's when Columbus stepped ashore—British influence came first with the Eleutherian Adventurers of 1647—After the Revolutions, American Loyalists fled from the newly independent states and settled in the Bahamas. Confederate blockade-runners used the island as a haven during the War between the States, and after the War, a number of Southerners moved to the Bahamas . . .

There it is again. Something proclaims itself a legitimate history and all it does is track white Mr. Columbus to the British Eleutherians through the Confederate Southerners as they barge into New World surf, land on New World turf, and nobody saying one word about the Bahamian people, the Black peoples, to whom the only thing new in their island world was this weird succession of crude intruders and its colonial consequences.

This is my consciousness of race as I unpack my bathing suit in the Sheraton British Colonial. Neither this hotel nor the British nor the long ago Italians nor the white Delta airline pilots belong here, of course. And every

---

From: June Jordan, *On Call: Political Essays* (Boston: South End Press, 1985), pp. 39–49. Reprinted by permission of the author.

time I look at the photograph of that fool standing in the water with his shoes on I'm about to have a West Indian fit, even though I know he's no fool; he's a middle-aged Black man who needs a job and this is his job—pretending himself a servile ancillary to the pleasures of the rich. (Compared to his options in life, I am a rich woman. Compared to most of the Black Americans arriving for this Easter weekend on a three nights four days' deal of bargain rates, the middle-aged waiter is a poor Black man.)

We will jostle along with the other (white) visitors and join them in the tee shirt shops or, laughing together, learn ruthless rules of negotiation as we, Black Americans as well as white, argue down the price of handwoven goods at the nearby straw market while the merchants, frequently toothless Black women seated on the concrete in their only presentable dress, humble themselves to our careless games:

"Yes? You like it? Eight dollar."

"Five."

"I give it to you. Seven."

And so it continues, this weird succession of crude intruders that, now, includes me and my brothers and my sisters from the North.

This is my consciousness of class as I try to decide how much money I can spend on Bahamian gifts for my family back in Brooklyn. No matter that these other Black women incessantly weave words and flowers into the straw hats and bags piled beside them on the burning dusty street. No matter that these other Black women must work their sense of beauty into these things that we will take away as cheaply as we dare, or they will do without food.

We are not white, after all. The budget is limited. And we are harmlessly killing time between the poolside rum punch and "The Native Show on the Patio" that will play tonight outside the hotel restaurant.

This is my consciousness of race and class and gender identity as I notice the fixed relations between these other Black women and myself. They sell and I buy or I don't. They risk not eating. I risk going broke on my first vacation afternoon.

We are not particularly women anymore; we are parties to a transaction designed to set us against each other.

"Olive" is the name of the Black woman who cleans my hotel room. On my way to the beach I am wondering what "Olive" would say if I told her why I chose The Sheraton British Colonial; if I told her I wanted to swim. I wanted to sleep. I did not want to be harassed by the middle-aged waiter, or his nephew. I did not want to be raped by anybody (white or Black) at all and I calculated that my safety as a Black woman alone would best be assured by a multinational hotel corporation. In my experience, the big guys take customer complaints more seriously than the little ones. I would suppose that's one reason why they're big; they don't like to lose money anymore than I like to be bothered when I'm trying to read a goddamned book underneath a palm tree I paid $264 to get next to. A Black woman seeking refuge in a multinational corporation may seem like a contradiction to some, but there you are. In this case it's a coincidence of entirely different self-interests: Sheraton/cash = June Jordan's short run safety.

Anyway, I'm pretty sure "Olive" would look at me as though I came from someplace as far away as Brooklyn. Then she'd probably allow herself one indignant query before righteously removing her vacuum cleaner from my room; "and why in the first place you come down you without your husband?"

I cannot imagine how I would begin to answer her.

My "rights" and my "freedom" and my "desire" and a slew of other New World values; what would they sound like to this Black woman described on the card atop my hotel bureau as "Olive the Maid"? "Olive" is older than I am and I may smoke a cigarette while she changes the sheets on my bed. Whose rights? Whose freedom? Whose desire?

And why should she give a shit about mine unless I do something, for real, about hers?

It happens that the book that I finished reading under a palm tree earlier today was the novel, *The Bread Givers*, by Anzia Yezierska. Definitely autobiographical, Yezierska lays out the difficulties of being both female and "a person" inside a traditional Jewish family at the start of the 20th century. That any Jewish woman became anything more than the abused servant of her father or her husband is really an improbable piece of news. Yet Yezierska managed such an unlikely outcome for her own life. In *The Bread Givers*, the heroine also manages an important, although partial, escape from traditional Jewish female destiny. And in the unpardonable, despotic father, the Talmudic scholar of that Jewish family, did I not see my own and hate him twice, again? When the heroine, the young Jewish child, wanders the streets with a filthy pail she borrows to sell herring in order to raise the ghetto rent and when she cries, "Nothing was before me but the hunger in our house, and no bread for the next meal if I didn't sell the herring. No longer like a fire engine, but like a houseful of hungry mouths my heart cried, 'herring—herring! Two cents apiece!' " who would doubt the ease, the sisterhood of conversation possible between that white girl and the Black women selling straw bags on the streets of paradise because they do not want to die? And is it not obvious that the wife of that Talmudic scholar and "Olive," who cleans my room here at the hotel, have more in common than I can claim with either one of them?

This is my consciousness of race and class and gender identity as I collect wet towels, sunglasses, wristwatch, and head towards a shower.

I am thinking about the boy who loaned this novel to me. He's white and he's Jewish and he's pursuing an independent study project with me, at the State University where I teach whether or not I feel like it, where I teach without stint because, like the waiter, I am no fool. It's my job and either I work or I do without everything you need money to buy. The boy loaned me the novel because he thought I'd be interested to know how a Jewish-American writer used English so that the syntax, and therefore the cultural habits of mind expressed by the Yiddish language, could survive translation. He did this because he wanted to create another connection between us on the basis of language, between his knowledge/his love of Yiddish and my knowledge/my love of Black English.

He has been right about the forceful survival of the Yiddish. And I had become excited by this further evidence of the written voice of spoken language protected from the monodrone of "standard" English, and so we had grown closer on this account. But then our talk shifted to student affairs more generally, and I had learned that this student does not care one way or the other about currently jeopardized Federal Student Loan Programs because, as he explained it to me, they do not affect him. He does not need financial help outside his family. My own son, however, is Black. And I am the only family help available to him and that means, if Reagan succeeds in eliminating Federal programs to aid minority students, he will have to forget about furthering his studies, or he or I or both of us will have to hit the numbers pretty big. For these reasons of difference, the student and I had moved away from each other, even while we continued to talk.

My consciousness turned to race, again, and class.

Sitting in the same chair as the boy, several weeks ago, a graduate student came to discuss her grade. I praised the excellence of her final paper; indeed it had seemed to me an extraordinary pulling together of recent left brain/right brain research with the themes of transcendental poetry.

She told me that, for her part, she'd completed her reading of my political essays. "You are so lucky!" she exclaimed.

"What do you mean by that?"

"You have a cause. You have a purpose to your life."

I looked carefully at this white woman; what was she really saying to me?

"What do you mean?" I repeated.

"Poverty. Police violence. Discrimination in general."

(Jesus Christ, I thought: Is that her idea of lucky?)

"And how about you?" I asked.

"Me?"

"Yeah, you. Don't you have a cause?"

"Me? I'm just a middle-aged woman: a housewife and a mother. I'm a nobody."

For a while, I made no response.

First of all, speaking of race and class and gender in one breath, what she said meant that those lucky preoccupations of mine, from police violence to nuclear wipe-out, were not shared. They were mine and not hers. But here she sat, friendly as an old stuffed animal, beaming good will or more "luck" in my direction.

In the second place, what this white woman said to me meant that she did not believe she was "a person" precisely because she had fulfilled the traditional female functions revered by the father of that Jewish immigrant, Anzia Yezierska. And the woman in front of me was not a Jew. That was not the connection. The link was strictly female. Nevertheless, how should that woman and I, another female connect, beyond this bizarre exchange?

If she believed me lucky to have regular hurdles of discrimination then why shouldn't I insist that she's lucky to be a middle class white Wasp female who lives in such well-sanctioned and normative comfort that she even has the luxury to deny the power of the privileges that paralyze her life?

If she deserts me and "my cause" where we differ, if, for example, she abandons me to "my" problems of race, then why should I support her in "her" problems of housewifely oblivion?

Recollection of this peculiar moment brings me to the shower in the bathroom cleaned by "Olive." She reminds me of the usual Women's Studies curriculum because it has nothing to do with her or her job: you won't find "Olive" listed anywhere on the reading list. You will likewise seldom hear of Anzia Yezierska. But yes, you will find, from Florence Nightingale to Adrienne Rich, a white procession of independently well-to-do women writers. (Gertrude Stein/Virginia Woolf/Hilda Doolittle are standard names among the "essential" women writers.)

In other words, most of the women of the world—Black and First World and white who work because we must—most of the women of the world persist far from the heart of the usual Women's Studies syllabus.

Similarly, the typical Black History course will slide by the majority experience it pretends to represent. For example, Mary McLeod Bethune will scarcely receive as much attention as Nat Turner, even though Black women who bravely and efficiently provided for the education of Black people hugely outnumber those few Black men who led successful or doomed rebellions against slavery. In fact, Mary McLeod Bethune may not receive even honorable mention because Black History too often apes those ridiculous white history courses which produce such dangerous gibberish as The Sheraton British Colonial "history" of the Bahamas. Both Black and white history courses exclude from their central consideration those people who neither killed nor conquered anyone as the means to new identity, those people who took care of every one of the people who wanted to become "a person," those people who still take care of the life at issue: the ones who wash and who feed and who teach and who diligently decorate straw hats and bags with all of their historically unrequired gentle love: the women.

Oh the old rugged cross
on a hill far away
Well I cherish the old rugged cross

It's Good Friday in the Bahamas. Seventy-eight degrees in the shade. Except for Sheraton territory, everything's closed.

It so happens that for truly secular reasons I've been fasting for three days. My hunger has now reached nearly violent proportions. In the hotel sandwich shop, the Black woman handling the counter complains about the tourists; why isn't the shop closed and why don't the tourists stop eating for once in their lives. I'm famished and I order chicken salad and cottage cheese and lettuce and tomato and a hard boiled egg and a hot cross bun and apple juice.

She eyes me with disgust.

To be sure, the timing of my stomach offends her serious religious practices. Neither one of us apologizes to the other. She seasons the chicken salad to the peppery max while I listen to the loud radio gospel she plays to console herself. It's a country Black version of "The Old Rugged Cross."

As I heave much chicken into my mouth tears start. It's not the pepper. I am, after all, a West Indian daughter. It's the Good Friday music that dominates the humid atmosphere.

Well I cherish the old rugged cross

And I am back, faster than a 747, in Brooklyn, in the home of my parents where we are wondering, as we do every year, if the sky will darken until Christ has been buried in the tomb. The sky should darken if God is in His heavens. And then, around 3 p.m., at the conclusion of our mournful church service at the neighborhood St. Phillips, and even while we dumbly stare at the black cloth covering the gold altar and the slender unlit candles, the sun should return through the high gothic windows and vindicate our waiting faith that the Lord will rise again, on Easter.

How I used to bow my head at the very name of Jesus: ecstatic to abase myself in deference to His majesty.

My mouth is full of salad. I can't seem to eat quickly enough. I can't think how I should lessen the offense of my appetite. The other Black woman on the premises, the one who disapprovingly prepared this very tasty break from my fast, makes no remark. She is no fool. This is a job that she needs. I suppose she notices that at least I included a hot cross bun among my edibles. That's something in my favor. I decide that's enough.

I am suddenly eager to walk off the food. Up a fairly steep hill I walk without hurrying. Through the pastel desolation of the little town, the road brings me to a confectionary pink and white plantation house. At the gates, an unnecessarily large statue of Christopher Columbus faces me down, or tries to. His hand is fisted to one hip. I look back at him, laugh without deference, and turn left.

It's time to pack it up. Catch my plane. I scan the hotel room for things not to forget. There's that white report card on the bureau.

"Dear Guests:" it says, under the name "Olive." "I am your maid for the day. Please rate me: Excellent. Good. Average. Poor. Thank you."

I tuck this momento from the Sheraton British Colonial into my notebook. How would "Olive" rate *me*? What would it mean for us to seem "good" to each other? What would that rating require?

But I am hastening to leave. Neither turtle soup nor kidney pie nor any conch shell delight shall delay my departure. I have rested, here, in the Bahamas, and I'm ready to return to my usual job, my usual work. But the skin on my body has changed and so has my mind. On the Delta flight home I realize I am burning up, indeed.

So far as I can see, the usual race and class concepts of connection, or gender assumptions of unity, do not apply very well. I doubt that they ever did. Otherwise why would Black folks forever bemoan our lack of solidarity when the deal turns real. And if unity on the basis of sexual oppression is something natural, then why do we women, the majority people on the planet, still have a problem?

The plane's ready for takeoff. I fasten my seatbelt and let the tumult inside my head run free. Yes: race and class and gender remain as real as the weather. But what they must mean about the contact between two individuals is less obvious and, like the weather, not predictable.

And when these factors of race and class and gender absolutely collapse is whenever you try to use them as automatic concepts of connection. They may serve well as indicators of commonly felt conflict, but as elements of connection they seem about as reliable as precipitation probability for the day after the night before the day.

It occurs to me that much organizational grief could be avoided if people understood that partnership in misery does not necessarily provide for partnership for change: *When we get the monsters off our backs all of us may want to run in very different directions.*

And not only that: even though both "Olive" and "I" live inside a conflict neither one of us created, and even though both of us therefore hurt inside that conflict, I may be one of the monsters she needs to eliminate from her universe and, in a sense, she may be one of the monsters in mine.

I am reaching for the words to describe the difference between a common identity that has been imposed and the individual identity any one of us will choose, once she gains that chance.

That difference is the one that keeps us stupid in the face of new, specific information about somebody else with whom we are supposed to have a connection because a third party, hostile to both of us, has worked it so that the two of us, like it or not, share a common enemy. *What happens beyond the idea of that enemy and beyond the consequences of that enemy?*

I am saying that the ultimate connection cannot be the enemy. The ultimate connection must be the need that we find between us. It is not only who you are, in other words, but what we can do for each other that will determine the connection.

I am flying back to my job. I have been teaching contemporary women's poetry this semester. One quandary I have set myself to explore with my students is the one of taking responsibility without power. We had been wrestling ideas to the floor for several sessions when a young Black woman, a South African, asked me for help, after class.

Sokutu told me she was "in a trance" and that she'd been unable to eat for two weeks.

"What's going on?" I asked her, even as my eyes startled at her trembling and emaciated appearance.

"My husband. He drinks all the time. He beats me up. I go to the hospital. I can't eat. I don't know what/anything."

In my office, she described her situation. I did not dare to let her sense my fear and horror. She was dragging about, hour by hour, in dread. Her husband, a young Black South African, was drinking himself into more and more deadly violence against her.

Sokutu told me how she could keep nothing down. She weighed 90 lbs. at the outside, as she spoke to me. She'd already been hospitalized as a result of her husband's battering rage.

I knew both of them because I had organized a campus group to aid the liberation struggles of Southern Africa.

Nausea rose in my throat. What about this presumable connection: this husband and this wife fled from that homeland of hatred against them, and now what? He was destroying himself. If not stopped, he would certainly murder his wife.

She needed a doctor, right away. It was a medical emergency. She needed protection. It was a security crisis. She needed refuge for battered wives and personal therapy and legal counsel. She needed a friend.

I got on the phone and called every number in the campus directory that I could imagine might prove helpful. Nothing worked. There were no institutional resources designed to meet her enormous, multifaceted, and ordinary woman's need.

I called various students. I asked the Chairperson of the English Department for advice. I asked everyone for help.

Finally, another one of my students, Cathy, a young Irish woman active in campus IRA activities, responded. She asked for further details. I gave them to her.

"Her husband," Cathy told me, "is an alcoholic. You have to understand about alcoholics. It's not the same as anything else. And it's a disease you can't treat any old way."

I listened, fearfully. Did this mean there was nothing we could do?

"That's not what I'm saying," she said. "But you have to keep the alcoholic part of the thing central in everybody's mind, otherwise her husband will kill her. Or he'll kill himself."

She spoke calmly. I felt there was nothing to do but to assume she knew what she was talking about.

"Will you come with me?" I asked her, after a silence. "Will you come with me and help us figure out what to do next?"

Cathy said she would but that she felt shy: Sokutu comes from South Africa. What would she think about Cathy?

"I don't know," I said. "But let's go."

We left to find a dormitory room for the young battered wife.

It was late, now, and dark outside.

On Cathy's VW that I followed behind with my own car, was the sticker that reads BOBBY SANDS FREE AT LAST. My eyes blurred as I read and reread the words. This was another connection: Bobby Sands and Martin Luther King Jr. and who would believe it? I would not have believed it; I grew up terrorized by Irish kids who introduced me to the word "nigga."

And here I was following an Irish woman to the room of a Black South African. We were going to that room to try to save a life together.

When we reached the little room, we found ourselves awkward and large. Sokutu attempted to treat us with utmost courtesy, as though we were honored guests. She seemed surprised by Cathy, but mostly Sokutu was flushed with relief and joy because we were there, with her.

I did not know how we should ever terminate her heartfelt courtesies and address, directly, the reason for our visit: her starvation and her extreme physical danger.

Finally, Cathy sat on the floor and reached out her hands to Sokutu.

"I'm here," she said quietly, "Because June has told me what has happened to you. And I know what it is. Your husband is an alcoholic. He has a disease. I know what it is. My father was an alcoholic. He killed himself. He almost killed my mother. I want to be your friend."

"Oh," was the only small sound that escaped from Sokutu's mouth. And then she embraced the other student. And then everything changed and I watched all of this happen so I know that this happened: this connection.

And after we called the police and exchanged phone numbers and plans were made for the night and for the next morning, the young South African woman walked down the dormitory hallway, saying goodbye and saying thank you to us.

I walked behind them, the young Irish woman and the young South African, and I saw them walking as sisters walk, hugging each other, and whispering and sure of each other and I felt how it was not who they were but what they both know and what they were both preparing to do about what they know that was going to make them both free at last.

And I look out the windows of the plane and I see clouds that will not kill me and I know that someday soon other clouds may erupt to kill us all.

And I tell the stewardess No thanks to the cocktails she offers me. But I look about the cabin at the hundred strangers drinking as they fly and I think even here and even now I must make the connection real between me and these strangers everywhere before those other clouds unify this ragged bunch of us, too late.

ANGRY WOMEN ARE BUILDING: 4

Issues and Struggles Facing American Indian Women Today

Paula Gunn Allen

The central issue that confronts American Indian women throughout the hemisphere is survival, *literal survival*, both on a cultural and biological level. According to the 1980 census, population of American Indians is just over one million. This figure, which is disputed by some American Indians, is probably a fair estimate, and it carries certain implications.

Some researchers put our pre-contact population at more than 45 million, while others put it at around 20 million. The U.S. government long put it at

From: Paula Gunn Allen, *The Sacred Hoop: Recovering the Feminism in American Indian Traditions* (Boston: Beacon Press, 1986), pp. 189–193. © 1986, 1992 by Paula Gunn Allen. Reprinted by permission of Beacon Press.

450,000—a comforting if imaginary figure, though at one point it was put at around 270,000. If our current population is around one million; if, as some researchers estimate, around 25 percent of Indian women and 10 percent of Indian men in the United States have been sterilized without informed consent; if our average life expectancy is, as the best-informed research presently says, 55 years; if our infant mortality rate continues at well above national standards; if our average unemployment for all segments of our population—male, female, young, adult, and middle-aged—is between 60 and 90 percent; if the U.S. government continues its policy of termination, relocation, removal, and assimilation along with the destruction of wilderness, reservation land, and its resources, and severe curtailment of hunting, fishing, timber harvesting and water-use rights—then existing tribes are facing the threat of extinction which for several hundred tribal groups has already become fact in the past five hundred years.

In this nation of more than 200 million, the Indian people constitute less than one-half of one percent of the population. In a nation that offers refuge, sympathy, and billions of dollars in aid from federal and private sources in the form of food to the hungry, medicine to the sick, and comfort to the dying, the indigenous subject population goes hungry, homeless, impoverished, cut out of the American deal, new, old, and in between. Americans are daily made aware of the worldwide slaughter of native peoples such as the Cambodians, the Palestinians, the Armenians, the Jews—who constitute only a few groups faced with genocide in this century. We are horrified by South African apartheid and the removal of millions of indigenous African black natives to what is there called "homelands"—but this is simply a replay of nineteenth-century U.S. government removal of American Indians to reservations. Nor do many even notice the parallel or fight South African apartheid by demanding an end to its counterpart within the borders of the United States. The American Indian people are in a situation comparable to the imminent genocide in many parts of the world today. The plight of our people north and south of us is no better; to the south it is considerably worse. Consciously or unconsciously, deliberately, as a matter of national policy, or accidentally as a matter of "fate," *every single government*, right, left, or centrist in the western hemisphere is consciously or subconsciously dedicated to the extinction of those tribal people who live within its borders.

Within this geopolitical charnel house, American Indian women struggle on every front for the survival of our children, our people, our self-respect, our value systems, and our way of life. The past five hundred years testify to our skill at waging this struggle: for all the varied weapons of extinction pointed at our heads, we endure.

We survive war and conquest; we survive colonization, acculturation, assimilation; we survive beating, rape, starvation, mutilation, sterilization, abandonment, neglect, death of our children, our loved ones, destruction of our land, our homes, our past, and our future. We survive, and we do more than just survive. We bond, we care, we fight, we teach, we nurse, we bear, we feed, we earn, we laugh, we love, we hang in there, no matter what.

Of course, some, many of us, just give up. Many are alcoholics, many are addicts. Many abandon the children, the old ones. Many commit suicide. Many become violent, go insane. Many go "white" and are never seen or heard from again. But enough hold on to their traditions and their ways so that even after almost five hundred brutal years, we endure. And we even write songs and poems, make paintings and drawings that say "We walk in beauty. Let us continue."

Currently our struggles are on two fronts: physical survival and cultural survival. For women this means fighting alcoholism and drug abuse (our own and that of our husbands, lovers, parents, children);[1] poverty; affluence—a destroyer of people who are not traditionally socialized to deal with large sums of money; rape, incest, battering by Indian men; assaults on fertility and other health matters by the Indian Health Service and the Public Health Service; high infant mortality due to substandard medical care, nutrition, and health information; poor educational opportunities or education that takes us away from our traditions, language, and communities; suicide, homicide, or similar expressions of self-hatred; lack of economic opportunities; substandard housing; sometimes violent and always virulent racist attitudes and behaviors directed against us by an entertainment and educational system that wants only one thing from Indians: our silence, our invisibility, and our collective death.

A headline in the *Navajo Times* in the fall of 1979 reported that rape was the number one crime on the Navajo reservation. In a professional mental health journal of the Indian Health Services, Phyllis Old Dog Cross reported that incest and rape are common among Indian women seeking services and that their incidence is increasing. "It is believed that at least 80 percent of the Native Women seen at the regional psychiatric service center (5 state area) have experienced some sort of sexual assault."[2] Among the forms of abuse being suffered by Native American women, Old Dog Cross cites a recent phenomenon, something called "training." This form of gang rape is "a punitive act of a group of males who band together and get even or take revenge on a selected woman."[3]

These and other cases of violence against women are powerful evidence that the status of women within the tribes has suffered grievous decline since contact, and the decline has increased in intensity in recent years. The amount of violence against women, alcoholism, and violence, abuse, and neglect by women against their children and their aged relatives have all increased. These social ills were virtually unheard of among most tribes fifty years ago, popular American opinion to the contrary. As Old Dog Cross remarks:

> Rapid, unstable and irrational change was required of the Indian people if they were to survive. Incredible loss of all that had meaning was the norm. Inhuman treatment, murder, death, and punishment was a typical experience for all the tribal groups and some didn't survive.
>
> The dominant society devoted its efforts to the attempt to change the Indian into a white-Indian. No inhuman pressure to effect this change was overlooked. These pressures included starvation, incarceration and enforced education. Religious and healing customs were banished.
>
> In spite of the years of oppression, the Indian and the Indian spirit survived. Not, however, without adverse effect. One of the major effects was the loss of cultured values and the concomitant loss of personal identity . . . The Indian was taught to be ashamed of being Indian and to emulate the non-Indian. In short, "white was right." For the Indian male, the only route to be successful, to be good, to be right, and to have an identity was to be as much like the white man as he could.[4]

Often it is said that the increase of violence against women is a result of various sociological factors such as oppression, racism, poverty, hopelessness, emasculation of men, and loss of male self-esteem as their own place within traditional society has been systematically destroyed by increasing urbanization, industrialization, and institutionalization, but seldom do we notice that for the past forty to fifty years, American popular media have depicted American Indian men as bloodthirsty savages devoted to treating women cruelly. While traditional Indian men seldom did any such thing—and in fact among most tribes abuse of women was simply unthinkable, as was abuse of children or the aged—the lie about "usual" male Indian behavior seems to have taken root and now bears its brutal and bitter fruit.

Image casting and image control constitute the central process that American Indian women must come to terms with, for on that control rests our sense of self, our claim to a past and to a future that we define and that we build. Images of Indians in media and educational materials profoundly influence how we act, how we relate to the world and to each other, and how we value ourselves. They also determine to a large extent how our men act toward us, toward our children, and toward each other. The popular American media image of Indian people as savages with no conscience, no compassion, and no sense of the value of human life and human dignity was hardly true of the tribes—however true it was of the invaders. But as Adolf Hitler noted a little over fifty years ago, if you tell a lie big enough and often enough, it will be believed. Evidently, while Americans and people all over the world have been led into a deep and unquestioned belief that American Indians are cruel savages, a number of American Indian men have been equally deluded into internalizing that image and acting on it. Media images, literary images, and artistic images, particularly those embedded in popular culture, must be changed before Indian women will see much relief from the violence that destroys so many lives.

To survive culturally, American Indian women must often fight the United States government, the tribal governments, women and men of their tribe or their urban community who are virulently misogynist or who are threatened by attempts to change the images foisted on us over the centuries by whites. The colonizers' revisions of our lives, values, and histories have devastated us at the most critical level of all—that of our own minds, our own sense of who we are.

Many women express strong opposition to those who would alter our life supports, steal our tribal lands, colonize our cultures and cultural expressions, and revise our very identities. We must strive to maintain tribal status; we must make certain that the tribes continue to be legally recognized entities, sovereign nations within the larger United States, and we must wage this struggle in many ways—political, educational, literary, artistic, individual, and communal. We are doing all we can: as mothers and grandmothers; as family members and tribal members; as professionals, workers, artists, shamans, leaders, chiefs, speakers, writers, and organizers, we daily demonstrate that we have no intention of disappearing, of being silent, or of quietly acquiescing in our extinction.

NOTES

1. It is likely, say some researchers, that fetal alcohol syndrome, which is serious among many Indian groups, will be so serious among the White Mountain Apache and the Pine Ridge Sioux that if present trends continue, by the year 2000 some people estimate that almost one half of all children born on those reservations will in some way be affected by FAS. (Michael Dorris, Native American Studies, Dartmouth College, private conversation. Dorris has done extensive research into the syndrome as it affects native populations in the United States as well as in New Zealand.)

2. Phyllis Old Dog Cross, "Sexual Abuse, a New Threat to the Native American Woman: An Overview," *Listening Post: A Periodical of the Mental Health Programs of Indian Health Services*, vol. 6, no. 2 (April 1982), p. 18.

3. Old Dog Cross, p. 18.

4. Old Dog Cross, p. 20.

OPPRESSION

5

Marilyn Frye

It is a fundamental claim of feminism that women are oppressed. The word "oppression" is a strong word. It repels and attracts. It is dangerous and dangerously fashionable and endangered. It is much misused, and sometimes not innocently.

The statement that women are oppressed is frequently met with the claim that men are oppressed too. We hear that oppressing is oppressive to those who oppress as well as to those they oppress. Some men cite as evidence of their oppression their much-advertised inability to cry. It is tough, we are told, to be masculine. When the stresses and frustrations of being a man are cited as evidence that oppressors are oppressed by their oppressing, the word "oppression" is being stretched to meaninglessness; it is treated as though its scope includes any and all human experience of limitation or suffering, no matter the cause, degree or consequence. Once such usage has been put over on us, then if ever we deny that any person or group is oppressed, we seem to imply that we think they never suffer and have no feelings. We are accused of insensitivity; even of bigotry. For women, such accusation is particularly intimidating, since sensitivity is one of the few virtues that has been assigned to us. If we are found insensitive, we may fear we have no redeeming traits at all and perhaps are not real women. Thus are we silenced before we begin: the name of our situation drained of meaning and our guilt mechanisms tripped.

But this is nonsense. Human beings can be miserable without being oppressed, and it is perfectly consistent to deny that a person or group is oppressed without denying that they have feelings or that they suffer. . . .

The root of the word "oppression" is the element "press." *The press of the crowd; pressed into military service; to press a pair of pants; printing press; press the button.* Presses are used to mold things or flatten them or reduce them in bulk, sometimes to reduce them by squeezing out the gases or liquids in them. Something pressed is something caught between or among forces and barriers

---

From: Marilyn Frye, *The Politics of Reality* (Trumansburg, N.Y.: The Crossing Press, 1983), pp. 1–16. Reprinted by permission.

which are so related to each other that jointly they restrain, restrict or prevent the thing's motion or mobility. Mold. Immobilize. Reduce.

The mundane experience of the oppressed provides another clue. One of the most characteristic and ubiquitous features of the world as experienced by oppressed people is the double bind—situations in which options are reduced to a very few and all of them expose one to penalty, censure or deprivation. For example, it is often a requirement upon oppressed people that we smile and be cheerful. If we comply, we signal our docility and our acquiescence in our situation. We need not, then, be taken note of. We acquiesce in being made invisible, in our occupying no space. We participate in our own erasure. On the other hand, anything but the sunniest countenance exposes us to being perceived as mean, bitter, angry or dangerous. This means, at the least, that we may be found "difficult" or unpleasant to work with, which is enough to cost one one's livelihood; at worst, being seen as mean, bitter, angry or dangerous has been known to result in rape, arrest, beating and murder. One can only choose to risk one's preferred form and rate of annihilation.

Another example: It is common in the United States that women, especially younger women, are in a bind where neither sexual activity nor sexual inactivity is all right. If she is heterosexually active, a woman is open to censure and punishment for being loose, unprincipled or a whore. The "punishment" comes in the form of criticism, snide and embarrassing remarks, being treated as an easy lay by men, scorn from her more restrained female friends. She may have to lie and hide her behavior from her parents. She must juggle the risks of unwanted pregnancy and dangerous contraceptives. On the other hand, if she refrains from heterosexual activity, she is fairly constantly harassed by men who try to persuade her into it and pressure her to "relax" and "let her hair down"; she is threatened with labels like "frigid," "uptight," "man-hater," "bitch" and "cocktease." The same parents who would be disapproving of her sexual activity may be worried by her inactivity because it suggests she is not or will not be popular, or is not sexually normal. She may be charged with lesbianism. If a woman is raped, then if she has been heterosexually active she is subject to the presumption that she liked it (since her activity is presumed to show that she likes sex), and if she has not been heterosexually active, she is subject to the presumption that she liked it (since she is supposedly "repressed and frustrated"). Both heterosexual activity and heterosexual nonactivity are likely to be taken as proof that you wanted to be raped, and hence, of course, weren't *really* raped at all. You can't win. You are caught in a bind, caught between systematically related pressures.

Women are caught like this, too, by networks of forces and barriers that expose one to penalty, loss or contempt whether one works outside the home or not, is on welfare or not, bears children or not, raises children or not, marries or not, stays married or not, is heterosexual, lesbian, both or neither. Economic necessity; confinement to racial and/or sexual job ghettos; sexual harassment; sex discrimination; pressures of competing expectations and judgments about *women, wives* and *mothers* (in the society at large, in racial and ethnic subcultures and in one's own mind); dependence (full or partial) on husbands, parents or the state; commitment to political ideas; loyalties to racial or ethnic or other "minority" groups; the demands of self-respect and responsibilities to others. Each of these factors exists in complex tension with every other, penalizing or prohibiting all of the apparently available options. And nipping at one's heels, always, is the endless pack of little things. If one dresses one way, one is subject to the assumption that one is advertising one's sexual availability; if one dresses another way, one appears to "not care about oneself" or to be "unfeminine." If one uses "strong language," one invites categorization as a whore or slut; if one does not, one invites categorization as a "lady"—one too delicately constituted to cope with robust speech or the realities to which it presumably refers.

The experience of oppressed people is that the living of one's life is confined and shaped by forces and barriers which are not accidental or occasional and hence avoidable, but are systematically related to each other in such a way as to catch one between and among them and restrict or penalize motion in any direction. It is the experience of being caged in: all avenues, in every direction, are blocked or booby trapped.

Cages. Consider a birdcage. If you look very closely at just one wire in the cage, you cannot see the other wires. If your conception of what is before you is determined by this myopic focus, you could look at that one wire, up and down the length of it, and be unable to see why a bird would not just fly around the wire any time it wanted to go somewhere. Furthermore, even if, one day at a time, you myopically inspected each wire, you still could not see why a bird would have trouble going past the wires to get anywhere. There is no physical property of any one wire, *nothing* that the closest scrutiny could discover, that will reveal how a bird could be inhibited or harmed by it except in the most accidental way. It is only when you step back, stop looking at the wires one by one, microscopically, and take a macroscopic view of the whole cage, that you can see why the bird does not go anywhere; and then you will see it in a moment. It will require no great subtlety of mental powers. It is perfectly *obvious* that the bird is surrounded by a network of systematically related barriers, no one of which would be the least hindrance to its flight, but which, by their relations to each other, are as confining as the solid walls of a dungeon.

It is now possible to grasp one of the reasons why oppression can be hard to see and recognize: one can study the elements of an oppressive structure with great care and some good will without seeing the structure as a whole, and hence without seeing or being able to understand that one is looking at a cage and that there are people there who are caged, whose motion and mobility are restricted, whose lives are shaped and reduced.

The arresting of vision at a microscopic level yields such common confusion as that about the male door-opening ritual. This ritual, which is remarkably widespread across classes and races, puzzles many people, some of whom do and some of whom do not find it offensive. Look at the scene of the two people approaching a door. The male steps slightly ahead and opens the door. The male holds the door open while the female glides through. Then the male goes through. The door closes after them. "Now how," one innocently asks, "can those crazy womenslibbers say that is oppressive? The guy *removed* a barrier to the lady's smooth and unruffled progress." But each repetition of this ritual has a place in a pattern, in fact in several patterns. One has to shift the level of one's perception in order to see the whole picture.

The door-opening pretends to be a helpful service, but the helpfulness is false. This can be seen by noting that it will be done whether or not it makes any practical sense. Infirm men and men burdened with packages will open doors for able-bodied women who are free of physical burdens. Men will impose themselves awkwardly and jostle everyone in order to get to the door first. The act is not determined by convenience or grace. Furthermore, these very numerous acts of unneeded or even noisome "help" occur in counterpoint to a pattern of men not being helpful in many practical ways in which women might welcome help. What *women* experience is a world in which gallant princes charming commonly make a fuss about being helpful and providing small services when help and services are of little or no use, but in which there are rarely ingenious and adroit princes at hand when substantial assistance is really wanted either in mundane affairs or in situations of threat, assault or terror. There is no help with the (his) laundry; no help typing a report at 4:00 a.m.; no help in mediating disputes among relatives or children. There is nothing but advice that women should stay indoors after dark, be chaperoned by a man, or when it comes down to it, "lie back and enjoy it."

The gallant gestures have no practical meaning. Their meaning is symbolic. The door-opening and similar services provided are services which really are needed by people who are for one reason or another incapacitated— unwell, burdened with parcels, etc. So the message is that women are incapable. The detachment of the acts from the concrete realities of what women need and do not need is a vehicle for the message that women's actual needs and interests are unimportant or irrelevant. Finally, these gestures imitate the behavior of servants toward masters and thus mock women, who are in most respects the servants and caretakers of men. The message of the false helpfulness of male gallantry is female dependence, the invisibility or insignificance of women, and contempt for women.

One cannot see the meanings of these rituals if one's focus is riveted upon the individual event in all its particularity, including the particularity of the individual man's present conscious intentions and motives and the individual woman's conscious perception of the event in the moment. It seems sometimes that people take a deliberately myopic view and fill their eyes with things seen microscopically in order not to see macroscopically. At any rate, whether it is deliberate or not, people can and do fail to see the oppression of women because they fail to see macroscopically and hence fail to see the various elements of the situation as systematically related in larger schemes.

As the cageness of the birdcage is a macroscopic phenomenon, the oppressiveness of the situations in which women live our various and different lives is a macroscopic phenomenon. Neither can be *seen* from a microscopic perspective. But when you look macroscopically you can see it—a network of forces and barriers which are systematically related and which conspire to the immobilization, reduction and molding of women and the lives we live. . . .

A DIFFERENT MIRROR

6

Ronald T. Takaki

I had flown from San Francisco to Norfolk and was riding in a taxi to my hotel to attend a conference on multiculturalism. Hundreds of educators from across the country were meeting to discuss the need for greater cultural diversity in the curriculum. My driver and I chatted about the weather and the tourists. The sky was cloudy, and Virginia Beach was twenty minutes away. The rearview mirror reflected a white man in his forties. "How long have you been

From: Ronald T. Takaki, *A Different Mirror: A History of Multicultural America* (Boston: Little, Brown and Company, 1993), pp. 1–17. Copyright © 1993 by Ronald Takaki. Reprinted by permission of the publisher.

in this country?" he asked. "All my life," I replied, wincing. "I was born in the United States." With a strong southern drawl, he remarked: "I was wondering because your English is excellent!" Then, as I had many times before, I explained: "My grandfather came here from Japan in the 1880s. My family has been here, in America, for over a hundred years." He glanced at me in the mirror. Somehow I did not look "American" to him; my eyes and complexion looked foreign.

Suddenly, we both became uncomfortably conscious of a racial divide separating us. An awkward silence turned my gaze from the mirror to the passing landscape, the shore where the English and the Powhatan Indians first encountered each other. Our highway was on land that Sir Walter Raleigh had renamed "Virginia" in honor of Elizabeth I, the Virgin Queen. In the English cultural appropriation of America, the indigenous peoples themselves would become outsiders in their native land. Here, at the eastern edge of the continent, I mused, was the site of the beginning of multicultural America. Jamestown, the English settlement founded in 1607, was nearby: the first twenty Africans were brought here a year before the Pilgrims arrived at Plymouth Rock. Several hundred miles offshore was Bermuda, the "Bermoothes" where William Shakespeare's Prospero had landed and met the native Caliban in *The Tempest*. Earlier, another voyager had made an Atlantic crossing and unexpectedly bumped into some islands to the south. Thinking he had reached Asia, Christopher Columbus mistakenly identified one of the islands as "Cipango" (Japan). In the wake of the admiral, many peoples would come to America from different shores, not only from Europe but also Africa and Asia. One of them would be my grandfather. My mental wandering across terrain and time ended abruptly as we arrived at my destination. I said good-bye to my driver and went into the hotel, carrying a vivid reminder of why I was attending this conference.

Questions like the one my taxi driver asked me are always jarring, but I can understand why he could not see me as American. He had a narrow but widely shared sense of the past—a history that has viewed American as European in ancestry. "Race," Toni Morrison explained, has functioned as a "metaphor" necessary to the "construction of Americanness": in the creation of our national identity, "American" has been defined as "white."[1]

But America has been racially diverse since our very beginning on the Virginia shore, and this reality is increasingly becoming visible and ubiquitous. Currently, one-third of the American people do not trace their origins to Europe; in California, minorities are fast becoming a majority. They already predominate in major cities across the country—New York, Chicago, Atlanta, Detroit, Philadelphia, San Francisco, and Los Angeles.

This emerging demographic diversity has raised fundamental questions about America's identity and culture. In 1990, *Time* published a cover story on "America's Changing Colors." "Someday soon," the magazine announced, "white Americans will become a minority group." How soon? By 2056, most Americans will trace their descent to "Africa, Asia, the Hispanic world, the Pacific Islands, Arabia—almost anywhere but white Europe." This dramatic change in our nation's ethnic composition is altering the way we think about ourselves. "The deeper significance of America's becoming a majority non-white society is what it means to the national psyche, to individuals' sense of themselves and their nation—their idea of what it is to be American." . . .[2]

What is fueling the debate over our national identity and the content of our curriculum is America's intensifying racial crisis. The alarming signs and symptoms seem to be everywhere—the killing of Vincent Chin in Detroit, the black boycott of a Korean grocery store in Flatbush, the hysteria in Boston over the Carol Stuart murder, the battle between white sportsmen and Indians over tribal fishing rights in Wisconsin, the Jewish-black clashes in Brooklyn's Crown Heights, the black-Hispanic competition for jobs and educational resources in Dallas, which *Newsweek* described as "a conflict of the have-nots," and the Willie Horton campaign commercials, which widened the divide between the suburbs and the inner cities.[3]

This reality of racial tension rudely woke America like a fire bell in the night on April 29, 1992. Immediately after four Los Angeles police officers were found not guilty of brutality against Rodney King, rage exploded in Los Angeles. Race relations reached a new nadir. During the nightmarish rampage, scores of people were killed, over two thousand injured, twelve thousand arrested, and almost a billion dollars' worth of property destroyed. The live televised images mesmerized America. The rioting and the murderous melee on the streets resembled the fighting in Beirut and the West Bank. The thousands of fires burning out of control and the dark smoke filling the skies brought back images of the burning oil fields of Kuwait during Desert Storm. Entire sections of Los Angeles looked like a bombed city. "Is this America?" many shocked viewers asked. "Please, can we get along here," pleaded Rodney King, calling for calm. "We all can get along. I mean, we're all stuck here for a while. Let's try to work it out."[4]

But how should "we" be defined? Who are the people "stuck here" in America? One of the lessons of the Los Angeles explosion is the recognition of the fact that we are a multiracial society and that race can no longer be defined in the binary terms of white and black. "We" will have to include Hispanics and Asians. While blacks currently constitute 13 percent of the Los Angeles population, Hispanics represent 40 percent. The 1990 census revealed that South Central Los Angeles, which was predominantly black in 1965 when the Watts rebellion occurred, is now 45 percent Hispanic. A majority of the first 5,438 people arrested were Hispanic, while 37 percent were black. Of the fifty-eight people who died in the riot, more than a third were Hispanic, and about 40 percent of the businesses destroyed were Hispanic-owned. Most of the other shops and stores were Korean-owned. The dreams of many Korean immigrants went up in smoke during the riot: two thousand Korean-owned businesses were damaged or demolished, totaling about $400 million in losses. There is evidence indicating they were targeted. "After all," explained a black gang member, "we didn't burn our community, just *their* stores."[5]

"I don't feel like I'm in America anymore," said Denisse Bustamente as she watched the police protecting the firefighters. "I feel like I am far away." Indeed, Americans have been witnessing ethnic strife erupting around the world—the rise of neo-Nazism and the murder of Turks in Germany, the ugly "ethnic cleansing" in Bosnia, the terrible and bloody clashes between Muslims and Hindus in India. Is the situation here different, we have been nervously wondering, or do ethnic conflicts elsewhere represent a prologue for America? What is the nature of malevolence? Is there a deep, perhaps primordial, need for group identity rooted in hatred for the other? Is ethnic pluralism possible for America? But answers have been limited. Television reports have been little more than thirty-second sound bites. Newspaper articles have been mostly superficial descriptions of racial antagonisms and the current urban malaise. What is lacking is historical context; consequently, we are left, feeling bewildered.[6]

How did we get to this point, Americans everywhere are anxiously asking. What does our diversity mean, and where is it leading us? *How* do we work it out in the post–Rodney King era?

Certainly one crucial way is for our society's various ethnic groups to develop a greater understanding of each other. For example, how can African Americans and Korean Americans work it out unless they learn about each other's cultures, histories, and also economic situations? This need to share knowledge about our ethnic diversity has acquired new importance and has given new urgency to the pursuit for a more accurate history. . . .

While all of America's many groups cannot be covered in one book, the English immigrants and their descendants require attention, for they possessed inordinate power to define American culture and make public policy. What men like John Winthrop, Thomas Jefferson, and Andrew Jackson thought as well as did mattered greatly to all of us and was consequential for everyone. A broad range of groups has been selected: African Americans, Asian Americans, Chicanos, Irish, Jews, and Indians. While together they help to explain general patterns in our society, each has contributed to the making of the United States.

African Americans have been the central minority throughout our country's history. They were initially brought here on a slave ship in 1619. Actually, these first twenty Africans might not have been slaves; rather, like most of the white laborers, they were probably indentured servants. The transformation of Africans into slaves is the story of the "hidden" origins of slavery. How and when was it decided to institute a system of bonded black labor? What happened, while freighted with racial significance, was actually conditioned by class conflicts within white society. Once established, the "peculiar institution" would have consequences for centuries to come. During the nineteenth century, the political storm over slavery almost destroyed the nation. Since the Civil War and emancipation, race has continued to be largely defined in relation to African Americans—segregation, civil rights, the underclass, and affirmative action. Constituting the largest minority group in our society, they have been at the cutting edge of the Civil Rights Movement. Indeed, their struggle has been a constant reminder of America's moral vision as a country committed to the principle of liberty. Martin Luther King clearly understood this truth when he wrote from a jail cell: "We will reach the goal of freedom in Birmingham and all over the nation, because the goal of America is freedom. Abused and scorned though we may be, our destiny is tied up with America's destiny."[7]

Asian Americans have been here for over one hundred and fifty years, before many European immigrant groups. But as "strangers" coming from a "different shore," they have been stereotyped as "heathen," exotic, and unassimilable. Seeking "Gold Mountain," the Chinese arrived first, and what happened to them influenced the reception of the Japanese, Koreans, Filipinos, and Asian Indians as well as the Southeast Asian refugees like the Vietnamese and the Hmong. The 1882 Chinese Exclusion Act was the first law that prohibited the entry of immigrants on the basis of nationality. The Chinese condemned this restriction as racist and tyrannical. "They call us 'Chink,'" complained a Chinese immigrant, cursing the "white demons." "They think we no good! America cuts us off. No more come now, too bad!" This precedent later provided a basis for the restriction of European immigrant groups such as Italians, Russians, Poles, and Greeks. The Japanese painfully discovered that their accomplishments in America did not lead to acceptance, for during World War II, unlike Italian Americans and German Americans, they were placed in internment camps. Two-thirds of them were citizens by birth. "How could I as a 6-month-old child born in this country," asked Congressman Robert Matsui years later, "be declared by my own

Government to be an enemy alien?" Today, Asian Americans represent the fastest-growing ethnic group. They have also become the focus of much mass media attention as "the Model Minority" not only for blacks and Chicanos, but also for whites on welfare and even middle-class whites experiencing economic difficulties.[8]

Chicanos represent the largest group among the Hispanic population, which is projected to outnumber African Americans. They have been in the United States for a long time, initially incorporated by the war against Mexico. The treaty had moved the border between the two countries, and the people of "occupied" Mexico suddenly found themselves "foreigners" in their "native land." As historian Albert Camarillo pointed out, the Chicano past is an integral part of America's westward expansion, also known as "manifest destiny." But while the early Chicanos were a colonized people, most of them today have immigrant roots. Many began the trek to El Norte in the early twentieth century. "As I had heard a lot about the United States," Jesus Garza recalled, "it was my dream to come here." "We came to know families from Chihuahua, Sonora, Jalisco, and Durango," stated Ernesto Galarza. "Like ourselves, our Mexican neighbors had come this far moving step by step, working and waiting, as if they were feeling their way up a ladder." Nevertheless, the Chicano experience has been unique, for most of them have lived close to their homeland—a proximity that has helped reinforce their language, identity, and culture. This migration to El Norte has continued to the present. Los Angeles has more people of Mexican origin than any other city in the world, except Mexico City. A mostly mestizo people of Indian as well as African and Spanish ancestries, Chicanos currently represent the largest minority group in the Southwest, where they have been visibly transforming culture and society.[9]

The Irish came here in greater numbers than most immigrant groups. Their history has been tied to America's past from the very beginning. Ireland represented the earliest English frontier: the conquest of Ireland occurred before the colonization of America, and the Irish were the first group that the English called "savages." In this context, the Irish past foreshadowed the Indian future. During the nineteenth century, the Irish, like the Chinese, were victims of British colonialism. While the Chinese fled from the ravages of the Opium Wars, the Irish were pushed from their homeland by "English tyranny." Here they became construction workers and factory operatives as well as the "maids" of America. Representing a Catholic group seeking to settle in a fiercely Protestant society, the Irish immigrants were targets of American nativist hostility. They were also what historian Lawrence J. McCaffrey called "the pioneers of the American urban ghetto," "previewing" experiences that would later be shared by the Italians, Poles, and other groups from southern and eastern Europe. Furthermore, they offer contrast to the immigrants from Asia. The Irish came about the same time as the Chinese, but they had a distinct advantage: the Naturalization Law of 1790 had reserved citizenship for "whites" only. Their compatible complexion allowed them to assimilate by blending into American society. In making their journey successfully into the mainstream, however, these immigrants from Erin pursued an Irish "ethnic" strategy: they promoted "Irish" solidarity in order to gain political power and also to dominate the skilled blue-collar occupations, often at the expense of the Chinese and blacks.[10]

Fleeing pogroms and religious persecution in Russia, the Jews were driven from what John Cuddihy described as the "Middle Ages into the Anglo-American world of the *goyim* 'beyond the pale.' " To them, America represented the Promised Land. This vision led Jews to struggle not only for themselves but also for other oppressed groups, especially blacks. After the 1917 East St. Louis race riot, the Yiddish *Forward* of New York compared this anti-black violence to a 1903 pogrom in Russia: "Kishinev and St. Louis—the same soil, the same people." Jews cheered when Jackie Robinson broke into the Brooklyn Dodgers in 1947. "He was adopted as the surrogate hero by many of us growing up at the time," recalled Jack Greenberg of the NAACP Legal Defense Fund. "He was the way we saw ourselves triumphing against the forces of bigotry and ignorance." Jews stood shoulder to shoulder with blacks in the Civil Rights Movement: two-thirds of the white volunteers who went south during the 1964 Freedom Summer were Jewish. Today Jews are considered a highly successful "ethnic" group. How did they make such great socioeconomic strides? This question is often reframed by neoconservative intellectuals like Irving Kristol and Nathan Glazer to read: if Jewish immigrants were able to lift themselves from poverty into the mainstream through self-help and education without welfare and affirmative action, why can't blacks? But what this thinking overlooks is the unique history of Jewish immigrants, especially the initial advantages of many of them as literate and skilled. Moreover, it minimizes the virulence of racial prejudice rooted in American slavery.[11]

Indians represent a critical contrast, for theirs was not an immigrant experience. The Wampanoags were on the shore as the first English strangers arrived in what would be called "New England." The encounters between Indians and whites not only shaped the course of race relations, but also influenced the very culture and identity of the general society. The architect of Indian removal, President Andrew Jackson told Congress: "Our conduct toward these people is deeply interesting to the national character." Frederick Jackson Turner understood the meaning of this observation when he identified the frontier as our transforming crucible. At first, the European newcomers had to wear Indian moccasins and shout the war cry. "Little by little," as they subdued the wilderness, the pioneers became "a new product" that was "American." But Indians have had a different view of this entire process. "The white man," Luther Standing Bear of the Sioux explained, "does not understand the Indian for the reason that he does not understand America." Continuing to be "troubled with primitive fears," he has "in his consciousness the perils of this frontier continent. . . . The man from Europe is still a foreigner and an alien. And he still hates the man who questioned his path across the continent." Indians questioned what Jackson and Turner trumpeted as "progress." For them, the frontier had a different "significance": their history was how the West was lost. But their story has also been one of resistance. As Vine Deloria declared, "Custer died for your sins."[12]

By looking at these groups from a multicultural perspective, we can comparatively analyze their experiences in order to develop an understanding of their differences and similarities. Race, we will see, has been a social construction that has historically set apart racial minorities from European immigrant groups. Contrary to the notions of scholars like Nathan Glazer and Thomas Sowell, race in America has not been the same as ethnicity. A broad comparative focus also allows us to see how the varied experiences of different racial and ethnic groups occurred within shared contexts.

During the nineteenth century, for example, the Market Revolution employed Irish immigrant laborers in New England factories as it expanded cotton fields worked by enslaved blacks across Indian lands toward Mexico. Like blacks, the Irish newcomers were stereotyped as "savages," ruled by passions rather than "civilized" virtues such as self-control and hard work. The Irish saw themselves as the "slaves" of British oppressors, and during a visit to Ireland in the 1840s, Frederick Douglass found that the "wailing notes" of the Irish ballads reminded him of the "wild notes" of slave songs. The United States annexation of California, while incorporating Mexicans, led to trade with Asia and the migration of "strangers" from Pacific shores. In 1870, Chinese immigrant laborers were transported to Massachusetts as scabs to break an Irish immigrant strike; in response, the Irish recognized the need for interethnic working-class solidarity and tried to organize a Chinese lodge of the Knights of St. Crispin. After the Civil War, Mississippi planters recruited Chinese immigrants to discipline the newly freed blacks. During the debate over an immigration exclusion bill in 1882, a senator asked: If Indians could be located on reservations, why not the Chinese?[13]

Other instances of our connectedness abound. In 1903, Mexican and Japanese farm laborers went on strike together in California: their union officers had names like Yamaguchi and Lizarras, and strike meetings were conducted in Japanese and Spanish. The Mexican strikers declared that they were standing in solidarity with their "Japanese brothers" because the two groups had toiled together in the fields and were now fighting together for a fair wage. Speaking in impassioned Yiddish during the 1909 "uprising of twenty thousand" strikers in New York, the charismatic Clara Lemlich compared the abuse of Jewish female garment workers to the experience of blacks: "[The bosses] yell at the girls and 'call them down' even worse than I imagine the Negro slaves were in the South." During the 1920s, elite universities like Harvard worried about the increasing numbers of Jewish students, and new admissions criteria were instituted to curb their enrollment. Jewish students were scorned for their studiousness and criticized for their "clannishness," Recently, Asian-American students have been the targets of similar complaints: they have been called "nerds" and told there are "too many" of them on campus.[14]

Indians were already here, while blacks were forcibly transported to America, and Mexicans were initially enclosed by America's expanding border. The other groups came here as immigrants: for them, America represented liminality—a new world where they could pursue extravagant urges and do things they had thought beyond their capabilities. Like the land itself, they found themselves "betwixt and between all fixed points of classification." No longer fastened as fiercely to their old countries, they felt a stirring to become new people in a society still being defined and formed.[15]

These immigrants made bold and dangerous crossings, pushed by political events and economic hardships in their homelands and pulled by America's demand for labor as well as by their own dreams for a better life. "By all means let me go to America," a young man in Japan begged his parents. He had calculated that in one year as a laborer here he could save almost a thousand yen—an amount equal to the income of a governor in Japan. "My dear Father," wrote an immigrant Irish girl living in New York, "Any man or woman without a family are fools that would not venture and come to this plentyful Country where no man or woman ever hungered." In the shtetls of Russia, the cry "To America!" roared like "wild-fire." "America was in everybody's mouth," a Jewish immigrant recalled. "Businessmen talked [about] it over their accounts; the market women made up their quarrels that they might discuss it from stall to stall; people who had relatives in the famous land went around reading their letters." Similarly, for Mexican immigrants crossing the border in the early twentieth century, El Norte became the stuff of overblown hopes. "If only you could see how nice the United States is," they said, "that is why the Mexicans are crazy about it."[16]

The signs of America's ethnic diversity can be discerned across the continent—Ellis Island, Angel Island, Chinatown, Harlem, South Boston, the Lower East Side, places with Spanish names like Los Angeles and San Antonio or Indian names like Massachusetts and Iowa. Much of what is familiar in America's cultural landscape actually has ethnic origins. The Bing cherry was developed by an early Chinese immigrant named Ah Bing. American Indians were cultivating corn, tomatoes, and tobacco long before the arrival of Columbus. The term *okay* was derived from the Choctaw word *oke*, meaning "it is so." There is evidence indicating that the name *Yankee* came from Indian terms for the English—from *eankke* in Cherokee and *Yankwis* in Delaware. Jazz and blues as well as rock and roll have African-American origins. The "Forty-Niners" of the Gold Rush learned mining techniques from the Mexicans; American cowboys acquired herding skills from Mexican *vaqueros* and adopted their range terms—such as *lariat* from *la reata*, *lasso* from *lazo*, and *stampede* from *estampida*. Songs like "God Bless America," "Easter Parade," and "White Christmas" were written by a Russian-Jewish immigrant named Israel Baline, more popularly known as Irving Berlin.[17]

Furthermore, many diverse ethnic groups have contributed to the building of the American economy, forming what Walt Whitman saluted as "a vast, surging, hopeful army of workers." They worked in the South's cotton fields, New England's textile mills, Hawaii's canefields, New York's garment factories, California's orchards, Washington's salmon canneries, and Arizona's copper mines. They built the railroad, the great symbol of America's industrial triumph. . . .

Moreover, our diversity was tied to America's most serious crisis: the Civil War was fought over a racial issue—slavery. . . .

. . . The people in our study have been actors in history, not merely victims of discrimination and exploitation. They are entitled to be viewed as subjects— as men and women with minds, wills, and voices.

> *In the telling and retelling*
> *of their stories,*
> *They create communities*
> *of memory.*

They also re-vision history. "It is very natural that the history written by the victim," said a Mexican in 1874, "does not altogether chime with the story of the victor." Sometimes they are hesitant to speak, thinking they are only "little people." "I don't know why anybody wants to hear my history," an Irish maid said apologetically in 1900. "Nothing ever happened to me worth the tellin'."[18]

But their stories are worthy. Through their stories, the people who have lived America's history can help all of us, including my taxi driver, understand that Americans originated from many shores, and that all of us are entitled to dignity. "I hope this survey do a lot of good for Chinese people," an immigrant told an interviewer from Stanford University in the 1920s. "Make American people realize that Chinese people are humans. I think very few American people really know anything about Chinese." But the remembering is also for the sake of the children. "This story is dedicated to the descendants of Lazar and Goldie Glauberman," Jewish immigrant Minnie Miller wrote in her autobiography. "My history is bound up in their history and the generations that follow should know where they came from to know better who they are." Similarly, Tomo Shoji, an elderly Nisei woman, urged Asian Americans to learn more about their roots: "We got such good, fantastic stories to tell. All our stories are different." Seeking to know how they fit into America, many young people have become listeners; they are eager to learn about the hardships and humiliations experienced by their parents and grandparents. They want to hear their stories, unwilling to remain ignorant or ashamed of their identity and past.[19]

The telling of stories liberates. By writing about the people on Mango Street, Sandra Cisneros explained, "the ghost does not ache so much." The place no longer holds her with "both arms. She sets me free." Indeed, stories may not be as innocent or simple as they seem to be. Native-American novelist Leslie Marmon Silko cautioned:

> I will tell you something about stories . . .
> They aren't just entertainment.
>   Don't be fooled.

Indeed, the accounts given by the people in this study vibrantly re-create moments, capturing the complexities of human emotions and thoughts. They also provide the authenticity of experience. After she escaped from slavery, Harriet Jacobs wrote in her autobiography: "[My purpose] is not to tell you what I have heard but what I have seen—and what I have suffered." In their sharing of memory, the people in this study offer us an opportunity to see ourselves reflected in a mirror called history.[20]

In his recent study of Spain and the New World, *The Buried Mirror*, Carlos Fuentes points out that mirrors have been found in the tombs of ancient Mexico, placed there to guide the dead through the underworld. He also tells us about the legend of Quetzalcoatl, the Plumed Serpent: when this god was given a mirror by the Toltec deity Tezcatlipoca, he saw a man's face in the mirror and realized his own humanity. For us, the "mirror" of history can guide the living and also help us recognize who we have been and hence are. In *A Distant Mirror*, Barbara W. Tuchman finds "phenomenal parallels" between the "calamitous 14th century" of European society and our own era.

We can, she, observes, have "greater fellow-feeling for a distraught age" as we painfully recognize the "similar disarray," "collapsing assumptions," and "unusual discomfort."[21]

But what is needed in our own perplexing times is not so much a "distant" mirror, as one that is "different." While the study of the past can provide collective self-knowledge, it often reflects the scholar's particular perspective or view of the world. What happens when historians leave out many of America's peoples? What happens, to borrow the words of Adrienne Rich, "when someone with the authority of a teacher" describes our society, and "you are not in it"? Such an experience can be disorienting—"a moment of psychic disequilibrium, as if you looked into a mirror and saw nothing."[22]

Through their narratives about their lives and circumstances, the people of America's diverse groups are able to see themselves and each other in our common past. They celebrate what Ishmael Reed has described as a society "unique" in the world because "the world is here"—a place "where the cultures of the world crisscross." Much of America's past, they point out, has been riddled with racism. At the same time, these people offer hope, affirming the struggle for equality as a central theme in our country's history. At its conception, our nation was dedicated to the proposition of equality. What has given concreteness to this powerful national principle has been our coming together in the creation of a new society. "Stuck here" together, workers of different backgrounds have attempted to get along with each other.

> People harvesting
> Work together unaware
> Of racial problems,

wrote a Japanese immigrant describing a lesson learned by Mexican and Asian farm laborers in California.[23]

Finally, how do we see our prospects for "working out" America's racial crisis? Do we see it as through a glass darkly? Do the televised images of racial hatred and violence that riveted us in 1992 during the days of rage in Los Angeles frame a future of divisive race relations—what Arthur Schlesinger, Jr., has fearfully denounced as the "disuniting of America"? Or will Americans of diverse races and ethnicities be able to connect themselves to a larger narrative? Whatever happens, we can be certain that much of our society's future will be influenced by which "mirror" we choose to see ourselves. America does not belong to one race or one group, the people in this study remind us, and Americans have been constantly redefining their national identity from the moment of first contact on the Virginia shore. By sharing their stories, they invite us to see ourselves in a different mirror.[24]

NOTES

1. Toni Morrison, *Playing in the Dark: Whiteness in the Literary Imagination* (Cambridge, Mass., 1992), p. 47.

2. William A. Henry III, "Beyond the Melting Pot," in "America's Changing Colors," *Time*, vol. 135, no. 15 (April 9, 1990), pp. 28–31.

3. "A Conflict of the Have-Nots," *Newsweek*, December 12, 1988, pp. 28–29.

4. Rodney King's statement to the press, *New York Times*, May 2, 1992, p. 6.

5. Tim Rutten, "A New Kind of Riot," *New York Review of Books*, June 11, 1992, pp. 52–53; Maria Newman, "Riots Bring Attention to Growing Hispanic Presence in South-Central Area," *New York Times*, May 11, 1992, p. A10; Mike Davis, "In L.A. Burning All Illusions," *The Nation*, June 1, 1992, pp. 744–745; Jack Viets and Peter Fimrite, "S.F. Mayor Visits Riot-Torn Area to Buoy Businesses," *San Francisco Chronicle*, May 6, 1992, p. A6.

6. Rick DelVecchio, Suzanne Espinosa, and Carl Nolte, "Bradley Ready to Lift Curfew," *San Francisco Chronicle*, May 4, 1992, p. A1.

7. Abraham Lincoln, "The Gettysburg Address," in *The Annals of America*, vol. 9, *1863–1865: The Crisis of the Union* (Chicago, 1968), pp. 462–463; Martin Luther King, *Why We Can't Wait* (New York, 1964), pp. 92–93.

8. Interview with old laundryman, in "Interviews with Two Chinese," circa 1924, Box 326, folder 325, Survey of Race Relations, Stanford University, Hoover Institution Archives; Congressman Robert Matsui, speech in the House of Representatives on the 442 bill for redress and reparations, September 17, 1987, *Congressional Record* (Washington, D.C., 1987), p. 7584.

9. Camarillo, *Chicanos in a Changing Society*, p. 2; Juan Nepornuceno Seguín, in David J. Weber (ed.), *Foreigners in Their Native Land: Historical Roots of the Mexican Americans* (Albuquerque, N. Mex., 1973), p. vi; Jesus Garza, in Manuel Garnio, *The Mexican Immigrant: His Life Story* (Chicago, 1931), p. 15; Ernesto Galarza, *Barrio Boy: The Story of a Boy's Acculturation* (Notre Dame, Ind., 1986), p. 200.

10. Lawrence J. McCaffrey, *The Irish Diaspora in America* (Washington, D.C., 1984), pp. 6, 62.

11. John Murray Cuddihy, *The Ordeal of Civility: Freud, Marx, Levi Strauss, and the Jewish Struggle with Modernity* (Boston, 1987), p. 165; Jonathan Kaufman, *Broken Alliance: The Turbulent Times between Blacks and Jews in America* (New York, 1989), pp. 28, 82, 83–84, 91, 93, 106.

12. Andrew Jackson, First Annual Message to Congress, December 8, 1829, in James D. Richardson (ed.), *A Compilation of the Messages and Papers of the Presidents, 1789–1897*

(Washington, D.C., 1897), vol. 2, p. 457; Frederick Jackson Turner, "The Significance of the Frontier in American History," in *The Early Writings of Frederick Jackson Turner* (Madison, Wis., 1938), pp. 185ff.; Luther Standing Bear, "What the Indian Means to America," in Wayne Moquin (ed.), *Great Documents in American Indian History* (New York, 1973), p. 307; Vine Deloria, Jr., *Custer Died for Your Sins: An Indian Manifesto* (New York, 1969).

13. Nathan Glazer, *Affirmative Discrimination: Ethnic Inequality and Public Policy* (New York, 1978); Thomas Sowell, *Ethnic America: A History* (New York, 1981); David R. Roediger, *The Wages of Whiteness: Race and the Making of the American Working Class* (London, 1991), pp. 134–136; Dan Caldwell, "The Negroization of the Chinese Stereotype in California," *Southern California Quarterly*, vol. 33 (June 1971), pp. 123–131.

14. Thomas Almaguer, "Racial Domination and Class Conflict in Capitalist Agriculture: The Oxnard Sugar Beet Workers' Strike of 1903," *Labor History*, vol. 25, no. 3 (summer 1984), p. 347; Howard M. Sachar, *A History of the Jews in America* (New York, 1992), p. 183.

15. For the concept of liminality, see Victor Turner, *Dramas, Fields, and Metaphors: Symbolic Action in Human Society* (Ithaca, N.Y., 1974), pp. 232, 237; and Arnold Van Gennep, *The Rites of Passage* (Chicago, 1960). What I try to do is to apply liminality to the land called America.

16. Kazuo Ito, *Issei: A History of Japanese Immigrants in North America* (Seattle, 1973), p. 33; Arnold Schrier, *Ireland and the American Emigration, 1850–1900* (New York, 1970), p. 24; Abraham Cahan, *The Rise of David Levinsky* (New York, 1960; originally published in 1917), pp. 59–61; Mary Antin, quoted in Howe, *World of Our Fathers* (New York, 1983), p. 27; Lawrence A. Cardoso, *Mexican Emigration to the United States, 1897–1931* (Tucson, Ariz., 1981), p. 80.

17. Ronald Takaki, *Strangers from a Different Shore: A History of Asian Americans* (Boston, 1989), pp. 88–89; Jack Weatherford, *Native Roots: How the Indians Enriched America* (New York, 1991), pp. 210, 212; Carey McWilliams, *North from Mexico: The Spanish-Speaking People of the United States* (New York, 1968), p. 154; Stephan Themstrom (ed.), *Harvard Encyclopedia of American Ethnic Groups* (Cambridge, Mass., 1980), p. 22; Sachar, *A History of the Jews in America*, p. 367.

18. Weber (ed.), *Foreigners in Their Native Land*, p. vi; Hamilton Holt (ed.), *The Life Stories of Undistinguished Americans as Told by Themselves* (New York, 1906), p. 143.

19. "Social Document of Pany Lowe, interviewed by C. H. Burnett, Seattle, July 5, 1924," p. 6, Survey of Race Relations, Stanford University, Hoover Institution Archives; Minnie Miller, "Autobiography," private manuscript, copy from Richard

Balkin; Tomo Shoji, presentation, Obana Cultural Center, Oakland, California, March 4, 1988.

20. Sandra Cisneros, *The House on Mango Street* (New York, 1991), pp. 109–110; Leslie Marmon Silko, *Ceremony* (New York, 1978), p. 2; Harriet A. Jacobs, *Incidents in the Life of a Slave Girl, written by herself* (Cambridge, Mass., 1987; originally published in 1857), p. xiii.

21. Carlos Fuentes, *The Buried Mirror: Reflections on Spain and the New World* (Boston, 1992), pp. 10, 11, 109; Barbara W. Tuchman, *A Distant Mirror: The Calamitous 14th Century* (New York, 1978), pp. xiii, xiv.

22. Adrienne Rich, *Blood, Bread, and Poetry: Selected Prose, 1979–1985* (New York, 1986), pp. 199.

23. Ishmael Reed, "America: The Multinational Society," in Rick Simonson and Scott Walker (eds.), *Multi-cultural Literacy* (St. Paul, 1988), p. 160; Ito, *Issei*, p. 497.

24. Arthur M. Schlesinger, Jr., *The Disuniting of America: Reflections on a Multicultural Society* (Knoxville, Tenn., 1991); Carlos Bulosan, *America Is in the Heart: A Personal History* (Seattle, 1981), pp. 188–189.

Conceptualizing Race, Class, and Gender

Picture a typical college basketball game. It all seems pretty familiar—the players on the court, the cheerleaders arrayed at the side, the band playing, fans cheering, boosters watching from the best seats, and if the team is ranked, perhaps a television crew. Everybody seems to have a place in the game. Everybody seems to be following the rules. But what are the "rules" of this game? What explains the patterns that we see and don't see?

Race clearly matters. The predominance of young African American males on many college basketball teams is noticeable. Why do so many young Black men play basketball? Some argue that African Americans are better in areas requiring physical skills such as sports and are less capable of doing intellectual work in fields such as physics, law, and medicine. Others look to Black culture for explanations, suggesting that African Americans would be perfectly happy just playing ball and partying. But these perspectives fail to take into account the continuing effects of institutional racism. Lack of access to decent jobs, adequate housing, quality education, and adequate health care have resulted in higher rates of poverty for African Americans. For example, in 1991, 33 percent of African Americans were poor, as compared to 9 percent of Whites (O'Hare 1992: 6). For young Black men growing up in communities with few opportunities, sports like basketball become attractive as a way out of poverty. But despite the importance of institutionalized racism, does a racial analysis fully explain the "rules" of the game of college basketball?

Not really, because Black men are not the only players. White men also play college basketball, and their backgrounds in some ways mirror those of the Black players. This leads to the significance of social class in explaining a basketball game. Who benefits from college basketball? Some argue that because the players get scholarships and are offered a chance to earn college degrees, the players reap the rewards, but this misses the point of who really benefits. College athletics is big business, and the players make far less from it than we typically believe. As amateur athletes, the players themselves are forbidden to take any payment for their skills. They are offered the hope of an NBA contract when they turn pro, or at least a college degree should they manage to graduate. But, as Messner shows in his "Masculinities and Athletic Careers," few actually turn pro; moreover, 47 percent of male college athletes actually graduate, compared to 54 percent of college students overall (*The Chronicle of Higher Education*, May 26, 1993, p. A33). These trends are accentuated among Black athletes. Interestingly, among women college athletes, 61 percent graduate.

So, who actually benefits from college basketball? The colleges that recruit these athletes certainly benefit. Winning teams garner increased alumni giving, corporate support, and television revenues. In athletics as a business, corporate sponsors want their names and products identified with winning teams and athletes; advertisers want their products promoted by members of winning teams. Even though the athletes themselves are forbidden to promote the products, corporations create and market products that can be sold in conjunction with the excitement about basketball sustained by the players' achievements. Products such as athletic shoes, workout clothing, cars, and beer are all targeted toward the consumer dollars of males who enjoy watching basketball. Also, consider how many full-time jobs are supported by the revenues generated by the enterprise of college basketball. Referees, sports reporters both at the actual games and on local news stations, athletic trainers, coaches, and health personnel all benefit. Unlike the players, these people all get paid for their contributions to college basketball.

Do race and class fully explain the "rules" of basketball? Sometimes what we *don't* see can be just as revealing as what we do see. One other feature of the game on the court may escape attention because it is so familiar and, therefore, so unquestioned. All of the players are men. All of the coaches and support personnel are men. Probably, most of the camera crew and announcers are men. Where are the women? Those closest to the action on the court are probably cheerleaders, tumbling, dancing, and being thrown into the air in support of the exploits of the athletes. Others may be in the band. Some women are in the stands, cheering on the team—many of them accompanied by their husbands, boyfriends, parents, and children. Many work in the concession stands, fulfilling women's traditional roles of serving others. Still others are even more invisible, left to clean the restrooms, locker rooms, and stands after the crowd goes home. Women remain on the sidelines in other ways as well. The treatment of women basketball players is markedly different than that of their male counterparts; women have many fewer opportunities for scholarships and professional careers in athletics. The centrality of men's activities in basketball mirrors the centrality afforded men's activities in the society as a whole; thus, women's seeming invisibility in basketball ironically highlights the salience of gender.

This brief discussion of a college basketball game demonstrates how race, class, and gender each provide an important, yet partial, perspective on the action on the court. Race, class, and gender are so inextricably intertwined that gaining a comprehensive understanding of one basketball game requires that we think inclusively. If something as familiar and widespread as college basketball is embedded in race, class, and gender relations, we might ask how other social practices, institutions, relations, and social issues are similarly structured.

Although some people believe that race, gender, and class divisions are relics of the past, we see them as deeply embedded in the structure of social institutions such as work, family, education, the state, and sports. Moreover, the politics of race, class, and gender in turn shape both the social issues that emerge from within and shape these institutional contexts. Despite their interlocking nature, however, in Part Two, we focus on race, class, and gender as separate, institutionalized systems of oppression. Our purpose is to conceptualize how each system structures and intersects with the others. For example, we want readers to understand how race and class structure gender relations, as well as how gender and race structure class relations. We look at each distinctly only to provide a foundation for understanding their interlocking nature.

RACE

In this volume we examine race, class, and gender relations from an institutional or structural perspective. Locating racial oppression in the structure of social institutions provides a different frame of analysis than does analysis of individuals only. *Individual racism* is one person's belief in the superiority of one race over another. Individual racism is related to *prejudice*, a hostile attitude toward a person who is presumed to have the negative characteristics associated with a group to which he or she belongs.

Institutional racism is more systemic than this. Institutions such as work, family, education, the state, and sports are "fairly stable social arrangements and practices" that persist over time (Knowles and Prewitt 1970: 5). *Institutional racism* is a system of beliefs and behaviors by which a group defined as a race is oppressed, controlled, and exploited because of presumed cultural or biological characteristics (Blauner 1972). Racism is not the same thing as prejudice, although prejudice is one manifestation of racism. The concept of institutional racism reminds us that racism is not just a matter of belief but is also embedded in a system of power relations.

In this definition of institutional racism, note first that racism is systemic—it is part of society's structure, not just present in individual bigots. As Gloria Yamato discusses in "Something About the Subject Makes It Hard to Name," racism can be intentional or unintentional. In a racist system, well-meaning White people benefit from racism even if they have no intentions of acting or thinking like a "racist." Thus, institutionalized racism creates a built-in system of privilege. As Yamato suggests, different groups internalize it in different forms of consciousness. Peggy McIntosh's essay, "White Privilege and Male Privilege," describes how the system of racial privilege becomes invisible to those who benefit from it, even though it structures the everyday life of both White people and people of color.

Second, institutionalized racism shapes everyday social relations. In other words, despite its systemic nature, institutionalized racism depends on techniques of everyday racism carried out by individuals in order to continue. According to sociologist Philomena Essed, it is important to identify practices, rather than individuals, as racist. Essed identifies the numerous practices engaged in by White people as examples of everyday racism. For example, White people deny racism in many ways. Failing to take a stand against racism, refusing to admit racism, acknowledging only extreme racism, and expressing anger at Black people who point out racism are all strategies of denial (Essed 1991). While each act of everyday racism in and of itself may be relatively benign, institutionalized racism operates in our everyday lives as a sustained pattern of everyday actions such as these.

Practices of everyday racism are part of the edifice of institutionalized racism; yet, we often misread their meaning. Many people believe that to be nonracist means that we must be color-blind—that is, we must not recognize or place significance on a person's racial background and identity. But to ignore the significance of race in a society where racial groups have distinct historical and contemporary experiences is to deny the reality of their group experience. Being color-blind in a society structured on racial privilege means everybody is assumed to be "White," so that people of color might be offended by friends who say, for example, "But, I never think of you as Black." Such practices of everyday racism are so powerful because, instead of seeing them as part of patterns of institutionalized racism, we experience these interactions as ordinary occurrences.

Another dimension of institutionalized racism is that its forms change over time. For example, the meaning of *race* itself reflects institutionalized racist practices and beliefs. Race is a social-historical-political concept (Omi and Winant 1986). Societies construct race, and racial meanings constantly change because different groups contest prevailing racial definitions that empower some groups at the expense of others. Consider, for example, the changing definitions of race in the U.S. Census. In 1860, only three "races" existed—whites, blacks, and mulattoes, yet by 1890, these original three "races" had been joined by five others—quadroon, octoroon, Chinese, Japanese, and Indian. In the following ten short years, this list shrank to five "races"—White, Black, Chinese, Japanese, and Indian—a situation reflecting the growth of strict segregation in the South (O'Hare 1992).

Recent decades have seen additional revisions to definitions of race. In particular, the experiences of Latino groups in the United States challenge long-standing racial categories of "Black" and "White." In the 1990 census, a person could be classified in one of five "racial" groups—White; Black or African American; Asian or Pacific Islander; American Indian, Eskimo, or Aleut; and other—and one of two ethnicities, Hispanic or non-Hispanic. Race and ethnic definitions overlap because Hispanics may be of any race, and many Hispanics do not identify with any of the race categories identified by the census. For example, in the 1990 census, 43 percent of Hispanics identified their race as "other"; moreover, most people (96 percent) in the "other" race category were Hispanic (O'Hare 1992: 6).

In addition to the shifting definitions of race grounded in shifting relations of power, as Michael Thornton points out in "Is Multiracial Status Unique?", we can also no longer think of race in terms of mutually exclusive categories. Persons of mixed racial heritage, a small but growing group, present a challenge to the current system. Because race is self-reported, individuals choose the racial category with which they most identify. How do we account for groups that have long resisted inclusion in existing racial categories?

This overarching structure of racial power relations means that placement in this structure leads to differences in outlook concerning the very presence of racism and what can be done about it. The reappearance of racial hostilities on college campuses is certainly evidence of the continuation of racist practices and beliefs; yet, despite this and other evidence, Whites continue to be optimistic in their assessment of racial progress. They say they are tired of hearing about racism and that they have done all they can to eliminate racial discrimination. Blacks are less sanguine about racial progress and are more aware of the nuances of racism. Although overt discrimination has been lessened, marked differences by race are still evident in comparing employment, political representation, schooling, and other basic measures of group well-being.

Racism does not exist in a vacuum. As we have said, race, gender, and class, are intersecting systems—experienced simultaneously, not separately. Do not think of any one category in the absence of the others. People's experiences with race and racism are framed by their location in this overarching system of race, class, and gender privileges and penalties. Race possesses not only objective dimensions resulting from institutionalized racism; it simultaneously has subjective dimensions of how people experience it. For example, some people of color grow up with relative class privilege; yet, this does not eliminate racism, as Michael Dyson's experience of having his credit card nearly cut in half by a bank manager shows (in "The Plight of Black Men"). Moreover, the experiences of poor Black men in Michael Dyson's discussion differ from those of the Black women in Elizabeth Higginbotham and Lynn Weber's analysis of differing mobility experiences ("Moving Up with Kin and Community"). All people of color encounter institutional racism, but their actual experiences with racism vary, depending on social class, gender, age, sexuality, and other markers of social position.

CLASS

Like institutionalized racism, the social class system is grounded in social institutions and practices. Rather than thinking of social class as an identity owned by an individual, think of social class as a series of relations that pervade the entire society and shape our social institutions and relationships with each other. In the United States, the social class system differentially structures group access to material resources, including economic, political, and social resources. This system results from patterns of capitalist development, especially as those patterns intersect with institutionalized racism and gender oppression.

In the United States, the social class system is marked by striking differences in wealth. For example, people of color lag far behind Whites in measures of accumulated wealth, such as savings or property. Accumulated wealth is a critical dimension of social class position. It helps pay for college costs for children and down payments on houses, or it can cushion the impact of emergencies, such as unexpected unemployment or sudden health problems. The median net worth of White households is approximately ten times that of African American or Latino households. White households had an accumulated wealth of $44,400 as compared to $3,800 for African Americans and $5,500 for Latinos. Household income data show similar patterns. In

1992, median income was $32,368 for Whites, $18,660 for Blacks, and $22,848 for Latinos (U.S. Bureau of the Census, 1993a: 2–4).

While such figures demonstrate wide differences in economic status between Whites and people of color, they can simultaneously obscure the workings of the social class system. We must be careful not to generalize and see, for example, all Whites and Asians as wealthy and all African Americans and Latinos as poor. Consider the range of social class experiences just among Whites. While White households on the average possess higher accumulated wealth and have higher incomes, large numbers of White households are not so fortunate. Even though 9 percent of White households were in poverty in 1992—a figure lower than the 31 percent of African Americans, 14 percent of Asians, 32 percent of Native Americans, and 26 percent of Latinos classified in poverty—more White people are in poverty than all four of the other groups combined. In 1992, 19,012,000 people of color lived in poverty. In contrast, an astonishing 24,523,000 White people lived in poverty (U.S. Bureau of the Census, 1993b: 1, 4). Imagine what the range in White wealth must actually be to produce such a high White median income while so many White people remain in poverty.

In the United States, the social class system is also marked by differences in power. Social class is not just a matter of material difference. It is a pattern of domination in which some groups have more power than others. Power is the ability to influence and in some cases dominate others. This means not just interpersonal power but also the structural power that some groups hold because of their position of domination over others. Power is, of course, influenced by material well-being. Groups with vast amounts of wealth, for example, have the ability to influence systems like the media and the political process in ways that less powerful groups do not have. The power held by different groups varies enormously, depending on a group's position in the class system. Social class is then both a position of material advantage and the ability to control and influence others.

Like the system of institutionalized racism, the system of social class undergoes change. The changes in class are intimately linked to patterns of economic transformation in the American political economy. Specific economic changes include a shift from a manufacturing economy to a service economy with corresponding changes in the types of jobs available and the wages attached to these jobs. Mechanization, job export, deskilling of work, and other changes in the workplace have meant that fewer skilled, well-paying, manufacturing jobs exist today than in the past. This situation in turn has benefited some but left many people in jobs that do not pay enough, in part-time work, or without any work at all.

We examine these changes further in the Part Three, but point out here that, as a result of such changes, social class divisions in the United States are becoming more marked. In the nation's cities and towns, homelessness has become increasingly apparent even to casual observers. Middle-class people feel that their way of life is slipping away—a feeling supported by data showing increasing economic polarization. As Barbara Ehrenreich points out in "Are You Middle Class?" despite popular perceptions fueled by the media that the middle class remains secure, changes in the economy are affecting everyone, including the middle class. Moreover, poverty remains a pressing social concern, as Holly Sklar informs us in "The Upperclass and Mothers N the Hood."

If social class is so important in shaping life chances, why don't more people realize its significance? The answer lies in how dominant groups use ideology to explain the social class system and other systems of inequality. *Ideology* created by dominant groups refers to a system of beliefs that distort reality at the same time that they justify the status quo. As we learn in "Tired of Playing Monopoly?" by Donna Langston, the class system in the United States has been supported through the myth that we live in a classless society.

This myth serves the dominant class, making class privilege seem like something one earns, not something that is deeply embedded in the institutions of society. Langston also suggests that the system of privilege and inequality (by race, class, and gender) is least visible to those who are most privileged and who, in turn, control the resources to define the dominant cultural belief systems. This is perhaps why the privileged, not the poor, are more likely to believe that one gets ahead through hard work. Similarly, this is perhaps why men more than women deny that patriarchy exists, and why Whites more than Blacks believe racism is disappearing.

In addition to giving some groups more privileges than others, class shapes social relationships. As Elizabeth Higginbotham and Lynn Weber point out in "Moving Up with Kin and Community," one of the best ways of seeing how social class shapes relationships among members of the same class and relationships among people of different social classes is to examine the experiences of people who have undergone social mobility. But as they also observe, social class mobility experiences are quite different for men and women, and for people of color and Whites.

The social class system intersects with the system of institutionalized racism in shaping the experiences of men and women. The social class system also intersects with the system of gender oppression by placing men and women differently within this same overarching structure. For example, in 1992, White men's median income of $31,737 far outstripped the income of all other groups of year-round, full-time workers. Arrayed below were Black men who earned $22,942; White women, $22,423; Latino men, $20,312; Black women, $20,299; and Latino women, $17,743—all figures substantially below those for White men (U.S. Bureau of the Census 1993a: 95). Overall, race, class, and gender work together in crosscutting ways. In some ways, women of color share a class position with men of color—features such as rates of poverty, geographic location, and the racial gap in earnings in the labor market. In other ways, women as a group share a class position as women—

women still make less than men of their racial group. Together, these trends point to the need for inclusive thinking in revisioning social class as a concept.

GENDER

Gender, like race, is a socially constructed experience, not a biological imperative. Sociologists distinguish between the terms sex and gender to emphasize this point. *Sex* refers to one's biological identity as male or female; *gender* refers to the systematic structuring of relationships between women and men in social institutions. Gender is a learned identity, but as with race, it cannot be understood at the individual level alone. Gender is structured in social institutions, including work, families, mass media, and education, and, again, changing gender relations is not just a matter of changing individual attitudes. Transformation of institutional structures requires change in consciousness and collective activism.

The ideology of sexism supports the system of gender relations, just as the ideology of racism supports the system of race relations. Like racism, *sexism* is a system of beliefs and behaviors by which a group is oppressed, controlled, and exploited because of presumed gender differences. Sexism is manifested both in individually held beliefs and social behaviors, and it is embedded in cultural symbols. It supports the gender inequalities that are structured into social institutions.

Gender oppression also works with another major system of oppression, that of sexuality. *Homophobia*—the fear and hatred of homosexuality—is part of the system of social control that legitimates and enforces gender oppression. It supports the system of compulsory heterosexuality, the institutionalized power and privilege given to heterosexual behavior and identification. If only heterosexual forms of gender identity are "normal," then gays, lesbians, and bisexuals become ostracized, oppressed, and defined as "socially deviant."

Homophobia affects heterosexuals as well because it is part of gender ideology used to define "normal" men and women from those deemed deviant.

Ideologies claiming that women are now equal to men have led many people to believe that sexism is disappearing. Seeing more women in the labor force, especially some in jobs traditionally reserved for men, can give the impression that women are doing well. But how real have these gains been? The gap between women's and men's income did close slightly through the 1980s; however, most analysts agree that the decreased gap reflects a drop in male wages. The most dramatic gains in economic status have been for younger women in professional jobs. The majority of women remain concentrated in gender-segregated occupations with low wages, little opportunity for mobility, and stressful conditions. This is particularly true for women of color who are more likely to be in occupations that are race and gender segregated. And for women heading their own households, poverty persists at alarmingly high rates. Close to half of Black and Latino families and 30 percent of White families headed by women are officially categorized as poor (U.S. Bureau of the Census, 1993a). Income and occupational data, however, do not tell the full story for women. High rates of violence against women, whether in the home, on campus, in the workplace, or on the streets, indicate the continuing devaluation of and danger for women in this society.

Despite their similar status as women, women are not oppressed in the same ways by gender. Race and class create significant differences among women. As Johnnetta Cole notes in "Commonalities and Differences," women share many features of gender oppression, but the specific form of their experience depends on race and class as well. Tracy Lai in "Asian American Women" writes of the specific forms of gender oppression that Asian American women experience. Rosemary Bray also explores this in "Taking Sides Against Ourselves"—an analysis of race and gender in the treatment of Anita Hill during the confirmation hearings of now Supreme Court Justice Clarence

Thomas. During the hearings, Hill was defined as a traitor to her race, although clearly her race *and* her gender were part of her experience as an African American woman. Peggy McIntosh in "White Privilege and Male Privilege" reminds us that although White women are oppressed by gender, they enjoy privileges of race and, in some cases, class and heterosexual identification. Evelyn Torton Beck in "From 'Kike' to 'JAP' " also shows the ways in which anti-Semitism, hatred of Jewish people, mingles with sexism, racism, and homophobia.

Institutionalized gender relations shape men's, as well as women's, experiences. But, as Peter Blood, Alan Tuttle, and George Lakey show in "Understanding and Fighting Sexism," not all men benefit equally from patriarchy. Depending on their race and class, men experience gender differently. Michael Messner in "Masculinities and Athletic Careers" shows how men's experiences and choices in sports are conditioned by race, class, and gender.

Throughout Part Two, keep the concept of social structure in mind. Race, class, and gender are often discussed in terms of cultural difference, but they are also part of the institutional framework of society. A structural analysis studies the intersections of race, class, and gender within institutions and within individual's experiences in those institutions. While we have divided the articles into three sections of race, class, and gender, many theoretically fit in multiple categories. Thinking inclusively means that practices rarely belong to only one category—race or class or gender. A more accurate way of viewing the world and the articles in Part Two is to see them as interconnected.

References

Blauner, Robert. 1972. *Racial Oppression in America*. New York: Harper & Row.

Essed, Philomena. 1991. *Understanding Everyday Racism: An Interdisciplinary Theory*. Newbury Park, CA: Sage.

Knowles, Louis L., and Kenneth Prewitt. 1970. *Institutional Racism in America*. Englewood Cliffs, NJ: Prentice-Hall.

O'Hare, William P. 1992. "America's Minorities—The Demographics of Diversity." *Population Bulletin* 47, no. 4 (December). Washington, DC: Population Reference Bureau.

Omi, Michael, and Howard Winant. 1986. *Racial Formation in the United States: From the 1960s to the 1980s*. New York: Routledge and Kegan Paul.

U.S. Bureau of the Census. 1993a. *Money Income of Households, Families, and Persons in the United States: 1992*, Series P-60, #184. Washington, DC: U.S. Government Printing Office.

U.S. Bureau of the Census. 1993b. *Poverty in the United States: 1992*, Series P-60, #185. Washington, DC: U.S. Government Printing Office.

Race and Racism

SOMETHING ABOUT THE SUBJECT MAKES IT HARD TO NAME

7

Gloria Yamato

Racism—simple enough in structure, yet difficult to eliminate. Racism—pervasive in the U.S. culture to the point that it deeply affects all the local town folk and spills over, negatively influencing the fortunes of folk around the world. Racism is pervasive to the point that we take many of its manifestations for granted, believing "that's life." Many believe that racism can be dealt with effectively in one hellifying workshop, or one hour-long heated discussion. Many actually believe this monster, racism, that has had at least a few hundred years to take root, grow, invade our space and develop subtle variations . . . this mind-funk that distorts thought and action, can be merely wished away. I've run into folks who really think that we can beat this devil, kick this habit, be healed of this disease in a snap. In a sincere blink of a well-intentioned eye,

From: Jo Whitehorse Cochran, Donna Langston, and Carolyn Woodward (eds.), *Changing Our Power: An Introduction to Women's Studies* (Dubuque, Iowa: Kendall-Hunt, 1988), pp. 3–6. Reprinted by permission.

presto—poof—racism disappears. "I've dealt with my racism . . . (envision a laying on of hands) . . . Hallelujah! Now I can go to the beach." Well, fine. Go to the beach. In fact, why don't we all go to the beach and continue to work on the sucker over there? Cuz you can't even shave a little piece off this thing called racism in a day, or a weekend, or a workshop.

When I speak of *oppression*, I'm talking about the systematic, institution-alized mistreatment of one group of people by another for whatever reason. The oppressors are purported to have an innate ability to access economic resources, information, respect, etc., while the oppressed are believed to have a corresponding negative innate ability. The flip side of oppression is *internal-ized oppression*. Members of the target group are emotionally, physically, and spiritually battered to the point that they begin to actually believe that their oppression is deserved, is their lot in life, is natural and right, and that it doesn't even exist. The oppression begins to feel comfortable, familiar enough that when mean ol' Massa lay down de whip, we got's to pick up and whack ourselves and each other. Like a virus, it's hard to beat racism, because by the time you come up with a cure, it's mutated to a "new cure-resistant" form. One shot just won't get it. Racism must be attacked from many angles.

The forms of racism that I pick up on these days are 1) aware/blatant racism, 2) aware/covert racism, 3) unaware/unintentional racism, and 4) un-aware/self-righteous racism. I can't say that I prefer any one form of racism over the others, because they all look like an itch needing a scratch. I've heard it said (and understandably so) that the aware/blatant form of racism is preferable if one must suffer it. Outright racists will, without apology or confusion, tell us that because of our color we don't appeal to them. If we so choose, we can attempt to get the hell out of their way before we get the sweat knocked out of us. Growing up, aware/covert racism is what I heard many of my elders bemoaning "up north," after having escaped the overt racism "down south." Apartments were suddenly no longer vacant or rents were outra-geously high, when black, brown, red, or yellow persons went to inquire about them. Job vacancies were suddenly filled, or we were fired for very vague reasons. It still happens, though the perpetrators really take care to cover their tracks these days. They don't want to get gummed to death or slobbered on by the toothless laws that supposedly protect us from such inequities.

Unaware/unintentional racism drives usually tranquil white liberals wild when they get called on it, and confirms the suspicions of many people of color who feel that white folks are just plain crazy. It has led white people to believe that it's just fine to ask if they can touch my hair (while reaching). They then exclaim over how soft it is, how it does not scratch their hand. It has led whites to assume that bending over backwards and speaking to me in high-pitched (terrified), condescending tones would make up for all the racist wrongs that distort our lives. This type of racism has led whites right to my doorstep, talking 'bout, "We're sorry/we love you and want to make things right," which is fine, and further, "We're gonna give you the opportunity to fix it while we sleep. Just tell us what you need. 'Bye!!"—which *ain't* fine. With the best of intentions, the best of educations, and the greatest generosity of heart, whites, operating on the misinformation fed to them from day one, will behave in ways that are racist, will perpetuate racism by being "nice" the way we're taught to be nice. You can just "nice" somebody to death with naïveté and lack of awareness of privilege. Then there's guilt and the desire to end racism and how the two get all tangled up to the point that people, morbidly fascinated with their guilt, are immobilized. Rather than deal with ending racism, they sit and ponder their guilt and hope nobody notices how awful they are. Meanwhile, racism picks up momentum and keeps on keepin' on.

Now, the newest form of racism that I'm hip to is unaware/self-righteous racism. The "good white" racist attempts to shame Blacks into being blacker, scorns Japanese-Americans who don't speak Japanese, and knows more about the Chicano/a community than the folks who make up the community. They assign themselves as the "good whites," as opposed to the "bad whites," and are often so busy telling people of color what the issues in the Black, Asian, Indian, Latino/a communities should be that they don't have time to deal with their errant sisters and brothers in the white community. Which means that people of color are still left to deal with what the "good whites" don't want to . . . racism.

Internalized racism is what really gets in my way as a Black woman. It influences the way I see or don't see myself, limits what I expect of myself or others like me. It results in my acceptance of mistreatment, leads me to believe that being treated with less than absolute respect, at least this once, is to be expected because I am Black, because I am not white. "Because I am (*you fill in the color*), you think, "Life is going to be hard." The fact is life may be hard, but the color of your skin is not the cause of the hardship. The color of your skin may be used as an excuse to mistreat you, but there is no reason or logic involved in the mistreatment. If it seems that your color is the reason; if it seems that your ethnic heritage is the cause of the woe, it's because you've been deliberately beaten down by agents of a greedy system until you swallowed the garbage. That is the internalization of racism.

Racism is the systematic, institutionalized mistreatment of one group of people by another based on racial heritage. Like every other oppression, racism can be internalized. People of color come to believe misinformation about their particular ethnic group and thus believe that their mistreatment is justified. With that basic vocabulary, let's take a look at how the whole thing works together. Meet "the Ism Family," racism, classism, ageism, adultism, elitism, sexism, heterosexism, physicalism, etc. All these ism's are systematic, that is, not only are these parasites feeding off our lives, they are also dependent on one another for foundation. Racism is supported and reinforced by classism, which is given a foothold and a boost by adultism, which also feeds sexism, which is validated by heterosexism, and so it goes on. You cannot have the "ism" functioning without first effectively installing its flip-side, the internalized version of the ism. Like twins, as one particular form of the ism grows in potency, there is a corresponding increase in its internalized form within the population. Before oppression becomes a specific ism like racism, usually all hell breaks loose. War. People fight attempts to enslave them, or to subvert their will, or to take what they consider theirs, whether that is territory or dignity. It's true that the various elements of racism, while repugnant, would not be able to do very much damage, but for one generally overlooked key piece: power/privilege.

While in one sense we all have power we have to look at the fact that, in our society, people are stratified into various classes and some of these classes have more privilege than others. The owning class has enough power and privilege to not have to give a good whinney what the rest of the folks have on their minds. The power and privilege of the owning class provides the ability to pay off enough of the working class and offer that paid-off group, the middle class, just enough privilege to make it agreeable to do various and sundry oppressive things to other working-class and outright disenfranchised folk, keeping the lid on explosive inequities, at least for a minute. If you're at the bottom of this heap, and you believe the line that says you're there because that's all you're worth, it is at least some small solace to believe that there are others more worthless than you, because of their gender, race, sexual preference . . . whatever. The specific form of power that runs the show here is the power to intimidate. The power to take away the most lives the quickest, and back it up with legal and "divine" sanction, is the very bottom line. It makes the difference between who's holding the racism end of the stick and who's getting beat with it (or beating others as vulnerable as they are) on the internalized racism end of the stick. What I am saying is, while people of color are welcome to tear up their own neighborhoods and each other, everybody knows that you cannot do that to white folks without hell to pay. People of color can be prejudiced against one another and whites, but do not have an ice-cube's chance in hell of passing laws that will get whites sent to relocation camps "for their own protection and the security of the nation." People who have not thought about or refuse to acknowledge this imbalance of power/privilege often want to talk about the racism of people of color. But then that is one of the ways racism is able to continue to function. You look for someone to blame and you blame the victim, who will nine times out of ten accept the blame out of habit.

So, what can we do? Acknowledge racism for a start, even though and especially when we've struggled to be kind and fair, or struggled to rise above it all. It is hard to acknowledge the fact that racism circumscribes and pervades our lives. Racism must be dealt with on two levels, personal and societal, emotional and institutional. It is possible—and most effective—to do both at the same time. We must reclaim whatever delight we have lost in our own ethnic heritage or heritages. This so-called melting pot has only succeeded in turning us into fast food-gobbling "generics" (as in generic "white folks" who were once Irish, Polish, Russian, English, etc. and "black folks," who were once Ashanti, Bambara, Baule, Yoruba, etc.). Find or create safe places to actually *feel* what we've been forced to repress each time we were a victim of, witness to or perpetrator of racism, so that we do not continue, like puppets, to act out the past in the present and future. Challenge oppression. Take a stand against it. When you are aware of something oppressive going down, stop the show. At least call it. We become so numbed to racism that we don't even think twice about it, unless it is immediately life-threatening.

Whites who want to be allies to people of color: You can educate yourselves via research and observation rather than rigidly, arrogantly relying solely on interrogating people of color. Do not expect that people of color should teach you how to behave non-oppressively. Do not give into the pull to be lazy. Think, hard. Do not blame people of color for your frustration about racism, but do appreciate the fact that people of color will often help you get in touch with that frustration. Assume that your effort to be a good friend is appreciated, but don't expect or accept gratitude from people of color. Work on racism for your sake, not "their" sake. Assume that you are needed and capable of being a good ally. Know that you'll make mistakes and commit yourself to correcting them and continuing on as an ally, no matter what. Don't give up.

People of color, working through internalized racism: Remember always that you and others like you are completely worthy of respect, completely capable of achieving whatever you take a notion to do. Remember that the term "people of color" refers to a variety of ethnic and cultural backgrounds. These various groups have been oppressed in a variety of ways. Educate yourself about the ways different peoples have been oppressed and how they've resisted that oppression. Expect and insist that whites are capable of being good allies against racism. Don't give up. Resist the pull to give out the "people of color seal of approval" to aspiring white allies. A moment of appreciation is fine, but more than that tends to be less than helpful. Celebrate yourself. Celebrate yourself. Celebrate the inevitable end of racism.

WHITE PRIVILEGE AND MALE PRIVILEGE: *A Personal Account of Coming to See Correspondences Through Work in Women's Studies*

8

Peggy McIntosh

Through work to bring materials and perspectives from Women's Studies into the rest of the curriculum, I have often noticed men's unwillingness to grant that they are overprivileged in the curriculum, even though they may grant that women are disadvantaged. Denials that amount to taboos surround the subject of advantages that men gain from women's disadvantages. These denials protect male privilege from being fully recognized, acknowledged, lessened, or ended.

Thinking through unacknowledged male privilege as a phenomenon with a life of its own, I realized that since hierarchies in our society are interlocking, there was most likely a phenomenon of white privilege that was similarly denied and protected, but alive and real in its effects. As a white person, I realized I had been taught about racism as something that puts others at a disadvantage, but had been taught not to see one of its corollary aspects, white privilege, which puts me at an advantage.

I think whites are carefully taught not to recognize white privilege, as males are taught not to recognize male privilege. So I have begun in an untutored way to ask what it is like to have white privilege. This paper is a partial record of my personal observations and not a scholarly analysis. It is based on my daily experiences within my particular circumstances.

I have come to see white privilege as an invisible package of unearned assets that I can count on cashing in each day, but about which I was "meant"

I have appreciated commentary on this paper from the Working Papers Committee of the Wellesley College Center for Research on Women, from members of the Dodge seminar, and from many individuals, including Margaret Andersen, Sorel Berman, Joanne Braxton, Johnnella Butler, Sandra Dickerson, Marnie Evans, Beverly Guy-Sheftall, Sandra Harding, Eleanor Hinton Hoytt, Pauline Houston, Paul Lauter, Joyce Miller, Mary Norris, Gloria Oden, Beverly Smith, and John Walter.

to remain oblivious. White privilege is like an invisible weightless knapsack of special provisions, assurances, tools, maps, guides, codebooks, passports, visas, clothes, compass, emergency gear, and blank checks.

Since I have had trouble facing white privilege, and describing its results in my life, I saw parallels here with men's reluctance to acknowledge male privilege. Only rarely will a man go beyond acknowledging that women are disadvantaged to acknowledging that men have unearned advantage, or that unearned privilege has not been good for men's development as human beings, or for society's development, or that privilege systems might ever be challenged and *changed*.

I will review here several types or layers of denial that I see at work protecting, and preventing awareness about, entrenched male privilege. Then I will draw parallels, from my own experience, with the denials that veil the facts of white privilege. Finally, I will list forty-six ordinary and daily ways in which I experience having white privilege, by contrast with my African American colleagues in the same building. This list is not intended to be generalizable. Others can make their own lists from within their own life circumstances.

Writing this paper has been difficult, despite warm receptions for the talks on which it is based.[1] For describing white privilege makes one newly accountable. As we in Women's Studies work reveal male privilege and ask men to give up some of their power, so one who writes about having white privilege must ask, "Having described it, what will I do to lessen or end it?"

The denial of men's overprivileged state takes many forms in discussions of curriculum change work. Some claim that men must be central in the curriculum because they have done most of what is important or distinctive in life or in civilization. Some recognize sexism in the curriculum but deny that it makes male students seem unduly important in life. Others agree that certain *individual* thinkers are male oriented but deny that there is any *systemic* tendency in disciplinary frameworks or epistemology to overempower men as a group. Those men who do grant that male privilege takes institutionalized and embedded forms are still likely to deny that male hegemony has opened doors for them personally. Virtually all men deny that male overreward alone can explain men's centrality in all the inner sanctums of our most powerful institutions. Moreover, those few who will acknowledge that male privilege systems have overempowered them usually end up doubting that we could

---

1. This paper was presented at the Virginia Women's Studies Association conference in Richmond in April, 1986, and the American Educational Research Association conference in Boston in October, 1986, and discussed with two groups of participants in the Dodge seminars for Secondary School Teachers in New York and Boston in the spring of 1987.

dismantle these privilege systems. They may say they will work to improve women's status, in the society or in the university, but they can't or won't support the idea of lessening men's. In curricular terms, this is the point at which they say that they regret they cannot use any of the interesting new scholarship on women because the syllabus is full. When the talk turns to giving men less cultural room, even the most thoughtful and fair-minded of the men I know will tend to reflect, or fall back on, conservative assumptions about the inevitability of present gender relations and distributions of power, calling on precedent or sociobiology and psychobiology to demonstrate that male domination is natural and follows inevitably from evolutionary pressures. Others resort to arguments from "experience" or religion or social responsibility or wishing and dreaming.

After I realized, through faculty development work in Women's Studies, the extent to which men work from a base of unacknowledged privilege, I understood that much of their oppressiveness was unconscious. Then I remembered the frequent charges from women of color that white women whom they encounter are oppressive. I began to understand why we are justly seen as oppressive, even when we don't see ourselves that way. At the very least, obliviousness of one's privileged state can make a person or group irritating to be with. I began to count the ways in which I enjoy unearned skin privilege and have been conditioned into oblivion about its existence, unable to see that it put me "ahead" in any way, or put my people ahead, overrewarding us and yet also paradoxically damaging us, or that it could or should be changed.

My schooling gave me no training in seeing myself as an oppressor, as an unfairly advantaged person, or as a participant in a damaged culture. I was taught to see myself as an individual whose moral state depended on her individual moral will. At school, we were not taught about slavery in any depth; we were not taught to see slaveholders as damaged people. Slaves were seen as the only group at risk of being dehumanized. My schooling followed the pattern which Elizabeth Minnich has pointed out: whites are taught to think of their lives as morally neutral, normative, and average, and also ideal, so that when we work to benefit others, this is seen as work that will allow "them" to be more like "us." I think many of us know how obnoxious this attitude can be in men.

After frustration with men who would not recognize male privilege, I decided to try to work on myself at least by identifying some of the daily effects of white privilege in my life. It is crude work, at this stage, but I will give here a list of special circumstances and conditions I experience that I did not earn but that I have been made to feel are mine by birth, by citizenship, and by virtue of being a conscientious law-abiding "normal" person of goodwill. I have chosen those conditions that I think in my case *attach somewhat more to skin-color privilege* than to class, religion, ethnic status, or geographical location, though these other privileging factors are intricately intertwined. As far as I can see, my Afro-American co-workers, friends, and acquaintances with whom I come into daily or frequent contact in this particular time, place, and line of work cannot count on most of these conditions.

1. I can, if I wish, arrange to be in the company of people of my race most of the time.
2. I can avoid spending time with people whom I was trained to mistrust and who have learned to mistrust my kind or me.
3. If I should need to move, I can be pretty sure of renting or purchasing housing in an area which I can afford and in which I would want to live.
4. I can be reasonably sure that my neighbors in such a location will be neutral or pleasant to me.
5. I can go shopping alone most of the time, fairly well assured that I will not be followed or harassed by store detectives.
6. I can turn on the television or open to the front page of the paper and see people of my race widely and positively represented.
7. When I am told about our national heritage or about "civilization," I am shown that people of my color made it what it is.
8. I can be sure that my children will be given curricular materials that testify to the existence of their race.
9. If I want to, I can be pretty sure of finding a publisher for this piece on white privilege.
10. I can be fairly sure of having my voice heard in a group in which I am the only member of my race.
11. I can be casual about whether or not to listen to another woman's voice in a group in which she is the only member of her race.
12. I can go into a book shop and count on finding the writing of my race represented, into a supermarket and find the staple foods that fit with my cultural traditions, into a hairdresser's shop and find someone who can deal with my hair.
13. Whether I use checks, credit cards, or cash, I can count on my skin color not to work against the appearance that I am financially reliable.
14. I could arrange to protect our young children most of the time from people who might not like them.
15. I did not have to educate our children to be aware of systemic racism for their own daily physical protection.
16. I can be pretty sure that my children's teachers and employers will tolerate them if they fit school and workplace norms; my chief worries about them do not concern others' attitudes toward their race.
17. I can talk with my mouth full and not have people put this down to my color.

18. I can swear, or dress in secondhand clothes, or not answer letters, without having people attribute these choices to the bad morals, the poverty, or the illiteracy of my race.
19. I can speak in public to a powerful male group without putting my race on trial.
20. I can do well in a challenging situation without being called a credit to my race.
21. I am never asked to speak for all the people of my racial group.
22. I can remain oblivious to the language and customs of persons of color who constitute the world's majority without feeling in my culture any penalty for such oblivion.
23. I can criticize our government and talk about how much I fear its policies and behavior without being seen as a cultural outsider.
24. I can be reasonably sure that if I ask to talk to "the person in charge," I will be facing a person of my race.
25. If a traffic cop pulls me over or if the IRS audits my tax return, I can be sure I haven't been singled out because of my race.
26. I can easily buy posters, postcards, picture books, greeting cards, dolls, toys, and children's magazines featuring people of my race.
27. I can go home from most meetings of organizations I belong to feeling somewhat tied in, rather than isolated, out of place, outnumbered, un-heard, held at a distance, or feared.
28. I can be pretty sure that an argument with a colleague of another race is more likely to jeopardize her chances for advancement than to jeopardize mine.
29. I can be fairly sure that if I argue for the promotion of a person of another race, or a program centering on race, this is not likely to cost me heavily within my present setting, even if my colleagues disagree with me.
30. If I declare there is a racial issue at hand, or there isn't a racial issue at hand, my race will lend me more credibility for either position than a person of color will have.
31. I can choose to ignore developments in minority writing and minority activist programs, or disparage them, or learn from them, but in any case, I can find ways to be more or less protected from negative consequences of any of these choices.
32. My culture gives me little fear about ignoring the perspectives and powers of people of other races.
33. I am not made acutely aware that my shape, bearing, or body odor will be taken as a reflection on my race.
34. I can worry about racism without being seen as self-interested or self-seek-ing.

35. I can take a job with an affirmative action employer without having my co-workers on the job suspect that I got it because of my race.
36. If my day, week, or year is going badly, I need not ask of each negative episode or situation whether it has racial overtones.
37. I can be pretty sure of finding people who would be willing to talk with me and advise me about my next steps, professionally.
38. I can think over many options, social, political, imaginative, or professional, without asking whether a person of my race would be accepted or allowed to do what I want to do.
39. I can be late to a meeting without having the lateness reflect on my race.
40. I can choose public accommodation without fearing that people of my race cannot get in or will be mistreated in the places I have chosen.
41. I can be sure that if I need legal or medical help, my race will not work against me.
42. I can arrange my activities so that I will never have to experience feelings of rejection owing to my race.
43. If I have low credibility as a leader, I can be sure that my race is not the problem.
44. I can easily find academic courses and institutions that give attention only to people of my race.
45. I can expect figurative language and imagery in all of the arts to testify to experiences of my race.
46. I can choose blemish cover or bandages in "flesh" color and have them more or less match my skin.

I repeatedly forgot each of the realizations on this list until I wrote it down. For me, white privilege has turned out to be an elusive and fugitive subject. The pressure to avoid it is great, for in facing it I must give up the myth of meritocracy. If these things are true, this is not such a free country; one's life is not what one makes it; many doors open for certain people through no virtues of their own. These perceptions mean also that my moral condition is not what I had been led to believe. The appearance of being a good citizen rather than a troublemaker comes in large part from having all sorts of doors open automatically because of my color.

A further paralysis of nerve comes from literary silence protecting privilege. My clearest memories of finding such analysis are in Lillian Smith's unparalleled *Killers of the Dream* and Margaret Andersen's review of Karen and Mamie Fields' *Lemon Swamp*. Smith, for example, wrote about walking toward black children on the street and knowing they would step into the gutter; Andersen contrasted the pleasure that she, as a white child, took on summer driving trips to the south with Karen Fields' memories of driving in a closed car stocked with all necessities lest, in stopping, her black family should suffer "insult, or worse." Adrienne Rich also recognizes and writes about daily experiences of privilege, but in my observation, white women's writing in this area is far more often on systemic racism than on our daily lives as light-skinned women.[2]

In unpacking this invisible knapsack of white privilege, I have listed conditions of daily experience that I once took for granted, as neutral, normal, and universally available to everybody, just as I once thought of a male-focused curriculum as the neutral or accurate account that can speak for all. Nor did I think of any of these perquisites as bad for the holder. I now think that we need a more finely differentiated taxonomy of privilege, for some of these varieties are only what one would want for everyone in a just society, and others give license to be ignorant, oblivious, arrogant, and destructive. Before proposing some more finely tuned categorization, I will make some observations about the general effects of these conditions on my life and expectations.

In this potpourri of examples, some privileges make me feel at home in the world. Others allow me to escape penalties or dangers that others suffer. Through some, I escape fear, anxiety, insult, injury, or a sense of not being welcome, not being real. Some keep me from having to hide, to be in disguise, to feel sick or crazy, to negotiate each transaction from the position of being an outsider or, within my group, a person who is suspected of having too close links with a dominant culture. Most keep me from having to be angry.

I see a pattern running through the matrix of white privilege, a pattern of assumptions that were passed on to me as a white person. There was one main piece of cultural turf; it was my own turf, and I was among those who could control the turf. I could measure up to the cultural standards and take advantage of the many options I saw around me to make what the culture would call a success of my life. *My skin color was an asset for any move I was educated to want to make.* I could think of myself as "belonging" in major ways and of making social systems work for me. I could freely disparage, fear, neglect, or be oblivious to anything outside of the dominant cultural forms. Being of the main culture, I could also criticize it fairly freely. My life was reflected back to me frequently enough so that I felt, with regard to my race, if not to my sex, like one of the real people.

Whether through the curriculum or in the newspaper, the television, the economic system, or the general look of people in the streets, I received daily

---

2. Andersen, Margaret, "Race and the Social Science Curriculum: A Teaching and Learning Discussion." *Radical Teacher*, November, 1984, pp. 17–20. Smith, Lillian, *Killers of the Dream*, New York: W. W. Norton, 1949.

signals and indications that my people counted and that others *either didn't exist or must be trying, not very successfully, to be like people of my race.* I was given cultural permission not to hear voices of people of other races or a tepid cultural tolerance for hearing or acting on such voices. I was also raised not to suffer seriously from anything that darker-skinned people might say about my group, "protected," though perhaps I should more accurately say *prohibited*, through the habits of my economic class and social group, from living in racially mixed groups or being reflective about interactions between people of differing races.

In proportion as my racial group was being made confident, comfortable, and oblivious, other groups were likely being made unconfident, uncomfortable, and alienated. Whiteness protected me from many kinds of hostility, distress, and violence, which I was being subtly trained to visit in turn upon people of color.

For this reason, the word "privilege" now seems to me misleading. Its connotations are too positive to fit the conditions and behaviors which "privilege systems" produce. We usually think of privilege as being a favored state, whether earned, or conferred by birth or luck. School graduates are reminded they are privileged and urged to use their (enviable) assets well. The word "privilege" carries the connotation of being something everyone must want. Yet some of the conditions I have described here work to systemically overempower certain groups. Such privilege simply *confers dominance*, gives permission to control, because of one's race or sex. The kind of privilege that gives license to some people to be, at best, thoughtless and, at worst, murderous should not continue to be referred to as a desirable attribute. Such "privilege" may be widely desired without being in any way beneficial to the whole society.

Moreover, though "privilege" may confer power, it does not confer moral strength. Those who do not depend on conferred dominance have traits and qualities that may never develop in those who do. Just as Women's Studies courses indicate that women survive their political circumstances to lead lives that hold the human race together, so "underprivileged" people of color who are the world's majority have survived their oppression and lived survivors' lives from which the white global minority can and must learn. In some groups, those dominated have actually become strong through *not* having all of these unearned advantages, and this gives them a great deal to teach the others. Members of so-called privileged groups can seem foolish, ridiculous, infantile, or dangerous by contrast.

I want, then, to distinguish between earned strength and unearned power conferred systemically. Power from unearned privilege can look like strength when it is, in fact, permission to escape or to dominate. But not all of the privileges on my list are inevitably damaging. Some, like the expectation that neighbors will be decent to you, or that your race will not count against you in court, should be the norm in a just society and should be considered as the entitlement of everyone. Others, like the privilege not to listen to less powerful people, distort the humanity of the holders as well as the ignored groups. Still others, like finding one's staple foods everywhere, may be a function of being a member of a numerical majority in the population. Others have to do with not having to labor under pervasive negative stereotyping and mythology.

We might at least start by distinguishing between positive advantages that we can work to spread, to the point where they are not advantages at all but simply part of the normal civic and social fabric, and negative types of advantage that unless rejected will always reinforce our present hierarchies. For example, the positive "privilege" of belonging, the feeling that one belongs within the human circle, as Native Americans say, fosters development and should not be seen as privilege for a few. It is, let us say, an entitlement that none of us should have to earn; ideally it is an *unearned entitlement*. At present, since only a few have it, it is an *unearned advantage* for them. The negative "privilege" that gave me cultural permission not to take darker-skinned Others seriously can be seen as arbitrarily conferred dominance and should not be desirable for anyone. This paper results from a process of coming to see that some of the power that I originally saw as attendant on being a human being in the United States consisted in *unearned advantage* and *conferred dominance*, as well as other kinds of special circumstance not universally taken for granted.

In writing this paper I have also realized that white identity and status (as well as class identity and status) give me considerable power to choose whether to broach this subject and its trouble. I can pretty well decide whether to disappear and avoid and not listen and escape the dislike I may engender in other people through this essay, or interrupt, answer, interpret, preach, correct, criticize, and control to some extent what goes on in reaction to it. Being white, I am given considerable power to escape many kinds of danger or penalty as well as to choose which risks I want to take.

There is an analogy here, once again, with Women's Studies. Our male colleagues do not have a great deal to lose in supporting Women's Studies, but they do not have a great deal to lose if they oppose it either. They simply have the power to decide whether to commit themselves to more equitable distributions of power. They will probably feel few penalties whatever choice they make; they do not seem, in any obvious short-term sense, the ones at risk, though they and we are all at risk because of the behaviors that have been rewarded in them.

Through Women's Studies work I have met very few men who are truly distressed about systemic, unearned male advantage and conferred dominance. And so one question for me and others like me is whether we will be like them, or whether we will get truly distressed, even outraged, about unearned race advantage and conferred dominance and if so, what we will do to lessen them. In any case, we need to do more work in identifying how they actually affect our daily lives. We need more down-to-earth writing by people about these taboo subjects. We need more understanding of the ways in which white "privilege" damages white people, for these are not the same ways in which it damages the victimized. Skewed white psyches are an inseparable part of the picture, though I do not want to confuse the kinds of damage done to the holders of special assets and to those who suffer the deficits. Many, perhaps most, of our white students in the United States think that racism doesn't affect them because they are not people of color; they do not see "whiteness" as a racial identity. Many men likewise think that Women's Studies does not bear on their own existences because they are not female; they do not see themselves as having gendered identities. Insisting on the universal "effects" of "privilege" systems, then, becomes one of our chief tasks, and being more explicit about the *particular* effects in particular contexts is another. Men need to join us in this work.

In addition, since race and sex are not the only advantaging systems at work, we need to similarly examine the daily experience of having age advantage, or ethnic advantage, or physical ability, or advantage related to nationality, religion, or sexual orientation. Professor Marnie Evans suggested to me that in many ways the list I made also applies directly to heterosexual privilege. This is a still more taboo subject than race privilege: the daily ways in which heterosexual privilege makes some persons comfortable or powerful, providing supports, assets, approvals, and rewards to those who live or expect to live in heterosexual pairs. Unpacking that content is still more difficult, owing to the deeper imbeddedness of heterosexual advantage and dominance and stricter taboos surrounding these.

But to start such an analysis I would put this observation from my own experience: The fact that I live under the same roof with a man triggers all kinds of societal assumptions about my worth, politics, life, and values and triggers a host of unearned advantages and powers. After recasting many elements from the original list I would add further observations like these:

1. My children do not have to answer questions about why I live with my partner (my husband).
2. I have no difficulty finding neighborhoods where people approve of our household.

3. Our children are given texts and classes that implicitly support our kind of family unit and do not turn them against my choice of domestic partnership.
4. I can travel alone or with my husband without expecting embarrassment or hostility in those who deal with us.
5. Most people I meet will see my marital arrangements as an asset to my life or as a favorable comment on my likability, my competence, or my mental health.
6. I can talk about the social events of a weekend without fearing most listeners' reactions.
7. I will feel welcomed and "normal" in the usual walks of public life, institutional and social.
8. In many contexts, I am seen as "all right" in daily work on women because I do not live chiefly with women.

Difficulties and dangers surrounding the task of finding parallels are many. Since racism, sexism, and heterosexism are not the same, the advantages associated with them should not be seen as the same. In addition, it is hard to isolate aspects of unearned advantage that derive chiefly from social class, economic class, race, religion, region, sex, or ethnic identity. The oppressions are both distinct and interlocking, as the Combahee River Collective statement of 1977 continues to remind us eloquently.[3]

One factor seems clear about all of the interlocking oppressions. They take both active forms that we can see and embedded forms that members of the dominant group are taught not to see. In my class and place, I did not see myself as racist because I was taught to recognize racism only in individual acts of meanness by members of my group, never in invisible systems conferring racial dominance on my group from birth. Likewise, we are taught to think that sexism or heterosexism is carried on only through intentional, individual acts of discrimination, meanness, or cruelty, rather than in invisible systems conferring unsought dominance on certain groups. Disapproving of the systems won't be enough to change them. I was taught to think that racism could end if white individuals changed their attitudes; many men think sexism can be ended by individual changes in daily behavior toward women. But a man's sex provides advantage for him whether or not he approves of the way in which dominance has been conferred on his group. A "white" skin in the United States opens many doors for whites whether or not we approve of the way dominance has been conferred on us. Individual acts can palliate, but

---

3. "A Black Feminist Statement," The Combahee River Collective, pp. 13–22 in G. Hull, P. Scott, B. Smith, Eds., *All the Women Are White, All the Blacks Are Men, But Some of Us Are Brave: Black Women's Studies*, Old Westbury, NY: The Feminist Press, 1982.

cannot end, these problems. To redesign social systems, we need first to acknowledge their colossal unseen dimensions. The silences and denials surrounding privilege are the key political tool here. They keep the thinking about equality or equity incomplete, protecting unearned advantage and conferred dominance by making these taboo subjects. Most talk by whites about equal opportunity seems to me now to be about equal opportunity to try to get into a position of dominance while denying that *systems* of dominance exist.

Obliviousness about white advantage, like obliviousness about male advantage, is kept strongly inculturated in the United States so as to maintain the myth of meritocracy, the myth that democratic choice is equally available to all. Keeping most people unaware that freedom of confident action is there for just a small number of people props up those in power and serves to keep power in the hands of the same groups that have most of it already. Though systemic change takes many decades, there are pressing questions for me and I imagine for some others like me if we raise our daily consciousness on the perquisites of being light-skinned. What will we do with such knowledge? As we know from watching men, it is an open question whether we will choose to use unearned advantage to weaken invisible privilege systems and whether we will use any of our arbitrarily awarded power to try to reconstruct power systems on a broader base.

FROM "KIKE" TO "JAP": 9
How Misogyny, Anti-Semitism, and Racism
Construct the "Jewish American Princess"

Evelyn Torton Beck

The stereotyping of the Jewish American woman as the JAP, which stands for Jewish American Princess, is an insult, an injury, and violence that is done to Jewish women. The term is used widely by both men and women, by both Jews and non-Jews. When gentiles use it, it is a form of anti-Semitism. When Jews use it, it is a form of self-hating or internalized anti-Semitism. It is a

From: Evelyn Torton Beck, "From 'Kike' to 'JAP,' " in *Sojourner: The Women's Forum*, September 1988, pp. 18–20. Reprinted by permission.

way of thinking that allows some Jewish women to harm other Jewish women who are just like them except for the fact that one is okay—she's *not* a JAP. The other is not okay—she's too JAPie. The seriousness of this term becomes evident when we substitute the words "too Jewish" for "too JAPie," and feel ourselves becoming considerably less comfortable.

When I speak on college campuses, young women frequently tell me that when someone calls them a Jew they are insulted because they know it's being said with a kind of hostility, but if someone calls them a JAP they don't mind because they frequently use this term themselves. They think the "J" in JAP really doesn't mean anything—it's just there. While everyone seems to know what the characteristics of a "Jewish American Princess" are, no one ever seems to think about what they are saying when they use the term. How is it that you don't have to be Jewish to be a JAP? If this is so, why is the word "Jewish" in the acronym at all? Words are not meaningless unless we choose to close our ears and pretend not to hear.

This subject is frequently trivialized, but when it is not, when we take it seriously, it makes us extremely tense. Why is that? I think it's because it takes us into several "war zones." It brings us in touch with Jew-hating, or anti-Semitism. It brings us in touch with misogyny, or woman-hating. And it brings us in touch with class-hatred, old money vs. nouveau riche. (Jews have classically been seen as intruders in the United States and have been resented for "making it.") It also puts us strongly in touch with racism. It is no accident that the acronym JAP is also the word used for our worst enemies in World War II—who were known as "the Japs." During World War II, posters and slogans saying "Kill Japs" were everywhere. It was a period in which slang terms were readily used in a pejorative way to identify many different minorities: "Japs," "Kikes," "Spics," "Wops," "Chinks" were commonplace terms used unthinkingly. And women were—and, unfortunately, still are—easily named "bitches," "sluts," and "cunts."

In such a climate, negative stereotypes easily overlap and elide. For example, in the popular imagination, Jews, "Japs," women and homosexuals have all been viewed as devious, unreliable, and power hungry. What has happened in the decades following World War II is that the "Japs," whom we de-humanized when we dropped our atom bomb on them, have subliminally merged in the popular imagination with "kikes" and other foreign undesirables. (The fact that in the 1980s Japan poses a serious economic threat to the United States should not be overlooked either.) While efforts to eradicate slurs against ethnic minorities have made it not okay to use explicitly ethnic epithets, women still provide an acceptable target, especially when the misogyny is disguised as supposedly "good-natured" humor. In this insidious and circuitous way, the Jewish American woman carries the stigmas of the "kikes" and "Japs" of a previous era. And that is very serious.

The woman, the Jewish woman as JAP has replaced the male Jew as the scapegoat, and the Jewish male has not only participated, but has, in fact, been instrumental in creating and perpetuating that image. I want to show how some of the images of Jewish women created in American culture by Jewish men provided the roots of the "Jewish American Princess." But first I want to provide a context for understanding the development of this image. I want to look at anti-Semitism in the United States, and at misogyny, and show how the merging of anti-Semitism and misogyny creates the Jewish American Princess.

Between 1986 and 1987 there was a 17 percent rise in anti-Semitic incidents in this country. Of these incidents, 48 percent occurred in the Northeast, and the highest rates of increase were in New York State; California, particularly Los Angeles; and Florida. These are all areas where there are high concentrations of Jews. On November 9 and 10, 1987, the anniversary of *Kristalnacht*, the Night of Broken Crystal, when Goebbels staged a mass "spontaneous pogrom" in Austria and Germany in 1938, swastikas were painted on entrances to synagogues in a number of different cities in the United States: Chicago, Yonkers, and others dotted across the country. Windows were smashed—not simply of Jewish-owned stores, but of identifiably Jewish businesses such as kosher meat markets, a kosher fish store, a Jewish book store—in five different neighborhoods, particularly in a Chicago suburb largely populated by Holocaust survivors. Having grown up in Vienna and having lived under the Nazis, the horror of that night resonated for me in a way that it might not for those who are much younger. But that these pogrom-like episodes happened on the eve of *Kristalnacht* cannot have been an accident, and the timing of these incidents should not be lost upon us.

In response to these attacks many members of the Jewish community wanted to hide the facts, and one member of the Jewish community in Chicago actually said, "The swastikas could have meant general white supremacy; they were not necessarily aimed at Jews." They just "happened" to be placed on synagogues, right? In the same way, it just "happens" that the word "Jewish" is lodged in the very negative image of this hideous creature known as the Jewish American Princess. Anyone who is aware of Jewish history and knows about the ridicule, defamation, and violence to which Jews have been subject will not be able to write this off so easily.

The Jewish American Princess phenomenon is not new; I (as well as other Jewish feminists) have been talking about it for at least ten years now, but only recently has it been given wide public attention. One reason for this is that it is beginning to be seen in the light of increased anti-Semitism and racism, particularly on college campuses. Dr. Gary Spencer, who is a *male* professor of sociology at Syracuse University (and it is unfortunate that his being male gives him credibility over women saying the same things) closely examined the library and bathroom graffiti of his school and interviewed hundreds of students on his campus and has concluded that "JAP"-baiting is widespread, virulent, and threatening to all Jews, not "just" Jewish women (which we gather might have been okay or certainly considerably less serious).

Spencer discovered that nasty comments about "JAPS" led to more generally anti-Semitic graffiti that said among other slogans, "Hitler was right!" "Give Hitler a second chance!" and "I hate Jews." He also discovered that there were certain places in which Jewish women—JAPS—were not welcome: for example, certain cafes where Jewish women were hassled if they entered. He also found that certain areas of the University were considered "JAP-free zones" and other areas (particular dorms) that were called "Jew havens." At The American University in Washington, D.C., largely Jewish residence halls are called "Tokyo Towers," making the racial overtones of "JAP" explicit. But let the parallels to Nazi occupied Europe not be lost upon us. Under the Nazis, movements of Jews were sharply restricted: there were many areas which Jews could not enter, and others (like ghettos and concentration camps) that they could not leave.

What I want to do now is to show how characteristics that have historically been attributed to Jews, primarily Jewish men, have been reinterpreted in terms of women: how misogyny combined with Jew-hating creates the Jewish American Princess. And I want you to remember that Jewish men have not only participated in this trashing, but they have not protected Jewish women when other men and women have talked about "JAPs" in this way. And this fact, I think, has made this an arena into which anyone can step—an arena that becomes a minefield when Jews step into it.

Jews have been said to be materialistic, money-grabbing, greedy, and ostentatious. Women have been said to be vain, trivial and shallow; they're only interested in clothing, in show. When you put these together you get the Jewish-woman type who's only interested in designer clothes and sees her children only as extensions of herself. The Jew has been seen as manipulative, crafty, untrustworthy, unreliable, calculating, controlling, and malevolent. The Jewish Princess is seen as manipulative, particularly of the men in her life, her husband, her boyfriend, her father. And what does she want? Their *money!* In addition, she's lazy—she doesn't work inside or outside the home. She is the female version of the Jew who, according to anti-Semitic lore, is a parasite on society; contradictorily, the Jew has been viewed both as dangerous "communist" as well as non-productive "capitalist." The cartoon vision of the

Jewish American Princess is someone who sucks men dry: she is an "unnatural mother" who refuses to nurture her children (the very opposite of the "Jewish mother" whose willingness to martyr herself makes *her* ludicrous). And she doesn't "put out" except in return for goods; she isn't really interested in either sexuality or lovingness. We live in a world climate and culture in which materialism is rampant, and Jewish women are taking the rap for it. The irony is they are taking the rap from non-Jews and Jewish men alike—even from some Jewish women.

Another way in which Jewish women are carrying the anti-Semitism that was directed in previous eras at Jewish men is in the arena of sexuality. Jews have been said to be sexually strange, exotic. There are many stereotypes of Jewish men as lechers. The Jewish American Princess is portrayed as both sexually frigid (withholding) and as a nymphomaniac. Here we again see the familiar anti-Semitic figure of the Jew as controlling and insatiably greedy, always wanting more, combining with the misogynist stereotype of the insatiable woman, the woman who is infinitely orgasmic, who will destroy men with her desire. Like the Jew of old, the Jewish woman will suck men dry. But she is worse than "the Jew"—she will also turn on her own kind.

There are physical stereotypes as well: the Jew with the big hook nose, thick lips, and frizzy hair. The Jewish American Princess has had a nose job and her hair has been straightened, but she too has large lips (an image we immediately also recognize as racist). Jews are supposed to be loud, pushy, and speak with unrefined accents. Jewish American Princesses are said to come from Long Island and speak with funny accents: "Oh my Gawd!" The accent has changed from the lampooned immigrant speech of previous generations, but assimilation into the middle class hasn't helped the Jewish American Princess get rid of her accent. It doesn't matter how she speaks, because if it's Eastern and recognizably Jewish, it's not okay.

I also want to give you some idea of how widespread and what a money-making industry the Jewish American Princess phenomenon has become. There are greeting cards about the "JAP Olympics," with the JAP doing things like "bank-vaulting" instead of pole-vaulting. Or cross-country *"kvetching"* instead of skiing. In this card the definition of the Yiddish term *"kvetch"* reads: "an irritable whine made by a three-year-old child or a JAP at any age." So in addition to the all-powerful monster you also have the infantilization of the Jewish woman. And there are the Bunny Bagelman greeting cards: Bunny has frizzy hair, big lips, is wearing ostentatious jewelry—and is always marked as a Jew in some way. One of her cards reads, "May God Bless you and keep you . . . rich!" Or Bunny Bagelman is a professional, dressed in a suit carrying a briefcase, but this image is undermined by the little crown she incongruously wears on her head bearing the initials "JAP." There is also a Halloween card with a grotesque female figure; the card reads, "Is it a vicious vampire? No, it's Bunny Bagelman with PM syndrome!" In analyzing these kinds of cartoons, you begin to see how sexism is absolutely intertwined with anti-Semitism.

Such attacks devalue Jewish women and keep them in line. An incident reported by Professor Spencer at Syracuse University makes this quite evident. At a basketball game, when women who were presumed to be "JAPs" stood up and walked across the floor at half-time (and it happened to Jewish and non-Jewish women), 2,000 students stood up, accusingly pointed their fingers at them, and repeatedly yelled, "JAP, JAP, JAP, JAP, JAP" in a loud chorus. This was so humiliating and frightening that women no longer got out of their seats to go to the bathroom or to get a soda. This is a form of public harassment that is guaranteed to control behavior and parallels a phenomenon called "punching" at the University of Dar El Salaam, Tanzania. Here, when women were "uppity" or otherwise stepped out of line, huge posters with their pictures on them were put all over campus, and no one was to speak to them. If you spoke to these women, you were considered to be like them. This is a very effective way of controlling people.

The threat of physical violence against Jewish women (in the form of "Slap-a-JAP" T-shirts and contests at bars) is evident on many Eastern college campuses. A disc jockey at The American University went so far as to sponsor a "fattest JAP-on-campus" contest. That this kind of unchecked verbal violence can lead to murder is demonstrated by lawyer Shirley Frondorf in a recent book entitled *Death of a "Jewish American Princess": The True Story of a Victim on Trial* (Villard Books, 1988). Frondorf shows how the murder of a Jewish woman by her husband was exonerated and the victim placed on trial because she was someone who was described by her husband as "materialistic, who shopped and spent, nagged shrilly and bothered her husband at work"—in other words, she was a "JAP" and therefore deserved what she got. This account demonstrates the dangers inherent in stereotyping and the inevitable dehumanization that follows.

One of the most aggressively sexual forms of harassment of Jewish women, which amounted to verbal rape, were signs posted at a college fair booth at Cornell University that read, "Make her prove she's not a JAP, make her swallow." Part of the mythology is that the Jewish woman will suck, but she won't swallow. So you see that as the degradation of woman *as woman* escalates, the anti-Semitism also gets increasingly louder. In a recent Cornell University student newspaper, a cartoon offered advice on how to "exterminate" JAPS by setting up a truck offering bargains, collecting the JAPS as they scurried in, and dropping them over a cliff. While the word "Jew" was not specifically mentioned, the parallels to the historical "rounding up" of Jews and herding them into trucks to be exterminated in the camps during World War II can hardly be ignored. This cartoon was created by a Jewish man.

This leads me directly to the third thing I want to discuss, namely, how and why Jewish men have participated in constructing and perpetuating the image of the Jewish American Princess as monster. How is it that the Jewish Mother (a mildly derogatory stereotype that nonetheless contained some warmth) has become the grotesque that is the Jewish American Princess, who, unlike the Jewish Mother, has absolutely no redeeming features? Exactly how the Jewish Mother (created entirely by second generation American men who had begun to mock the very nurturance they had relied upon for their success) gave birth to the Jewish American Princess is a long and complex story. This story is intertwined with the overall economic success of Jews as a class in the United States, the jealousy others have felt over this success, and the discomfort this success creates in Jews who are fearful of living out the stereotype of the "rich Jew." It is also a likely conjecture that middle-class American Jewish men view the large numbers of Jewish women who have successfully entered the work force as professionals as a serious economic and ego threat.

We find the origins of the Jewish American Princess in the fiction of American Jewish males of the last three decades. In the '50s, Herman Wouk's *Marjory Morningstar* (nee Morgenstern) leaves behind her immigrant background, takes a new name (one that is less recognizably Jewish), manipulates men, has no talent, and is only interested in expensive clothing. The postwar Jewish male, who is rapidly assimilating into American middle-class culture and leaving behind traditional Jewish values, is creating the Jewish woman— the materialistic, empty, manipulative Jewish woman, the Americanized daughter who fulfills the American Dream for her parents but is, at the same time, punished for it. It looks as if the Jewish woman was created in the image of the postwar Jewish male but viewed by her creator as grotesque. All the characteristics he cannot stand in himself are displaced onto the Jewish woman.

In the '60s, Philip Roth created the spoiled and whiny Brenda Potemkin in *Goodbye, Columbus* at the same time that Shel Silverstein created his image of the perfect Jewish mother as martyr. Some of you may remember this popular story from your childhood. A synopsis goes something like this: "Once there was a tree and she loved the little boy. And he slept in her branches, and loved the tree and the tree was happy. And as the boy grew older, he needed things from her. He needed apples, so she gave him apples, and she was happy. Then she cut off her branches because the boy needed them to build a house. And she was happy. Then finally he needed her trunk because he wanted to build a big boat for himself. And she was happy. The tree gave and gave of herself, and finally the tree was alone and old when the boy returned one more time. By now, the tree had nothing to give. But the boy/man is himself old now, and he doesn't need much except a place on which to sit. And the tree said, 'An old stump is good for sitting and resting on. Come boy and sit and rest on me.' And the boy did, and the tree was happy." This "positive" entirely self-*less* mother, created as a positive wish fantasy by a Jewish man, very easily tips over into its opposite, the monstrous woman, the self-absorbed "JAP" who is negatively self-less. She has no center. She *is* only clothes, money, and show.

In concluding, I want to bring these strands together and raise some questions. Obviously Jews need to be as thoughtful about consumerism as others, but we need to ask why the Jewish woman is taking the rap for the consumerism which is rampant in our highly materialistic culture in general. We need to think about the image of the Jewish American Princess and the father she tries to manipulate. What has happened to the Jewish Mother? Why has she dropped out of the picture? If (as is likely true of all groups) some middle-class Jewish women (and men) are overly focused on material things, what is the other side of that? What about the middle-class fathers who measure their own success by what material goods they are able to provide to their wives and children and who don't know how to show love in any other way? Someone who doesn't know how to give except through material goods could easily create a child who comes to expect material goods as a proof of love and self-worth, especially if sexist gender expectations limit the options for women. We need to look more closely at the relationship between the "monster" daughter and the father who helped create her.

This brings up another uncomfortable subject—incest in Jewish families. We have to look carefully at the image of the "little princess" who sits on Daddy's lap and later becomes this monstrous figure. (My father thought it appropriate for me to sit on his lap until the day he died, well into his '80s, and I do not believe he was unusual in his expectations.) There are enough stories of incest in which we know that the father who sexually abuses his daughter when she is a child becomes quite distant when she reaches adolescence and may continue to abuse her in psychological ways. And the JAP image is a real form of psychological abuse. We need to look at these things to understand that this phenomenon is not trivial, and to understand how it undermines *all* Jewish women and particularly harms young women coming of age. It cannot do Jewish men such good either to think of their sisters, daughters, mothers, and potential girlfriends with such contempt.

Last, I want to say that we have many false images of Jewish families. There *is* violence in Jewish families, just as there is violence in families of all groups. It is time to put the whole question of the Jewish American Princess into the context of doing away with myths of all kinds. The Jewish family is no more nor less cohesive than other families, although there is great pressure on Jewish families to pretend they are. Not all Jewish families are non-alcoholic; not all Jewish families are heterosexual; not all Jews are upper or middle class; and not all are urban or Eastern. It's important that the truth of Jewish women's (and also Jewish men's) lives be spoken. Beginning to take apart this image of the Jewish American Princess can make us look more closely at what it is that we, in all of our diversity as Jews, are; what we are striving towards; and what we hope to become.

IS MULTIRACIAL STATUS UNIQUE? THE PERSONAL AND SOCIAL EXPERIENCE

10

Michael C. Thornton

Most of us use the term *race* with little difficulty. We can even describe the various races, usually by referring to physical and other overt qualities. Nevertheless, the apparent consensus over the meaning and use of the term remains so only on the surface, for when asked for details beyond physical differences, we are at a loss for words, often citing cultural artifacts to explain racial differences. Ultimately, there is no real consensus on the meaning of the term. This puzzlement is due to a misunderstanding of what race is, how it is commonly perceived by society, and how racial minorities experience it. This confusion about race is most visible in how we view multiracial people.

Race and racial differences seem somehow self-evident and distinct. However, what we perceive as race is often intertwined with culture and biology. Physical qualities, the belief goes, reflect biological influences, which seem tied to attitudinal, life-style, and socioeconomic factors (in toto, culture).

From: Michael C. Thornton, "Is Multiracial Status Unique: The Personal and Social Experience," in *Racially Mixed People in America*, ed. Maria P. P. Root (Newbury Park, CA: Sage Publications, 1992), pp. 321–325. Copyright © 1992 Sage Publications, Inc. Reprinted by permission of Sage Publications, Inc.

Biological factors still underpin explanations of racial differences observed in our society.

But what is race, and what does it explain about group differences? Biologists and geneticists talk of races possessing dissimilar gene pools, revealed as obvious physical disparity. Furthermore, these gene pools are not distinct and in fact overlap. Because there are no such things as pure races, technically we are all interracial. Therefore, the significant distinctions among racial groups are socially constructed artifacts. Racial differences explain how society parcels out resources and privileges and have little to do with biological deviations of any note. Race is important because we say it is.

Thus societies have developed subjective social measures to identify racial groups. In the United States, these groups are said to be distinct and well-defined entities. For example, one is considered "Black" (i.e., has a Black identity) if one has an African heritage or "Black" blood. Nevertheless, there are significant cases that deviate from this rule. By the above definition, Hispanics and Native Americans should be but are not considered Black. Further, we are all regarded as members of only one racial group. You are, for example, Black *or* Asian *or* White. For a number of people such a designation is problematic. Multiracial people present a quandary for the either/or, dichotomous racial classifications used in this society. This is reflected in a lack of labels for those of multiple racial heritage and in our ambiguous treatment of non-Black-White combinations. If you are Asian and White are you White or Asian?

The subjective nature of racial labels is emphasized by a debate that appeared in a recent *Los Angeles Times* article titled "Call for Census Category Creates Interracial Debate" (Njeri, 1991). In it, a self-proclaimed multiracial female (Asian, Black, and Native American heritage) argued that multiracial people deserve to have their distinctive status legitimated by the creation of a unique census category. In opposition, a Black male (who, while not defining himself as multiracial, also has Native American ancestry) suggested that the present categories are sufficient. A new grouping would negatively affect Black representation.

Often racial labels become a political tool to measure strength of identity. This tendency can be seen in an allegation that the woman described above desires a separate category in part to deny her African heritage. It is acceptable to make statements about one's multiple ancestry (I am part Asian, Polish, or whatever), but not about identity; a private but not public affirmation of multiple heritage is acceptable. As Jack Forbes (1990) has noted, in embracing White notions of racial cataloging, some Blacks are trapped by a system that blocks self-identification and serves to reinforce what racists have long advocated. Nevertheless, it is understandable why they take this stance. The campaign by the United States to subordinate and oppress people of color has forced them to forge a united front to generate pride in what Whites have declared to be inferior. In this environment many multiracial people pass or take pride in their mixed heritage for the wrong reasons, exalting in their White (or higher-status) ancestry and denying their ethnic origins. An authentic identity cannot be developed while an individual is in a state of intellectual and emotional subservience to racism. Still worse is to accept only one's majority and not one's minority ancestry. To lean toward the former is to undermine the struggle for self-determination, while aligning with the latter at least maintains an alliance with the oppressed.

The man in the debate described above assumes, with little evidence, that his opponent reveals latent racism. While he ostensibly attempts to support the idea of Black unity, for the woman to do as he suggests would be subversive to self-identification, an antithesis to the primary thrust of the Black power movement—self-determination. In this regard, pressure is from both Black and White society, and in the same direction—for both seem to accept White definitions of Whiteness and of race—the presence of African ancestry is the sole determinant of racial status.

These underlying assumptions can be seen in the categories selected for the U.S. Census. Interestingly, although in the census racial and national labels are supposedly measures of self-identity, it is clear that the universe of self-designations is limited to socially decreed categories. Racial labels are not objective configurations, but are political in nature. African heritage equals being Black because we as a society say it is so, although Blacks are realistically a number of diverse people—Jamaican, Ethiopian, Nigerian, and so on.

The process of labeling of course oversimplifies how race is experienced. There is no intrinsic personality related to race; what develops is a complex interaction between individual and social definitions. Individuals are expected to locate themselves "accurately" within established racial structures (such as current census categories), finding where society places them and reconciling this placement with what they want to be. Ideally, individual choice and social designations are psychologically satisfying for the individual and socially approved. The woman in the debate described above has not found that ideal, for society denies the validity of her perceptions and experience. For her, a census category would legitimate an otherwise socially ambiguous status.

Thus individual experience often differs from social expectations of that experience. The secular existence of multiracial people overlaps but differs from other racial minorities. Socially they are often treated like their monoracial compatriots, but psychologically they encounter life in unique ways. The contrast is between, as Jack Forbes describes it, fruits and apples; fruits are the general class, apples are the specific. Fruits are not the same as apples, yet the two overlap. In the present context, an Asian American's experience and a

Black-Asian American's experience, for example, will overlay one another, yet they are not the same. If self-identification were the goal, then a mixed racial category would be part of the census and would serve to legitimate this status.

Such a new category would suggest that there is an experience that ties all multiracial people together; any label utilized implies important continuity in experience. If all multiracial people have some basic background in common, what is it and how does it bind them more closely to one another than to their parental racial groups? What is the unifying theme? There is little evidence indicating other than some superficial basis for commonality between groups of different racial mixture. The two protagonists in the *L.A. Times* piece show the complexity in combining people of all racial mixtures into one group. Both are of mixed racial heritage, but one identifies as mixed and the other does not. It is predictable that many people would exclude themselves from this category, a trend perhaps more pronounced among those of particular combinations, Black-other than, say, Asian-White, because of the virulent nature of racism against Blacks. What would the new designation mean if one were allowed to choose among a number of possibilities?

Further, the other groups that have designated census categories have more clear-cut bases on which to expect similar experiences, both because society identifies and treats them as separate groups and because they have common heritages (i.e., cultures). None of these categories exists simply because of common experience. Do multiracials have a core (cultural) heritage, or are they viewed as alike by others in society? Are multiracials seen as a different (racial/ethnic) group from, say, Blacks? In fact, what seems to bind multiracial people is not race or culture, but living with an ambiguous status, an experience similar to that of all people of color. Facing a different set of dilemmas does not make one an ethnic or racial group, or signify a culture. As a group, multiracials are too diverse to categorize. This group is more biologically diverse than others, and has no common ancestry and little community. These are things one cannot say about the other census groupings.

In comparing people of so-called multiracial status to monoracials, it is clear that the former have a different flavor to their lives. Society defines race as distinctive and homogeneous; multiracial people experience it as multidimensional. That background should and must be appreciated in and of itself. But that experience is irrelevant in a social sense. Race is a social and political tool not meant to reflect the range of experience or the diversity found within any group to any great extent. By definition, racial labels are tools used to categorize and to separate and/or exclude. There remains insufficient social justification to exclude multiracials from other groups. While there is experiential rationale for identifying a unique experience, that alone is not reason enough, and does not provide a consistent basis, to describe multiracials as one sort of people.

REFERENCES

Forbes, J. D. 1990. "The Manipulation of Race, Caste, and Identity: Classifying Afro-Americans, Native Americans, and red-black people." *Journal of Ethnic Studies* 17: 1–51.

Njeri, I. 1991. "Call for Census Category Creates Interracial Debate." *Los Angeles Times*, January 13: pp. E1, E9–11.

Class and Inequality

TIRED OF PLAYING MONOPOLY? **11**

Donna Langston

I. Magnin, Nordstrom, The Bon, Sears, Penneys, K mart, Goodwill, Salvation Army. If the order of this list of stores makes any sense to you, then we've begun to deal with the first question which inevitably arises in any discussion of class here in the U.S.—huh? Unlike our European allies, we in the U.S. are reluctant to recognize class differences. This denial of class divisions functions to reinforce ruling class control and domination. America is, after all, the supposed land of equal opportunity where, if you just work hard enough, you can get ahead, pull yourself up by your bootstraps. What the old bootstraps theory overlooks is that some were born with silver shoe horns. Female-headed households, communities of color, the elderly, disabled and children find themselves, disproportionately, living in poverty. If hard work were the sole determinant of your ability to support yourself and your family, surely we'd have a different outcome for many in our society. We also, however, believe in luck and, on closer examination, it certainly is quite a coincidence that the "unlucky" come from certain race, gender and class backgrounds. In order to perpetuate racist, sexist and classist outcomes, we also have to believe that the current economic distribution is unchangeable, has always existed, and probably exists in this form throughout the known universe, i.e., it's

From: Jo Whitehorse Cochran, Donna Langston, and Carolyn Woodward (eds.), *Changing Our Power: An Introduction to Women's Studies* (Dubuque, IA: Kendall-Hunt, 1988). Reprinted by permission.

"natural." Some people explain or try to account for poverty or class position by focusing on the personal and moral merits of an individual. If people are poor, then it's something they did or didn't do; they were lazy, unlucky, didn't try hard enough, etc. This has the familiar ring of blaming the victims. Alternative explanations focus on the ways in which poverty and class position are due to structural, systematic, institutionalized economic and political power relations. These power relations are based firmly on dynamics such as race, gender, and class.

In the myth of the classless society, ambition and intelligence alone are responsible for success. The myth conceals the existence of a class society, which serves many functions. One of the main ways it keeps the working-class and poor locked into a class-based system in a position of servitude is by cruelly creating false hope. It perpetuates the false hope among the working-class and poor that they can have different opportunities in life. The hope that they can escape the fate that awaits them due to the class position they were born into. Another way the rags-to-riches myth is perpetuated is by creating enough visible tokens so that oppressed persons believe they, too, can get ahead. The creation of hope through tokenism keeps a hierarchical structure in place and lays the blame for not succeeding on those who don't. This keeps us from resisting and changing the class-based system. Instead, we accept it as inevitable, something we just have to live with. If oppressed people believe in equality of opportunity, then they won't develop class consciousness and will internalize the blame for their economic position. If the working-class and poor do not recognize the way false hope is used to control them, they won't get a chance to control their lives by acknowledging their class position, by claiming that identity and taking action as a group.

The myth also keeps the middle class and upper class entrenched in the privileges awarded in a class-based system. It reinforces middle- and upper-class beliefs in their own superiority. If we believe that anyone in society really can get ahead, then middle- and upper-class status and privileges must be deserved, due to personal merits, and enjoyed—and defended at all costs. According to this viewpoint, poverty is regrettable but acceptable, just the outcome of a fair game: "There have always been poor people, and there always will be."

Class is more than just the amount of money you have; it's also the presence of economic security. For the working class and poor, working and eating are matters of survival, not taste. However, while one's class status can be defined in important ways in terms of monetary income, class is also a whole lot more—specifically, class is also culture. As a result of the class you are born into and raised in, class is your understanding of the world and where you fit in; it's composed of ideas, behavior, attitudes, values, and language; class is how you think, feel, act, look, dress, talk, move, walk; class is what stores you shop at, restaurants you eat in; class is the schools you attend, the education you attain; class is the very jobs you will work at throughout your adult life. Class even determines when we marry and become mothers. Working-class women become mothers long before middle-class women receive their bachelor's degrees. We experience class at every level of our lives; class is who our friends are, where we live and work even what kind of car we drive, if we own one, and what kind of health care we receive, if any. Have I left anything out? In other words, class is socially constructed and all-encompassing. When we experience classism, it will be because of our lack of money (i.e., choices and power in this society) and because of the way we talk, think, act, move—because of our culture.

Class affects what we perceive as and what we have available to us as choices. Upon graduation from high school, I was awarded a scholarship to attend any college, private or public, in the state of California. Yet it never occurred to me or my family that it made any difference which college you went to. I ended up just going to a small college in my town. It never would have occurred to me to move away from my family for school, because no one ever had and no one would. I was the first person in my family to go to college. I had to figure out from reading college catalogs how to apply—no one in my family could have sat down and said, "Well, you take this test and then you really should think about . . . " Although tests and high school performance had shown I had the ability to pick up white middle-class lingo, I still had quite an adjustment to make—it was lonely and isolating in college. I lost my friends from high school—they were at the community college, vo-tech school, working, or married. I lasted a year and a half in this foreign environment before I quit college, married a factory worker, had a baby and resumed living in a community I knew. One middle-class friend in college had asked if I'd like to travel to Europe with her. Her father was a college professor and people in her family had actually travelled there. My family had seldom been able to take a vacation at all. A couple of times my parents were able—by saving all year—to take the family over to the coast on their annual two-week vacation. I'd seen the time and energy my parents invested in trying to take a family vacation to some place a few hours away; the idea of how anybody ever got to Europe was beyond me.

If class is more than simple economic status but one's cultural background, as well, what happens if you're born and raised middle-class, but spend some of your adult life with earnings below a middle-class income bracket—are you then working-class? Probably not. If your economic position changes, you still have the language, behavior, educational background, etc., of the middle class, which you can bank on. You will always have choices. Men who consciously try to refuse male privilege are still male; whites who want to challenge white privilege are still white. I think those who come from middle-class backgrounds need to recognize that their class privilege does not float out with the rinse water. Middle-class people can exert incredible power just by being nice and polite. The middle-class way of doing things is the standard—they're always right, just by being themselves. Beware of middle-class people who deny their privilege. Many people have times when they struggle to get shoes for the kids, when budgets are tight, etc. This isn't the same as long-term economic conditions without choices. Being working-class is also generational. Examine your family's history of education, work, and standard of living. It may not be a coincidence that you share the same class status as your parents and grandparents. If your grandparents were professionals, or your parents were professionals, it's much more likely you'll be able to grow up to become a yuppie, if your heart so desires, or even if you don't think about it.

How about if you're born and raised poor or working-class, yet through struggle, usually through education, you manage to achieve a different economic level: do you become middle class? Can you pass? I think some working class people may successfully assimilate into the middle class by learning to dress, talk, and act middle-class—to accept and adopt the middle-class way of doing things. It all depends on how far they're able to go. To succeed in the middle-class world means facing great pressures to abandon working-class friends and ways.

Contrary to our stereotype of the working class—white guys in overalls—the working class is not homogeneous in terms of race or gender. If you are a person of color, if you live in a female-headed household, you are much more likely to be working-class or poor. The experience of Black, Latino, American Indian or Asian American working classes will differ significantly from the white working classes, which have traditionally been able to rely on white privilege to provide a more elite position within the working class. Working-class people are often grouped together and stereotyped, but distinctions can be made among the working-class, working-poor and poor. Many working-class families are supported by unionized workers who possess marketable skills. Most working-poor families are supported by non-unionized, unskilled men and women. Many poor families are dependent on welfare for their income.

Attacks on the welfare system and those who live on welfare are a good example of classism in action. We have a "dual welfare" system in this country whereby welfare for the rich in the form of tax-free capital gain, guaranteed loans, oil depletion allowances, etc., is not recognized as welfare. Almost everyone in America is on some type of welfare; but, if you're rich, it's in the form of tax deductions for "business" meals and entertainment, and if you're poor, it's in the form of food stamps. The difference is the stigma and humiliation connected to welfare for the poor, as compared to welfare for the rich, which is called "incentives." Ninety-three percent of AFDC (Aid to Families with Dependent Children, our traditional concept of welfare) recipients are women and children. Eighty percent of food stamp recipients are single mothers, children, the elderly and disabled. Average AFDC payments are $93 per person, per month. Payments are so low nationwide that in only three states do AFDC benefits plus foodstamps bring a household *up to* the poverty level. Food stamp benefits average $10 per person, per week (Sar Levitan, *Programs in Aid of the Poor for the 1980s*). A common focal point for complaints about "welfare" is the belief that most welfare recipients are cheaters—goodness knows there are no middle-class income tax cheaters out there. Imagine focusing the same anger and energy on the way corporations and big business cheat on their tax revenues. Now, there would be some dollars worth quibbling about. The "dual welfare" system also assigns a different degree of stigma to programs that benefit women and children, such as AFDC, and programs whose recipients are primarily male, such as veterans' benefits. The implicit assumption is that mothers who raise children do not work and therefore are not deserving of their daily bread crumbs.

Anti-union attitudes are another prime example of classism in action. At best, unions have been a very progressive force for workers, women and people of color. At worst, unions have reflected the same regressive attitudes which are out there in other social structures: classism, racism, and sexism. Classism exists within the working class. The aristocracy of the working class—unionized, skilled workers—have mainly been white and male and have viewed themselves as being better than unskilled workers, the unemployed and poor, who are mostly women and people of color. The white working class must commit itself to a cultural and ideological transformation of racist attitudes. The history of working people, and the ways we've resisted many types of oppressions, are not something we're taught in school. Missing from our education is information about workers and their resistance.

Working-class women's critiques have focused on the following issues:

Education: White middle-class professionals have used academic jargon to rationalize and justify classism. The whole structure of education is a classist system. Schools in every town reflect class divisions: like the store list at the beginning of this article, you can list schools in your town by what classes of kids attend, and in most cities you can also list by race. The classist system is perpetuated in schools with the tracking system, whereby the "dumbs" are tracked into homemaking, shop courses and vocational school futures, while the "smarts" end up in advanced math, science, literature, and college-prep courses. If we examine these groups carefully, the coincidence of poor and working-class backgrounds with "dumbs" is rather alarming. The standard measurement of supposed intelligence is white middle-class English. If you're other than white middle-class, you have to become bilingual to succeed in the educational system. If you're white middle-class, you only need the language and writing skills you were raised with, since they're the standard. To do well in society presupposes middle-class background, experiences and learning for everyone. The tracking system separates those from the working class who can potentially assimilate to the middle class from all our friends, and labels us "college bound."

After high school, you go on to vocational school, community college, or college—public or private—according to your class position. Apart from the few who break into middle-class schools, the classist stereotyping of the working class as being dumb and inarticulate tracks most into vocational and low-skilled jobs. A few of us are allowed to slip through to reinforce the idea that equal opportunity exists. But for most, class position is destiny—determining our educational attainment and employment. Since we must overall abide by middle-class rules to succeed, the assumption is that we go to college in order to "better ourselves"—i.e., become more like them. I suppose it's assumed we have "yuppie envy" and desire nothing more than to be upwardly mobile individuals. It's assumed that we want to fit into their world. But many of us remain connected to our communities and families. Becoming college-educated doesn't mean we have to, or want to, erase our first and natural language and value system. It's important for many of us to remain in and return to our communities to work, live, and stay sane.

Jobs: Middle-class people have the privilege of choosing careers. They can decide which jobs they want to work, according to their moral or political commitments, needs for challenge or creativity. This is a privilege denied the working-class and poor, whose work is a means of survival, not choice (see Hartsock). Working-class women have seldom had the luxury of choosing between work in the home or market. We've generally done both, with little ability to purchase services to help with this double burden. Middle- and upper-class women can often hire other women to clean their houses, take care of their children, and cook their meals. Guess what class and race those "other" women are? Working a double or triple day is common for working-class women. Only middle-class women have an array of choices such as: parents put you through school, then you choose a career, then you choose when and if to have babies, then you choose a support system of working-class women to take care of your kids and house if you choose to resume your career. After the birth of my second child, I was working two part-time jobs—one loading trucks at night—and going to school during the days. While I was quite privileged because I could take my colicky infant with me to classes and the day-time job, I was in a state of continuous semi-consciousness. I had to work to support my family; the only choice I had was between school or sleep: Sleep became a privilege. A white middle-class feminist instructor at the university suggested to me, all sympathetically, that I ought to hire someone to clean my house and watch the baby. Her suggestion was totally out of my reality, both economically and socially. I'd worked for years cleaning other peoples' houses. Hiring a working-class woman to do the shit work is a middle-class woman's solution to any dilemma which her privileges, such as a career, may present her.

Mothering: The feminist critique of families and the oppressive role of mothering has focused on white middle-class nuclear families. This may not be an appropriate model for communities of class and color. Mothering and families may hold a different importance for working class women. Within this context, the issue of coming out can be a very painful process for working-class lesbians. Due to the homophobia of working-class communities, to be a lesbian is most often to be excommunicated from your family, neighborhood, friends and the people you work with. If you're working-class, you don't have such clearly demarcated concepts of yourself as an individual, but instead see yourself as part of a family and community that forms your survival structure. It is not easy to be faced with the risk of giving up ties which are so central to your identity and survival.

Individualism: Preoccupation with one's self—one's body, looks, relationships—is a luxury working-class women can't afford. Making an occupation out of taking care of yourself through therapy, aerobics, jogging, dressing for success, gourmet meals and proper nutrition, etc., may be responses that are directly rooted in privilege. The middle-class have the leisure time to be preoccupied with their own problems, such as their waistlines, planning their vacations, coordinating their wardrobes, or dealing with what their mother said to them when they were five—my!

The white middle-class women's movement has been patronizing to working-class women. Its supporters think we don't understand sexism. What we don't understand is white middle-class feminism. They act as though they invented the truth, the light, and the way, which they merely need to pass along to us lower-class drudges. What they invented is a distorted form of what working-class women already know—if you're female, life sucks. Only at least we were smart enough to know that it's not just being female, but also being a person of color or class, which makes life a quicksand trap. The class system weakens all women. It censors and eliminates images of female strength. The idea of women as passive, weak creatures totally discounts the strength, self-dependence and inter-dependence necessary to survive as work-ing-class and poor women. My mother and her friends always had a less-than-passive, less-than-enamoured attitude toward their spouses, male bosses, and men in general. I know from listening to their conversations, jokes and what they passed on to us, their daughters, as folklore. When I was five years old, my mother told me about how Aunt Betty had hit Uncle Ernie over the head with a skillet and knocked him out because he was raising his hand to hit her, and how he's never even thought about doing it since. This story was told to me with a good amount of glee and laughter. All the men in the neighborhood were told of the event as an example of what was a very acceptable response in the women's community for that type of male behavior. We kids in the neighborhood grew up with these stories of women giving husbands, bosses, the welfare system, schools, unions and men in general—hell, whenever they deserved it. For me there were many role models of women taking action, control and resisting what was supposed to be their lot. Yet many white middle-class feminists continue to view feminism like math homework, where there's only supposed to be one answer. Never occurs to them that they might be talking algebra while working-class women might be talking metaphysics.

Women with backgrounds other than white middle-class experience compounded, simultaneous oppressions. We can't so easily separate our experiences by categories of gender, or race, or class, i.e., "I remember it well: on Saturday, June 3, I was experiencing class oppression, but by Tuesday, June 6, I was caught up in race oppression, then all day Friday, June 9, I was in the middle of gender oppression. What a week!" Sometimes, for example, gender and class reinforce each other. When I returned to college as a single parent after a few years of having kids and working crummy jobs—I went in for vocational testing. Even before I was tested, the white middle-class male vocational counselor looked at me, a welfare mother in my best selection from the Salvation Army racks, and suggested I quit college, go to vo-tech school and become a grocery clerk. This was probably the highest paying female working-class occupation he could think of. The vocational test results sug-gested I become an attorney. I did end up quitting college once again, not because of his suggestion, but because I was tired of supporting my children in ungenteel poverty. I entered vo-tech school for training as an electrician and, as one of the first women in a non-traditional field, was able to earn a living wage at a job which had traditionally been reserved for white working-class males. But this is a story for another day. Let's return to our little vocational counselor example. Was he suggesting the occupational choice of grocery clerk to me because of my gender or my class? Probably both. Let's imagine for a moment what this same vocational counselor might have advised, on sight only, to the following people:

HOW TO CHALLENGE CLASSISM

If you're middle-class you can begin to challenge classism with the following:

1. Confront classist behavior in yourself, others and society. Use and share the privileges, like time or money, which you do have.
2. Make demands on working-class and poor communities' issues—anti-racism, poverty, unions, public housing, public transportation, literacy and day care.
3. Learn from the skills and strength of working people—study working and poor people's history; take some Labor Studies, Ethnic Studies, Women Studies classes. Challenge elitism. There are many different types of intelligence: white middle-class, academic, professional intellectualism being one of them (reportedly). Finally, educate yourself, take responsibility and take action.

If you're working-class, just some general suggestions (it's cheaper than therapy—free, less time-consuming and I won't ask you about what your mother said to you when you were five):

1. Face your racism! Educate yourself and others, your family, community, any organizations you belong to; take responsibility and take action. Face your classism, sexism, heterosexism, ageism, able-bodiness, adultism. . . .
2. Claim your identity. Learn all you can about your history and the history and experience of all working and poor peoples. Raise your children to be anti-racist, anti-sexist and anti-classist. Teach them the language and culture of working peoples. Learn to survive with a fair amount of anger and lots of humor, which can be tough when this stuff isn't even funny.
3. Work on issues which will benefit your community. Consider remaining in or returning to your communities. If you live and work in white middle-class environments, look for working-class allies to help you survive with your humor and wits intact. How do working-class people spot each other? We have antenna.

We need not deny or erase the differences of working class cultures but can embrace their richness, their variety, their moral and intellectual heritage. We're not at the point yet where we can celebrate differences—not having money for a prescription for your child is nothing to celebrate. It's not time yet to party with the white middle class, because we'd be the entertainment ("Aren't they quaint? Just love their workboots and uniforms and the way they cuss!"). We need to overcome divisions among working people, not by ignoring the multiple oppressions many of us encounter, or by oppressing each other, but by becoming committed allies on all issues which affect working people: racism, sexism, classism, etc. An injury to one is an injury to all. Don't play by ruling class rules, hoping that maybe you can live on Connecticut Avenue instead of Baltic, or that you as an individual can make it to Park Place and Boardwalk. Tired of Monopoly? Always ending up on Mediterranean Avenue? How about changing the game?

THE PLIGHT OF BLACK MEN   **12**

Michael Dyson

On a recent trip to Knoxville, I visited Harold's barber shop, where I had gotten my hair cut during college, and after whenever I had the chance. I had developed a friendship with Ike, a local barber who took great pride in this work. I popped my head inside the front door, and after exchanging friendly greetings with Harold, the owner, and noticing Ike missing, I inquired about his return. It had been nearly two years since Ike had cut my hair, and I was hoping to receive the careful expertise that comes from familiarity and repetition.

"Man, I'm sorry to tell you, but Ike got killed almost two years ago," Harold informed me. "He and his brother, who was drunk, got into a fight, and he stabbed Ike to death."

I was shocked, depressed, and grieved, these emotions competing in rapid-fire fashion for the meager psychic resources I was able to muster. In a daze of retreat from the fierce onslaught of unavoidable absurdity, I half-consciously slumped into Harold's chair, seeking solace through his story of Ike's untimely and brutal leave-taking. Feeling my pain, Harold filled in the details of Ike's last hours, realizing that for me Ike's death had happened only yesterday. Harold proceeded to cut my hair with a methodical precision that was itself a temporary and all-too-thin refuge from the chaos of arbitrary

From: Michael Dyson, *Reflecting Black: African-American Cultural Criticism* (Minneapolis: University of Minnesota Press, 1993), pp. 182–194. Reprinted by permission of the author.

death, a protest against the nonlinear progression of miseries that claim the lives of too many black men. After he finished, I thanked Harold, both of us recognizing that we would not soon forget Ike's life, or his terrible death.

This drama of tragic demise, compressed agony, nearly impotent commiseration, and social absurdity is repeated countless times, too many times, in American culture for black men. Ike's death forced to the surface a painful awareness that provides the chilling sound track to most black men's lives: it is still hazardous to be a human being of African descent in America.

Not surprisingly, much of the ideological legitimation for the contemporary misery of African-Americans in general, and black men in particular, derives from the historical legacy of slavery, which continues to assert its brutal presence in the untold suffering of millions of everyday black folk. For instance, the pernicious commodification of the black body during slavery was underwritten by the desire of white slave owners to completely master black life. The desire for mastery also fueled the severe regulation of black sexual activity, furthering the telos of southern agrarian capital by reducing black men to studs and black women to machines of production. Black men and women became sexual and economic property. Because of the arrangement of social relations, slavery was also the breeding ground for much of the mythos of black male sexuality that survives to this day: that black men are imagined as peripatetic phalluses with unrequited desire for their denied object—white women.

Also crucial during slavery was the legitimation of violence toward blacks, especially black men. Rebellion in any form was severely punished, and the social construction of black male image and identity took place under the disciplining eye of white male dominance. Thus healthy black self-regard and self-confidence were outlawed as punitive consequences were attached to their assertion in black life. Although alternate forms of resistance were generated, particularly those rooted in religious praxis, problems of self-hatred and self-abnegation persisted. The success of the American political, economic, and social infrastructure was predicated in large part upon a squelching of black life by white modes of cultural domination. The psychic, political, economic, and social costs of slavery, then, continue to be paid, but mostly by the descendants of the oppressed. The way in which young black men continue to pay is particularly unsettling.

Black men are presently caught in a web of social relations, economic conditions, and political predicaments that portray their future in rather bleak terms.[1] For instance, the structural unemployment of black men has reached virtually epidemic proportions, with black youth unemployment double that of white youth. Almost one-half of young black men have had no work experience at all. Given the permanent shift in the U.S. economy from manufacturing and industrial jobs to high-tech and service employment and the flight of these jobs from the cities to the suburbs, the prospects for eroding the stubborn unemployment of black men appear slim.[2]

The educational front is not much better. Young black males are dropping out of school at alarming rates, due to a combination of severe economic difficulties, disciplinary entanglements, and academic frustrations. Thus the low level of educational achievement by young black men exacerbates their already precarious employment situation. Needless to say, the pool of high-school graduates eligible for college has severely shrunk, and even those who go on to college have disproportionate rates of attrition.

Suicide, too, is on the rise, ranking as the third leading cause of death among young black men. Since 1960, the number of black men who have died from suicide has tripled. The homicide rate of black men is atrocious. For black male teenagers and young adults, homicide ranks as the leading cause of death. In 1987, more young black men were killed within the United States in a single year than had been killed abroad in the entire nine years of the Vietnam War. A young black man has a one-in-twenty-one lifetime chance of being killed, most likely at the hands of another black man, belying the self-destructive character of black homicide.

Even with all this, a contemporary focus on the predicament of black males is rendered problematical and ironic for two reasons. First, what may be termed the "Calvin Klein" character of debate about social problems—which amounts to a "designer" social consciousness—makes it very difficult for the concerns of black men to be taken seriously. Social concern, like other commodities, is subject to cycles of production, distribution, and consumption. With the dwindling of crucial governmental resources to address a range of social problems, social concern is increasingly relegated to the domain of private philanthropic and nonprofit organizations. Furthermore, the selection of which problems merit scarce resources is determined, in part, by such philanthropic organizations, which highlight special issues, secure the services of prominent spokespersons, procure capital for research, and distribute the benefits of their information.

Unfortunately, Americans have rarely been able to sustain debate about pressing social problems over long periods of time. Even less have we been able to conceive underlying structural features that bind complex social issues together. Such conceptualization of the intricate relationship of social problems would facilitate the development of broadly formed coalitions that address a range of social concern. As things stand, problems like poverty, racism, and sexism go in and out of style. Black men, with the exception of star athletes and famous entertainers, are out of style.

Second, the the irony of the black male predicament is that it has reached its nadir precisely at the point when much deserved attention has been devoted to the achievements of black women like Alice Walker, Toni Morrison, and Terry McMillan.

The identification and development of the womanist tradition in African-American culture has permitted the articulation of powerful visions of black female identity and liberation. Michele Wallace, bell hooks, Alice Walker, Audre Lorde, and Toni Morrison have written in empowering ways about the disenabling forms of racism and patriarchy that persist in white and black communities. They have expressed the rich resources for identity that come from maintaining allegiances to multiple kinship groups defined by race, gender, and sexual orientation, while also addressing the challenges that arise in such membership.

Thus discussions about black men should not take place in an ahistorical vacuum, but should be informed by sensitivity to the plight of black women. To isolate and examine the pernicious problems of young black men does not privilege their perspectives or predicament. Rather, it is to acknowledge the decisively deleterious consequences of racism and classism that plague black folk, particularly young black males.

The aim of my analysis is to present enabling forms of consciousness that may contribute to the reconstitution of the social, economic, and political relations that continually consign the lives of black men to psychic malaise, social destruction, and physical death. It does not encourage or dismiss the sexism of black men, nor does it condone the patriarchal behavior that sometimes manifests itself in minority communities in the form of misdirected machismo. Above all, African-Americans must avoid a potentially hazardous situation that plays musical chairs with scarce resources allocated to black folk and threatens to inadvertently exacerbate already deteriorated relations between black men and women. The crisis of black inner-city communities is so intense that it demands our collective resources to stem the tide of violence and catastrophe that has besieged them.

I grew up as a young black male in Detroit in the 1960s and 1970s. I witnessed firsthand the social horror that is entrenched in inner-city communities, the social havoc wreaked from economic hardship. In my youth, Detroit had been tagged the "murder capital of the world," and many of those murders were of black men, many times by other black men. Night after night, the news media in Detroit painted the ugly picture of a homicide-ridden city caught in the desperate clutches of death, depression, and decay. I remember having recurring nightmares of naked violence, in which Hitchcockian vertigo emerged in Daliesque perspective to produce gun-wielding perpetrators of doom seeking to do me in.

And apart from those disturbing dreams, I was exercised by the small vignettes of abortive violence that shattered my circle of friends and acquaintances. My next-door neighbor, a young black man, was stabbed in the jugular vein by an acquaintance and bled to death in the midst of a card game. (Of course, one of the ugly statistics involving black-on-black crime is that many black men are killed by those whom they know.) Another acquaintance murdered a businessman in a robbery; another executed several people in a gangland-style murder.

At fourteen, I was at our corner store at the sales counter, when suddenly a jolt in the back revealed a young black man wielding a sawed-off, double-barreled shotgun, requesting, along with armed accomplices stationed throughout the store, that we hit the floor. We were being robbed. At the age of eighteen I was stopped one Saturday night at 10:30 by a young black man who ominously materialized out of nowhere, much like the .357 Magnum revolver that he revealed to me in an robbery attempt. Terror engulfed my entire being in the fear of imminent death. In desperation I hurled a protest against the asphyxiating economic hardships that had apparently reduced him to desperation, too, and appealed to the conscience I hoped was buried beneath the necessity that drove him to rob me. I proclaimed, "Man, you don't look like the type of brother that would be doin' something like this."

"I wouldn't be doin' this, man," he shot back, "but I got a wife and three kids, and we ain't got nothin' to eat. And besides, last week somebody did the same thing to me that I'm doin' to you." After convincing him that I really only had one dollar and thirty-five cents, the young man permitted me to leave with my life intact.

The terrain for these and so many other encounters that have shaped the lives of black males was the ghetto. Much social research and criticism has been generated in regard to the worse-off inhabitants of the inner city, the so-called underclass. From the progressive perspective of William Julius Wilson to the archconservative musings of Charles Murray, those who dwell in ghettos, or enclaves of civic, psychic, and social terror, have been the object of recrudescent interest within hallowed academic circles and governmental policy rooms.[3] In most cases they have not fared well and have borne the brunt of multifarious "blame the victim" social logics and policies.

One of the more devastating developments in inner-city communities is the presence of drugs and the criminal activity associated with their production, marketing, and consumption. Through the escalation of the use of the rocklike form of cocaine known as "crack" and intensified related gang activity, young black men are involved in a vicious subculture of crime. This subculture is sustained by two potent attractions: the personal acceptance and affirmation gangs offer and the possibility of enormous economic reward.

U.S. gang life had its genesis in the Northeast of the 1840s, particularly in the depressed neighborhoods of Boston and New York, where young

Irishmen developed gangs to sustain social solidarity and to forge a collective identity based on common ethnic roots.[4] Since then, youth of every ethnic and racial origin have formed gangs for similar reasons, and at times have even functioned to protect their own ethnic or racial group from attack by harmful outsiders. Overall, a persistent reason for joining gangs is the sense of absolute belonging and unsurpassed social love that results from gang membership. Especially for young black men whose life is at a low premium in America, gangs have fulfilled a primal need to possess a sense of social cohesion through group identity. Particularly when traditional avenues for the realization of personal growth, esteem, and self-worth, usually gained through employment and career opportunities, have been closed, young black men find gangs a powerful alternative.

Gangs also offer immediate material gratification through a powerful and lucrative underground economy. This underground economy is supported by exchanging drugs and services for money, or by barter. The lifestyle developed and made possible by the sale of crack presents often irresistible economic alternatives to young black men frustrated by their own unemployment. The death that can result from involvement in such drug- and gang-related activity is ineffective in prohibiting young black men from participating.

To understand the attraction such activity holds for black men, one must remember the desperate economic conditions of urban black life. The problems of poverty and joblessness have loomed large for African-American men, particularly in the Rust Belt, including New York, Chicago, Philadelphia, Detroit, Cleveland, Indianapolis, and Baltimore. From the 1950s to the 1980s, there was severe decline in manufacturing and in retail and wholesale trade, attended by escalating unemployment and a decrease in labor-force participation by black males, particularly during the 1970s.

During this three-decade decline of employment, however, there was not an expansion of social services or significant increase in entry-level service jobs. As William Julius Wilson rightly argues, the urban ghettos then became more socially isolated than at any other time. Also, with the mass exodus of black working and middle-class families from the ghetto, the inner city's severe unemployment and joblessness became even more magnified. With black track from the inner city mimicking earlier patterns of white flight, severe class changes have negatively affected black ghettos. Such class changes have depleted communities of service establishments, local businesses, and stores that could remain profitable enough to provide full-time employment so that persons could support families, or even to offer youths part-time employment in order to develop crucial habits of responsibility and work. Furthermore, ghetto residents are removed from job networks that operate in more affluent neighborhoods. Thus, they are deprived of the informal contact with employers that results in finding decent jobs. All of these factors create a medium for the development of criminal behavior by black men in order to survive, ranging from fencing stolen goods to petty thievery to drug dealing. For many black families, the illegal activity of young black men provides their only income.

Predictably, then, it is in these Rust Belt cities, and other large urban and metropolitan areas, where drug and gang activity has escalated in the past decade. Detroit, Philadelphia, and New York have had significant gang and drug activity, but Chicago and Los Angeles have dominated of late. Especially in regard to gang-related criminal activity such as homicides, Chicago and L.A. form a terrible one-two punch. Chicago had 47 gang-related deaths in 1987, 75 in 1986, and 60 in 1988. L.A.'s toll stands at 400 for 1988.

Of course, L.A.'s gang scene has generated mythic interpretation in the Dennis Hopper film *Colors*. In the past decade, gang membership in L.A. has risen from 15,000 to almost 60,000 (with some city officials claiming as much as 80,000), as gang warfare claims one life per day. The ethnic composition of the groups include Mexicans, Armenians, Samoans, and Fijians. But gang life is dominated by South Central L.A. black gangs, populated by young black men willing to give their lives in fearless fidelity to their group's survival. The two largest aggregates of gangs, composed of several hundred microgangs, are the Bloods and the Crips, distinguished by the colors of their shoelaces, T-shirts, and bandannas.

The black gangs have become particularly dangerous because of their association with crack. The gangs control more than 150 crack houses in L.A., each of which does over five thousand dollars of business per day, garnering over half a billion dollars per year. Crack houses, which transform powdered cocaine into crystalline rock form in order to be smoked, offer powerful material rewards to gang members. Even young teens can earn almost a thousand dollars a week, often outdistancing what their parents, if they work at all, can earn in two months.

So far, most analyses of drug gangs and the black youth who comprise their membership have repeated old saws about the pathology of black culture and weak family structure, without accounting for the pressing economic realities and the need for acceptance that help explain such activity. As long as the poverty of young black men is ignored, the disproportionate number of black unemployed males is overlooked, and the structural features of racism and classism are avoided, there is room for the proliferation of social explanations that blame the victim. Such social explanations reinforce the misguided efforts of public officials to stem the tide of illegal behavior by state repression aimed at young blacks, such as the sweeps of L.A. neighborhoods resulting in mass arrests of more than four thousand black men, more than at any time since the Watts rebellion of 1965.

Helpful remedies must promote the restoration of job training (such as Neighborhood Youth Corps [NYC] and the Comprehensive Employment and Training Act [CETA]); the development of policies that support the family, such as child-care and education programs; a full employment policy; and dropout prevention in public schools. These are only the first steps toward the deeper structural transformation necessary to improve the plight of African-American men, but they would be vast improvements over present efforts.

Not to be forgotten, either, are forms of cultural resistance that are developed and sustained within black life and are alternatives to the crack gangs. An example that springs immediately to mind is rap music. Rap music provides space for cultural resistance to the criminal-ridden ethos that pervades segments of many underclass communities. Rap was initially a form of musical play that directed the creative urges of its producers into placing often humorous lyrics over the music of well-known black hits.

As it evolved, however, rap became a more critical and conscientious forum for visiting social criticism upon various forms of social injustice, especially racial and class oppression. For instance, Grandmaster Flash and the Furious Five pioneered the social awakening of rap with two rap records, "The Message" and "New York, New York." These rap records combined poignant descriptions of social misery and trenchant criticism of social problems as they remarked upon the condition of black urban America. They compared the postmodern city of crime, deception, political corruption, economic hardship, and cultural malaise to a "jungle." These young savants portrayed a chilling vision of life that placed them beyond the parameters of traditional African-American cultural resources of support: religious faith, communal strength, and familial roots. Thus, they were creating their own aesthetic of survival, generated from the raw material of their immediate reality, the black ghetto. This began the vocation of the rap artist, in part, as urban griot dispensing social and cultural critique.

Although rap music has been saddled with a reputation for creating violent outbursts by young blacks, especially at rap concerts, most of rap's participants have repeatedly spurned violence and all forms of criminal behavior as useless alternatives for black youth. Indeed rap has provided an alternative to patterns of identity formation provided by gang activity and has created musical vehicles for personal and cultural agency. A strong sense of self-confidence permeates the entire rap genre, providing healthy outlets for young blacks to assert, boast, and luxuriate in a rich self-conception based upon the achievement that their talents afford them. For those reasons alone, it deserves support. Even more, rap music, although its increasing expansion means being influenced by the music industry's corporate tastes and decisions, presents an economic alternative to the underground economy of crack gangs and the illegal activity associated with them.

However, part of the enormous difficulty in discouraging illegal activity among young black men has to do, ironically, with their often correct perception of the racism and classism still rampant in employment and educational opportunities open to upwardly mobile blacks. The subtle but lethal limits continually imposed upon young black professionals, for instance, as a result of the persistence of racist ideologies operating in multifarious institutional patterns and personal configurations, send powerful signals to young black occupants of the underclass that education and skill do not ward off racist, classist forms of oppression.

This point was reiterated to me upon my son's recent visit with me at Princeton near Christmas. Excited about the prospect of spending time together catching up on new movies, playing video games, reading, and the like, we dropped by my bank to get a cash advance on my MasterCard. I presented my card to the young service representative, expecting no trouble since I had just paid my bill a couple of weeks before. When he returned, he informed me that not only could I not get any money, but that he would have to keep my card. When I asked for an explanation, all he could say was that he was following the instructions of my card's bank, since my MasterCard was issued by a different bank.

After we went back and forth a few times about the matter, I asked to see the manager. "He'll just tell you the same thing that I've been telling you," he insisted. But my persistent demand prevailed, as he huffed away to the manager's office, resentfully carrying my request to his boss. My son, sitting next to me the whole time, asked what the problem was, and I told him that there must be a mistake, and that it would all be cleared up shortly. He gave me that confident look that says, "My dad can handle it." After waiting for about seven or eight minutes, I caught the manager's figure peripherally, and just as I turned, I saw him heading with the representative to an empty desk, opening the drawer and pulling out a pair of scissors. I could feel the blood begin to boil in my veins as I beseeched the manager, "Sir, if you're about to do what I fear you will, can we please talk first?" Of course, my request was to no avail, as he sliced my card in two before what had now become a considerable crowd. I immediately jumped up and followed him into his office, my son trailing close behind, crying now, tearfully pumping me with "Daddy, what's going on?"

I rushed into the manager's office and asked for the privacy of a closed door, to which he responded, "Don't let him close the door," as he beckoned three other employees into his office. I angrily grabbed the remnants of my card from his hands and proceeded to tell him that I was a reputable member of the community and a good customer of the bank and that if I had been wearing a three-piece suit (instead of the black running suit I was garbed in) and if I had been a white male (and not a black man) I would have been at least accorded the respect of a conversation, prior to a private negotiation of an embarrassing situation, which furthermore was the apparent result of a mistake on the bank's part.

His face flustered, the manager then prominently positioned his index finger beneath his desk drawer, and pushed a button, while declaring, "I'm calling the police on you." My anger now piqued, I was tempted to vent my rage on his defiant countenance, arrested only by the vision terrible that flashed before my eyes as a chilling premonition of destruction: I would assault the manager's neck; his co-workers would join the fracas, as my son stood by horrified by his helplessness to aid me; the police would come, and abuse me even further, possibly harming my son in the process. I retreated under the power of this proleptic vision, grabbing my son's hand as I marched out of the bank. Just as we walked through the doors, the policemen were pulling up.

Although after extensive protests, phone calls, and the like, I eventually received an apology from the bank's board and a MasterCard from their branch, this incident seared an indelible impression onto my mind, reminding me that regardless of how much education, moral authority, or personal integrity a black man possesses, he is still a "nigger," still powerless in many ways to affect his destiny.

The tragedy in all of this, of course, is that even when articulate, intelligent black men manage to rise above the temptations and traps of "the ghetto," they are often subject to continuing forms of social fear, sexual jealousy, and obnoxious racism. More pointedly, in the 1960s, during a crucial stage in the development of black pride and self-esteem, highly educated, deeply consci-entious black men were gunned down in cold blood. This phenomenon finds paradigmatic expression in the deaths of Medgar Evers, Malcolm X, and Martin Luther King, Jr. These events of public death are structured deep in the psyches of surviving black men, and the ways in which these horrible spectacles of racial catastrophe represent and implicitly sanction lesser forms of social evil against black men remains hurtful to black America.

I will never forget the effect of King's death on me as a nine-year-old boy in Detroit. For weeks I could not be alone at night before an open door or window without fearing that someone would kill me, too. I thought that if they killed this man who taught justice, peace, forgiveness, and love, then they would kill all black men. For me, Martin's death meant that no black man in America was safe, that no black man could afford the gift of vision, that no black man could possess an intelligent fire that would sear the fierce edges of ignorance and wither to ashes the propositions of hate without being extin-guished. Ultimately Martin's death meant that all black men, in some way, are perennially exposed to the threat of annihilation.

As we move toward the last decade of this century, the shadow of Du Bois's prophetic declaration that the twentieth century's problem would be the color line continues to extend itself in foreboding manner. The plight of black men, indeed, is a microcosmic reflection of the problems that are at the throat of all black people, an idiomatic expression of hurt drawn from the larger discourse of racial pain. Unless, however, there is vast reconstitution of our social, economic, and political policies and practices, most of which target black men with vicious specificity, Du Bois's words will serve as the frontispiece to the racial agony of the twenty-first century, as well.

NOTES

1. For a look at the contemporary plight of black men, especially black juvenile males, see *Young, Black, and Male in America: An Endangered Species,* ed. Jewelle Taylor Gibbs (Dover, Mass.: Auburn House, 1988).

2. See William Julius Wilson, *The Truly Disadvantaged: The Inner City, the Underclass, and Public Policy* (Chicago: University of Chicago Press, 1987).

3. See Wilson, *The Truly Disadvantaged.* For Charles Murray's views on poverty, welfare, and the ghetto underclass, see his influential book, *Losing Ground: American Social Policy, 1950–1980* (New York: Basic Books, 1984).

4. This section on gangs is informed by the work of Mike Davis in his *City of Quartz* (New York: Verso Press, 1991).

ARE YOU MIDDLE CLASS?   13

Barbara Ehrenreich

In the 1980s, the Reagan administration redistributed the wealth upward by cutting the taxes of the rich and cutting services for the poor and middle class.

Excerpted from: Barbara Ehrenreich, "The Hour Glass Society," *New Perspectives Quarterly* 7 (Fall 1990): 44–46. Reprinted by permission.

Figures from Citizens for Tax Justice show that 9 out of 10 Americans now face higher taxes than before Reagan. It's just the top 10 percent that pays lower taxes. This from a decade in which the rallying cry was "no new taxes."

But that's not all. Another factor, which has not been sufficiently addressed, is the effect of the Reagan years on what is traditionally called the "working class." Many liberal and moderate observers hedge on this issue, writing off the wage concessions of the '80s to the natural operation of the global economy—the effect of wage equilibration worldwide.

However, the story of the '80s in the United States is of one beaten strike after another, of givebacks rather than wage gains. Clearly, in some cases wage concessions and givebacks were results of capital mobility: When producers could move operations to northern Mexico or Thailand and pay a few dollars a day, the bargaining power of domestic employees diminished significantly.

One can't make the same argument about airplane pilots or about clerical workers in the insurance industry. The Reagan administration busted PATCO, the air traffic controllers' union, in 1981, then proceeded to all but eviscerate the Occupational Safety and Health Administration and the National Labor Relations Board, which guarantee minimal legal protections to labor.

Between 1979 and 1984 alone, 11.5 million American workers lost their jobs because of plant shutdowns or relocations. Only 60 percent were able to find new jobs, and nearly half of the new jobs were at lower pay than the lost jobs. One study found that laid-off steelworkers in Chicago saw their incomes fall, on the average, from $22,000 a year to $12,500 a year, just slightly above the official poverty level. Nonetheless, during the '80s no one seemed to be overly concerned that the continued assault on the wages of the working and middle class threatened underconsumption of goods and services. Indeed, it appeared that business could profit by selling more to a smaller and smaller number of people with lots of money. Until quite recently, luxury department stores like Saks Fifth Avenue did very well. And, at the other end, the Kmarts did well. The middle dropped out of the retail market. In New York, for example, middle-class emporiums like Gimbels and Korvettes closed.

This polarization in the retail market reflects what happened to the middle class in general. Moreover, as the extremes moved further apart, it became much harder for those in the middle range of income to survive. As the rich got richer, they were able to bid up the cost of goods that middle-income people also consumed, particularly housing. Wildly inflated housing costs hurt the affluent upper fifth, too, but they were far more likely than middle-income people to be able to command salary increases to match their escalating cost of living.

The middle class is still a statistical reality, of course. A graph of income distribution in America still comes out as a bell-shaped curve, with most people hovering near the middle rather than at either extreme. But today, the people in the middle of the American population in terms of income are not necessarily in the middle class, at least not in the way we have traditionally thought of it. In the '80s, the after-tax median family income of $20,000 became insufficient for the attainment of what we traditionally looked on as a middle-class lifestyle. We have a certain cultural sense of what it means to be middle class, its key elements being home ownership and the ability to put one's kids through college and take a family vacation during the summer. Families getting by on $20,000 are probably not taking the kids to Yellowstone or paying tuition at State U without any loans.

Class polarization becomes a vicious cycle, especially in cities where the polarization is most acute and where the extremes are pressed together geographically. In these areas, the affluent do what they can to avoid contact with the poor and the downwardly mobile. They abandon public services and public spaces—schools, parks, mass transportation—which then deteriorate. One result of this is that the living conditions and opportunities available to the poor, and many in the middle range of income, worsen. And, of course, as the poor sink lower, the affluent have all the more reason to withdraw into their own "good" neighborhoods and private services.

As the more affluent cease to use public services, they also tend to withdraw political support for public spending designed to benefit the community as a whole. If one sends one's children to private school and commutes to work by taxi, one is likely to prefer a tax cut to an expansion of government services. The affluent don't want to pay taxes for schools they don't use. And they are not going to use them because they find the poor too frightening to be around.

As the upper class flees public services like schools, those schools lose influential advocates—a dynamic that takes its toll on the poor. Another factor is that today issues such as education, which used to be very important to low-income families in New York City, for instance, have been subsumed by the more pressing survival considerations—drugs, violence, AIDS, teenage pregnancy, poverty.

We've been fixated on the stressed-out, overworked yuppie in the '80s, but another phenomenon was taking place concurrently—the poor person with two jobs and a long commute, since many poor people couldn't afford to live where their jobs were. Indeed, the "working poor" now make up over 40 percent of the poverty population over the age of 14. Their hard work isn't getting them anywhere. When hard work is no longer sufficient to lift families out of poverty, we are asking for mass cynicism and despair.

Even as this great divide opens between the affluent class and everyone else, most people think of it in terms of race or merit or skills or luck—not class. Unfortunately, Americans are notorious for their lack of class consciousness or even class awareness. We have a much greater consciousness of race and gender issues than we do of class. Race and gender are immediate—and they are irreversible. Class is different: The American myth is that we can escape our class. We can work our way up and out of it.

The anti-communism of the '50s is as much to blame as anything for this kind of aversion to "class." In the '50s, the notion of class—which was well articulated throughout the '20s and '30s in the United States—became suspect and un-American, part of a left-wing heritage that mainstream intellectuals found important to repudiate. Classlessness became part of America's official ideology in the '50s, and those who believed otherwise risked being driven from their teaching posts and rejected by publishers.

As the Cold War dissolves, I hope that the United States will avail itself of the opportunity to become a bit more introspective as a society and look at its problems, especially those arising from class polarization.

Some foresee that in the '90s, our class problems will be addressed in an increasingly liberal political policy trend. Though I am greatly heartened to see conservative analysts such as Kevin Phillips employ historic trends to herald a leftward turn in American politics, I am skeptical. Cyclical theories—that the '30s, '60s, and now the '90s, are left-wing decades—discount the need for human intervention. The '90s will not be progressive unless we make them progressive. Social transformation does not occur through the operation of some sort of natural law. It requires human effort and leadership at both the community and the national level.

THE UPPERCLASS AND MOTHERS N THE HOOD

14

Holly Sklar

The reality is that most poor Americans are white, many married couples are poor, and even if there were no nonwhite children and no single mother

From: Holly Sklar, "The Upperclass and Mothers N the Hood," *Z Magazine* 6 (March 1993): 22–31. Reprinted by permission of the author.

families, the United States would have one of the highest child poverty rates among the capitalist powers. But that doesn't stop liberals and conservatives alike from blaming poverty on single mothers, especially Black single mothers, and accusing them of breeding a pathological underclass culture of poverty, drug abuse, sloth, and savagery.

In a 1992 speech to Yale University, slandering single mothers and affirmative action, neoliberal Massachusetts Senator John Kerry recycled the refuted, racist Black matriarchy myth popularized by neoconservative Daniel Patrick Moynihan in a 1965 report released by the White House shortly after the Watts riots: "Twenty-seven years ago, my Senate colleague Daniel Patrick Moynihan warned that: 'from the wild Irish slums of the 19th century eastern seaboard, to the riot-torn suburbs of Los Angeles, there is one unmistakable lesson in American history: A society that allows a large number of young men to grow up in broken families . . . never acquiring any stable relationship to . . . authority, never acquiring any rational expectations about the future—that society asks for and gets chaos. Crime, violence, unrest, disorder—more particularly, the furious unrestrained lashing out at the whole social structure—that is not only to be expected; it is very near inevitable.' " [ellipses Kerry's] (See Z, May/June 1992.)

As films like *Boys N the Hood* show, you don't have to be a neoconservative (Black or white) to equate Black female-headed families with disorder, savagery, and death and male-headed families with discipline, salvation, and success. Once again, children are stigmatized from birth as the pathological bastards of their mother's presence and their father's absence. Once again, misogynist myths are used to perpetuate racial, gender, and class discrimination.

MAMMYS, MATRIARCHS, AND PATRIARCHY

Culture of poverty theories are neither new nor true, but back they come to mask cultures of greed, racism, and sexism. "It's clear women have been viewed as the breeders of poverty, juvenile delinquency, criminality, and other social problems," says Mimi Abramovitz, professor of Social Work at Hunter College, "from the 'tenement class' of the mid 1800s and the 'dangerous classes' of the 1880s, to Social Darwinism and eugenics, to Freudian theories of motherhood, to today's 'underclass'."

Stereotypes reflect power relations, as some past generations of poor white European immigrants could attest. As Oscar Handlin writes in *Boston's Immigrants*, "the Irish were the largest components of the state poorhouse population and a great majority of all paupers . . . after 1845." They were economically exploited and socially stereotyped as immoral, drunkards, and criminals (recall the term "Paddy wagon" for police wagon). Alcoholism was once recorded as a cause of death for Irish immigrants in the Massachusetts registry, not for Protestant Anglo-Saxons. A century later, Oscar Lewis coined the phrase "culture of poverty," first for Mexicans in 1959, then Puerto Ricans and African-Americans.

Imagine labeling married-couple families as pathological breeding grounds of patriarchal domestic violence, or suggesting that women should never marry, because they are more likely to be beaten and killed by a spouse than a stranger. In Massachusetts during the first half of 1992, nearly three out of four women whose murderers are known were killed by husbands, boyfriends, or ex-partners. Misogynist domestic violence is so rampant that over 50,000 Massachusetts women have taken out restraining orders against former mates. Violence is the leading cause of injuries to women ages 15 to 44, "more common than automobile accidents, muggings, and cancer deaths combined." It is estimated that a woman has between a one in five and a one in three chance of being physically assaulted by a partner or ex-partner during her lifetime. More than 90 women were murdered every week in 1991. In the words of an October 1992 Senate Judiciary Committee report, "Every week is a week of terror for at least 21,000 American women" of all races, regions, educational, and economic backgrounds, whose "domestic assaults, rapes and murders were reported to the police." As many as three million more domestic violence crimes may go unreported.

Stephanie Coontz writes in her myth-busting study of families, *The Way We Never Were*, "families whose members are police officers or who serve in the military have much higher rates of divorce, family violence, and substance abuse than do other families, but we seldom accuse them of constituting an 'underclass' with a dysfunctional culture."

In *The Negro Family*, published the year following the 1964 Civil Rights Act, Moynihan embellished sociologist E. Franklin Frazier's thesis of the Black matriarch in whom "neither economic necessity nor tradition had instilled the spirit of subordination to masculine authority." Moynihan's notion that matriarchal families are at the core of a Black "tangle of pathology" was the perfect divisive response to the Black liberation movement, feminism, and the welfare rights movement.

African-American women have been stereotyped since slavery as "mammies, matriarchs, and other controlling images," explains Patricia Hill Collins in *Black Feminist Thought*. The mammy was "the faithful, obedient domestic servant. Created to justify the economic exploitation of house slaves and sustained to explain Black women's long-standing restriction to domestic service, the mammy represents the normative yardstick to evaluate all Black women's behavior. By loving, nurturing, and caring for her white children and 'family' better than her own, the mammy symbolizes the dominant group's perceptions of the ideal Black female relationship to elite white male power. . . . She has accepted her subordination." While "the mammy represents the 'good' Black mother, the matriarch symbolizes the 'bad' Black mother. . . . Spending too much time away from home, these working mothers ostensibly cannot properly supervise their children and are a major contributing factor to their children's school failure. As overly aggressive, unfeminine women, Black matriarchs allegedly emasculate their lovers and husbands." . . .

"WELFARE QUEENS" AND WORKER BEES

The third controlling image of Black women, explains Patricia Hill Collins, is the welfare mother. "Essentially an updated version of the breeder woman image created during slavery, this image provides an ideological justification for efforts to harness Black women's fertility to the needs of a changing political economy. . . . Slaveowners wanted enslaved Africans to 'breed' because every slave child born represented a valuable unit of property, another unit of labor, and, if female, the prospects for more slaves." The welfare mother is labeled a bad mother, like the matriarch, but "while the matriarch's unavailability contributed to her children's poor socialization, the welfare mother's accessibility is deemed the problem." Blacks made up a higher percentage of the U.S. population in 1850 than in 1950 or any time in the twentieth century.

In the postwar period, as the percentage of births to unmarried women rose, especially among white women, and Aid to Dependent Children was opened to their offspring, both Black and white women were viewed as breeders. . . .

Aid to Families with Dependent Children (AFDC) expanded for many reasons, among them the inclusion of mothers (and not just their children) as recipients after 1950, higher rates of female-headed households due to divorce and unmarried births, and later the mobilization of poor people in the National Welfare Rights Organization. Black women were "blamed" though only about 16 percent of nonwhite unwed mothers received welfare grants while 30 percent of the unwed mothers who did not give their children up for adoption received grants in 1959. (In 1960, about 94 percent of Black and 29 percent of white "illegitimate" babies lived with natural parents or relatives.) As Piven and Cloward point out in *Regulating the Poor*, the proportion of Blacks on

AFDC rose after 1948 because of two often-neglected factors: the displacement of Blacks from southern agriculture by mechanization and their migration to northern cities (where jobs and low-cost housing became scarcer) and the lessening of eligibility discrimination. While the proportion of AFDC parents who are white (non-Hispanic) was the same in 1973 (38 percent) as 1990, the proportion who are Black declined from 45.8 percent to 39.7 percent in the same period.

The stereotype "welfare queen" lazily collects government checks and reproduces poverty by passing on her pathologies to her many children. Two decades ago, Senator Russell Long of Louisiana referred to welfare mothers as "brood mares." The slaveowners' control of fertility is mirrored again in the present economy which wants Black women's reproduction further reduced because Black workers and therefore Black children are increasingly seen as surplus. Norplant contraceptive implants, which can cause bleeding and other side effects, have become a eugenics weapon for judges and politicians.

The myth of an intergenerational matriarchy of "welfare queens" is particularly disgusting since Black women were enslaved workers for over two centuries and have always had a high labor force participation rate and a disproportionate share of low wages and poverty. In 1900, Black women's labor force participation rate was 40.7 percent, white women's 16 percent. The 1960 rates were 42.2 percent for Black women and 33.6 percent for whites; in 1970, 49.5 percent and 42.6 percent; in 1980, 53.2 percent and 51.2 percent; and in 1991 they converged at 57 percent. . . .

In 1990, there were about 3.4 million women, 374,000 men, and 7.7 million children under 18 receiving AFDC. The number of AFDC child recipients as a percent of children in poverty fell from 80.5 percent in 1973 to 59.9 percent in 1990. About 38 percent of AFDC families are white, 40 percent are Black, 17 percent are Latino, 3 percent are Asian, and 1 percent are Native American. There are disproportionately more people of color on welfare because disproportionately more people of color are poor and, as discussed below, they have disproportionately less access to other government benefits such as Social Security and Unemployment Insurance.

Contrary to image, most daughters in families who received welfare do not become welfare recipients as adults. And, women receiving welfare don't have more children than others. Most families on AFDC have one child (42 percent) or two children (30 percent); only 10 percent have more than three children. . . .

The welfare system minimizes help and maximizes humiliation. When Barbara Sobel, head of the New York City Human Resources administration, posed as a welfare applicant to experience the system first hand, she was misdirected, mistreated, and so "depersonalized," she says, "I ceased to be." She remained on welfare, with a mandatory part-time job as a clerk in a city office, despite repeated pleas for full-time work, and learned that most recipients desperately want work (*New York Times*, February 5, 1993). . . .

Below-subsistence welfare payments are governmental child abuse. The child-abusing budget cutters hide behind budget deficits (in 1991, AFDC accounted for less than 1 percent of federal outlays and states spent 2.2 percent of their revenues on AFDC) and their stereotypes of cheating "welfare queens." When California reduced its monthly AFDC payment for a mother and two children in 1991 from $694 (which was $2,645 below the annual poverty line) to $663, Governor Pete Wilson said it meant "one less six-pack per week" (*Equal Means*, Spring 1992).

Women turn to AFDC to support them and their children after divorce (when their incomes plummet because of no or low wages and no or low child support), after losing a job, after childbirth outside marriage, or while completing their education or job training. While most families receiving AFDC do so for two years or less, a minority of families become long-term recipients. As the 1992 House Committee on Ways and Means *Green Book* noted, "the typical recipient is a short-term user."

Long-term recipients have greater obstacles to getting off welfare such as lacking prior work experience, a high school degree or child care, or having poor health. Many women leave welfare—though often not poverty—after finding jobs and/or marrying men. Black women have a harder time doing either than white women, not because of a self-perpetuating "cycle of dependency," but a cycle of discrimination and demographics. It's fashionable to point to a "dearth of marriageable Black men," e.g. employed men earning above-poverty wages, without mentioning the dearth of Black men, period, as racism-fueled mortality takes its toll. The Black female-male ratio between the ages of 25 and 44, for example, was 100 to 87 in 1989, while it was 100 to 101 for whites.

Being married is neither necessary nor sufficient to avoid poverty. The 1991 poverty rates for married-couple families with children were 7.7 percent for whites, 14.3 percent for Blacks, and 23.5 percent for Latinos.

A new wave of policies is being enacted to address the "behavioral roots of poverty" and reinforce an old patriarchy with a "new paternalism." They punish unmarried women who have additional children—and punish those children—by denying women any increased benefits for new dependents and they reward women who marry. . . .

Although two-thirds of AFDC recipients are children, critics make it sound like most recipients should be employed (then again, child labor is on the rise). Many women work or seek work outside the home while receiving welfare in spite of the near dollar-for-dollar reductions in benefits for wages and insufficient allowance for child care and other work expenses. In recent years, state and federal policy has imposed mandatory work and training programs. In 1990, nearly two-thirds of adult recipients were exempt from registration in work programs, most commonly because they had very young children to care for. Nearly 40 percent of AFDC families had at least one child two years old or younger. In a discriminatory, dangerous move to expand day care for AFDC recipients many states are exempting child care providers from health and safety regulations or loosening them. And prevailing "workfare" programs by whatever name do not help women transcend the growing ranks of the working poor.

WHOSE "CULTURE OF POVERTY"?

The myth of a "culture of poverty" masks the reality of an economy of impoverishment. A lot of single mother families are broke, they aren't broken. . . .

It's not surprising that many single parent households are poor since the U.S. government neither assures affordable child care nor provides the universal child supports common in Western Europe. France, Britain, Denmark, and Sweden, for example, have similar or higher proportions of births to unmarried women without U.S. proportions of poverty.

It shouldn't be surprising that Black and Latino single parent families have higher rates of poverty than white families since the earnings and job opportunities of people of color reflect continued educational and employment discrimination. The overall poverty rates of Black (28.5 percent) and Latino (26.2 percent) males, are much closer to the poverty rate of all female-headed households with and without children, of any race, 35.6 percent, than of white males (9.8 percent). And, it shouldn't be surprising that single mother families are the poorest of all since women are the lowest paid and women of color are doubly discriminated against. The fact that many female-headed households are poorer because women earn less than men is taken as a given in much welfare reform discussion, as if pay equity was a pipe dream not even worth mentioning. . . .

UNDER THE UPPERCLASS

Terms like "underclass" and "persistent poverty" imply that poverty persists in spite of society's commitment to eliminate it. In reality, the socioeconomic system reproduces poverty no matter how persistently people are trying to get out or stay out of poverty.

As Adolph Reed Jr. writes in *Radical America* (January 1992) in a critique of various underclass theories, "behavioral tendencies supposedly characterizing the underclass exist generally throughout the society. Drug use, divorce, educational underattainment, laziness, and empty consumerism exist no less in upper status suburbs than in inner-city bantustans. The difference lies not in the behavior but in the social position of those exhibiting it" and in their access to safety nets. And in their imprisonment rates.

A 1990 study in the *New England Journal of Medicine* found that substance abuse rates are slightly higher for white women than nonwhite women, but nonwhite women are ten times more likely to be reported to authorities. And, while mothers are increasingly prosecuted for drug use during pregnancy, the doors of most drug treatment centers remain closed to pregnant women. Similarly, Black kids are less likely to use drugs than white kids (according to government studies), but much more likely to be stigmatized and jailed for it. If the irrational drug laws were applied equally, we'd see a lot more handcuffed white movie stars, rich teenagers, politicians, doctors, stockbrokers, and CIA officers on the TV news, trying to hide their faces.

Low-income people and communities, like middle- and upper-class people and communities, have a mix of strengths and weaknesses, needs and capacities. But poor communities are uniquely portrayed as the negative sum of their needs and "risks." All people and communities need services. In higher-income communities, people needing doctors or psychologists, lawyers or drug treatment, birth control or abortion, tutors or child care, can afford pricey private practitioners and avoid the stigma that often accompanies stingy public social services. In lower-income communities they cannot. This problem is especially bad in the United States because it lags far behind all other industrialized democracies in assuring the basic human needs of its people.

When health care is a privilege, not a right, children die. But it's cheaper to blame their mothers than provide real universal health care (not Clinton's "managed competition"). Seventy countries provide prenatal care to *all* pregnant women and many have policies requiring *paid* maternity leave. The United States does not. An Ohio study found that a woman on pregnancy leave is 10 times more likely to lose her job than one on medical leave for other reasons (*New York Times*, January 12, 1993). More children die before their first birthday in the United States per capita than in 21 other countries. Nationally, the infant mortality rate for Black babies is more than twice as high as for whites—the widest gap since 1940, when race specific data were first collected. In Boston, it is three times higher. . . .

One out of four children is born into poverty in the United States—the highest official rate of any industrialized nation. The official 1991 child poverty rate was 21.8 percent (25.5 percent for children under age three). For white children, it was 16.8 percent; for Latino children, 40.4 percent; and for Black children, 45.9 percent. Poverty rates would be even higher if they counted families whose incomes fell below the poverty line after taxes, and if the poverty threshold was adjusted upward to reflect, not just an inflation-multiplied out of date standard, but the real cost of living. The last time the Department of Labor compiled a "lower family budget," in 1981, it was 65 percent above that year's official poverty line for the same size family. . . .

Many countries provide a children's allowance or other universal public benefit for families raising children. The United States does not. Throughout the 1980s, the U.S. government preached family values without valuing families. "Under our tax laws," Colorado Congresswoman Pat Schroeder was quoted in a *Time* feature on children (October 8, 1990), "the deduction for a Thoroughbred horse is greater than that for children."

Everything from prenatal care to college is rationed by money in a country where income inequality has grown so much that the top 4 percent of Americans earned as much in wages and salaries in 1989 as the bottom 51 percent; in 1959, the top 4 percent earned as much as the bottom 35 percent. The average chief executive officer (CEO) of a large corporation earned as much in salary as 42 factory workers in 1980 and 104 factory workers in 1991 (Japan's CEOs earn about as much as 18 factory workers).

The top 1 percent of families now have a net worth much greater than that of the bottom 90 percent. In 1989, reports the Economic Policy Institute, the top 1 percent of families had 37.7 percent of total net worth (assets minus debt) and the bottom 90 percent has 29.2 percent; the bottom 95 percent had 40.7 percent. The top fifth of families had 83.6 percent of net worth; the upper middle fifth, 12.3 percent; the middle fifth, 4.9 percent; the lower middle fifth, 0.8 percent; and the bottom fifth, –1.7 percent. Looking at family income, the top fifth (families with pretax incomes of $61,490 and above in 1990) had 55.5 percent; the upper middle fifth, 20.7 percent; the middle fifth, 13.3 percent; the lower middle fifth, 7.6 percent; and the bottom fifth, 3.1 percent.

U.S. wealth concentration is now more extreme than any time since 1929, and getting worse. For many Americans there's an endless economic depression. The shrinking middle class is misled into thinking those below them on the economic ladder are pulling them down, when in reality those at the top of the ladder are pushing everyone down.

The stereotype of deadbeat poor people masks the growing reality of dead-end jobs. It is fashionable to point to the so-called breakdown of the family as a cause of poverty and ignore the breakdown in wages. The average inflation-adjusted earnings of nonsupervisory workers crashed 19 percent between 1973 and 1990. Minimum wage is 23 percent below its average value during the 1970s. For more and more Americans and their children, work is not a ticket out of poverty, but a condition of poverty.

Living standards are falling for younger generations, despite the fact that many households have two wage earners. The inflation-adjusted median income for families with children headed by persons younger than 30 plummeted 32 percent between 1973 and 1990. Forty percent of all children in families headed by someone younger than 30 were living in poverty in 1990—including one out of four children in white young families.

The entry-level wage for high school graduates fell 22 percent between 1979 and 1991, a reflection, reports the Economic Policy Institute, of "the shift toward lower-paying industries, the lower value of the minimum wage, less unionization" and other trends. Entry level wages for college graduates fell slightly overall (–0.2 percent) between 1979 and 1991, but Black college graduates lost over 3 percent and Latino college graduates lost nearly 15 percent. Between 1979 and 1990, the proportion of full-time, year-round workers, ages 18 to 24, paid low wages (below $12,195 in 1990) jumped from 23 percent in 1979 to over 43 percent in 1990. Among young women workers, the figure is nearly one in two workers. And low-wage jobs are often dead-end jobs with low or no benefits (e.g. health insurance, paid vacation, pension), round-the-clock shifts, and little prospect of advancement.

During the 1960s and 1970s, Blacks were about twice as likely to be unemployed as whites, according to official, undercounting statistics. In the 1980s, the gap widened: when white unemployment was 8.4 percent in 1983, Black unemployment was 19.5 percent. When white unemployment was 4.1 percent in 1990, Black unemployment was 2.76 times higher at 11.3 percent. Black college graduates had a jobless rate 2.24 times that of white college graduates. As the Urban Institute documented in a 1990 study using carefully matched and trained pairs of white and Black young men applying for entry-level jobs, discrimination against Black job seekers is "entrenched and widespread."

To make matters worse, most unemployed people do not receive unemployment insurance benefits. An average one third of the officially-counted unemployed nationwide received benefits from 1984 to 1989; the figure rose to 42 percent in the severe recession year of 1991 (76 percent received benefits during the 1975 recession). Eligibility varies by state and unemployment insurance typically lasts only a maximum of 26 weeks whether or not you've found a job.

Low wage workers, disproportionately women and people of color, are less likely than other workers to qualify for unemployment benefits (they may not earn enough or meet work history requirements) and, when they do qualify, their unemployment payments are only a portion of their meager wages. When the New Deal era Unemployment Insurance and Social Security programs were established, the occupations excluded from coverage—such as private domestic workers, agricultural laborers, government and nonprofit employees—were ones with large numbers of women and people of color. A recent General Accounting Office study reported in the *New York Times* (May 11, 1992) found that after accounting for such factors as age, education, and types of disability, "blacks with serious ailments have been much more likely than whites to be rejected for benefits" under the Social Security Disability Insurance and Supplementary Security Income programs. . . .

In the unusually blunt words of *Time* magazine (September 28, 1992), "Official statistics fail to reveal the extent of the pain. Unemployment stands at 7.6 percent . . . but more people are experiencing distress. A comprehensive tally would include workers who are employed well below their skill level, those who cannot find more than a part-time job, people earning poverty-level wages, workers who have been jobless for more than four weeks at a time and all those who have grown discouraged and quit looking. Last year those distressed workers totaled 36 million, or 40 percent of the American labor force, according to the Washington-based Economic Policy Institute."

"We never meant to quit our jobs. They quit on us," says a former Rath Meatpacking employee from Waterloo, Iowa, quoted by Jacqueline Jones in *The Dispossessed*. Corporations are permanently downsizing their workforces and shifting more operations (including service sector jobs such as data processing) to countries where workers have even lower wages and few or no rights. The newer U.S. jobs not only pay less than disappearing unionized jobs, but employers are replacing full-time workers with part-time and temporary workers with even lower benefits and job security. . . .

In the words of the Children's Defense Fund, "The slow, grinding violence of poverty takes an American child's life every 53 minutes. The deadly, quick violence of guns takes an American child's life every three hours." Single mothers do not direct the economy—legal or underground. They don't direct the drug war, the National Rifle Association, the military, or the television and movie industries which teach children violence through entertainment and government action.

Pointing fingers at an "underclass culture of poverty" diverts attention and anger from the poverty-reproducing upperclass culture of greed. While subsidizing the luxury lifestyle of corporate kingpins and bailing out wealthy bank speculators, politicians pretend that below-subsistence subsidies for poor women and children are destroying the family and bringing down the American economy. Upperclass white America has been built on centuries of discriminatory subsidy and violence, from slavery to segregated suburbanization, Indian removal to "urban renewal," redbaiting, redlining, and union-busting. It's way past time to break upperclass dependency on the cycle of unequal opportunity.

MOVING UP WITH KIN AND COMMUNITY: *Upward Social Mobility for Black and White Women* 15

Elizabeth Higginbotham and Lynn Weber

. . . When women and people of color experience upward mobility in America, they scale steep structural as well as psychological barriers. The long process of moving from a working-class family of origin to the professional-managerial class is full of twists and turns: choices made with varying degrees of information and varying options; critical junctures faced with support and encouragement or disinterest, rejection, or active discouragement; and interpersonal relationships in which basic understandings are continuously negotiated and renegotiated. It is a fascinating process that profoundly shapes the lives of those who experience it, as well as the lives of those around them. Social mobility is also a process engulfed in myth. One need only pick up any newspaper or turn on the television to see that the myth of upward mobility remains firmly entrenched in American culture: With hard work, talent, determination, and some luck, just about anyone can "make it." . . .

The image of the isolated and detached experience of mobility that we have inherited from past scholarship is problematic for anyone seeking to understand the process for women or people of color. Twenty years of

Authors' note: The research reported here was supported by National Institute for Mental Health Grant MH38769. All the names used in this article are pseudonyms.

From: *Gender & Society*, vol. 6, no. 3 (September 1992):416–440. © 1992 Sociologists for Women in Society. Reprinted by permission of Sage Publications, Inc. and the authors.

scholarship in the study of both race and gender has taught us the importance of interpersonal attachments to the lives of women . . . and a commitment to racial uplift among people of color. . . .

. . . Lacking wealth, the greatest gift a Black family has been able to give to its children has been the motivation and skills to succeed in school. Aspirations for college attendance and professional positions are stressed as *family* goals, and the entire family may make sacrifices and provide support. . . . Black women have long seen the activist potential of education and have sought it as a cornerstone of community development—a means of uplifting the race. When women of color or white women are put at the center of the analysis of upward mobility, it is clear that different questions will be raised about social mobility and different descriptions of the process will ensue. . . .

Research Design

These data are from a study of full-time employed middle-class women in the Memphis metropolitan area. This research is designed to explore the processes of upward social mobility for Black and white women by examining differences between women professionals, managers, and administrators who are from working- and middle-class backgrounds—that is, upwardly mobile and middle-class stable women. In this way, we isolate subjective processes shared among women who have been upwardly mobile from those common to women who have reproduced their family's professional-managerial class standing. Likewise, we identify many experiences in the attainment process that are shared by women of the same race, be they upwardly mobile or stable middle class. Finally, we specify some ways in which the attainment process is unique for each race-class group. . . .

. . . We rely on a model of social class basically derived from the work of Poulantzas (1974), Braverman (1974), Ehrenreich and Ehrenreich (1979), and elaborated in Vanneman and Cannon (1987). These works explicate a basic distinction between social class and social status. Classes represent bounded categories of the population, groups set in a relation of opposition to one another by their roles in the capitalist system. The middle class, or professional-managerial class, is set off from the working class by the power and control it exerts over workers in three realms: economic (power through ownership), political (power through direct supervisory authority), and ideological (power to plan and organize work; Poulantzas 1974; Vanneman and Cannon 1987).

In contrast, education, prestige, and income represent social statuses—hierarchically structured relative rankings along a ladder of economic success and social prestige. Positions along these dimensions are not established by social relations of dominance and subordination but, rather, as rankings on scales representing resources and desirability. In some respects, they represent both the justification for power differentials vested in classes and the rewards for the role that the middle class plays in controlling labor.

Our interest is in the process of upward social class mobility, moving from a working-class family of origin to a middle-class destination—from a position of working-class subordination to a position of control over the working class. Lacking inherited wealth or other resources, those working-class people who attain middle-class standing do so primarily by obtaining a college education and entering a professional, managerial, or administrative occupation. Thus we examine carefully the process of educational attainment not as evidence of middle-class standing but as a necessary part of the mobility process for most working-class people.

Likewise, occupation alone does not define the middle class, but professional, managerial, and administrative occupations capture many of the supervisory and ideologically based positions whose function is to control workers' lives. Consequently, we defined subjects as *middle class* by virtue of their employment in either a professional, managerial, or administrative occupation. . . . Classification of subjects as either professional or managerial-administrative was made on the basis of the designation of occupations in the U.S. Bureau of the Census's (1983) "Detailed Population Characteristics: Tennessee." Managerial occupations were defined as those in the census categories of managers and administrators; professionals were defined as those occupations in the professional category, excluding technicians, whom Braverman (1974) contends are working class.

Upwardly mobile women were defined as those women raised in families where neither parent was employed as a professional, manager, or administrator. Typical occupations for working-class fathers were postal clerk, craftsman, semi-skilled manufacturing worker, janitor, and laborer. Some working-class mothers had clerical and sales positions, but many of the Black mothers also worked as private household workers. *Middle-class stable* women were defined as those women raised in families where *either* parent was employed as a professional, manager, or administrator. Typical occupations of middle-class parents were social worker, teacher, and school administrator as well as high-status professionals such as attorneys, physicians, and dentists. . . .

Family Expectations for Educational Attainment

Four questions assess the expectations and support among family members for the educational attainment of the subjects. First, "Do you recall your father or mother stressing that you attain an education?" Yes was the response of 190 of the 200 women. Each of the women in this study had obtained a college degree, and many have graduate degrees. It is clear that for Black and white women, education was an important concern in their families. . . .

The comments of Laura Lee, a 39-year-old Black woman who was raised middle class, were typical:

> Going to school, that was never a discussable issue. Just like you were born to live and die, you were going to go to school. You were going to prepare yourself to do something.

It should be noted, however, that only 86 percent of the white working-class women answered yes, compared to 98 percent of all other groups. Although this difference is small, it foreshadows a pattern where white women raised in working-class families received the least support and encouragement for educational and career attainment.

"When you were growing up, how far did your father expect you to go in school?" While most fathers expected college attendance from their daughters, differences also exist by class of origin. Only 70 percent of the working-class fathers, both Black and white, expected their daughters to attend college. In contrast, 94 percent of the Black middle-class and 88 percent the white middle-class women's fathers had college expectations for their daughters.

When asked the same question about mother's expectations, 88 percent to 92 percent of each group's mothers expected their daughters to get a college education, except the white working-class women, for whom only 66 percent of mothers held such expectations. In short, only among the white working-class women did a fairly substantial proportion (about one-third) of both mothers and fathers expect less than a college education from their daughters. About 30 percent of Black working-class fathers held lower expectations for their daughters, but not the mothers; virtually all middle-class parents expected a college education for their daughters.

Sara Marx is a white, 33-year-old director of counseling raised in a rural working-class family. She is among those whose parents did not expect a college education for her. She was vague about the roots of attending college:

> It seems like we had a guest speaker who talked to us. Maybe before our exams somebody talked to us. I really can't put my finger on anything. I don't know where the information came from exactly.

"Who provided emotional support for you to make the transition from high school to college?" While 86 percent of the Black middle-class women indicated that family provided that support, 70 percent of the white middle class, 64 percent of the Black working class, and only 56 percent of the white working class received emotional support from family.

"Who paid your college tuition and fees?" Beyond emotional support, financial support is critical to college attendance. There are clear class differences in financial support for college. Roughly 90 percent of the middle-class respondents and only 56 percent and 62 percent of the Black and white working-class women, respectively, were financially supported by their families. These data also suggest that working-class parents were less able to give emotional or financial support for college than they were to hold out the expectation that their daughters should attend.

Family Expectations for Occupation or Career

When asked, "Do you recall your father or mother stressing that you should have an occupation to succeed in life?", racial differences appear. Ninety-four percent of all Black respondents said yes. In the words of Julie Bird, a Black woman raised-middle-class junior high school teacher:

> My father would always say, "You see how good I'm doing? Each generation should do more than the generation before." He expects me to accomplish more than he has.

Ann Right, a 36-year-old Black attorney whose father was a janitor, said:

> They wanted me to have a better life than they had. For all of us. And that's why they emphasized education and emphasized working relationships and how you get along with people and that kind of thing.

Ruby James, a Black teacher from a working-class family, said:

> They expected me to have a good-paying job and to have a family and be married. Go to work every day. Buy a home. That's about it. Be happy.

In contrast, only 70 percent of the white middle-class and 56 percent of the white working-class women indicated that their parents stressed that an occupation was needed for success. Nina Pentel, a 26-year-old white medical social worker, expressed a common response: "They said 'You're going to get married but get a degree, you never know what's going to happen to you.' They were pretty laid back about goals."

When the question focuses on a career rather than an occupation, the family encouragement is lower and differences were not significant, but similar patterns emerged. We asked respondents, "Who, if anyone, encouraged you to think about a career?" Among Black respondents, 60 percent of the middle-class and 56 percent of the working-class women answered that family encouraged them. Only 40 percent of the white working-class women indicated that their family encouraged them in their thinking about a career, while 52 percent of the white middle-class women did so. . . .

When working-class white women seek to be mobile through their own attainments, they face conflicts. Their parents encourage educational attainment, but when young women develop professional career goals, these same parents sometimes become ambivalent. This was the case with Elizabeth Marlow, who is currently a public interest attorney—a position her parents never intended her to hold. She described her parents' traditional expectations and their reluctance to support her career goals fully.

> My parents assumed that I would go college and meet some nice man and finish, but not necessarily work after. I would be a good mother for my children. I don't think that they ever thought I would go to law school. Their attitude about my interest in law school was, "You can do it if you want to, but we don't think it is a particularly practical thing for a woman to do."

Elizabeth is married and has three children, but she is not the traditional housewife of her parents' dreams. She received more support outside the family for her chosen life-style.

Although Black families are indeed more likely than white families to encourage their daughters to prepare for careers, like white families, they frequently steer them toward highly visible traditionally female occupations, such as teacher, nurse, and social worker. Thus many mobile Black women are directed toward the same gender-segregated occupations as white women. . . .

Marriage

Although working-class families may encourage daughters to marry, they recognize the need for working-class women to contribute to family income or to support themselves economically. To achieve these aims, many working-class girls are encouraged to pursue an education as preparation for work in gender-segregated occupations. Work in these fields presumably allows women to keep marriage, family, and child rearing as life goals while contributing to the family income and to have "something to fall back on" if the marriage does not work out. This interplay among marriage, education, financial need, and class mobility is complex (Joslin 1979).

We asked, "Do you recall your mother or father emphasizing that marriage should be your primary life goal?" While the majority of all respondents did not get the message that marriage was the *primary life goal,* Black and white women's parents clearly saw this differently. Virtually no Black parents stressed marriage as the primary life goal (6 percent of the working class and 4 percent of the middle class), but significantly more white parents did (22 percent of the working class and 18 percent of the middle class).

Some white women said their families expressed active opposition to marriage, such as Clare Baron, a raised-working-class nursing supervisor, who said, "My mother always said, 'Don't get married and don't have children!' "

More common responses recognized the fragility of marriage and the need to support oneself. For example, Alice Page, a 31-year-old white raised-middle-class librarian, put it this way:

> I feel like I am really part of a generation that for the first time is thinking, "I don't want to have to depend on somebody to take care of me because what if they say they are going to take care of me and then they are not there? They die, or they leave me or whatever." I feel very much that I've got to be able to support myself and I don't know that single women in other eras have had to deal with that to the same degree.

While white working-class women are often raised to prepare for work roles so that they can contribute to family income and, if necessary, support themselves, Black women face a different reality. Unlike white women, Black women are typically socialized to view marriage separately from economic security, because it is not expected that marriage will ever remove them from the labor market. As a result, Black families socialize all their children—girls and boys—for self-sufficiency (Clark 1986; Higginbotham and Cannon 1988). . . .

. . . Fairly substantial numbers of each group had never married by the time of the interview, ranging from 20 percent of the white working-class to 34 percent of the Black working-class and white middle-class respondents. Some of the women were pleased with their singlehood, like Alice Page, who said:

> I am single by choice. That is how I see myself. I have purposely avoided getting into any kind of romantic situation with men. I have enjoyed going out but never wanted to get serious. If anyone wants to get serious, I quit going out with him.

Other women expressed disappointment and some shock that they were not yet married. When asked about her feeling about being single, Sally Ford, a 32-year-old white manager, said:

> That's what I always wanted to do: to be married and have children. To me, that is the ideal. I want a happy, good marriage with children. I do not like being single at all. It is very, very lonesome. I don't see any advantages to being single. None! . . .

Subjective Sense of Debt to Kin and Friends

McAdoo (1978) reports that upwardly mobile Black Americans receive more requests to share resources from their working-class kin than do middle-class Black Americans. Many mobile Black Americans feel a "social debt" because their families aided them in the mobility process and provided emotional support. When we asked the white women in the study the following question: "Generally, do you feel you owe a lot for the help given to you by your family and relatives?" many were perplexed and asked what the question meant. In contrast, both the working- and middle-class Black women tended to respond immediately that they felt a sense of obligation to family and friends in return for the support they had received. Black women, from both the working class and the middle class, expressed the strongest sense of debt to family, with 86 percent and 74 percent, respectively, so indicating. White working-class women were least likely to feel that they owed family (46 percent), while 68 percent of white middle-class women so indicated. In short, upwardly mobile Black women were almost twice as likely as upwardly mobile white women to express a sense of debt to family.

Linda Brown, an upwardly mobile Black woman, gave a typical response, "Yes, they are there when you need them." Similar were the words of Jean Marsh, "Yes, because they have been supportive. They're dependable. If I need them I can depend upon them."

One of the most significant ways in which Black working-class families aided their daughters and left them with a sense of debt related to care for their children. Dawn March expressed it thus:

> They have been there more so during my adult years than a lot of other families that I know about. My mother kept all of my children until they were old enough to go to day care. And she not only kept them, she'd give them a bath for me during the daytime and feed them before I got home from work. Very, very supportive people. So, I really would say I owe them for that.

Carole Washington, an upwardly mobile Black woman occupational therapist, also felt she owed her family. She reported:

> I know the struggle that my parents have had to get me where I am. I know the energy they no longer have to put into the rest of the family even though they want to put it there and they're willing. I feel it is my responsibility to give back some of that energy they have given to me. It's self-directed, not required.

White working-class women, in contrast, were unlikely to feel a sense of debt and expressed their feelings in similar ways. Irma Cox, part owner of a computer business, said, "I am appreciative of the values my parents instilled in me. But I for the most part feel like I have done it on my own." Carey Mink, a 35-year-old psychiatric social worker, said, "No, they pointed me in a direction and they were supportive, but I've done a lot of the work myself." Debra Beck, a judge, responded, "No, I feel that I've gotten most places on my own." . . .

Commitment to Community

The mainstream "model of community stresses the rights of individuals to make decisions in their own self interest, regardless of the impact on the larger society" (Collins 1990, 52). This model may explain relations to community of origin for mobile white males but cannot be generalized to other racial and gender groups. In the context of well-recognized structures of racial oppression, America's racial-ethnic communities develop collective survival strategies that contrast with the individualism of the dominant culture but ensure the community's survival. . . . Widespread community involvement enables mobile people of color to confront and challenge racist obstacles in credentialing institutions, and it distinguishes the mobility process in racial-ethnic communities from mobility in the dominant culture. For example, Lou Nelson, now a librarian, described the support she felt in her southern segregated inner-city school. She said:

> There was a closeness between people and that had a lot to do with neighborhood schools. I went to Tubman High School with people that lived in the Tubman area. I think that there was a bond, a bond between parents, the PTA . . . I think that it was just that everybody felt that everybody knew everybody. And that was special.

Family and community involvement and support in the mobility process means that many Black professionals and managers continue to feel linked to their communities of origin. Lillian King, a high-ranking city official who was raised working class, discussed her current commitment to the Black community. She said:

> Because I have more opportunities, I've got an obligation to give more back and to set a positive example for Black people and especially for Black women. I think we've got to do a tremendous job in building self-esteem and giving people the desire to achieve.

Judith Moore is a 34-year-old single parent employed as a health investigator. She has been able to maintain her connection with her community, and that is a source of pride.

> I'm proud that I still have a sense of who I am in terms of Black people. That's very important to me. No matter how much education or professional status I get, I do not want to lose touch with where I've come from. I think that you need to look back and that kind of pushes you forward. I think the degree and other things can make you lose sight of that, especially us Black folks, but I'm glad that I haven't and I try to teach that [commitment] to my son.

For some Black women, their mobility has enabled them to give to an even broader community. This is the case with Sammi Lewis, a raised-working-class woman who is a director of a social service agency. She said, "I owe a responsibility to the entire community, and not to any particular group." . . .

Crossing the Color Line

Mobility for people of color is complex because in addition to crossing class lines, mobility often means crossing racial and cultural ones as well. Since the 1960s, people of color have increasingly attended either integrated or predominantly white schools. Only mobile white ethnics have a comparable experience of simultaneously crossing class and cultural barriers, yet even this experience is qualitatively different from that of Black and other people of color. White ethnicity can be practically invisible to white middle-class school peers and co-workers, but people of color are more visible and are subjected to harsher treatment. Our research indicates that no matter when people of color first encounter integrated or predominantly white settings, it is always a shock. The experience of racial exclusion cannot prepare people of color to deal with the racism in daily face-to-face encounters with white people.

For example, Lynn Johnson was in the first cohort of Black students at Regional College, a small private college in Memphis. The self-confidence and stamina Lynn developed in her supportive segregated high school helped her withstand the racism she faced as the first female and the first Black to graduate in economics at Regional College. Lynn described her treatment:

> I would come into class and Dr. Simpson (the Economics professor) would al-phabetically call the roll. When he came to my name, he would just jump over it. He would not ask me any questions, he would not do anything. I stayed in that class. I struggled through. When it was my turn, I'd start talking. He would say, "Johnson, I wasn't talking to you" [because he never said *Miss* Johnson). I'd say, "That's all right, Dr. Simpson, it was my turn. I figured you just over-looked me. I'm just the littlest person in here. Wasn't that the right answer?" He would say, "Yes, that was the right answer." I drove him mad, I really did. He finally got used to me and started to help me.

In southern cities, where previous interaction between Black and white people followed a rigid code, adjustments were necessary on both sides. It was clear to Lynn Johnson and others that college faculty and students had to adapt to her small Black cohort at Regional College.

Wendy Jones attended a formerly predominantly white state university that had just merged with a formerly predominantly Black college. This new institution meant many adjustments for faculty and students. As a working-class person majoring in engineering, she had a rough transition. She recalled:

> I had never gone to school with white kids. I'd always gone to all Black schools all my life and the Black kids there [at the university] were snooty. Only one friend from high school went there and she flunked out. The courses were harder and all my teachers were men and white. Most of the kids were white. I was in classes where I'd be the only Black and woman. There were no similarities to grasp for. I had to adjust to being in that situation. In about a year I was comfortable where I could walk up to people in my class and have conversations.

For some Black people, their first significant interaction with white people did not come until graduate school. Janice Freeman described her experiences:

> I went to a Black high school, a Black college and then worked for a Black man who was a former teacher. Everything was comfortable until I had to go to State University for graduate school. I felt very insecure. I was thrown into an environment that was very different—during the 1960s and 1970s there was so much unrest anyway—so it was extremely difficult for me.

It was not in graduate school but on her first job as a social worker that Janice had to learn to work *with* white people. She said, "After I realized that

I could hang in school, working at the social work agency allowed me to learn how to work *with* white people. I had never done that before and now I do it better than anybody."

Learning to live in a white world was an additional hurdle for all Black women in this age cohort. Previous generations of Black people were more likely to be educated in segregated colleges and to work within the confines of the established Black community. They taught in segregated schools, provided dental and medical care to the Black communities, and provided social services and other comforts to members of their own communities. They also lived in the Black community and worshiped on Sunday with many of the people they saw in different settings. As the comments of our respondents reveal, both Black and white people had to adjust to integrated settings, but it was more stressful for the newcomers.

SUMMARY AND CONCLUSIONS

Our major aim in this research was to reopen the study of the subjective experience of upward social mobility and to begin to incorporate race and gender into our vision of the process. In this exploratory work, we hope to raise issues and questions that will cast a new light on taken-for-granted assumptions about the process and the people who engage in it. The experiences of these women have certainly painted a different picture from the one we were left some twenty years ago. First and foremost, these women are not detached, isolated, or driven solely by career goals. Relationships with family of origin, partners, children, friends, and the wider community loom large in the way they envision and accomplish mobility and the way they sustain themselves as professional and managerial women.

Several of our findings suggest ways that race and gender shape the mobility process for baby boom Black and white women. Education was stressed as important in virtually all of the families of these women; however, they differed in how it was viewed and how much was desired. The upwardly mobile women, both Black and white, shared some obstacles to attainment. More mobile women had parents who never expected them to achieve a college education. They also received less emotional and financial support for college attendance from their families than the women in middle-class families received. Black women also faced the unique problem of crossing racial barriers simultaneously with class barriers.

There were fairly dramatic race differences in the messages that the Black and white women received from family about what their lives should be like as adults. Black women clearly received the message that they needed an occupation to succeed in life and that marriage was a secondary concern. Many Black women also expressed a sense that their mobility was connected to an entire racial uplift process, not merely an individual journey.

White upwardly mobile women received less clear messages. Only one-half of these women said that their parents stressed the need for an occupation to succeed, and 20 percent said that marriage was stressed as the primary life goal. The most common message seemed to suggest that an occupation was necessary, because marriage could not be counted on to provide economic survival. Having a career, on the other hand, could even be seen as detrimental to adult happiness.

Upward mobility is a process that requires sustained effort and emotional and cognitive, as well as financial, support. The legacy of the image of mobility that was built on the white male experience focuses on credentialing institutions, especially the schools, as the primary place where talent is recognized and support is given to ensure that the talented among the working class are mobile. Family and friends are virtually invisible in this portrayal of the mobility process.

Although there is a good deal of variation in the roles that family and friends play for these women, they are certainly not invisible in the process. Especially among many of the Black women, there is a sense that they owe a great debt to their families for the help they have received. Black upwardly mobile women were also much more likely to feel that they give more than they receive from kin. Once they have achieved professional managerial employment, the sense of debt combines with their greater access to resources to put them in the position of being asked to give and of giving more to both family and friends. Carrington (1980) identifies some potential mental health hazards of such a sense of debt in upwardly mobile Black women's lives.

White upwardly mobile women are less likely to feel indebted to kin and to feel that they have accomplished alone. Yet even among this group, connections to spouses and children played significant roles in defining how women were mobile, their goals, and their sense of satisfaction with their life in the middle class.

These data are suggestive of a mobility process that is motivated by a desire for personal, but also collective, gain and that is shaped by interpersonal commitments to family, partners and children, community, and the race. Social mobility involves competition, but also cooperation, community support, and personal obligations. Further research is needed to explore fully this new image of mobility and to examine the relevance of these issues for white male mobility as well.

REFERENCES

Braverman, Harry. 1974. *Labor and monopoly capital.* New York: Monthly Review Press.

Carrington, Christine. 1980. Depression in Black women: A theoretical appraisal. In *The Black woman,* edited by La Frances Rodgers Rose. Beverly Hills, CA: Sage.

Clark, Reginald. 1986. *Family life and school achievement.* Chicago: University of Chicago Press.

Collins, Patricia Hill. 1990. *Black feminist thought: Knowledge, consciousness, and the politics of empowerment.* Boston: Routledge.

Ehrenreich, Barbara, and John Ehrenreich. 1979. The professional-managerial class. In *Between labor and capital,* edited by Pat Walker. Boston: South End Press.

Higginbotham, Elizabeth, and Lynn Weber Cannon. 1988. *Rethinking mobility: Towards a race and gender inclusive theory.* Research Paper no. 8. Center for Research on Women, Memphis State University.

Joslin, Daphne. 1979. Working-class daughters, middle-class wives: Social identity and self-esteem among women upwardly mobile through marriage. Ph.D. diss., New York University, New York.

McAdoo, Harriette Pipes. 1978. Factors related to stability in upwardly mobile Black families. *Journal of Marriage and the Family* 40:761–76.

Poulantzas, Nicos. 1974. *Classes in contemporary capitalism.* London: New Left Books.

U.S. Bureau of the Census. 1983. Detailed population characteristics: Tennessee. Census of the Population, 1980. Washington, DC: GPO.

Vanneman, Reeve, and Lynn Weber Cannon. 1987. *The American perception of class.* Philadelphia: Temple University Press.

Gender and Sexism

COMMONALITIES AND DIFFERENCES   **16**

Johnnetta B. Cole

If you see one woman, have you seen them all? Does the heavy weight of patriarchy level all differences among US women? Is it the case, as one woman put it, that "there isn't much difference between having to say 'Yes suh Mr. Charlie' and 'Yes dear'?" Does "grandmother" convey the same meaning as "abuela," as "buba," as "gran'ma"? Is difference a part of what we share, or is it, in fact, *all* that we share? As early as 1970, Toni Cade Bambara asked: "How relevant are the truths, the experiences, the findings of white women to black women? Are women after all simply women?" (Bambara 1970: 9).

Are US women bound by our similarities or divided by our differences? The only viable response is *both*. To address our commonalities without dealing with our differences is to misunderstand and distort that which separates as well as that which binds us as women. Patriarchal oppression is not limited to women of one race or of one particular ethnic group, women in one class, women of one age group or sexual preference, women who live in one part of the country, women of any one religion, or women with certain physical abilities or disabilities. Yet, while oppression of women knows no such limitations, we cannot, therefore, conclude that the oppression of all women is identical.

Among the things which bind women together are the assumptions about the way that women think and behave, the myths—indeed the stereotypes—about what is common to all women. For example, women will be asked nicely in job interviews if they type, while men will not be asked such a question. In response to certain actions, the expression is used: "Ain't that just like a woman?" Or during a heated argument between a man and a woman, as the voice of each rises and emotions run high, the woman makes a particularly good point. In a voice at the pitch of the ongoing argument, the man screams at her: "You don't have to get hysterical!"

In an interesting form of "what goes around comes around," as Malcolm X put it, there is the possibility that US women are bound together by our assumptions, attitudes toward, even stereotypes of the other gender. Folklorist Rayna Green, referring to women of the Southern setting in which she grew up, says this:

> Southern or not, women everywhere talk about sex. . . . In general men are more often the victims of women's jokes than not. Tit for tat, we say. Usually the subject for laughter is men's boasts, failures, or inadequacies ("comeuppance for lack of upcommance," as one of my aunts would say). Poking fun at a man's sexual ego, for example, might never be possible in real social situations with the men who have power over their lives, but it is possible in a joke. (Green 1984: 23–24)

That which US women have in common must always be viewed in relation to the particularities of a group, for even when we narrow our focus to one particular group of women it is possible for differences within that group to challenge the primacy of what is shared in common. For example, what have we said and what have we failed to say when we speak of "Asian American women"? As Shirley Hune notes (1982), Asian American women as a group share a number of characteristics. Their participation in the work force is higher than that of women in any other ethnic group. Many Asian American women live life supporting others, often allowing their lives to be subsumed by the needs of the extended family. And they are subjected to stereotypes by the dominant society: the sexy but "evil dragon lady," the "neuter gender," the "passive/demure" type, and the "exotic/erotic" type.

However, there are many circumstances when these shared experiences are not sufficient to accurately describe the condition of particular Asian American women. Among Asian American women there are those who were born in the United States, fourth and fifth generation Asian American women with firsthand experience of no other land, and there are those who recently arrived in the United States. Asian American women are diverse in their heritage or country of origin: China, Japan, the Philippines, Korea, India,

Vietnam, Cambodia, Thailand, or another country in Asia. If we restrict ourselves to Asian American women of Chinese descent, are we referring to those women who are from the People's Republic of China or those from Taiwan, those from Hong Kong or those from Vietnam, those from San Francisco's Chinatown or those from Mississippi? Are we subsuming under "Asian American" those Pacific Island women from Hawaii, Samoa, Guam, and other islands under U.S. control? Although the majority of Asian American women are working-class—contrary to the stereotype of the "ever successful" Asians—there are poor, "middle-class," and even affluent Asian American women (Hune 1982: 1–2, 13–14).

It has become very common in the United States today to speak of "Hispanics," putting Puerto Ricans, Chicanos, Dominicans, Cubans, and those from every Spanish-speaking country in the Americas into one category of people, with the women referred to as Latinas or Hispanic women. Certainly there is a language, or the heritage of a language, a general historical experience, and certain cultural traditions and practices which are shared by these women. But a great deal of harm can be done by sweeping away differences in the interest of an imposed homogeneity.

Within one group of Latinas there is, in fact, considerable variation in terms of self-defined ethnic identity, such that some women refer to themselves as Mexican Americans, others as *Chicanas*, others as Hispanics, and still others as Americans. Among this group of women are those who express a commitment to the traditional roles of women and others who identify with feminist ideals. Some Chicanas are monolingual—in Spanish or English—and others are bilingual. And there are a host of variations among Chicanas in terms of educational achievements, economic differences, rural or urban living conditions, and whether they trace their ancestry from women who lived in this land well before the United States forcibly took the northern half of Mexico, or more recently arrived across the border that now divides the nations called Mexico and the United States.

Women of the Midwest clearly share a number of experiences which flow from living in the U.S. heartland, but they have come from different places, and they were and are today part of various cultures.

> Midwestern women are the Native American women whose ancestors were brought to the plains in the mid-nineteenth century to be settled on reservations, the black women whose fore-bears emigrated by the thousands from the South after Reconstruction. They are the descendants of the waves of Spanish, French, Norwegian, Danish, Swedish, Bohemian, Scottish, Welsh, British, Irish, German, and Russian immigrants who settled the plains, the few Dutch, Italians, Poles, and Yugoslavs who came with them. (Boucher 1982: 3)

There is another complexity: when we have identified a commonality among women, cutting across class, racial, ethnic, and other major lines of difference, the particular ways that commonality is acted out and its consequences in the larger society may be quite diverse. Ostrander makes this point in terms of class:

> When women stroke and soothe men, listen to them and accommodate their needs, men of every class return to the workplace with renewed energies. When women arrange men's social lives and relationships, men of every class are spared investing the time and energy required to meet their social needs. When women run the households and keep family concerns in check, men of every class are freer than women to pursue other activities, including work, outside the home. But upper-class women perform these tasks for men at the very top of the class structure. . . . Supporting their husbands as individuals, they support and uphold the very top of the class structure. In this way they distinguish themselves from women of other social classes. (Ostrander 1984: 146)

Suppose that we can accurately and exclusively identify the characteristics shared by one particular group of women. For each of the women within that group, into how many other groups does she want to, or is she forced to, fit? Or can we speak of similarities only with respect to a group such as Puerto Rican women who are forty-three years old, were born in San Juan, Puerto Rico, migrated to New York City when they were five years old, work as eighth-grade school teachers, attend a Catholic church, are heterosexual, married, with two male and two female children, and have no physical disabilities?

Then there is that unpredictable but often present quality of individuality, the idiosyncrasies of a particular person. Shirley Abbott, describing experiences of growing up in the South, contrasts her mother's attitude and behavior toward the black woman who was her maid with what was the usual stance of "Southern white ladies."

> I don't claim that my mother's way of managing her black maid was typical. Most white women did not help their laundresses hang the washing on the line. . . . Compulsive housewifery had some part in it. So did her upbringing. . . . There was another motive too. . . . Had she used Emma in just the right way, Mother could have become a lady. But Mother didn't want to be a lady. Something in her was against it, and she couldn't explain what frightened her, which was why she cried when my father ridiculed her. (Abbott 1983: 78–79)

Once we have narrowed our focus to one specific group of women (Armenian American women, or women over sixty-five, or Arab American women, or black women from the Caribbean, or Ashkenazi Jewish women), the oppression that group of women experiences may take different forms at different times. Today, there is no black woman in the United States who is the legal slave of a white master: "chosen" for that slave status because of her race, forced to give her labor power without compensation because of the class arrangements of the society, and subjected to the sexual whims of her male master because of her gender. But that does not mean that black women today are no longer oppressed on the basis of race, class, and gender.

There are also groups of women who experience intense gender discrimination today, but in the past had a radically different status in their society. Contrary to the popular image of female oppression as being both universal and as old as human societies, there is incontestable evidence of egalitarian societies in which men and women related in ways that did not involve male dominance and female subjugation. Eleanor Leacock is the best known of the anthropologists who have carried out the kind of detailed historical analysis which provides evidence on gender relations in precolonial North American societies. In discussing the debate on the origins and spread of women's oppression, Leacock points out that women's oppression is a reality today in virtually every society, and while socialist societies have reduced it, they have not eliminated gender inequality. However, it does not follow that women's oppression has always existed and will always exist. What such arguments about universal female subordination do is to project onto the totality of human history the conditions of today's world. Such an argument also "affords an important ideological buttress for those in power" (Leacock 1979: 10–11).

Studies of precolonial societies indicate considerable variety in terms of gender relations.

> Women retained great autonomy in much of the pre-colonial world, and related to each other and to men through public as well as private procedures as they carried out their economic and social responsibilities and protected their rights. Female and male modalities of various kinds operated reciprocally within larger kin and community contexts before the principle of male dominance within individual families was taught by missionaries, defined by legal status, and solidified by the economic relations of colonialism. (Leacock 1979: 10–11)

Even when there is evidence of female oppression among women of diverse backgrounds, it is important to listen to the individual assessment which each women makes of her own condition, rather than assume that a synonymous experience of female oppression exists among all women. As a case in point, Sharon Burmeister Lord, in describing what it was like to grow up "Appalachian style," speaks of the influence of female role models in shaping the conditions of her development. In Williamson, West Virginia, she grew up knowing women whose occupations were Methodist preacher, elementary school principal, county sheriff, and university professor. Within her own family, her mother works as a secretary, writes poetry and songs, and "swims faster than any boy"; her aunt started her own seed and hardware store; one grandmother is a farmer and the other runs her own boarding house. Summarizing the effect of growing up among such women, Lord says:

> When a little girl has had a chance to learn strength, survival tactics, a firm grasp of reality, and an understanding of class oppression from the women around her, it doesn't remove oppression from her life, but it does give her a fighting chance. And that's an advantage! (Lord 1979: 25)

Finally, if it is agreed that today, to some extent, all women are oppressed, to what extent can a woman, or a group of women, also act as oppressor? Small as the numbers may be, there are some affluent black women. . . . Is it not possible that among this very small group of black women there are those who, while they experience oppression because of their race, act in oppressive ways toward other women because of their class? Does the experience of this society's heterosexism make a Euro-American lesbian incapable of engaging in racist acts toward women of color? The point is very simply that privilege can and does coexist with oppression (Bulkin et al. 1984: 99) and being a victim of one form of discrimination does not make one immune to victimizing someone else on a different basis. . . .

REFERENCES CITED

Abbott, S. 1983. *Womenfolks: Growing Up Down South*. New York: Ticknor and Fields.

Bambara, T. C., ed. 1970. *The Black Woman: An Anthology*. New York: Signet.

Boucher, S. 1982. *Heartwomen: An Urban Feminist Odyssey Home*. New York: Harper & Row.

Bulkin, E., M. E. Pratt, and B. Smith, eds. 1984. *Yours in Struggle*. Brooklyn, N.Y.: Long Haul Press.

Green, R. 1984. "Magnolias Grow in Dirt: The Bawdy Lore of Southern Women." In *Speaking for Ourselves*, M. Alexander, ed., pp. 20–28. New York: Pantheon.

Hune, S. 1982. Asian American Women: Past and Present, Myth and Reality. Unpublished manuscript prepared for conference on Black Women's Agenda for the Feminist Movement in the 80's, Williams College, Williamstown, Mass., November 12–14, 1982.

Leacock, E. 1979. "Women, Development and Anthropological Facts and Fictions." In *Women in Latin America: An Anthology from Latin American Perspectives*, pp. 7–16. Riverside, Calif.: Latin American Perspectives.

Lord, S. B. 1979. "Growin' Up—Appalachian, Female, and Feminist." In *Appalachian Women: A Learning/Teaching Guide*, S. B. Lord and C. Patton-Crowder, eds., pp. 22–25. Knoxville, Tenn.: University of Tennessee.

Ostrander, S. A. 1984. *Women of the Upper Class*. Philadelphia: Temple University Press.

UNDERSTANDING AND FIGHTING SEXISM: *A Call to Men*

17

Peter Blood, Alan Tuttle, and George Lakey

PART 1: UNDERSTANDING THE ENEMY: HOW SEXISM WORKS IN THE U.S.A.

What Is Sexism?

Sexism is much more than a problem with the language we use, our personal attitudes, or individual hurtful acts toward women. Sexism in our country is a complex mesh of practices, institutions, and ideas which have the overall effect of giving more power to men than to women. By "power" we mean the ability to influence important decisions—political decisions of govern-

From: *Off Their Backs . . . and on Our Own Two Feet* (Philadelphia: New Society Publishers, 1983), pp. 1–8. Reprinted by permission.

ment on every level, economic decisions (jobs, access to money, choice of priorities), and a wide variety of other life areas down to the most personal concerns, such as whether two people are going to make love on a given night or not. The word "patriarchy" is sometimes used to refer to the actual power structure built around men's domination of women. Two key areas where women are denied power are the area of jobs and the area of violence directed toward women.

Women have much less earning power in our labor market than men do. Reasons for this include the fact that much of women's labor is unwaged (housecleaning, childrearing, little services to please bosses or lovers); the low status and pay of most of the traditionally women's jobs that are waged (secretary, sales clerk, childcare, nursing home attendant); the non-union status of most women workers; and the discriminatory practices such as the recent Supreme Court decision allowing companies to exclude pregnancy from their medical insurance and sick leave benefits.

Women face a constant threat of physical violence and sexual aggression in our society. As men we are rarely aware of how pervasive this is or the powerful effect it has on women's outlook on themselves and the world. Actual rape or sadistic violence is the tip of the iceberg. Physical abuse of wives and lovers is common and rarely publicized. A majority of women have probably experienced some form of sexual abuse as children. The memory of these experiences often gets suppressed because they feel so humiliated and scared, and because adults deny repeatedly that such a thing could happen. Society is filled with messages pressuring women to provide men with sexual pleasure.

All of the above combine with differences in physical strength, voice, acculturated ways of dealing with anger, and the very concrete power men hold in other areas of life to keep many women intimidated, passive, and unable to even acknowledge their own fear openly. The rapist is the shock trooper for an overall system of unequal power.

Patriarchy is not just a power structure "out there"; it is mainly enforced by our own acceptance of its character ideals for our lives. The character ideal which is held up for men to reach toward is "masculinity." A masculine man is supposed to be tough, good at abstract reasoning, hard-working, unfeeling except for anger and sexual desire, and habitually taking the initiative. Masculinity exists only in contrast to femininity, the model for women. Feminine characteristics include cooperativeness, emotionality, patience, passivity, nurturance, and sexual appeal.

We all know that human characteristics are *not* distributed neatly between the sexes that way. A nursery school will often include girls who do abstract reasoning and get into fights and boys who cry easily. We also know that the culture does not leave them alone—the tomboy usually learns to become a lady, and the gentle boy develops armor to protect him from the jibes of his mates. They also learn that masculinity is valued in our culture more than femininity, especially when it comes to gaining power. In fact, the characteristics which are assigned to men by the patriarchy are the power-linked characteristics. In other words, by accepting masculinity as an ideal for ourselves, men buy into a system which keeps women down.

Sexism and Our Economic System

Why does sexism exist? Why is it so hard to root it out? Who really benefits from it? There is a variety of theories as to where sexism originated. Some say it started with the advent of class society; others say it preceded all other forms of oppression.

Regardless of its historic roots, it is clear that sexism today is intimately connected with our economic system, which is called "corporate capitalism."

In our society, a small group of people own and control almost all of the factories, financial institutions, networks of transport and sales of food and clothing. What this means, first of all, is that though we as men generally have more power than women do, the great majority of us have relatively little real power. Most of us are given very little chance to make the major decisions which affect our lives (for example, whether society will emphasize public transportation or private cars, how decisions are made at the factory where we work, what quality air we breathe, what is taught to our children in school). In fact, this kind of power is concentrated in the hands of a relatively small number of white, middle-aged to older men.

However much power the system has given to each of us, this power is only useful in keeping things going according to the present rules, not in changing things in any basic way. The moment one of us tries to use power conveyed by the system to make fundamental changes, we find it taken away from us by that system. (All of us do possess the power to influence the course of history and bring about fundamental changes if we work together. This power comes from a very different source: from our power to work collectively to change the world around us, by "grass roots" power.)

Corporate capitalism is supported by sexism in many ways. Perhaps the most important way is that women function as a surplus labor force, where they can be pushed in and out of employment in keeping with current needs of the economy (depression, expansion, wartime, or a period of union-busting). In effect, the prevalent attitude in this society might be: "A woman's place is in the home . . . *until she's needed in the factory.*" Women's low wages keep wages lower for all: our boss holds the implicit threat that he can replace us with lower-paid women workers if we get too uppity or demand too high wages. Job role stereotyping also helps get women to do unpleasant forms of work which men would rather not do.

There are other ways sexism helps maintain capitalism. Women's un-waged labor of reproducing and taking care of the work force is called "women's work" and taken for granted. Women's servicing of men at home keeps our feelings of alienation and frustration concerning our jobs in check, which might otherwise lead to rebellion or burnout. Men's role as sole breadwinner in many families makes them reluctant to take militant stands on safety, wages, or other issues at work. It is threatening for a man not to be filling the breadwinner role—and he knows his wife will have a very difficult time finding work that pays enough to support a family. Women and men are played off against each other in competing for scarce jobs rather than sharing the fruits of a non-competitive, non-wasteful economy. Finally, excess consumption is encouraged by the lifeless, unsatisfying sex roles which we get shunted into on the job and at home.

The capitalist economic-political structure works to preserve sexism through indoctrination in our schools, a constant flow of brainwashing (subtle or otherwise) in advertising, and through its control of the mass media (what news gets reported and what does not, what programs get chosen for airing . . .). It also maintains it through control of legislation (influence through money), through the policies of our large corporations, and through failure to make use of leadership positions to educate people about the ways sexism affects us and ways it can be overcome.

Sexism and capitalism are so intertwined, in fact, that we believe there is no way that sexism can be thoroughly uprooted from our society without the total remaking of our economic system. Changing over to a popularly controlled or socialist economy would be a great blow against sexism. Women in countries like Cuba, China, Mozambique, and the Soviet Union have made significant gains as compared with their status prior to the revolutions in those countries.

Setting up a socialist society in no way guarantees an end to sexism (as can also be seen in the persistence of sexism to one degree or another in all the above countries). The patriarchy is a power arrangement which has a life of its own independent of capitalism's indoctrination. The struggle against sexism, therefore, must be a high priority for any movement which is offering leadership in creating a new society. The style pervading socialist movements in this country and the groups which are leading socialist countries is all too often determined by the masculine conditioning of domination and competition. This conditioning needs to be struggled against much more forcefully than it has been if we are ever to achieve a world free from sexism.

Are We Men the Enemy?

Some people say that men are the enemy when it comes to fighting sexism. We do not agree; blame and guilt don't help in understanding why people function as they do or in getting them to change.

Does this mean we are not responsible for what is happening? Not at all! As men, we are all involved in the oppression women experience, and we benefit from it each day. Yet this is no reason to fix blame on ourselves as "the oppressor" (or, for that matter, to place blame on any woman for "failing to fight back"). Over many years society has forced men and women into these roles of domination and submission.

How did society do this? It is clear we were pushed into these roles when we were young children and especially vulnerable. At that time, all of us were hurt in many ways through the expectations put on our sex. These hurts came through ridicule and threats directed at us and also through having to watch others get hurt while feeling ourselves incapable (because of lack of information or strength) of stopping that hurt. This experience of powerlessness in turn has reduced our ability to see the world clearly and act accordingly. The ingrained fear from these experiences tends to lead to either of two responses: feeling powerless and playing the victim role, or turning the experience around and acting out an oppressor role.

These dynamics do not just happen between men and women, but also between races, age groups, gays and non-gays, healthy and physically challenged people, and a number of other groups. As a matter of fact, everyone in our society has probably ended up playing both of these roles at one time or another. Even the richest, most powerful man was once powerless as a child in relation to his parents. Even a poor black woman could experience some privilege in relation to a very young person or a gay person.

The point is that we were all taught very early not to go beyond the expected stereotyped behavior. The sooner we recognize the effects of this kind of conditioning on us, the sooner we can effectively change the way things happen. This may mean stopping our domination of others or ceasing to accept oppression ourselves. It is good to recognize that stepping out of the old roles usually feels uncomfortable and may well require years of painful struggle, but to give up before the process is over is to miss the rich rewards that are down the road. We can travel that road successfully if we join with others for strength and support.

Why Change?

What rewards are there for us in this process? Much of what society thinks of as being manly is really a way of hardening up so we can dominate or coerce other people. We pay a high price for this hardening, however; it means giving up much of our humanity. We men are conditioned to suppress our feelings and don't learn how to give and receive support, nurturance, and affection awarely. We are taught to take all the responsibility for a situation on our own shoulders. We are taught to have heavy expectations of accomplishment for ourselves and for others. The direct result of this is a high level of tension and anxiety; the indirect result is a high disease rate and early death.

So, as men we have a lot to gain by fighting sexism. From what we have seen and experienced, men (at least in the long run) feel relief and joy just from being freed from the roles that lead to the oppression of women.

But this in no way means that we change easily. Conditioning is far too strong, and the temptation to keep the privilege too great. Nor does it mean, as some people suggest, that "men and women are both oppressed equally by sexism." However much men are hurt or limited by sex roles in this country, the fact remains that they are *not* systematically denied power simply because of being born a certain sex, as women are. Tremendous amounts of struggle will be required—as well as lots of loving support, especially from other men—to undergo the process of change.

Racism, Ageism, and the Oppression of Lesbians and Gay Men

All forms of oppression in our society are closely connected both to each other and to our economic system. Racism, for example, has many close parallels with sexism. Racial minorities, like women, function as a "reserve labor force," shunted into unattractive jobs with little reward or decision-making power, pushed in and out of employment as benefits the system. Racism, like sexism, divides working people from other, preventing them from looking at who holds most of the power in our country. Racial minorities, like women, are thought to be unintelligent and to have less motivation to achieve and work hard than white males. Both groups are kept in place by violent intimidation and are referred to with abusive words. Both frequently get deflected from militant struggle against the oppression of their group by cooptation of a few of their leaders into the lower levels of the power structure. In addition, sexual myths and paranoia about blacks and

Latinos play an especially vicious role in undergirding white people's racial fears and stereotypes.

People are just beginning to have a glimpse of what oppression based on age involves. The fact is that our society is almost totally blind to the dignity and capacities of the very young and the very old. Children are like women in being considered helpless, dependent, and cute—creatures to be cherished and taken care of, but not full human beings to be deeply respected and trusted with significant power. They experience 10–15 years of unpaid labor and brainwashing in our current form of education. Older people are looked at as children—except that they often find themselves without anyone interested in cherishing or taking care of them. For men, growing up is associated with taking on more and more of the hard masculine traits we mentioned before. Crying, acting afraid, and showing too much tenderness are all considered shameful because they are "childish" or "woman-like." ("You gotta stop that crying—don't be a sissy.")

Most of us know gay people, but usually we do not know who they are! Lesbians and gay men are so frequently hurt in this society that they usually do not feel safe to come out, even to their friends. An invisible colony of 20,000,000 people in our midst, lesbians and gay men generally work to service the status quo—even the Hollywood illusions of heterosexual romance—while their own dignity and security are denied.

People are still beaten and even killed for being homosexual in America, and the memory of the mass deaths of gays in Nazi concentration camps remains vivid. Now the "new right" is cutting back some recent gains lesbians and gay men have made in rights to jobs and housing. The fact that the same forces are opposing the Equal Rights Amendment and gay rights is a tip-off to the intimate connection between sexism and heterosexism. For one thing, gay men are mistakenly seen as taking the role of women in heterosexual relationships, and therefore not being "real men." For another, lesbians are seen as "uppity" because they act with a freedom not usually found when dependent upon men for loving.

Gay oppression is one of the ways the potential unity of all workers is prevented. It is also one of the cornerstones of the American nuclear family, which in turn is used to promote consumption and to teach sexist division of labor and sex roles. Lesbians and gay men are also exploited by being forced into ghettos and by the commercialized culture which profits from them.

Everyone is hurt by gay oppression. The fear of being considered gay limits and distorts everyone's life choices and relationships. Men are often afraid to get close to their male friends because it might imply gayness—and might even reveal a half-suspected gay dimension of themselves. An essential prop for sexism, in keeping people within their accustomed sex roles, is this fear of homosexuality, or homophobia. Because of this, women's liberation and men's liberation depend partly on gay liberation. . . .

TAKING SIDES AGAINST OURSELVES   **18**

Rosemary L. Bray

The Anita Hill–Clarence Thomas hearings are over; Judge Thomas is Justice Thomas now. Yet the memories linger on and on. Like witnesses to a bad accident, many of us who watched the three days of Senate hearings continue to replay the especially horrible moments. We compare our memories of cool accusation and heated denial; we weigh in our minds the hours of testimony, vacuous and vindictive by turn. In the end, even those of us who thought we were beyond surprise had underestimated the trauma.

"I have not been so wrenched since Dr. King was shot," says Jewell Jackson McCabe, the founder of the National Coalition of 100 Black Women, an advocacy group with chapters in twenty-one states and the District of Columbia. "I cannot begin to tell you; this thing has been unbelievable."

The near-mythic proportions that the event has already assumed in the minds of Americans are due, in part, to the twin wounds of race and gender that the hearings exposed. If gender is a troubling problem in American life and race is still a national crisis, the synergy of the two embodied in the life and trials of Anita Hill left most of America dumbstruck. Even black people who did not support Clarence Thomas's politics felt that Hill's charges, made public at the 11th hour, smacked of treachery. Feminist leaders embraced with enthusiasm a woman whose conservative political consciousness might have given them chills only a month earlier.

Even before the hearings began, the nomination of Clarence Thomas had taken on, for me, the quality of a nightmare. The particular dread I felt was one of betrayal—not a betrayal by President Bush, from whom I expected

From: Rosemary L. Bray, "Taking Sides Against Ourselves," *New York Times Magazine* 141 (November 17, 1991): 56ff. Copyright ©1991 by The New York Times Company. Reprinted by permission.

nothing—but by Thomas himself, who not only was no Thurgood Marshall but also gradually revealed himself to be a man who rejoiced in burning the bridges that brought him over.

I felt the kind of heartbreak that comes only to those of us still willing to call ourselves race women and race men in the old and honorable sense, people who feel that African-Americans should live and work and succeed not only for ourselves but also for our people.

The heated debates about gender and race in America have occurred, for the most part, in separate spheres; the separation makes for neater infighting.

But black women can never skirt these questions; we are their living expression. The parallel pursuits of equality for African-Americans and for women have trapped black women between often conflicting agendas for more than a century. We are asked in a thousand ways, large and small, to take sides against ourselves, postponing a confrontation in one arena to address an equally urgent task in another. Black men and white women have often made claims to our loyalty and our solidarity in the service of their respective struggles for recognition and autonomy, understanding only dimly that what may seem like liberty to each is for us only a kind of parole. Despite the bind, more often than not we choose loyalty to the race rather than the uncertain allegiance of gender.

Ours is a complicity of guilty survivors. A black man's presence is often feared; a black woman's presence is at least tolerated. Because until recently so much of the work that black women were paid to do was work that white men and women would not do—cleaning, serving, tending, teaching, nursing, maintaining, caring—we seem forever linked to the needs of human life that are at once minor and urgent.

As difficult as the lives of black women often are, we know we are mobile in ways black men are not—and black men know that we know. They know that we are nearly as angered as they about their inability to protect us in the traditional and patriarchal way, even as many of us have moved beyond the need for such protection. And some black men know ways to use our anger, our sorrow, our guilt, against us.

In our efforts to make a place for ourselves and our families in America, we have created a paradigm of sacrifice. And in living out such lives, we have convinced even ourselves that no sacrifice is too great to insure what we view in a larger sense as the survival of the race.

That sacrifice has been an unspoken promise to our people; it has made us partners with black men in a way white women and men cannot know. Yet not all of us view this partnership with respect. There are those who would use black women's commitment to the race as a way to control black women. There are those who believe the price of solidarity is silence. It was that commitment that trapped Anita Hill. And it is a commitment we may come to rue.

As I watched Hill being questioned that Friday by white men, by turn either timid or incredulous, I grieved for her. The anguish in her eyes was recognizable to me. Not only did she dare speak about events more than one woman would regard as unspeakable, she did so publicly. Not only did she make public accusations best investigated in private, she made them against a man who was black and conservative, as she was—a man who in other ways had earned her respect.

"Here is a woman who went to Sunday school and took it seriously," says Cornel West, director of the African-American Studies department at Princeton University and a social critic who felt mesmerized by what he called "the travesty and tragedy" of the hearings. "She clearly is a product of social conservatism of a rural black Baptist community." For black women historically, such probity, hard-won and tenaciously held, was social salvation. For white onlookers, it suggested an eerie primness out of sync with contemporary culture.

In the quiet and resolute spirit she might very well have learned from Sunday school, Hill confronted and ultimately breached a series of taboos in the black community that have survived both slavery and the post-segregation life she and Clarence Thomas share. Anita Hill put her private business in the street, and she downgraded a black man in a room filled with white men who might alter his fate—surely a large enough betrayal for her to be read out of the race.

By Sunday evening, Anita Hill's testimony lay buried under an avalanche of insinuation and innuendo. Before the eyes of a nation, a tenured law professor beloved by her students was transformed into an evil, opportunistic harpy; a deeply religious Baptist was turned into a sick and delusional woman possessed by Satan and in need of exorcism; this youngest of thirteen children from a loving family became a frustrated spinster longing for the attentions of her fast-track superior, bent on exacting a cruel revenge for his rejection.

The skillful transformations of Anita Hill's character by some members of the Senate were effective because they were familiar, manageable images of African-American womanhood. What undergirds these images is the common terror of black women out of control. We are the grasping and materialistic Sapphire in an "Amos 'n' Andy" episode; the embodiment of a shadowy, insane sexuality; the raging, furious, rejected woman. In their extremity, these are images far more accessible and understandable than the polished and gracious dignity, the cool intelligence that Anita Hill displayed in the lion's den of the Senate chamber. However she found herself reconstituted, the result was the same. She was, on all levels, simply unbelievable.

Anita Hill fell on the double-edged sword of African-American womanhood. Her privacy, her reputation, her integrity—all were casualties of an ignorance that left her unseen by and unknown to most of those who meant either to champion or abuse her. As credible, as inspiring, as impressive as she was, most people who saw her had no context in which to judge her. The signs and symbols that might have helped place Hill were long ago appropriated by officials of authentic (male) blackness, or by representatives of authentic (white) womanhood. Quite simply, a woman like Anita Hill couldn't possibly exist. . . .

. . . The issues of race and sex illuminated by the hearings remain. So, too, do the myriad ways in which race and gender combine to confuse us. But for the first time in decades, the country has been turned, for a time, into a mobile social laboratory. A level of discussion between previously unaligned groups may have begun with new vigor and candor.

Segments of the feminist movement have been under attack for their selective wooing of black women. Yet many of these same women rallied to Hill with impressive speed. Some black women who had never before considered sexism as an issue serious enough to merit collective concern have begun to organize, including a group of black female academics known as African American Women in Defense of Ourselves. And even in brusque New York, people on opposite sides of this issue, still traumatized by the televised spectacle, seem eager to listen, to be civil, to talk things over.

"I am so pleased people are starting to ask questions, not only about race and gender, but about the America that has frustrated and disappointed them," says Jewell Jackson McCabe. "People who had become cynical, people who have not talked about issues in their lives are talking now. I think the experience was so bad, it was so raw. I don't know a woman who watched those hearings whose life hasn't been changed."

"It was an international drama," says Michael Eric Dyson, assistant professor of ethics, philosophy, and cultural criticism at Chicago Theological Seminary. "Anita Hill has put these issues on the American social agenda. She has allowed black men and women to talk freely for the first time about a pain that has been at the heart of our relationships since slavery. Black wives are beginning to tell their husbands about the kind of sexism they have faced not at the hands of white men, but black men."

What was most striking about the hearings, in the end, was the sense of destiny that surrounded them. There was something rewarding about seeing what began as a humiliating event become gradually transformed only in its aftermath. Two African-Americans took center stage in what became a national referendum on many of our most cherished values. In the midst of their shattering questions, Anita Hill and Clarence Thomas each made us ask questions that most of us had lost the heart to ask.

They are exactly the kinds of questions that could lead us out of the morass of cynicism and anger in which we've all been stuck. That is an immensely satisfying measurement of the Hill-Thomas hearings. It would not be the first time that African-Americans have used tragedy and contradiction as catalysts to make America remember its rightful legacy.

MASCULINITIES AND ATHLETIC CAREERS

19

Michael Messner

The growth of women's studies and feminist gender studies has in recent years led to the emergence of a new men's studies (Brod 1987; Kimmel 1987). But just as feminist perspectives on women have been justifiably criticized for falsely universalizing the lives and issues of white, middle-class, U.S. women (Hooks 1984; Zinn, Cannon, Higginbotham, and Dill 1986), so, too, men's studies has tended to focus on the lives of relatively privileged men. As Brod (1983–1984) points out in an insightful critique of the middle-class basis and bias of the men's movement, if men's studies is to be relevant to minority and working-class men, less emphasis must be placed on personal lifestyle trans-formations, and more emphasis must be placed on developing a structural critique of social institutions. Although some institutional analysis has begun in men's studies, very little critical scrutiny has been focused on that very masculine institution, organized sports (Messner 1985; Sabo 1985; Sabo and Runfola 1980). Not only is the institution of sports an ideal place to study men and masculinity, careful analysis would make it impossible to ignore the realities of race and class differences.

In the early 1970s, Edwards (1971, 1973) debunked the myth that the predominance of blacks in sports to which they have access signaled an end to

Author's note: Parts of this article were presented as papers at the American Sociological Association Annual Meeting, Chicago, in August 1987, and at the North American Society for the Sociology of Sport Annual Meeting in Edmonton, Alberta, in November 1987. I thank Maxine Baca Zinn, Bob Blauner, Bob Dunn, Pierrette Hondagneu-Sotelo, Carol Jacklin, Michael Kimmel, Judith Lorber, Don Sabo, Barrie Thorne, and Carol Warren for constructive comments on earlier versions of this article.

From: *Gender & Society* 3 (March 1989): 71–88. Reprinted by permission of Sage Publications, Inc.

institutionalized racism. It is now widely accepted in sport sociology that social institutions such as the media, education, the economy, and (a more recent and controversial addition to the list) the black family itself all serve to systematically channel disproportionately large numbers of young black men into football, basketball, boxing, and baseball, where they are subsequently "stacked" into low-prestige and high-risk positions, exploited for their skills, and finally, when their bodies are used up, excreted from organized athletics at a young age with no transferable skills with which to compete in the labor market (Edwards 1984; Eitzen and Purdy 1986; Eitzen and Yetman 1977).

While there are racial differences in involvement in sports, class, age, and educational differences seem more significant. Rudman's (1986) initial analysis revealed profound differences between whites' and blacks' orientations to sports. Blacks were found to be more likely than whites to view sports favorably, to incorporate sports into their daily lives, and to be affected by the outcome of sporting events. However, when age, education, and social class were factored into the analysis, Rudman found that race did not explain whites' and blacks' different orientations. Black's affinity to sports is best explained by their tendency to be clustered disproportionately in lower-income groups.

The 1980s has ushered in what Wellman (1986, p. 43) calls a "new political linguistics of race," which emphasize cultural rather than structural causes (and solutions) to the problems faced by black communities. The advocates of the cultural perspective believe that the high value placed on sports by black communities has led to the development of unrealistic hopes in millions of black youths. They appeal to family and community to bolster other choices based upon a more rational assessment of "reality." Visible black role models in many other professions now exist, they say, and there is ample evidence which proves that sports careers are, at best, a bad gamble.

Critics of the cultural perspective have condemned it as conservative and victim blaming. But it can also be seen as a response to the view of black athletes as little more than unreflexive dupes of an all-powerful system, which ignores the importance of agency. Gruneau (1983) has argued that sports must be examined within a theory that views human beings as active subjects who are operating within historically constituted structural constraints. Gruneau's reflexive theory rejects the simplistic views of sports as either a realm of absolute oppression or an arena of absolute freedom and spontaneity. Instead, he argues, it is necessary to construct an understanding of how and why participants themselves actively make choices and construct and define meaning and a sense of identity within the institutions that they find themselves.

None of these perspectives considers the ways that gender shapes men's definitions of meaning and choices. Within the sociology of sport, gender as a process that interacts with race and class is usually ignored or taken for granted—except when it is *women* athletes who are being studied. Sociologists who are attempting to come to grips with the experiences of black men in general, and in organized sports in particular, have almost exclusively focused their analytic attention on the variable "black," while uncritically taking "men" as a given. Hare and Hare (1984), for example, view masculinity as a biologically determined tendency to act as a provider and protector that is thwarted for black men by socioeconomic and racist obstacles. Staples (1982) does view masculinity largely as a socially produced script, but he accepts this script as a given, preferring to focus on black men's blocked access to male role fulfillment. These perspectives on masculinity fail to show how the male role itself, as it interacts with a constricted structure of opportunity, can contribute to locking black men into destructive relationships and life-styles (Franklin 1984; Majors 1986).

This article will examine the relationships among male identity, race, and social class by listening to the voices of former athletes. I will first briefly describe my research. Then I will discuss the similarities and differences in the choices and experiences of men from different racial and social class backgrounds. Together, these choices and experiences help to construct what Connell (1987) calls "the gender order." Organized sports, it will be suggested, is a practice through which men's separation from and power over women is embodied and naturalized at the same time that hegemonic (white, heterosexual, professional-class) masculinity is clearly differentiated from marginalized and subordinated masculinities.

DESCRIPTION OF RESEARCH

Between 1983 and 1985, I conducted 30 open-ended, in-depth interviews with male former athletes. My purpose was to add a critical understanding of male gender identity to Levinson's (1978) conception of the "individual lifecourse"—specifically, to discover how masculinity develops and changes as a man interacts with the socially constructed world of organized sports. Most of the men I interviewed had played the U.S. "major sports"—football, basketball, baseball, track. At the time of the interview, each had been retired from playing organized sports for at least 5 years. Their ages ranged from 21 to 48, with the median, 33. Fourteen were black, 14 were white, and 2 were Hispanic. Fifteen of the 16 black and Hispanic men had come from poor or working-class families, while the majority (9 of 14) of the white men had come from middle-class or professional families. Twelve had played organized sports through high school, 11 through college, and 7 had been professional athletes. All had at some time in their lives based their identities largely on their roles as athletes and could therefore be said to have had athletic careers.

MALE IDENTITY AND ORGANIZED SPORTS

Earlier studies of masculinity and sport argued that sports socialize boys to be men (Lever 1976; Schafer 1975). Here, boys learn cultural values and behaviors, such as competition, toughness, and winning at all costs, that are culturally valued aspects of masculinity. While offering important insights, these early studies of masculinity and sports suffered from the limiting assumptions of a gender-role theory that seems to assume that boys come to their first athletic experience as blank slates onto which the values of masculinity are imprinted. This perspective oversimplifies a complex reality. In fact, young boys bring an already gendered identity to their first sports experiences, an identity that is struggling to work through the developmental task of individuation (Chodorow 1978; Gilligan 1982). Yet, as Benjamin (1988) has argued, individuation is accomplished, paradoxically, only through relationships with other people in the social world. So, although the major task of masculinity is the development of a "positional identity" that clarifies the boundaries between self and other, this separation must be accomplished through some form of connection with others. For the men in my study, the rule-bound structure of organized sports became a context in which they struggled to construct a masculine positional identity.

All of the men in this study described the emotional salience of their earliest experiences in sports in terms of relationships with other males. It was not winning and victories that seemed important at first; it was something "fun" to do with fathers, older brothers or uncles, and eventually with same-aged peers. As a man from a white, middle-class family said, "The most important thing was just being out there with the rest of the guys—being friends." A 32-year-old man from a poor Chicano family, whose mother had died when he was 9 years old, put it more succinctly:

> What I think sports did for me is it brought me into kind of an instant family. By being on a Little League team, or even just playing with kids in the neighborhood, it brought what I really wanted, which was some kind of closeness.

Though sports participation may have initially promised "some kind of closeness," by the ages of 9 or 10, the less skilled boys were already becoming alienated from—or weeded out of—the highly competitive and hierarchical system of organized sports. Those who did experience some early successes received recognition from adult males (especially fathers and older brothers) and held higher status among peers. As a result, they began to pour more and more of their energies into athletic participation. It was only after they learned that they would get recognition from other people for being a good athlete— indeed, that this attention was contingent upon *being a winner*—that performance and winning (the dominant values of organized sports) became extremely important. For some, this created pressures that served to lessen or eliminated the fun of athletic participation (Messner 1987a, 1987b).

While feminist psychoanalytic and developmental theories of masculinity are helpful in explaining boys' early attraction and motivations in organized sports, the imperatives of core gender identity do not fully determine the contours and directions of the life course. As Rubin (1985) and Levinson (1978) have pointed out, an understanding of the lives of men must take into account the processual nature of male identity as it unfolds through interaction between the internal (psychological ambivalences) and the external (social, historical, and institutional) contexts.

To examine the impact of the social contexts, I divided my sample into two comparison groups. In the first group were 10 men from higher-status backgrounds, primarily white, middle-class, and professional families. In the second group were 20 men from lower-status backgrounds, primarily minority, poor, and working-class families. While my data offered evidence for the similarity of experiences and motivations of men from poor backgrounds, independent of race, I also found anecdotal evidence of a racial dynamic that operates independently of social class. However, my sample was not large enough to separate race and class, and so I have combined them to make two status groups.

In discussing these two groups, I will focus mainly on the high school years. During this crucial period, the athletic role may become a master status for a young man, and he is beginning to make assessments and choices about his future. It is here that many young men make a major commitment to—or begin to back away from—athletic careers.

Men from Higher-Status Backgrounds

The boyhood dream of one day becoming a professional athlete—a dream shared by nearly all the men interviewed in this study—is rarely realized. The sports world is extremely hierarchical. The pyramid of sports careers narrows very rapidly as one climbs from high school, to college, to professional levels of competition (Edwards 1984; Harris and Eitzen 1978; Hill and Lowe 1978).

In fact, the chances of attaining professional status in sports are approximately 4/100,000 for a white man, 2/100,000 for a black man, and 3/100,000 for a Hispanic man in the United States (Leonard and Reyman 1988). For many young athletes, their dream ends early when coaches inform them that they are not big enough, strong enough, fast enough, or skilled enough to compete at the higher levels. But six of the higher-status men I interviewed did not wait for coaches to weed them out. They made conscious decisions in high school or in college to shift their attentions elsewhere—usually toward educational and career goals. Their decision not to pursue an athletic career appeared to them in retrospect to be a rational decision based on the growing knowledge of how very slim their chances were to be successful in the sports world. For instance, a 28-year-old white graduate student said:

> By junior high I started to realize that I was a good player—maybe even one of the best in my community—but I realized that there were all these people all over the country and how few will get to play pro sports. By high school, I still dreamed of being a pro—I was a serious athlete, I played hard—but I knew it wasn't heading anywhere. I wasn't going to play pro ball.

A 32-year-old white athletic director at a small private college had been a successful college baseball player. Despite considerable attention from professional scouts, he had decided to forgo a shot at a baseball career and to enter graduate school to pursue a teaching credential. As he explained this decision:

> At the time I think I saw baseball as pissing in the wind, really. I was married, I was 22 years old with a kid. I didn't want to spend 4 or 5 years in the minors with a family. And I could see I wasn't a superstar; so it wasn't really worth it. So I went to grad school. I thought that would be better for me.

Perhaps most striking was the story of a high school student body president and top-notch student who was also "Mr. Everything" in sports. He was named captain of his basketball, baseball, and football teams and achieved All-League honors in each sport. This young white man from a middle-class family received attention from the press and praise from his community and peers for his athletic accomplishments, as well as several offers of athletic scholarships from universities. But by the time he completed high school, he had already decided to quit playing organized sports. As he said:

> I think in my own mind I kind of downgraded the stardom thing. I thought that was small potatoes. And sure, that's nice in high school and all that, but on a broad scale, I didn't think it amounted to all that much. So I decided that my goal's to be a dentist, as soon as I can.

In his sophomore year of college, the basketball coach nearly persuaded him to go out for the team, but eventually he decided against it:

> I thought, so what if I can spend two years playing basketball? I'm not going to be a basketball player forever and I might jeopardize my chances of getting into dental school if I play.

He finished college in three years, completed dental school, and now, in his mid-30s, is again the epitome of the successful American man: a professional with a family, a home, and a membership in the local country club.

How and why do so many successful male athletes from higher-status backgrounds come to view sports careers as "pissing in the wind," or as "small potatoes"? How and why do they make this early assessment and choice to shift from sports and toward educational and professional goals? The white, middle-class institutional context, with its emphasis on education and income, makes it clear to them that choices exist and that the pursuit of an athletic career is not a particularly good choice to make. Where the young male once found sports to be a convenient institution within which to construct masculine status, the postadolescent and young adult man from a higher-status background simply *transfers* these same strivings to other institutional contexts: education and careers.

For the higher-status men who had chosen to shift from athletic careers, sports remained important on two levels. First, having been a successful high school or college athlete enhances one's adult status among other men in the community —but only as a badge of masculinity that is *added* to his professional status. In fact, several men in professions chose to be interviewed in their offices, where they publicly displayed the trophies and plaques that attested to their earlier athletic accomplishments. Their high school and college athletic careers may have appeared to them as "small potatoes," but many successful men speak of their earlier status as athletes as having "opened doors" for them in their present professions and in community affairs. Similarly, Farr's (1988) research on "Good Old Boys Sociability Groups" shows how sports, as part of the glue of masculine culture, continues to facilitate "dominance bonding" among privileged men long after active sports careers end. The college-educated, career-successful men in Farr's study rarely express overtly sexist, racist, or classist attitudes; in fact, in their relationships with women, they "often engage in expressive intimacies" and "make fun of exaggerated 'machismo' " (p. 276). But though they outwardly conform more to what Pleck (1982) calls "the modern male role," their informal relationships within their sociability groups, in effect, affirm their own gender and class status by constructing and clarifying the boundaries between themselves and women and lower-status men. This dominance bonding is based largely upon ritual forms of sociability (camaraderie, competition), "the superiority of which was first affirmed in the exclusionary play activities of young boys in groups" (Farr 1988, p. 265).

In addition to contributing to dominance bonding among higher-status adult men, sports remains salient in terms of the ideology of gender relations. Most men continued to watch, talk about, and identify with sports long after their own disengagement from athletic careers. Sports as a mediated spectacle provides an important context in which traditional conceptions of masculine superiority—conceptions recently contested by women—are shored up. As a 32-year-old white professional-class man said of one of the most feared professional football players today:

> A woman can do the same job as I can do—maybe even be my boss. But I'll be *damned* if she can go out on the football field and take a hit from Ronnie Lott.

Violent sports as spectacle provide linkages among men in the project of the domination of women, while at the same time helping to construct and clarify differences among various masculinities. The statement above is a clear identification with Ronnie Lott *as a man*, and the basis of the identification is the violent male body. As Connell (1987, p. 85) argues, sports is an important organizing institution for the embodiment of masculinity. Here, men's power over women becomes naturalized and linked to the social distribution of violence. Sports, as a practice, suppresses natural (sex) similarities, constructs differences, and then, largely through the media, weaves a structure of symbol and interpretation around these differences that naturalizes them (Hargreaves 1986, p. 112). It is also significant that the man who made the above statement about Ronnie Lott was quite aware that he (and perhaps 99 percent of the rest of the U.S. male population) was probably as incapable as most women of taking a "hit" from someone like Lott and living to tell of it. For middle-class men, the "tough guys" of the culture industry—the Rambos, the Ronnie Lotts who are fearsome "hitters," who "play hurt"—are the heroes who "prove" that "we men" are superior to women. At the same time, they play the role of the "primitive other," against whom higher-status men define themselves as "modern" and "civilized."

Sports, then, is important from boyhood through adulthood for men from higher-status backgrounds. But it is significant that by adolescence and early adulthood, most of these young men have concluded that sports *careers* are not for them. Their middle-class cultural environment encourages them to decide to shift their masculine strivings in more "rational" directions: education and nonsports careers. Yet their previous sports participation continues to be very important to them in terms of constructing and validating their status within privileged male peer groups and within their chosen professional careers. And organized sports, as a public spectacle, is a crucial locus around which ideologies of male superiority over women, as well as higher-status men's superiority over lower-status men, are constructed and naturalized.

Men from Lower-Status Backgrounds

For the lower-status young men in this study, success in sports was not an added proof of masculinity; it was often their only hope of achieving public masculine status. A 34-year-old black bus driver who had been a star athlete in three sports in high school had neither the grades nor the money to attend college, so he accepted an offer from the U.S. Marine Corps to play on their baseball team. He ended up in Vietnam, where a grenade blew four fingers off his pitching hand. In retrospect, he believed that his youthful focus on sports stardom and his concomitant lack of effort in academics made sense:

> You can go anywhere with athletics—you don't have to have brains. I mean, I didn't feel like I was gonna go out there and be a computer expert, or something that was gonna make a lot of money. The only thing I could do and live comfortably would be to play sports—just to get a contract—doesn't matter if you play second or third team in the pros, you're gonna make big bucks. That's all I wanted, a confirmed livelihood at the end of my ventures, and the only way I could do it would be through sports. So I tried. It failed, but that's what I tried.

Similar, and even more tragic, is the story of a 34-year-old black man who is now serving a life term in prison. After a career-ending knee injury at the age of 20 abruptly ended what had appeared to be a certain road to professional football fame and fortune, he decided that he "could still be rich and famous" by robbing a bank. During his high school and college years, he said, he was nearly illiterate:

> I'd hardly ever go to classes and they'd give me *Cs*. My coaches taught some of the classes. And I felt, "So what? They *owe* me that! I'm an *athlete!* I thought that was what I was born to do—to play sports—and everybody understood that.

Are lower-status boys and young men simply duped into putting all their eggs into one basket? My research suggested that there was more than "hope for the future" operating here. There were also immediate psychological reasons that they chose to pursue athletic careers. By the high school years, class and ethnic inequalities had become glaringly obvious, especially for those who attended socioeconomically heterogeneous schools. Cars, nice clothes, and other signs of status were often unavailable to these young men, and this contributed to a situation in which sports took on an expanded importance for them in terms of constructing masculine identities and status. A white, 36-year-old man from a poor, single-parent family who later played professional baseball had been acutely aware of his low-class status in his high school:

> I had one pair of jeans, and I wore them every day. I was always afraid of what people thought of me—that this guy doesn't have anything, that he's wearing the same Levi's all the time, he's having to work in the cafeteria for his lunch. What's going on? I think that's what made me so shy. . . . But boy, when I got into sports, I let it all hang out—[laughs]—and maybe that's why I became so good, because I was frustrated, and when I got into that element, they gave me my uniform in football, basketball, and baseball, and I didn't have to worry about how I looked, because then it was *me* who was coming out, and not my clothes or whatever. And I think that was the drive.

Similarly, a 41-year-old black man who had a 10-year professional football career described his insecurities as one of the few poor blacks in a mostly white, middle-class school and his belief that sports was the one arena in which he could be judged solely on his merit:

> I came from a poor family, and I was very sensitive about that in those days. When people would say things like "Look at him—he has dirty pants on," I'd think about it for a week. [But] I'd put my pants on and I'd go out on the football field with the intention that I'm gonna do a job. And if that calls on me to hurt you, I'm gonna do it. It's as simple as that. I demand respect just like everybody else.

"Respect" was what I heard over and over when talking with the men from lower-status backgrounds, especially black men. I interpret this type of respect to be a crystallization of the masculine quest for recognition through public achievement, unfolding within a system of structured constraints due to class and race inequities. The institutional context of education (sometimes with the collusion of teachers and coaches) and the constricted structure of opportunity in the economy made the pursuit of athletic careers appear to be the most rational choice to these young men.

The same is not true of young lower-status women. Dunkle (1985) points out that from junior high school through adulthood, young black men are far more likely to place high value on sports than are young black women, who are more likely to value academic achievement. There appears to be a gender dynamic operating in adolescent male peer groups that contributes toward their valuing sports more highly than education. Franklin (1986, p. 161) has argued that many of the normative values of the black male peer group (little respect for nonaggressive solutions to disputes, contempt for nonmaterial culture) contribute to the constriction of black men's views of desirable social positions, especially through education. In my study, a 42-year-old black man who did succeed in beating the odds by using his athletic scholarship to get a college degree and eventually becoming a successful professional said:

> By junior high, you either got identified as an athlete, a thug, or a book-worm. It's very important to be seen as somebody who's capable in some area. And you *don't* want to be identified as a bookworm. I was very good with books, but I was kind of covert about it. I was a closet bookworm. But with sports, I was *somebody;* so I worked very hard at it.

For most young men from lower-status backgrounds, the poor quality of their schools, the attitudes of teachers and coaches, as well as the antieducation environment within their own male peer groups, made it extremely unlikely that they would be able to succeed as students. Sports, therefore, became *the* arena in which they attempted to "show their stuff." For these lower-status men, as Baca Zinn (1982) and Majors (1986) argued in their respective studies of chicano men and black men, when institutional resources that signify masculine status and control are absent, physical presence, personal style, and expressiveness take on increased importance. What Majors (1986, p. 6) calls "cool pose" is black men's expressive, often aggressive, assertion of masculinity. This self-assertion often takes place within a social context in which the young man is quite aware of existing social inequities. As the black bus driver, referred to above, said of his high school years:

> See, the rich people use their money to do what they want to do. I use my ability. If you wanted to be around me, if you wanted to learn something about sports, I'd teach you. But you're gonna take me to lunch. You're gonna let me use your car. See what I'm saying? In high school I'd go where I wanted to go. I didn't have to be educated. I was well-respected. I'd go some-where, and they'd say, "Hey, that's Mitch Harris,[1] yeah, that's a bad son of a bitch!'

Majors (1986) argues that although "cool pose" represents a creative survival technique within a hostile environment, the most likely long-term effect of this masculine posturing is educational and occupational dead ends. As a result, we can conclude, lower-status men's personal and peer-group responses to a constricted structure of opportunity—responses that are rooted, in part, in the developmental insecurities and ambivalences of masculinity—serve to lock many of these young men into limiting activities such as sports.

SUMMARY AND CONCLUSIONS

This research has suggested that within a social context that is stratified by social class and by race, the choice to pursue—or not to pursue—an athletic career is explicable as an individual's rational assessment of the available means to achieve a respected masculine identity. For nearly all of the men from lower-status backgrounds, the status and respect that they received through sports was temporary—it did not translate into upward mobility. Nonetheless, a strategy of discouraging young black boys and men from involvement in sports is probably doomed to fail, since it ignores the continued existence of structural constraints. Despite the increased number of black role models in nonsports professions, employment opportunities for young black males have actually deteriorated in the 1980s (Wilson and Neckerman 1986), and nonathletic opportunities in higher education have also declined. While blacks constitute 14 percent of the college-aged (18–24 years) U.S. population, as a proportion of students in four-year colleges and universities, they have dropped to 8 percent. In contrast, by 1985, black men constituted 49 percent of all college basketball players and 61 percent of basketball players in institutions that grant athletic scholarships (Berghorn et al., 1988). For young black men, then, organized sports appears to be more likely to get them to college than their own efforts in nonathletic activities.

But it would be a mistake to conclude that we simply need to breed socioeconomic conditions that make it possible for poor and minority men to mimic the "rational choices" of white, middle-class men. If we are to build an appropriate understanding of the lives of all men, we must critically analyze white middle-class masculinity, rather than uncritically taking it as a normative standard. To fail to do this would be to ignore the ways in which organized sports serves to construct and legitimate gender differences and inequalities among men and women.

Feminist scholars have demonstrated that organized sports gives men from all backgrounds a means of status enhancement that is not available to young women. Sports thus serve the interests of all men in helping to construct and legitimize their control of public life and their domination of women (Bryson 1987; Hall 1987; Theberge 1987). Yet concrete studies are suggesting that men's experiences within sports are not all of a piece. Brian Pronger's (1990) research suggests that gay men approach sports differently than straight men do, with a sense of "irony." And my research suggests that although sports are important for men from both higher- and lower-status backgrounds, there are crucial differences. In fact, it appears that the meaning that most men give to their athletic strivings has more to do with competing for status among men than it has to do with proving superiority over women. How can we explain this seeming contradiction between the feminist claim that sports links all men in the domination of women and the research findings that different groups of men relate to sports in very different ways?

The answer to this question lies in developing a means of conceptualizing the interrelationships between varying forms of domination and subordination. Marxist scholars of sports often falsely collapse everything into a class analysis; radical feminists often see gender domination as universally fundamental. Concrete examinations of sports, however, reveal complex and multilayered systems of inequality: Racial, class, gender, sexual preference, and age dynamics are all salient features of the athletic context. In examining this reality, Connell's (1987) concept of the "gender order" is useful. The gender order is a dynamic process that is constantly in a state of play. Moving beyond static gender-role theory and reductionist concepts of patriarchy that view men as an undifferentiated group which oppresses women, Connell argues that at any given historical moment, there are competing masculinities—some hegemonic, some marginalized, some stigmatized. Hegemonic masculinity (that definition of masculinity which is culturally ascendant) is constructed in relation to various subordinated masculinities as well as in relation to femininities. The project of male domination of women may tie all men together, but men share very unequally in the fruits of this domination.

These are key insights in examining the contemporary meaning of sports. Utilizing the concept of the gender order, we can begin to conceptualize how hierarchies of race, class, age, and sexual preference among men help to construct and legitimize men's overall power and privilege over women. And how, for some black, working-class, or gay men, the false promise of sharing in the fruits of hegemonic masculinity often ties them into their marginalized and subordinate statuses within hierarchies of intermale dominance. For instance, black men's development of what Majors (1986) calls "cool pose" within sports can be interpreted as an example of creative resistance to one form of social domination (racism); yet it also demonstrates the limits of an agency that adopts other forms of social domination (masculinity) as its vehicle. As Majors (1990) points out:

> Cool Pose demonstrates black males' potential to transcend oppressive conditions in order to express themselves *as men*. [Yet] it ultimately does not

put black males in a position to live and work in more egalitarian ways with women, nor does it directly challenge male hierarchies.

Indeed, as Connell's (1990) analysis of an Australian "Iron Man" shows, the commercially successful, publicly acclaimed athlete may embody all that is valued in present cultural conceptions of hegemonic masculinity—physical strength, commercial success, supposed heterosexual virility. Yet higher-status men, while they admire the public image of the successful athlete, may also look down on him as a narrow, even atavistic, example of masculinity. For these higher-status men, their earlier sports successes are often status enhancing and serve to link them with other men in ways that continue to exclude women. Their decisions not to pursue athletic careers are equally important signs of their status vis-à-vis other men. Future examinations of the contemporary meaning and importance of sports to men might take as a fruitful point of departure that athletic participation, and sports as public spectacle serve to provide linkages among men in the project of the domination of women, while at the same time helping to construct and clarify differences and hierarchies among various masculinities.

NOTE

1. "Mitch Harris" is a pseudonym.

REFERENCES

Benjamin, J. 1988. *The Bonds of Love: Psychoanalysis, Feminism, and the Problem of Domination*. New York: Pantheon.

Berghorn, F. J. et al. 1988. "Racial Participation in Men's and Women's Intercollegiate Basketball: Continuity and Change, 1958–1985." *Sociology of Sport Journal* 5:107–24.

Brod, H. 1983–84. "Work Clothes and Leisure Suits: The Class Basis and Bias of the Men's Movement." *M: Gentle Men for Gender Justice* 11:10–12, 38–40.

Brod, H. (ed.). 1987. *The Making of Masculinities: The New Men's Studies*. Winchester, MA: Allen & Unwin.

Bryson, L. 1987. "Sport and the Maintenance of Masculine Hegemony." *Women's Studies International Forum* 10:349–60.

Chodorow, N. 1978. *The Reproduction of Mothering.* Berkeley: University of California Press.

Connell, R. W. 1987. *Gender and Power.* Stanford, CA: Stanford University Press.

———. 1990. "An Iron Man: The Body and Some Contradictions of Hegemonic Masculinity." In *Sport, Men, and the Gender Order: Critical Feminist Perspectives,* edited by M. A. Messner and D. S. Sabo. Champaign, IL: Human Kinetics.

Dunkle, M. 1985. "Minority and Low-Income Girls and Young Women in Athletics." *Equal Play* 5(Spring-Summer):12–13.

Duquin, M. 1984 "Power and Authority: Moral Consensus and Conformity in Sport." *International Review for Sociology of Sport* 19:295–304.

Edwards, H. 1971. "The Myth of the Racially Superior Athlete." *The Black Scholar* 3(November).

———. 1973. *The Sociology of Sport.* Homewood, IL: Dorsey.

———. 1984. "The Collegiate Athletic Arms Race: Origins and Implications of the 'Rule 48' Controversy." *Journal of Sport and Social Issues* 8:4–22.

Eitzen, D. S. and D. A. Purdy, 1986. "The Academic Preparation and Achievement of Black and White College Athletes." *Journal of Sport and Social Issues* 10:15–29.

Eitzen, D. S. and N. B. Yetman. 1977. "Immune From Racism?" *Civil Rights Digest* 9:3–13.

Farr, K. A. 1988. "Dominance Bonding Through the Good Old Boys Sociability Group." *Sex Roles* 18:259–77.

Franklin, C. W. II. 1984. *The Changing Definition of Masculinity.* New York: Plenum.

———. 1986. "Surviving the Institutional Decimation of Black Males: Causes, Consequences, and Intervention." Pp. 155–70 in *The Making of Masculinities: The New Men's Studies,* edited by H. Brod. Winchester, MA: Allen & Unwin.

Gilligan, C. 1982. *In a Different Voice: Psychological Theory and Women's Development.* Cambridge, MA: Harvard University Press.

Gruneau, R. 1983. *Class, Sports, and Social Development.* Amherst: University of Massachusetts Press.

Hall, M. A. (ed.). 1987. "The Gendering of Sport, Leisure, and Physical Education." *Women's Studies International Forum* 10:361–474.

Hare, N. and J. Hare. 1984. *The Endangered Black Family: Coping With the Unisexualization and Coming Extinction of the Black Race.* San Francisco, CA: Black Think Tank.

Hargreaves, J. A. 1986. "Where's the Virtue? Where's the Grace? A Discussion of the Social Production of Gender Through Sport." *Theory, Culture and Society* 3:109–21.

Harris, D. S. and D. S. Eitzen. 1978. "The Consequences of Failure in Sport." *Urban Life* 7:177–88.

Hill, P. and B. Lowe. 1978. "The Inevitable Metathesis of the Retiring Athlete." *International Review of Sport Sociology* 9:5–29.

Hooks, B. 1984. *Feminist Theory: From Margin to Center.* Boston: South End Press.

Kimmel, M. S. (ed.). 1987. *Changing Men: New Directions in Research on Men and Masculinity.* Newbury Park, CA: Sage.

Leonard, W. M. II and J. M. Reyman. 1988. "The Odds of Attaining Professional Athlete Status: Refining the Computations." *Sociology of Sport Journal* 5:162–69.

Lever, J. 1976. "Sex Differences in the Games Children Play." *Social Problems* 23:478–87.

Levinson, D. J. 1978. *The Seasons of a Man's Life.* New York: Ballantine.

Majors, R. 1986. "Cool Pose: The Proud Signature of Black Survival." *Changing Men: Issues in Gender, Sex, and Politics* 17:5–6.

———. 1990. "Cool Pose: Black Masculinity in Sports." In *Sport, Men, and the Gender Order: Critical Feminist Perspectives,* edited by M. A. Messner and D. S. Sabo. Champaign, IL: Human Kinetics.

Messner, M. 1985. "The Changing Meaning of Male Identity in the Lifecourse of the Athlete." *Arena Review* 9:31–60.

———. 1987a. "The Meaning of Success: The Athletic Experience and the Development of Male Identity." Pp. 193–209 in *The Making of Masculinities: The New Men's Studies,* edited by H. Brod. Winchester, MA: Allen & Unwin.

———. 1987b. "The Life of a Man's Seasons: Male Identity in the Lifecourse of the Athlete." Pp. 53–67 in *Changing Men: New Directions in Research on Men and Masculinity,* edited by M. S. Kimmel. Newbury Park, CA: Sage.

Pleck, J. H. 1982. *The Myth of Masculinity.* Cambridge: MIT Press.

Pronger, B. 1990. "Gay Jocks: A Phenomenology of Gay Men in Athletics." In *Sport, Men, and the Gender Order: Critical Feminist Perspectives,* edited by M. A. Messner and D. S. Sabo. Champaign, IL: Human Kinetics.

Rubin, L. B. 1985. *Just Friends: The Role of Friendship in Our Lives.* New York: Harper & Row.

Rudman, W. J. 1986. "The Sport Mystique in Black Culture." *Sociology of Sport Journal* 3:305–19.

Sabo, D. 1985. "Sport, Patriarchy, and Male Identity: New Questions About Men and Sport." *Arena Review* 9:1–30.

Sabo, D. and R. Runfola (eds.). 1980. *Jocks: Sports and Male Identity.* Englewood Cliffs, NJ: Prentice-Hall.

Schafer, W. E. 1975. "Sport and Male Sex Role Socialization." *Sport Sociology Bulletin* 4:17–54.

Staples, R. 1982. *Black Masculinity.* San Francisco, CA: Black Scholar Press.

Theberge, N. 1987. "Sport and Women's Empowerment." *Women's Studies International Forum* 10:387–93.

Wellman, D. 1986. "The New Political Linguistics of Race." *Socialist Review* 87/88:43–62.

Wilson, W. J. and K. M. Neckerman. 1986. "Poverty and Family Structure: The Widening Gap Between Evidence and Public Policy Issues." Pp. 232–59 in *Fighting Poverty*, edited by S. H. Danzinger and D. H. Weinberg. Cambridge, MA: Harvard University Press.

Zinn, M. Baca. 1982. "Chicano Men and Masculinity." *Journal of Ethnic Studies* 10:29–44.

Zinn, M. Baca, L. Weber Cannon, E. Higginbotham, and B. Thornton Dill. 1986. "The Costs of Exclusionary Practices in Women's Studies." *Signs: Journal of Women in Culture and Society* 11:290–303.

ASIAN AMERICAN WOMEN: NOT FOR SALE

20

Tracy Lai

Asian American women are not for sale. We will not be bought off, materially or otherwise. We say this to the white men who use the mail-order bride

From: Jo Whitehorse Cochran, Donna Langston, and Carolyn Woodward (eds.), *Changing Our Power: An Introduction to Women's Studies* (Dubuque, Iowa: Kendall-Hunt, 1988), pp. 120–127. Reprinted by permission.

catalogues, hoping to buy an obedient Asian wife. But we also say this to anyone who participates in the long history of stereotyping Asian peoples and cultures as inferior and exotic. Stereotypes dehumanize people and turn them into objects to be manipulated.

It is a struggle to be an Asian in America. Historically, Asians have been denied political, economic and social equality in America. The very term "Asian American" carries a political assertion that Asians have been and continue to be a legitimate and integral part of American society. We are not forever foreign with an identity attached only to an Asian country. The Asian American movement has its origins in the social uprising of the 1960s and '70s, inspired by the black liberation movement. Chinese Americans, Japanese Americans, Pilipino Americans and Korean Americans united in recognition of a common history of oppression and common goals of community empowerment and pride. Asian American women played a leading role, drawing from the strength of their Vietnamese sisters in the National Liberation Front. The racist war abroad in Vietnam had its counterpart back home. Asian American communities fought the enemy at home, raising such issues as bilingual-bicultural education, ethnic studies, and low-income housing. Today, that quest for equality and political power continues. And Asian American women continue to be an important part of that struggle. The urgency of this struggle is reflected in the rising incidence of anti-Asian violence. *Pacific Citizen*, newspaper of the Japanese American Citizens League, reported the tragic death in February, 1984, of Ly Yung Cheung, a seamstress in New York's Chinatown. She was pushed into the path of an oncoming subway train and was decapitated. Her attacker, John Cardinale, reportedly shouted, "we're even" and later based his defense on a "psychotic phobia about Orientals."

What makes this struggle even more complicated is the die-hard myth that Asians have made it. We are told we have overcome our oppression, and that therefore we are the model minority. *Model* refers to the cherished dictum of capitalism that "pulling hard on your bootstraps" brings due rewards. The lesson drawn is that if you work hard enough, you will succeed—and if you don't succeed, you must not be working hard enough. High-profile Asian American women such as Connie Chung, a national television newscaster with a six-figure salary, are promoted as examples of Asian American success stories. But such examples, while certainly remarkable, do little to illuminate the actual conditions of the majority of Asian Americans. Such examples conceal the more typical Asian American experience of unemployment, underemployment and struggle to survive.

The model minority myth thus classically scapegoats Asian Americans. It labels us in a way that dismisses the real problems that many do face, while at the same time pitting Asians against other oppressed people of color. The fact that Asian Americans lack political representation, power and community control conveniently disappears. Model minority labelling has also given rise to an insidious cultural hierarchy. The concept of cultural deprivation implies that black, Latino, and American Indian cultures lack the cultural reinforcements that lead to successful achievement in education and career advancement, and hence to higher socioeconomic levels. For example, Asians are claimed to value education more than other minorities and to have special intellectual affinities for math and science. In fact, this is a racist rationale implying the intellectual inferiority of other minorities, while ignoring important historical and class differences in backgrounds. The cultural deprivation and model minority analyses also fail to examine capitalism as an economic system that thrives on exploitation by class, sex and race. The lack of success of other minority groups and women of all races is deliberate and necessary under capitalism. Profits for capitalism come from the low wages justified in sexist and racist terms.

A MULTIFACETED COMMUNITY

The forces shaping the experience of Asian American women come out of the historical policies of the U.S. toward Asians. These policies were largely aimed at using and discarding Asians as a temporary, cheap labor pool. Thus, young Asian men were desirable while Asian women were not, especially since women would bear children who could legally claim citizenship rights in the United States. Asian Americans are united by our oppression. The racist claim that all Asians look alike could more accurately be stated, "treat all Asians as if they were alike." We are denied respect for our cultural identities, and in the American eye, Asia and the Pacific merge into a single cultural entity. No matter how many generations we have lived in America, we are assumed to be foreigners. The 1980 Census counted more than 20 separate Asian and Pacific nationalities in the United States. The total number of Asian Americans is approximately 3.7 million, or 1.6 percent of the population, mostly concentrated in the western United States. The largest nationality groups, in decreasing order, are Chinese, Pilipino, Japanese, Asian Indian, Korean, and Vietnamese. Others include Hmong, Laotian, (native) Hawaiian, Samoan, Guamanian, Marshallese, and Micronesian. Each of these nationalities has a unique history and culture. However, common to all has been the history of U.S. intervention. Beginning in the nineteenth century, the U.S. followed Britain's example, forcing China and then Japan to open their doors to trade and immigration. Eventually, the U.S. sought to dominate the Pacific and Asia for trade and military purposes. As a consequence, the U.S. annexed, occupied or otherwise dominated Hawaii, the Micronesian Islands, the Philippines, South Korea, Vietnam, and Kampuchea.

Immigration of Asian and Pacific people has been heavily shaped by the nature of U.S. involvement in their countries. Today, the highest rate of immigration for Asians is from South Korea and the Philippines, two countries that are heavily dominated by U.S. capital and are occupied by strategic U.S. military bases. According to the *Seattle Times* (November 15, 1983), the U.S. maintains 38,882 military personnel in South Korea, 15,123 in the Philippines, 48,496 in Japan, 8,959 in Guam, and 23,214 at sea near Asia. The U.S. military bases have a devastating impact on the local Asian women, one that spills over to Asian American women as American military men bring back stereotypes and expectations of conquest. For instance, U.S. bases generate a huge prostitution business, reinforcing the stereotype that Asian women seek fulfillment through serving men.

Historically, the U.S. has also maintained discriminatory immigration policies explicitly aimed at controlling and eliminating the Asian population in the United States. The first wave of Chinese immigrants arrived in the 1850s, but by 1882, an organized movement of trade unions and opportunistic politicians secured the Chinese Exclusion Act which prohibited immigration of Chinese laborers and their wives. Thus, families were curtailed, wives effectively abandoned in China, and an entire generation of Chinese "bachelors" were trapped in the United States. Violent killings and expulsions followed in the wake of this legislation, such as the 1885–86 expulsions of the Chinese communities from Tacoma and Seattle, Washington. The Chinese were wanted for their labor but they were considered undesirable and unassimilable for settlement. The Chinese and, later, other Asians were accused of stealing white workers' jobs and lowering the standard of living. In fact, Asians performed work that white workers refused to do, and Asians had to be recruited to fill the labor shortage of the westward expansion. Their labor laid the foundation for the industrial and agricultural wealth of the West today.

Japanese, Korean and Pilipino workers followed in successive waves, filling one another's footsteps in low-paying, low-status jobs. In turn, each faced similarly hostile accusations and blame for downturns in the economy. Recruited for the sugar plantations in the 1800s, the Japanese immigrated to Hawaii, and subsequently to the mainland. The anti-Chinese movement reorganized itself as the Asiatic Exclusion League [i]n 1905, dedicated to the preservation of the Caucasian race upon American soil. In 1924 the League successfully passed the National Origins Act which barred immigration of all Asians. Only the Pilipinos remained problematic. Pilipinos were considered nationals, since the U.S. had seized the Philippines during the Spanish-American War of 1898. As nationals, they were exempt from exclusionary immigration laws. After 1924, Pilipinos were a primary source of labor and, as their numbers increased, so did white hostility. In 1929 and 1930, anti-Pilipino riots erupted all along the west coast. The Tydings-McDuffie Act of 1934 ostensibly granted independence to the Philippines but its real purpose was to limit the immigration quota of Pilipinos to 50 per year.

In many ways, these exclusionary laws were aimed at Asian women to prevent the development of families and communities. By keeping Asian women from immigrating, and by passing anti-miscegenation laws, it was hoped that the largely male Asian labor force would eventually die out. A common theme in the anti-Asian propaganda threatened destruction of America through an invasion of the "yellow hordes" or the "yellow peril." Asian American women were described as breeding like rats. This stereotype continues into the present in the form of the "Oriental Beauty" who has extraordinary sexual powers. To raise families and to build communities under such conditions become acts of resistance.

A more recent example of U.S. foreign policy impacting Asian Americans is the resettlement in the United States of 760,854 Southeast Asians between 1975–1985. Over half are Vietnamese; the rest are nearly evenly divided between Laotians and Kampucheans. They are all refugees created by the U.S. imperialist war in Southeast Asia, which raged from the 1950s to 1975. These newer Asian Americans have borne the brunt of the anti-Asian violence, as well as a specifically anti-refugee/anti-Southeast Asian hostility. Refugee women are especially vulnerable to attacks such as robbery, rape and intimidation. The perpetrators seem to believe that these attacks are acceptable, even deserved, because the U.S. lost the war in Vietnam, and war can now be made on its victims/refugees. Refugee women are perceived as likely targets, easier to physically overpower and intimidate. Because refugees are unfamiliar with American behavior, language and laws, it is difficult for them to fight back. At the same time, the pressure to assimilate falls heavily on the women, even as the community's traditional values and roles are fundamentally undermined and their children become strangers to them.

Historically, the tendency has been to reject and exclude Asians from participating fully in American society. Besides the exclusionary immigration laws, in many states, Asians could not be naturalized, vote, own property, or marry persons of other races. Their lives and jobs were restricted in every way. Although most of the discriminatory laws have eventually changed, deep-rooted stereotypes and hostility towards Asians have not. Throughout there has been a clear pattern of violence used to intimidate and eliminate Asians. Asians, like other minorities, have been expendable: their lives have not been valued as highly as white lives. Burned into Asian American consciousness is the violent uprooting of more than 110,000 Japanese who were interned in

American concentration camps during World War II. Western Defense Commander General John L. DeWitt declared that "the Japanese race is an enemy race," thus providing the racist rationale that all Japanese on the west coast must be locked up as a military necessity (see notes). They were marched away at gunpoint and imprisoned behind barbed wire, arbitrarily and illegally stripped of all civil rights. They were stripped of their dignity and pride as a people and of their dreams. Most lost their businesses, homes and possessions, an estimated value of as much as $400 million (Weglyn, p. 276). Japanese American women fiercely resisted these attacks on family and community integrity, and the demoralization and shame. They turned the prisons into homes and transformed the anger into the will to survive. Today, the camps remain a warning that the old anti-Asian hysteria could strike at any time; you are not safe behind the yellow face.

ASIAN AMERICAN WOMEN: DANGERS WITHIN AND WITHOUT

We are triply oppressed: as Asian Americans, as Asian American women, and as Asian American women workers. Racism has been and continues to be a primary force in shaping our oppression as women and as workers. The model minority stereotype has ominous overtones when applied to Asian American women. Asian American women are described as being desirable because they are cute (as in doll-like), quiet rather than militant, and unassuming rather than assertive. In a word, non-threatening. This is the image being sold as a commodity on the front page of the *Wall Street Journal*, January 25, 1984, headlined: "American Men Find Asian Brides Fill the Unliberated Bill— Mailorder Firms Help Them Look for the Ideal Women They Didn't Find at Home."

There are about 50 mailorder bride services in the U.S., carrying names such as "Cherry Blossom" and "Love Overseas." For a fee, men receive photo catalogs of Asian women, primarily from poor families in Malaysia and the Philippines. Descriptions of the women include statements such as "They love to do things to make their husbands happy," and "Most, if not all, are very feminine, loyal, loving—and virgins!" (as quoted in *Pacific Citizen*). The men who use these services are often middle-aged or older, disillusioned and divorced. Many have served in the U.S. military overseas in Asia. They blame previous marriage failures on the women's liberation movement. The mail order services claim to sell unliberated women who will supposedly be satisfied homemakers and be subservient to their husbands. The reality is that this market is a by-product of the U.S. military and economic domination in Asia.

The mail order phenomenon is a threat not only to Asian and Asian American women, but to all women. It promotes a degrading view of women's roles and the acceptability of selling women.

The stereotypes also pay off in the form of inflated profits extracted by superexploiting Asian American women workers. Businesses want docile, subservient workers who will not complain, file grievances, or organize unions. Many businesses purposely seek immigrant workers with limited English skills as further insurance against backtalk. Asian American women are also stereotyped as having special dexterity and endurance for routine, thus making them fit for assembly work of various types. They are thought to be "loyal, diligent and attentive to detail," again good qualities for subordinates, but certainly not for supervisors. Asian American women continue to be hired mainly in low-profile, low-status, low-paying occupations, such as clerical and service work. Asian American families tend to have multiple wage earners, to support larger, often extended, families, and to live in urban areas with a relatively high cost of living. These factors skew the Asian American family median income upwards. This higher figure has been used to suggest wrongly that Asian American families are as or more successful than white families. Comparisons of median income levels within specific cities (instead of using national averages) reveal that Asian Americans, like other minorities, consistently earn less than whites.

The force of history and the conditions of our lives demand many battles, but Asian American women must also simultaneously wage an inner struggle with feudal and religious cultural baggage. In Asian cultures that have been heavily influenced by Confucianism, women are regarded as secondary to men, existing for their service. The Spanish imposed Catholicism on the Philippines with similar results. While the Asian experience in America has modified some of these ideas, every wave of immigration tends to revive the old cultures. Asian American women must be able to reject negative traditions without feeling like they are rejecting their whole Asian American identity. Assimilation appears to demand this same rejection, but it is from the standpoint of shame and self-hatred. No culture is static, and this struggle to consciously develop and redefine the best in Asian American culture is based on pride and love of our people.

Writers Merle Woo and Kitty Tsui have eloquently articulated this many-fronted struggle, becoming a voice for other Asian American lesbians. As lesbians, they have faced a painful rejection from the Asian American community. In "Letter to Ma," Merle directly addresses homophobia in the community: "If my reaction to being a Yellow Woman is different than yours was, please know that that is not a judgment on you, a criticism or denial of you, your worth" (*Bridge*, p. 146). Merle explains that being a Yellow Feminist does not mean " 'separatism,' either by cutting myself off from non-Asians or men . . . it means changing the economic class system and psychological forces (sexism, racism, and homophobia) that really hurt all of us" (p. 142). Kitty describes herself as a warrior who grapples with those same three many-headed demons. In "The Words of A Woman Who Breathes Fire," she affirms: "I am a woman who loves women, / I am a woman who loves myself" (p. 52). Self-affirmation as a source of boldness and vision becomes a strength for Asian American women, lesbian and straight.

Writers/activists Sasha Hohri, Miya Iwataki and Janice Mirikitani are a few more of the unsung Asian American women continuing the strong tradition of political activism and organizing. Their issues range from re-dress/reparations for the Japanese interned during World War II to racist violence against Asians and the Rainbow Coalition for political power. They continue in the spirit of earlier Korean and Pilipino women who organized and continue to organize in the U.S. for independence in their homelands. A Japanese American woman, Mitsuye Endo, had partial success in the Supreme Court in 1944, challenging the internment of concededly loyal American citizens (Weglyn, p. 227). More recently, some of the largest labor rallies and significant employer concessions have been won by striking Chinese American garment workers. In 1982, 10,000 Chinese American women garment workers rallied in New York, winning a union contract that addressed their substandard working conditions. In 1986, following the shutdown of P & L Sportswear, 300 Chinese American women garment workers forced the city of Boston and the state of Massachusetts to implement the required retraining.

But while the Chinese American community is celebrating these victories, the women's movement has yet to recognize the significance of this achievement. Asian American women are trying to organize one of the least organized sectors of labor. They are fighting for basic working conditions denied to them precisely because they are Asian American women. Asian American women are often most actively involved in their communities and workplaces because those conditions directly determine their future. If we do not fight our own battles, who will? In *East Wind: Focus on Asian Women*, Sasha Hohri and Sadie Lum analyze Asian American women's oppression as intrinsically linked to class and race issues. We cannot separate ourselves from any one part. Our liberation is linked to that of our communities and ultimately, to that of our whole society. Liberation requires revolution.

As yet, feminism has not provided sufficient analysis and direction to the basic struggle of survival facing Asian Americans in this country. Feminism appears to have a more limited agenda, one concerned primarily with women's oppression. However, women's oppression is irresolvable in a society which is inherently unequal. A capitalist society means that a few will profit while the majority will not. Feminism must deal with the structure of capitalism and its exploitation of people by race and class, as well as the way this exploitation parallels and compounds women's oppression. As feminists broaden their perspective on what issues are of priority to women of all colors, more unity can be forged with Asian American and other sisters of color who are moving ahead, organizing and surviving. We cannot choose to stop struggling. We can only choose how we work together.

NOTES

"Pilipino"; Filipino is the anglicized form and symbolizes the colonization and domination of the Philippines by foreign powers such as the United States. In Tagalog, the national language of the Philippines, the word is pronounced with a "p" sound, hence "Pilipino."

General DeWitt's statement is part of his February 1942 recommendation to Secretary Stimson on exclusion of the Japanese, as quoted in *Personal Justice Denied*, p. 6.

REFERENCES

"American Men Find Asian Brides Fill the Unliberated Bill," Raymond A. Joseph, *Wall Street Journal*, January 25/84, p. 1, 22.

"Chinese Garment Workers Shake Up New York" August 20/82 and "Chinese Garment Workers Win Retraining in Boston" September 12/86 in *Unity*, Unity Publications: Oakland, California.

East Wind: Politicians and Culture of Asians in the U.S. Focus: Asian Women, V. 2 N. 1, Spring/Summer 1983, Getting Together Publications: Oakland, California.

"JACL Report on Asian Bride Catalogs," *Pacific Citizen*, February 22/85, p. 10–11.

"Letter to Ma," Merle Woo, in *This Bridge Called My Back*, Cherríe Moraga and Gloria Anzaldúa, editors, Persephone Press: Watertown, Massachusetts, 1981.

Personal Justice Denied, Report of the Commission on Wartime Relocation and Internment of Civilians, U.S. Government Printing Office, Washington, D.C., 1982.

Recent Activities Against Citizens and Residents of Asian Descent, U.S. Commission on Civil Rights, Clearinghouse Publicatio.` No. 88.

Rethinking
Institutions

Social institutions exert a powerful influence on our everyday lives. They are also powerful channels for societal penalties and privileges. The type of work you do, the structure of your family, the kind of education you receive, and how you are treated by the state are all shaped by the institutional structure of society. Because institutions are patterned by race, class, and gender, their effect is different, depending on who you are. We rely on institutions to meet our needs, although they do so better for some groups than others. When a specific institution (such as the economy) fails us, we often appeal to another institution (such as the state) for redress. In this sense, institutions are both sources of support and sources of oppression.

The concept of an institution is an abstract one, since there is not a thing or an object that one can point to as an institution. *Social institutions* are the established societal patterns of behavior organized around particular purposes. The economy is an institution, as are the family, education, and the state—the four societal institutions examined here. Each is organized around a specific purpose, such as, in the case of the economy, the production, distribution, and consumption of goods and services. Within a given institution, there may be various patterns, such as different family structures, but as a whole, institutions are general patterns of behavior that emerge because of the specific societal conditions in which groups live. Institutions do change over time, both as societal conditions evolve and as groups challenge specific institutional structures, but they are also enduring and persistent, even when there are active efforts to change them. Institutions confront us from birth and live on after we die.

Social institutions are the fundamental conduits for race, class, and gender oppression in this society, even though they are often presented as entities far removed from these experiences. The American creed portrays institutions as neutral in their treatment of different groups; indeed, the liberal framework of the law sees things like access to education and employment opportunities as gender- and race-blind, for the most part. Still, institutions differentiate on the basis of race, class, and gender; as the articles included here show, institutions are actually structured based on race, class, and gender relations. As an example, think of the economy. Economic institutions in this society are founded on capitalism—an economic system based on the pursuit of profit and the principle of private ownership. Such a system creates class inequality since, in simple terms, the profits of some stem from the exploitation of the labor of others. The U.S. capitalist economy is further divided by race and class, resulting in a *split labor market.* The split labor market includes: (1) a primary labor market, where there are relatively high wages, opportunities for advancement, employee benefits, and rules of due process that protect workers' rights and (2) a secondary labor market, where most women and minorities are located and where there are low wages, little opportunity for advancement, few, if any, benefits, and little protection for workers.

The point is that the split labor market is embedded in a larger economic institution that rests on race, class, and gender inequality. This is what it means to say that institutions are structured by race, class, and gender. In other words, institutions are built from and then reflect the historical and contemporary patterns of race, class, and gender relations in society. Think also of the state. The *state* refers to the organized system of power and authority in society.

This includes the government, the police, the military, and the law (as reflected in social policy, as well as in the civil and criminal justice system). The state is supposed to protect all citizens, regardless of their race, class, or gender (as well as other protections, such as disability and age); yet, like other "gendered" institutions, the state can be described as "male" (MacKinnon 1982). This means not only that the majority of powerful people in the state are men (such as elected officials, judges, police, and the military) but also and, just as important, that the state works to protect men's interests. Policies about reproductive rights provide a good example—they are largely enacted by men, but they have a particularly profound effect on women. Similarly, welfare policies designed to encourage people to work are based on the model of men's experiences, since they presume that staying home to care for one's children is not working. In this sense, the state is a "gendered" institution.

A gendered institution is one that embeds the specific characteristics associated with gender into its structure (Acker 1992). The family provides a good example. Historically, the family has been presumed to be the world of women; the ideology of the family (that is, the dominant belief systems about the family) purports that families are places for nurturing, love, and support—characteristics that have been associated with women. This ideal identified women with the private world of the family and men with the public sphere of work. In this sense, the family ideal identified the family as a gendered institution. Family ideology, of course, only projected an ideal, since we know that few families actually fit the presumed ideal; nonetheless, the ideology of the family provided a standard against which all families were judged.

This example demonstrates an important point: that institutions are not only gendered, they are also structured by race and class. In the example above, the ideology of the family is class- and race-specific. That is, it ignores and distorts the family experiences of African Americans, Latinos, and most White families as well. Bonnie Thornton Dill's essay, "Our Mothers' Grief," shows that for African American, Chinese American, and Mexican American women, family structure is deeply shaped by the relationships of families to the structures of race, class, and gender.

Examining the institutional structure of society when thinking about race, class, and gender is different from the way that most people commonly think about institutions. The individualist framework of the dominant culture sees race, class, and gender as attributes of individuals, instead of seeing them as embedded in institutional structures. People do, of course, have race, class, and gender identities, and race, class, and gender have an enormous impact on individual experience; however, seeing race, class, and gender from an individualist viewpoint overlooks their profoundly embedded position in the structure of American institutions.

In this section of the book we examine how institutions structure race, class, and gender experience and, in turn, how race, class, and gender structure institutions. We look at four major institutions: work and the economic system, families, education, and the state and social policy. Each can be shown to have a unique impact on different groups (such as the discriminatory treatment of African American men by the criminal justice system). At the same time, each institution and its interrelationship with other institutions can be seen as specifically structured through the dynamics of race, class, and gender relations. Moving historically marginalized groups to the center of analysis clarifies the importance of social institutions as links between individual experience and larger structures of race, class, and gender.

WORK AND ECONOMIC TRANSFORMATION

Structural transformations in the economy have dramatically changed the conditions under which people work. As D. Stanley Eitzen and Maxine Baca Zinn show in "Structural Transformation and Systems of Inequality," there are four basic transformations currently affecting the character of work: the development of new technologies, global economic interdependence, capital flight, and the growth of the service sector. These changes have brought new opportunities for some groups—those positioned to benefit from these changes—while, for others, these changes have meant massive economic dislocation. For example, as Teresa Amott shows in "Shortchanged," women have been uniquely affected by these changes; furthermore, different groups of women have been differently affected. Her analysis shows the effect of each race, gender, and class in shaping the economic crisis for women. High unemployment rates among African American, Asian, and Latino male workers also demonstrate the impact of economic restructuring on different groups, according to race, class, and gender.

In "The Latino Population" Joan Moore and Raquel Pinderhughes document the specific effects of economic restructuring on Latinos. They point to the decline in traditional manufacturing and the growth of low-wage service and manufacturing industries (most notably, the electronics and garment industries) as most affecting the employment of Latino workers. These industries rely on the use of cheap labor, also encouraging migration among Latinos. In addition, the growth of "global cities," in which internationalization of the labor force occurs, creates a unique mix of high- and low-paid service jobs in major cities and encourages the development of labor markets in these cities that have a distinct race, class, and gender structure.

In the midst of these changes, numerous myths abound about who succeeds and why. Deborah Woo in "The Gap Between Striving and Achieving" refutes one such myth—the recurring claim that current economic changes allow Asian Americans, the so-called model minority, to escape racial discrimination. Woo shows that the myth of the model minority obscures knowledge about the large number of Asian Americans in working-class jobs and the structural barriers to Asian-American success. Taken together, these essays demonstrate the interactive relationship of race, class, and gender in shaping economic institutions.

FAMILIES

Families are another primary social institution profoundly influenced by systems of race, class, and gender. Bonnie Thornton Dill's historical analysis of racial-ethnic women and their families in "Our Mothers' Grief" examines diverse patterns of family organization directly influenced by a group's place-ment in the larger political economy. Just as the political economy of the nineteenth century affected women's experience in families, so does the political economy of the late twentieth century shape family relations for women and men of all races. People do not, however, remain passive victims of these forces. As Melba Sánchez-Ayéndez shows in "Puerto Rican Elderly Women," people shape viable family networks and support systems that enable them to survive even in the face of cultural and economic assaults.

The articles in this section illustrate several points that have emerged from feminist studies of families, as identified by Barrie Thorne (1992). First, the family is not monolithic. The now widely acknowledged diversity among families refutes the idea that there is a normative family: White, middle-class, with children, organized around a heterosexual married couple (preferably with the wife not employed), and needing little support from relatives or neighbors. As the experiences of African Americans, Latinos, lesbians, Asian Americans, and others reveal, this so-called normal family actually represents a minority experience.

Second, the best way to analyze families is to look at the underlying struc-tures of "gender, generation, sexuality, race, and class" (Thorne 1992: 5). As an example, Audre Lorde's insightful view of Black lesbian mothers in "Man Child" shows the way that families have been presumed to be formed around a hetero-sexual norm. Raising children in diverse family environments so that they appreciate not only their own families but also diverse ways of forming families is a challenge in a society that so devalues gays, lesbians, and people of color.

Third, Thorne points out that the dominant ideology of the family has glorified the family, hiding the underlying conflicts that are embedded in family systems. Although we do not examine family violence directly here, we know that violence stems from the gender, race, and class conflicts that are encouraged in a sexist, racist, and class-based society. In addition, we think of family violence not only as that which occurs between family members but also as what happens to families as they confront the institutional structures of race, class, and gender domination. In "Reports from the Front," Diana Dujon, Judy Gradford, and Dottie Stevens show how this affects welfare mothers, in particular, as they struggle to meet their families' needs.

Fourth, Thorne argues that the boundaries separating family, work, welfare, and other institutions are more fluid than previously believed. No institution is isolated from another. Changes, for example, in the economy affect family structures; moreover, this effect is reciprocal, since changes in the family also generate changes in the economy. We think the interrelationships among institutions are especially apparent when studying race, class, and gender. Race, class, and gender oppression rest on a network of interconnected social institutions; moreover, understanding the interconnections between institutions helps us see that we are all part of one historically created system that finds structural form in interconnected social institutions.

Finally, Thorne's fifth point, that the presumed dichotomy between the public and private spheres is false, is especially evident when considering the experience of people of color. Racial-ethnic families have rarely been provided the protection and privacy that the alleged split between public and private assumes. From welfare policy to reproductive rights, this dichotomy is based on the myth that families are insulated from the society around them. Each of the articles in this section exposes this myth by revealing the complex relationship of all families to systems of race, class, and gender oppression.

EDUCATION

Next we turn to education. Education has recently and frequently been described as an institution in crisis. People think that children are not learning in school. High school dropout rates, especially among the poor and working class, reveal deep problems in the system of education. School violence indicates that the conflicts of race, class, and gender in the society at large are also played out within the schools. The persistence of inequalities in educational opportunity because of race, class, and gender mean that many are less able to succeed. All groups whose experience is different from the White norm are ignored and distorted in the educational curriculum. However they are described, the problems in education are a logical outgrowth of the race, class, and gender inequities that exist in society at large.

In "Education and the Struggle Against Race, Class, and Gender Inequality," Roslyn Mickelson and Stephen Smith show how schools promote inequalities of race, class, and gender. With their analysis, we can see the interplay between school policies that perpetuate existing hierarchies, such as credential inflation, and belief systems about race, class, and gender that are embedded in the so-called hidden curriculum. Students encounter these structures and belief systems while young, when identity takes on major importance. Schools thus become key sites where the inequalities of race, class, and gender are fostered and resisted.

Ironically, however, while schools reproduce inequality, education is also a source of social mobility. This is well illustrated in "Reminiscence of a Post-Integration Kid" by Gaye Williams and in "Canto, Locura y Poesia" by Olivia Castellano. These compelling narratives explain how African American and Latino women experience their schooling, yet manage to cope and often excel. Both women identify their struggle for a self-defined consciousness as fundamental to their survival.

Recently contested discussions about the content of the curriculum also show how race, class, and gender are embedded in the politics of education. Henry Louis Gates, Jr., in "Integrating the American Mind" addresses the alienation students feel when they do not see themselves reflected in the content of their learning. Transformation of the curriculum is not just hollow-minded "political correctness"; it is fundamental to addressing the problems in education, since students otherwise "learn without focus." In sum, as these articles show, education cannot be improved without addressing the arrangements of race, class, and gender in the society as a whole.

THE STATE AND SOCIAL POLICY

Finally, we include a section on the state and social policy. As we defined it before, the state is the system of legitimated power and authority in society. Race, class, and gender are fundamental in the construction of the state and its policies, just as the state is fundamental in shaping and defining these categories. State policies determine people's rights and, as the history of Jim Crow legislation shows, the state has often been the basis for extreme exclusionary action. At the same time, as the history of civil rights legislation shows, the state can be an invaluable resource for addressing the wrongs of race, class, and gender injustice. In this sense, the state is both a source of oppression *and* an avenue for seeking justice.

For example, state policy has done much to eliminate earlier structures on inequality. At least in law (*de jure*), citizens' rights are race-, class-, and gender-blind; in practice (*de facto*), however, these inequities continue. How does the state negotiate the relationships of race, class, and gender? In "Black Males and Social Policy" Ronald Taylor examines this by looking at the role of the state in shaping policies affecting African American men. Disadvantaged by education, employment, high rates of crime, and poverty, African American men have been victims of federal policies; yet, Taylor also shows how social policy can be used on behalf of disadvantaged groups.

In a different vein, Alan Berkman and Tim Blunk's analysis of political prisoners, "Thoughts on Class, Race, and Prison," shows how the state controls those judged to be a threat to the status quo. Their discussion puts a context around the fact that disproportionate numbers of Black, Latino, and Native American men are confined in prison. In fact, the criminal justice system exemplifies the differential treatment that individuals receive based on race, class, and gender. Prisons are, in this context, systems of social control.

Race, class, and gender are fundamental determinants of organizational policies fostering social control. Cynthia Brown in "The Vanished Native Americans" discusses the process by which the state gives a "legitimate" identity to Native American groups. Her article demonstrates the significance of the state in defining legitimate citizenship—a fact that will also be familiar to immigrant workers. She also shows how powerful the state is in managing group rights and organizing people's access to state and societal resources. In "The Brutality of the Bureaucracy" Theresa Funiciello painfully describes the brutality of the welfare state in her discussion of welfare mothers confronting the state bureaucracy. Her description will feel familiar to anyone who has been affronted by an impersonal and obstinate state bureaucracy, but, as she shows, the indignity is even greater when compounded by the insult of presumed race, class, and gender inferiority.

In conclusion, institutions are powerful mechanisms for perpetuating the race, class, and gender inequities in society. Who controls them, who benefits from institutional resources, and who is best able to negotiate their way through institutional structures all reveal patterns of race, class, and gender inequity. Still, it is important to realize that power does not operate only in a "top-down" fashion. Challenges to institutions come in many ways, but change begins with analysis. The articles in this section provide an institutional analysis of race, class, and gender.

References:

Acker, Joan. 1992. "Gendered Institutions: From Sex Roles to Gendered Institutions." *Contemporary Sociology* 21 (September): 565–569.

MacKinnon, Catharine. 1982. "Feminism, Marxism, Method, and the State: An Agenda for Theory." *Signs* 7 (Spring): 515–544.

Thorne, Barrie with Marilyn Yalom. 1992. *Rethinking the Family: Some Feminist Questions*. Rev. ed. Boston: Northeastern University Press.

Work and Economic Transformation

STRUCTURAL TRANSFORMATION AND SYSTEMS OF INEQUALITY

21

D. Stanley Eitzen and Maxine Baca Zinn

The technological and economically based reorganization of society has created wide disparities in the distribution of economic resources. All people in the United States are affected by the economic changes. The magnitude of structural transformation, however, is different throughout society. The old inequalities of class, race, and gender are thriving. New and subtle forms of discrimination are becoming prevalent throughout society as the economic base shifts and settles.

Four factors are at work here: new technologies, global economic inter-dependence, capital flight, and the dominance of the information and service sectors over basic manufacturing industries. Together these factors have reinforced the unequal placement of individuals and families in the larger society. They have deepened patterns of social inequality, and they have formed new patterns of domination in which the affluent control the poor, whites control people of color, and men control women.

The disproportionate effects of economic and industrial change are most visible in three trends: (1) structural unemployment, (2) the changing

From: D. Stanley Eitzen and Maxine Baca Zinn (eds.), *The Reshaping of America: Social Consequences of the Changing Economy* (Englewood Cliffs, N.J.: Prentice-Hall, 1989), pp. 131–143, © 1989. Reprinted by permission of Prentice-Hall, Inc.

distribution and organization of jobs, and (3) the low income-generating capacity of jobs. The first two trends have their most obvious effects on the changing class structure and the new racial order, while the second and third trends have particular ramifications for women and for gender relations. Still, all three trends have significant consequences for the hierarchies of class, race, and gender.

CLASS

Two major developments stand out when we look at the emerging class structure. The first is the growing gap between the rich and the poor since 1970. The second is the decline of the middle class.

The distribution of income is very unequal and widening. From 1970 to 1986, the income share of the highest quintile rose from 43.3 to 46.1 percent, while the bottom one fifth fell from 4.1 percent to 3.8 percent of all income (Pear 1987). The growing disparity in income and wealth is directly related to the changing job structure as the economy shifts from manufacturing to service.

Many Americans have experienced a sharp slowdown of income growth. This has caused a shrinkage of the middle class that is related in large measure to the erosion of middle-income jobs and the emergence of a bipolar wage structure in high tech and service work. The middle portion of the workforce fell from 52.3 percent of the population in 1978 to 44.3 in 1986 (Rose 1986:9). To be fair, demographic conditions as well as labor market forces are responsible for middle class decline. Nevertheless, the current period is the first in American history where the rate of downward mobility exceeds the rate of upward mobility. Home ownership illustrates how the middle class is losing ground. The percentage of families owning their homes has declined each year in the 1980s (American Demographics 1986). Furthermore, changes have occurred in the population with a middle-class life style. As Barbara Ehrenreich has observed: "Middle class is a matter of status as well as income and is signaled by subtler cues: how we live, what we spend our money on, what expectations we have for the future. Since the post-war period, middle class status has been defined by home ownership, college education (at least for the children), and the ability to afford such amenities as a second car and family vacations" (Ehrenreich 1986:50). Using this colloquial understanding of "middle class," it is clear that fewer Americans will achieve this status because many of the old avenues to social mobility no longer exist in the new society.

RACE

Technology and the changing distribution of jobs are having devastating effects on minority communities across America. The employment status of minorities is falling in all regions. It is worse however, in areas of industrial decline. The labor market status of whites has also been lowered in these areas, but the level of racial inequality has increased to scandalous levels. "In cities such as Detroit, Buffalo, Chicago, and Cleveland, the gap between the labor market position of blacks, especially black males, and whites probably exceeds the highest levels that ever existed in the most racist of the South's cities" (Swinton 1987:68). Racial inequality is partly a class issue. Prominent sociologist William J. Wilson has argued that the long-term removal of job opportunities for minority skilled and semiskilled workers is the force most responsible for the growth of the Black underclass in America's inner cities (Wilson 1987). Furthermore, the racial underclass is expanding to include Hispanics whose poverty rate now exceeds that of Blacks and who are expected, shortly after the turn of the century, to surpass Blacks as the largest racial-minority group.

Hispanics and Blacks have suffered disproportionately from industrial job loss and declining manufacturing employment. Their concentration in industries that have deteriorated in recent years is well documented. . . . By every measure including employment rates, occupational standing, and wage rates, the labor market status of racial minorities has deteriorated relative to whites. While the official unemployment rate for March 1987 was 6.5 percent, the rate for Blacks was 13.9 percent and 9 percent for Hispanics. These government rates, of course, are misleading because they count as employed those 5.5 million who work part time because they cannot find full-time jobs, and they do not count as unemployed the 1.17 million discouraged workers who have given up their search for work (Hershey 1987).

GENDER

Women and men are affected differently by the transformation of the economy from its manufacturing base to a base in service and high technology. Industrial jobs, traditionally filled by men, are being replaced with service jobs that are increasingly filled by women. Since 1980, women have taken 80 percent of the new jobs created in the economy. If this pace continues, women will make up most of the work force by the turn of the century (Hacker 1986:26). Unlike other forms of inequality, sexism in American society has not become more intense as a result of economic transformations. Instead, sexism has taken new forms as women are propelled into the labor market. Women have continued to move from the private sphere (family) to public arenas, but they have done so under conditions of labor market discrimination that have always plagued women. . . . Today, the "typical" job is a non-union, service sector, low-paying job occupied by a woman. Despite the growing feminization of the work force, male domination has been more firmly entrenched in the social organization of work. The rise of the contingent work force (including part-time work, temporary agencies, and subcontracted work) offers many advantages to employers but at considerable cost to women workers. Technological innovations in the workplace offer women dubious gains:

> In any number of cases, the outcome is more work opportunities for women, even if that work is less skilled, less autonomous and less rewarding . . . for many other women, new office machines have made them develop greater skills even if the jobs they use them for have become more rather than less fragmented and alienated and are no better than they were in the first place. What was once professional-level work—typically reserved for men—is often eliminated, and the residue or more routine tasks is added to the responsibilities of clerical workers who are, of course, women. The new technologies may develop more skills, but they also introduce clear monitoring and a stepped-up work pace. (Smith 1987:6)

The full impact of economic restructuring on women must take into account the low wage levels and the limited opportunities for advancement that characterize their work in the new economy.

NEW INTERSECTIONS OF CLASS, RACE, AND GENDER

The hierarchies of class, race, and gender are simultaneous and interlocking systems. For this reason, they frequently operate with and through each other to produce social inequality. Not only are many existing inequalities being exacerbated by the structural transformation of the economy, but the combined efforts of class, race, and gender are producing new kinds of subordination and exclusion throughout society and especially in the workplace. For example, the removal of manufacturing jobs has severely increased Black and Hispanic male job loss. Just how much of this is due to class and how much is due to race remains an important question.

In contrast, many women of color have found their work opportunities expanded, albeit in marginal work settings in service jobs and high-tech jobs. Does such growth offer traditionally oppressed race/gender groups new mobility opportunities or does the expansion of new kinds of work reproduce existing forms of inequality? The impact of economic restructuring on race and gender varies considerably. In some cases, it generates no jobs at all and displaces minority women workers. In other cases, minority women benefit by the creation of new jobs. Third world immigrant women provide the bulk of high-tech productive labor force in Silicon Valley. Yet a growing "underclass" in high tech consists of low-paid immigrant women from Mexico, Vietnam, Korea, and the Philippines. These examples reveal new labor systems as well as new forms of racial control based on class, race, and gender. . . .

REFERENCES

American Demographics. 1986. "The Affordable Dream." *American Demographics* (July):12, 14.

Ehrenreich, Barbara. 1986. "Is the Middle Class Doomed?" *The New York Times Magazine* (Sept. 7):44, 50, 54, 62, 64.

Hacker, Andrew. 1986. "Women at Work," *New York Review of Books* 33(13) (August 14):26–32.

Hershey, Robert D. 1987. "Jobless Rate Down but Growth of Jobs Also Falls." *The New York Times* (April 4):7.

Moberg, David. 1986. "Middle Class May Be Losing the Economic War of Attrition." *In These Times* (Nov. 12, 16).

Pear, Robert. 1987. "Poverty Rate Dips as the Median Family Income Rises." *The New York Times* (July 31):8.

Rose, Stephen J. 1986. *The American Profile Poster.* New York: Pantheon.

Smith, Joan. 1987. "Terminal Illnesses," *The Women's Review of Books IV*, 7 (April): 6–7.

Swinton, David. 1987. "Economic Status of Blacks 1986." *The State of Black America 1987.* New York: National Urban League (Jan.):49–73.

Wilson, William J. 1987. *The Truly Disadvantaged.* Chicago: University of Chicago Press.

SHORTCHANGED: *Restructuring* *Women's Work*

22

Teresa Amott

. . . The [economic] crisis has had different effects on men and women. In some ways, and for some women, the economic crisis has not been as severe as it has been for men. In other ways, and for other women, the crisis has been far more severe.

Throughout the crisis women continued to join the labor force, both to support themselves and their families and to find satisfaction in working outside the home. Companies, seeking to bolster their profits, hired women in order to cut labor costs. In fact, hiring women was a central part of the corporate strategy to restore profitability because women were not only cheaper than men, but were also less likely to be organized into unions and more willing to accept temporary work and no benefits. Women were hired rather than men in a variety of industries and occupations, in the United States and abroad. This led to what has been called the "feminization" of the labor force, as women moved into jobs that had previously been held only by men and as jobs that were already predominantly female became even more so. . . . This "feminization" of the workforce, in which women substitute for men, explains why women's unemployment rates, on average, were lower than men's during the 1980s.

Each of the corporate and government responses to the crisis . . . had its most damaging effects on a particular group (or groups) of workers. Union-busting, for instance, took its toll *primarily* on the manufacturing jobs that have been dominated by white men (although a weakened labor movement damages the bargaining power of all workers, as we see below). Capital flight also *primarily* affected men's manufacturing jobs, although it also affected women in the service sector and women who work in primarily female manufacturing jobs, such as the garment industry.

SEGREGATION AND SEGMENTATION

The way women experience the economic crisis depends on where they are located in the occupational hierarchy. One concept that will help us understand this hierarchy is *occupational segregation*, which can be by gender—most jobs are held by either men *or* women and few are truly integrated. To take one extreme example, in 1990, 82 percent of architects were male; 95 percent of typists were female. Occupational segregation can also be by race-ethnicity, although this is sometimes more difficult to detect—since racial-ethnic workers are a minority, they rarely dominate a job category numerically, although they may be in the majority at a particular workplace or in a geographical region. At the national level, we have to look for evidence of occupational segregation by race-ethnicity by examining whether a particular group is over- or underrepresented relative to its percent of the total workforce. For instance, if African American women make up 5.1 percent of the total workforce, they are underrepresented in an occupation if they hold less than 5.1 percent of the jobs.[1] Law would be such an occupation: African American women made up only 1.8 percent of all lawyers in 1990. In contrast, they are overrepresented in licensed practical nursing, where they hold 16.9 percent of the jobs.

A second concept that will be helpful in our examination of women's situation during the economic crisis is *labor market segmentation*, a term that refers to the division of jobs into categories with distinct working conditions. Economists generally distinguish two such categories, which they call the *primary* and *secondary* sectors. The first includes high-wage jobs that provide good benefits, job security, and opportunities for advancement. The upper level of this sector includes elite jobs that require long years of training and certification and offer autonomy on the job and a chance to advance up the corporate ladder. Access to upper level jobs is by way of family connections, wealth, talent, education, and government programs (like the GI bill, which guaranteed higher education to veterans returning from World War II). The lower level includes those manufacturing jobs that offer relatively high wages and job security (as a result of unionization), but do not require advanced training or degrees. The fact that unionized workers are part of this lower level is the result of the capital-labor accord . . . through which employers offered some unionized workers better pay and working conditions in exchange for labor peace. In both levels of the primary sector, job turnover is relatively low because it is more difficult for employers to replace these workers. Both the upper and lower levels of the primary sector were for many years the preserve of white men, with women (mostly white women) confined to small niches, such as schoolteaching and nursing.

The secondary sector includes low-wage jobs with few fringe benefits and little opportunity for advancement. Here too, there is a predominantly white-collar upper level (which includes sales and clerical workers), where working conditions, pay, and benefits are better than in the blue-collar lower level (private household, laborer, and most service jobs). Turnover is high in both levels of this sector because these workers have relatively few marketable skills and are easily replaced. For decades, the majority of women of all racial-ethnic groups, along with most men of color, were found in the secondary sector. Mobility between the primary and secondary sectors is limited: no career ladder connects jobs in the secondary sector to jobs in the primary sector.

While most jobs fall into these two sectors . . . during the economic crisis a third sector began to grow rapidly. This is known as the *informal sector*, or the underground economy. This name is not entirely accurate, however, since these activities do not make up a separate, distinct *economy* but are linked in many ways to the formal, above-ground sectors. Journalist and economist Philip Mattera believes that economic activity can be lined up along a continuum of formality and regulation.[2] At one end there is formal, regulated, and measured activity, where laws are observed, taxes are paid, inspections are frequent, and the participants report their activities to the relevant government entities. At the other end is work "off the books," where regulations are not enforced, participants do not report their activities, and taxes are evaded. Many economic activities exist somewhere in the middle. As the economic crisis deepens, many large corporations, whose own jobs are in the primary sector, have subcontracted some of their work out to underground firms that hire undocumented workers and escape health and safety, minimum wage, and environmental regulations. For example, in El Paso, Texas, only half of all garment industry workers earn adequate wages in union shops.[3] The other half work in sweatshops that contract work from big-name brands, such as Calvin Klein and Jordache, pay the minimum wage, and sometimes fail to pay anything at all. In 1990, the Labor Department fined one contractor for owing $30,000 in back wages and forced him to pay up. Fortunately for the employees, there is some justice in this world: the International Ladies' Garment Workers Union (ILGWU) then won a contract at his shops. According to Mattera,

> operating a business off the books—i.e., without any state regulation or union involvement—is the logical conclusion of the restructuring process. It represents the ultimate goal of the profit-maximizing entrepreneur: proverbial *free enterprise*. . . . The type of restructuring that has taken place makes it possible for firms that cannot or do not want to go underground to take advantage of unprotected labor nonetheless.[4]

Another reason for the growth in the informal sector is that worsening wages and conditions in the two formal sectors lead people to seek additional work "off the books" to supplement their shrinking incomes and inadequate welfare or social security benefits. Thus the numbers of people suffering what Mattera calls "the nightmarish working conditions of unregulated capitalism" grow rapidly. Women and men of color, particularly immigrants, are those most likely to be found in the informal sector. . . .

CAPITAL FLIGHT: A FLIGHT TO WOMEN?

. . . Over the past twenty years, many U.S. corporations shifted manufacturing jobs overseas. The creation of this "global assembly line" became a crucial component of the corporate strategy to cut costs. In their new locations, these companies hired women workers at minimal wages, both in the third world and in such countries as Ireland. Poorly paid as these jobs were, they were attractive to the thousands of women who were moving from impoverished rural villages into the cities in search of a better life for their families.

But in the United States, millions of workers lost their jobs as the result of capital flight or corporate downsizing. When workers lose their jobs because their plants or businesses close down or move, or their positions or shifts are abolished, it is called worker *displacement*. Over 5 million workers were displaced between 1979 and 1983, and another 4 million between 1985 and 1989.[5] In both periods, women were slightly *less* likely to lose their jobs than men of the same racial-ethnic group. Women in secondary sector factory jobs were hit hardest, primarily because they lacked union protection and the education and skills to find better jobs. (In 1989, over 35 percent of the women in manufacturing operative jobs had less than a high school education.)[6]

The overall result was that even though women lost jobs to capital flight and corporate downsizing, they did so at a slower rate than men. In fact, the share of manufacturing jobs going to women *rose* between 1970 and 1990. Women, in other words, claimed a growing share of a shrinking pie. Sociologist Joan Smith studied this growing tendency to replace male workers with women (as well as a parallel tendency to replace white workers with African Americans and Latinos). In her research on heavy manufacturing industries such as steel and automobiles, Smith found that employers hired men and/or white workers only in those areas of manufacturing where profits were high, jobs were being created, and there was substantial investment in new plant and equipment. In contrast,

in sectors where profits were slipping, the obvious search for less expensive workers led to the use of Black and women workers as a substitute for white workers and for men. . . . Close to 70 percent of women in these sectors and well over two-fifths of Blacks held their jobs as either substitutes or replacements for whites or men.[7]

While manufacturing jobs were feminizing, the rapidly expanding service sector was also hiring women in larger and larger numbers—both women entering the labor market for the first time and women displaced from manufacturing. In fact, all the jobs created during the 1980s were in the service sector. . . .[8]

A large part of this service sector growth took place in what were already predominantly female jobs, such as nurses' aides, child care workers, or hotel chambermaids (jobs that men would not take), as employers took advantage of the availability of a growing pool of women workers who were excluded from male-dominated jobs.[9] Chris Tilly argues that these sectors were able to grow so rapidly during the 1980s precisely *because* they were able to use low-wage, part-time labor.[10] In other words, if no women workers had been available, the jobs would not have been filled by men; instead, service employment would not have grown as rapidly.

The availability of service sector jobs helped hold the average official unemployment rates for women below those for men. However, the overall figure for women masks important differences by race-ethnicity. . . .[11] In addition, the official unemployment rate doesn't tell the whole story. If we construct a measure of underemployment, we get a different sense of the relative hardships faced by women and men. The underemployed are those who are working part-time but would prefer full-time work, those who are so discouraged that they have given up looking for work, and those who want a job but can't work because home responsibilities— such as caring for children or aged parents—or other reasons keep them out of the labor force. If we look at underemployment rather than unemployment, the rankings change: in contrast to the official unemployment rates, the underemployment rates are higher for women than for men. . . .

Even though the entire service sector grew during this period, service workers still risked losing their jobs. According to the Census Bureau's study of displaced workers, nearly half of those who lost their jobs between 1981 and 1983, and nearly two-thirds of those between 1985 and 1989, were in the service sector. As the crisis dragged on, in other words, corporate downsizing hit services as well as manufacturing. The financial sector, where women make up 60 percent of the workforce, was particularly hard hit, as jobs in banking, insurance, and real estate were lost as a result of the savings and loan scandal and other financial troubles.

UNION-BUSTING

... A second corporate strategy to bolster profitability was to attack labor unions. As the assault took its toll, union membership and union representation declined until by 1990 only 14 percent of women and 22 percent of men were represented by a union—compared to 18 percent of women and 28 percent of men only six years earlier. Even though women are less likely to be represented by unions than men, the *drop* in unionization was not as severe for women, largely because they are concentrated in the service sector where there were some organizing victories. In addition, most women do not hold jobs in factories, the area most vulnerable to union-busting. The drop in unionization was highest among Latinas, who are over-represented in manufacturing, followed closely by African American women. Still, African American women have the highest rate of union representation among women (22 percent) because so many work in the public sector.[12]

The drop in unionized jobs is dangerous for women for several reasons. The most obvious is that unionized women earn an average weekly wage that is 1.3 times that of nonunionized women. The gap is especially large in the service sector, where unionized workers earn 1.8 times as much as nonunionized workers. ... [13]

The loss of unionized jobs also hurts women in nonunion jobs because of what economists call a "spillover" effect from union to nonunion firms in the area of wages and benefits. While it is difficult to estimate the extent of this spillover, Harvard economists Richard Freeman and James Medoff suggest that it raises wages in blue collar jobs in large nonunion firms (those with lower level primary sector jobs) by anywhere from 10 to 20 percent, and also improves benefits and working conditions.[14] This happens for two reasons: (1) nonunion firms must compete for labor with the union firms and therefore have to meet unionized rates, and (2) some firms will keep wages higher than necessary in order to keep unions out. Thus when union workers lose wages and benefits, these spillover effects diminish, lowering wages and benefits in nonunion jobs as well.

Higher wages are only part of what women achieve when there is a strong labor movement. Unionization has an important spillover effect in the political as well as economic arena. For instance, support from the labor movement is responsible for the passage of most of the major safety net legislation in the United States, including the minimum wage and the Social Security Act. When the labor movement is weakened, important items on labor's agenda—including national health care, national child care, improved enforcement of job safety and health regulations, and broadened unemployment insurance coverage—all become more difficult to achieve, even though they address the needs of *all* workers. Further evidence of the critical role played by the labor movement comes from Europe, where family policies (parental leave, child allowances, and national day care) were all enacted with the backing, and sometimes at the initiative, of the labor movement.

WOMEN AND THE RESTRUCTURING OF WORK

. . . Another component of restructuring, and one that particularly affects women, is the use of homework. While homework can provide incomes for women who are unable to locate affordable child care or who live in rural areas, far from other employment, it also exposes women to intense exploitation. Both men and women homeworkers typically earn much less than those who do the same work outside the home: according to the federal Office of Technology Assessment, the poverty risk for homeworkers is nearly double that for other workers. In addition, working conditions in the home can be dangerous. In semiconductor manufacturing homework, for instance, workers are exposed to hazardous substances that can also contaminate residential sewage systems.[15]

Many homeworkers are undocumented immigrants who work out of their homes in order to escape detection by the Immigration and Naturalization Service (INS). During the early 1970s, the majority of undocumented immigrants were male, but since then women have begun to arrive in ever larger numbers:

> Women's migration has traditionally been ignored by researchers. . . . The INS reported a "dramatic rise" in the number of women apprehended at the U.S.-Mexico border between 1984 and 1986. . . . Many authors describe a new spirit of "independence" among women choosing to cross the border alone to reunite with family already in the U.S., or to seek employment on their own to support family left behind.[16]

In the spring of 1990, the Coalition for Immigrant and Refugee Rights and Services in San Francisco surveyed over 400 undocumented women in the Bay Area. Nearly half of them reported employment discrimination by employers who abused them sexually, physically, or emotionally, paid them less than their documented co-workers, or failed to pay them at all. Most of those surveyed worked as domestic servants, in stores, restaurants, and factories.

WOMEN'S RESPONSES TO RESTRUCTURING

As wages fell and employers pushed more and more work into the secondary and informal sectors, women responded with a variety of individual strategies.

Some started small businesses. Others sought "nontraditional" jobs in areas formerly dominated by men, hoping to earn a man's wage. And a growing number took on multiple jobs, "moonlighting" in a desperate effort to make a living wage out of two, or even three, different jobs. Each of these individual strategies . . . held some promise, but all failed to deliver substantial gains except to the lucky few. . . .

. . . Self-employment did not solve women's economic problems. The vast majority of these businesses remained small in scale: although they made up nearly 33 percent of all the businesses in the United States, they earned only 14 percent of the receipts. Nearly 40 percent had total receipts of less than $5,000 a year; only 10 percent had any employees.[17] Even the most successful women entrepreneurs faced difficulties finding affordable health insurance and pension coverage, which they had to buy for themselves out of their profits. Despite conservative rhetoric about the glories of entrepreneurship, self-employment has not proved to be a cure-all for women's economic troubles. . . .

As the number of poverty-level jobs has increased, more and more women have been forced to turn to moonlighting to boost their incomes. By 1989, 3.5 times as many women were working two or more jobs as in 1973. In contrast, the number of men moonlighting only rose by a factor of 1.2.[18] And many more women were moonlighting because of economic hardship—to meet regular household expenses or pay off debts—not to save for something special, get experience, or help out a family member or friend. African American women and Latinas were the most likely to report that they were moonlighting to meet *regular* household expenses, while white women were more likely to report saving for the future or other reasons—a difference caused by the lower average income of women of color and the greater likelihood that they were raising children on their own.

But moonlighting is ultimately limited by the number of hours in the day. . . . As the crisis wears on and more and more women become fully employed, more families turn to children to help out. According to recent estimates, at least 4 million children under the age of 19 are employed legally, while at least 2 million more work "off the books," and the number is growing.[19]

NEW JOBS FOR WOMEN

During the 1980s, the female labor force grew by over 10 million. Most of the new entrants found traditionally female jobs in secondary sector service and administrative support occupations, since that is where the majority of job growth took place. . . . However, some women made inroads into traditionally male jobs in the highly paid primary sector, and these gains are likely to be maintained. In addition, graduate degrees show how women are increasingly willing to prepare themselves for male-dominated occupations. . . . There has been an occupational "trickle down" effect, as white women improved their occupational status by moving into male-dominated professions such as law and medicine, while African American women moved into the *female-dominated* jobs, such as social work and teaching, vacated by white women. There is some evidence that the improvement for white women was related to federal civil rights legislation, particularly the requirement that firms receiving federal contracts comply with affirmative action guidelines.[20]

The movement of women into highly skilled blue-collar work, such as construction and auto-making, was sharply limited by the very slow growth in those jobs. Not coincidentally, male resistance to letting women enter these trades stiffened during the 1980s, when layoffs were common and union jobs were under attack. Moreover, all women lost ground in secondary sector manufacturing jobs, such as machine operators and laborers. Latinas were particularly hard hit, losing ground in most manufacturing jobs. . . .

UP THE DOWN ESCALATOR

If the postwar economic boom had continued into the 1970s and 1980s, women's economic status today would be substantially improved. The crisis produced some gains for women, but many of these evaporate on close inspection. The wage gap narrowed, but partly because men's wages fell. The gap between men's and women's rates of unionization and access to fringe benefits fell, but again partly because men's rates fell. More women entered the workforce, but they also worked longer hours than ever before, held multiple jobs, and sought work in the informal economy in order to maintain their standard of living. Finally, women's gains were not evenly distributed: highly educated women moved even further ahead of their less-educated counterparts.

All this took place against a backdrop of rising family responsibilities. . . . The most serious stress faced by married women . . . was associated with the reduced standard of living their families faced as a result of the cutbacks: "Their continued need to reduce what their wages can buy for their families means that the conflicts they feel between work and family life intensify. Earning less makes it feel harder and harder for them to continue to work and take care of their families."[21]

For these women, who depended on manufacturing jobs, it is not surprising that the economic crisis took a heavy toll. What is surprising is that they experienced it most acutely in the home rather than on the job. Work for them was not a career, a satisfying route to self-actualization, but a fate they accepted in order to provide for their families. When their earnings fell, it was their work at home—the work of marketing, cooking, cleaning, caring—that became more difficult. An increasing number were the sole support of their households. Others found that their household's standard of living could only be maintained if they took on one—or more—paid jobs in addition to their homemaking. They were caught between shrinking incomes and growing responsibilities. . . .

NOTES

1. Unpublished data, Bureau of Labor Statistics.

2. Philip Mattera, *Off the Books: The Rise of the Underground Economy* (New York: St. Martin's Press, 1985), p. 38.

3. Colatosti, Camille, "A Job Without a Future," *Dollars and Sense* (May 1992), p. 10.

4. Philip Mattera, *Prosperity Lost* (Reading, MA: Addison-Wesley, 1992), pp. 34–35.

5. For instance, between 1985 and 1989 the displacement rate—the number of workers displaced for every 1,000 workers employed—was 6.3 for white women compared to 6.7 for white men; 6.1 for African American women compared to 7.3 for African American men; and 8.3 for Latinas compared to 9.0 for Latinos. See U.S. Department of Labor, Bureau of Labor Statistics, *Displaced Workers, 1985–89*, June 1991, Bulletin 2382, Table 4.

6. U.S. Bureau of the Census, *Statistical Abstract of the United States 1991*, Table 656, p. 400. African American women factory operatives are better educated than whites: only 29 percent lack a high school degree, compared to 36 percent of whites.

7. Joan Smith, "Impact of the Reagan Years: Race, Gender, and the Economic Restructuring," *First Annual Women's Policy Research Conference Proceedings* (Washington, DC: Institute for Women's Policy Research, 1989), p. 20.

8. U.S. Bureau of the Census, *Statistical Abstract of the United States 1991*, Table 658, p. 401. The number of jobs in durable manufacturing fell by 0.8 percent a year between 1980 and 1988 and in nondurable manufacturing (i.e., light manufacturing, such as food processing) fell by 0.2 percent; service jobs grew an average of 2.7 percent. Manufacturing jobs did see some growth in the last half of the 1970s, however, so that over the two decades there was a small amount of overall growth in manufacturing.

9. Because the number of jobs held by women increased more than the number held by men, the percent of service sector jobs held by women increased from 43 percent to 52 percent between 1970 and 1990.

10. Polly Callaghan and Heidi Hartmann, *Contingent Work* (Washington, DC: Economic Policy Institute, 1991), p. 24.

11. For Latinas, on the other hand, women's unemployment was higher than men's. While it is difficult to pinpoint the reason for this, it may be because of the relatively rapid growth of Latina participation in the labor force. Although Latinas have the lowest labor force participation rate of the three groups of women, their participation rates are growing the most rapidly. It may be that this growth in the Latina workforce outstripped job creation in the secondary sector, which had traditionally hired Latinas, while discrimination still barred their way in the primary sector—resulting in high unemployment. See National Council of La Raza, *State of Hispanic America 1991*, p. 26.

12. Paula Ries and Anne J. Stone, eds., *The American Woman 1992–93: A Status Report* (New York, W. W. Norton and Co., 1991), p. 369.

13. U.S. Bureau of Labor Statistics, *Employment and Earnings*, January 1992, Tables 59–60.

14. Richard Freeman and James Medoff, *What Do Unions Do?* (New York: Basic Books, 1981), p. 153.

15. Virginia DuRivage and David Jacobs, "Home-Based Work: Labor's Choices," in *Homework: Historical and Contemporary Perspectives on Paid Labor at Home*, ed. Eileen Boris and Cynthia R. Daniels (Urbana: University of Illinois Press, 1989), p. 259.

16. Chris Hogeland and Karen Rose, *Dreams Lost, Dreams Found: Undocumented Women in the Land of Opportunity* (San Francisco, CA: Coalition for Immigrant and Refugee Rights and Services, 1990), p. 4.

17. Ries and Stone, *The American Woman*, pp. 347–48.

18. Lawrence Mishel and David M. Frankel, *The State of Working America* (Armonk, NY: M. E. Sharpe, 1991), p. 142.

19. Gina Kolata, "More Children Are Employed, Often Perilously," *New York Times*, 21 June 1992, p. 1.

20. Barbara Bergman, *The Economic Emergence of Women* (New York: Basic Books, 1986), p. 147.

21. Ellen Israel Rosen, *Bitter Choices: Blue Collar Women in and out of Work* (Chicago: University of Chicago Press, 1987), p. 164.

THE GAP BETWEEN STRIVING AND ACHIEVING: *The Case of Asian American Women*

23

Deborah Woo

Much academic research on Asian Americans tends to underscore their success, a success which is attributed almost always to a cultural emphasis on education, hard work, and thrift. Less familiar is the story of potential not fully realized. For example, despite the appearance of being successful and highly educated, Asian American women do not necessarily gain the kind of recognition or rewards they deserve.

The story of unfulfilled dreams remains unwritten for many Asian Americans. It is specifically this story about the gap between striving and achieving that I am concerned with here. Conventional wisdom obscures the discrepancy by looking primarily at whether society is adequately rewarding individuals. By comparing how minorities as disadvantaged groups are doing relative to each other, the tendency is to view Asian Americans as a "model minority." This practice programs us to ignore structural barriers and inequities and to insist that any problems are simply due to different cultural values or failure of individual effort.

Myths about the Asian American community derive from many sources. All ethnic groups develop their own cultural myths. Sometimes, however, they create myths out of historical necessity, as a matter of subterfuge and survival. Chinese Americans, for example, were motivated to create new myths because institutional opportunities were closed off to them. Succeeding in America meant they had to invent fake aspects of an "Oriental culture," which became the beginning of the Chinatown tourist industry.

What has been referred to as the "model minority myth," however, essentially originated from without. The idea that Asian Americans have been a successful group has been a popular news media theme for the last twenty years. It has become a basis for cutbacks in governmental support for all ethnic

minorities—for Asian Americans because they apparently are already successful as a group; for other ethnic minorities because they are presumably not
working as hard as Asian Americans or they would not need assistance. Critics
of this view argue that the portrayal of Asian Americans as socially and
economically successful ignores fundamental inequities. That is, the question
"Why have Asians been successful vis-à-vis other minorities?" has been asked
at the expense of another equally important question: "What has kept Asians
from *fully* reaping the fruits of their education and hard work?"

The achievements of Asian Americans are part reality, part myth. Part of
the reality is that a highly visible group of Asian Americans are college-
educated, occupationally well-situated, and earning relatively high incomes.
The myth, however, is that hard work reaps commensurate rewards. This essay
documents the gap between the level of education and subsequent occupational or income gains.

THE ROOTS AND CONTOURS OF THE "MODEL MINORITY" CONCEPT

Since World War II, social researchers and news media personnel have been
quick to assert that Asian Americans excel over other ethnic groups in terms
of earnings, education, and occupation. Asian Americans are said to save more,
study more, work more, and so achieve more. The reason given: a cultural
emphasis on education and hard work. Implicit in this view is a social judgment
and moral injunction: if Asian Americans can make it on their own, why can't
other minorities?

While the story of Asian American women workers is only beginning to
be pieced together, the success theme is already being sung. The image prevails
that despite cultural and racial oppression, they are somehow rapidly assimilating into the mainstream. As workers, they participate in the labor force at
rates higher than all others, including Anglo women. Those Asian American
women who pursue higher education surpass other women, and even men, in
this respect. Moreover, they have acquired a reputation for not only being
conscientious and industrious but docile, compliant, and uncomplaining as
well.

In the last few decades American women in general have been demanding
"equal pay for equal work," the legitimation of housework as work that needs
to be recompensed, and greater representation in the professional fields.
These demands, however, have not usually come from Asian American
women. From the perspective of those in power, this reluctance to complain
is another feature of the "model minority." But for those who seek to uncover

employment abuses, the unwillingness to talk about problems on the job is itself a problem. The garment industry, for example, is a major area of exploitation, yet it is also one that is difficult to investigate and control. In a 1983 report on the Concentrated Employment Program of the California Department of Industrial Relations, it was noted:

> The major problem for investigators in San Francisco is that the Chinese community is very close-knit, and employers and employees cooperate in refusing to speak to investigators. In two years of enforcing the Garment Registration Act, the CEP has never received a complaint from an Asian employee. The few complaints received have been from Anglo or Latin workers.[1]

While many have argued vociferously either for or against the model minority concept, Asian Americans in general have been ambivalent in this regard. Asian Americans experience pride in achievement born of hard work and self-sacrifice, but at the same time, they resist the implication that all is well. Data provided here indicate that Asian Americans have not been successful in terms of benefitting fully, (i.e., monetarily), from their education. It is a myth that Asian Americans have proven the American Dream. How does this myth develop?

The Working Consumer: Income and Cost of Living

One striking feature about Asian Americans is that they are geographically concentrated in areas where both income and cost of living are very high. . . . [Immigration] has not only produced dramatic increases, especially in the Filipino and Chinese populations, but has also continued the overwhelming tendency for these groups to concentrate in the same geographical areas, especially those in California.[2] Interestingly enough, the very existence of large Asian communities in the West has stimulated among more recent refugee populations what is now officially referred to as "secondary migration," that is, the movement of refugees away from their sponsoring communities (usually places where there was no sizeable Asian population prior to their own arrival) to those areas where there are well-established Asian communities.[3]

This residential pattern means that while Asian Americans may earn more by living in high-income areas, they also pay more as consumers. The additional earning power gained from living in San Francisco or Los Angeles, say, is absorbed by the high cost of living in such cities. National income averages which compare the income of Asian American women with that of the more broadly dispersed Anglo women systematically distort the picture. Indeed, if we compare women within the same area, Asian American women are frequently less well-off than Anglo American females, and the difference between women pales when compared with Anglo males, whose mean income is much higher than that of any group of women.[4]

When we consider the large immigrant Asian population and the language barriers that restrict women to menial or entry-level jobs, we are talking about a group that not only earns minimum wage or less, but one whose purchasing power is substantially undermined by living in metropolitan areas of states where the cost of living is unusually high.

Another striking pattern about Asian American female employment is the high rate of labor force participation. Asian American women are more likely than Anglo American women to work full time and year round. The model minority interpretation tends to assume that mere high labor force participation is a sign of successful employment. One important factor motivating minority women to enter the work force, however, is the need to supplement family resources. For Anglo American women some of the necessity for working is partly offset by the fact that they often share in the higher incomes of Anglo males, who tend not only to earn more than all other groups but, as noted earlier, also tend to receive higher returns on their education. Moreover, once regional variation is adjusted for, Filipino and Chinese Americans had a median annual income equivalent to black males in four mainland SMSAs—Chicago, Los Angeles/Long Beach, New York, San Francisco/Oakland.[5] Census statistics point to the relatively lower earning capacity of Asian males compared to Anglo males, suggesting that Asian American women enter the work force to help compensate for this inequality. Thus, the mere fact of high employment must be read cautiously and analyzed within a larger context.

The Different Faces of Immigration

Over the last decade immigration has expanded the Chinese population by 85.3 percent, making it the largest Asian group in the country at 806,027, and has swelled the Filipino population by 125.8 percent, making it the second largest at 774,640. Hence at present the majority of Chinese American and Filipino American women are foreign-born. In addition the Asian American "success story" is misleading in part because of a select group of these immigrants: foreign-educated professionals.

Since 1965 U.S. immigration laws have given priority to seven categories of individuals. Two of the seven allow admittance of people with special occupational skills or services needed in the United States. Four categories facilitate family reunification, and the last applies only to refugees. While occupation is estimated to account for no more than 20 percent of all visas, professionals are not precluded from entering under other preference categories. Yet this select group is frequently offered as evidence of the upward mobility possible in America when Asian Americans who are born and raised in the United States are far less likely to reach the doctoral level in their education. Over two-thirds of Asians with doctorates in the United States are trained and educated abroad.[6]

Also overlooked in some analyses is a great deal of downward mobility among the foreign-born. For example, while foreign-educated health professionals are given preferential status for entry into this country, restrictive licensing requirements deny them the opportunity to practice or utilize their special skills. They are told that their educational credentials, experience, and certifications are inadequate. Consequently, for many the only alternatives are menial labor or unemployment.[7] Other highly educated immigrants become owner/managers of Asian businesses, which also suggests downward mobility and an inability to find jobs in their field of expertise.

"Professional" Obscures More Than It Reveals

Another major reason for the perception of "model minority" is that the census categories implying success, "professional-managerial" or "executive, administrative, managerial," frequently camouflage important inconsistencies with this image of success. As managers, Asian Americans, usually male, are concentrated in certain occupations. They tend to be self-employed in small-scale wholesale and retail trade and manufacturing. They are rarely buyers, sales managers, administrators, or salaried managers in large-scale retail trade, communications, or public utilities. Among foreign-born Asian women, executive-managerial status is limited primarily to auditors and accountants.[8]

In general, Asian American women with a college education are concentrated in narrow and select, usually less prestigious, rungs of the "professional-managerial" class. In 1970, 27 percent of native-born Japanese women were either elementary or secondary school teachers. Registered nurses made up the next largest group. Foreign-born Filipino women found this to be their single most important area of employment, with 19 percent being nurses. They were least represented in the more prestigious professions—physicians, judges, dentists, law professors, and lawyers.[9] In 1980 foreign-born Asian women with four or more years of college were most likely to find jobs in administrative support or clerical occupations.

Self-Help Through "Taking Care of One's Own"

Much of what is considered ideal or model behavior in American society is based on Anglo-Saxon, Protestant values. Chief among them is an ethic of individual self-help, of doing without outside assistance or governmental support. On the other hand, Asian Americans have historically relied to a large extent on family or community resources. Their tightly-knit communities tend to be fairly closed to the outside world, even when under economic hardship. Many below the poverty level do not receive any form of public assistance.[10] Even if we include social security benefits as a form of supplementary income, the proportion of Asian Americans who use them is again very low, much lower than that for Anglo Americans.[11] Asian American families, in fact, are more likely than Anglo American families to bear economic hardships on their own.

While Asian Americans appear to have been self-sufficient as communities, we need to ask, at what personal cost? Moreover, have they as a group reaped rewards commensurate with their efforts? The following section presents data which document that while Asian American women may be motivated to achieve through education, monetary returns for them are less than for other groups.

THE NATURE OF INEQUALITY

The decision to use white males as the predominant reference group within the United States is a politically charged issue. When women raise and push the issue of "comparable worth," of "equal pay for equal work," they argue that women frequently do work equivalent to men's, but are paid far less for it. . . .

Another way of thinking about comparable worth is not to focus only on what individuals do on the job, but on what they bring to the job as well. Because formal education is one measure of merit in American society and because it is most frequently perceived as the means to upward mobility, we would expect greater education to have greater payoffs.

Asian American women tend to be extraordinarily successful in terms of attaining higher education. Filipino American women have the highest college completion rate of all women and graduate at a rate 50 percent greater than that of majority males. Chinese American and Japanese American women follow closely behind, exceeding both the majority male and female rate of college completion.[12] Higher levels of education, however, bring lower returns for Asian American women than they do for other groups.

While education enhances earnings capability, the return on education for Asian American women is not as great as that for other women, and is well below parity with white males. . . .

The fact that Asian American women do not reap the income benefits one might expect given their high levels of educational achievement raises questions about the reasons for such inequality. To what extent is this discrepancy based on outright discrimination? On self-imposed limitations related to cultural modesty? The absence of certain social or interpersonal skills required for upper managerial positions? Or institutional factors beyond their control? It is beyond the scope of this paper to address such concerns. However, the fact of inequality is itself noteworthy and poorly appreciated.

In general, Asian American women usually are overrepresented in clerical or administrative support jobs. While there is a somewhat greater tendency for foreign-born college-educated Asian women to find clerical-related jobs, both native- and foreign-born women have learned that clerical work is the area where they are most easily employed. . . . In addition Asian American women tend to be overrepresented as cashiers, file clerks, office machine operators, and typists. They are less likely to get jobs as secretaries or receptionists. The former occupations not only carry less prestige but generally have "little or no decision-making authority, low mobility and low public contact."[13]

In short, education may improve one's chances for success, but it cannot promise the American Dream. For Asian American women education seems to serve less as an opportunity for upward mobility than as a protection against jobs as service or assembly workers, or as machine operatives—all areas where foreign-born Asian women are far more likely to find themselves.

CONCLUSION

In this essay I have attempted to direct our attention on the gap between achievement and reward, specifically the failure to reward monetarily those who have demonstrated competence. Asian American women, like Asian American men, have been touted as "model minorities," praised for their outstanding achievements. The concept of model minority, however, obscures the fact that one's accomplishments are not adequately recognized in terms of commensurate income or choice of occupation. By focusing on the achievements of one minority in relation to another, our attention is diverted from larger institutional and historical factors which influence a group's success. Each ethnic group has a different history, and a simplistic method of modeling which assumes the experience of all immigrants is the same ignores the sociostructural context in which a certain kind of achievement occurred. For example, World War II enabled many Asian Americans who were technically trained and highly educated to move into lucrative war-related industries.[14] More recently, Korean immigrants during the 1960s were able to capitalize on the fast-growing demand for wigs in the United States. It was not simply cultural ingenuity or individual hard work which made them successful in this enterprise, but the fact that Korean immigrants were in the unique position of being able to import cheap hair products from their mother country.[15]

Just as there are structural opportunities, so there are structural barriers. However, the persistent emphasis in American society on individual effort deflects attention away from such barriers and creates self-doubt among those who have not "made it." The myth that Asian Americans have succeeded as a group, when in actuality there are serious discrepancies between effort and achievement, and between achievement and reward, adds still further to this self-doubt.

While others have also pointed out the myth of the model minority, I want to add that myths do have social functions. It would be a mistake to dismiss the model minority concept as merely a myth. Asian Americans are—however inappropriately—thrust into the role of being models for other minorities.

A closer look at the images associated with Asians as a model minority group suggests competing or contradictory themes. One image is that Asian Americans exemplify a competitive spirit enabling them to overcome structural barriers through perseverance and ingenuity. On the other hand, they are also seen as complacent, content with their social lot, and expecting little in the way of outside help. A third image is that Asian Americans are experts at assimilation, demonstrating that this society still functions as a melting pot. Their values are sometimes equated with white, middle-class, Protestant values of hard work, determination, and thrift. Opposing this image, however, is still another, namely that Asian Americans have succeeded because they possess cultural values unique to them as a group—their family-centeredness and long tradition of reverence for scholarly achievement, for example.

Perhaps, then, this is why so many readily accept the myth, whose tenacity is due to its being vague and broad enough to appeal to a variety of different groups. Yet to the extent that the myth is based on misconceptions, we are called upon to reexamine it more closely in an effort to narrow the gap between striving and achieving.[16]

1. Ted Bell, "Quiet Loyalty Keeps Shops Running," *Sacramento Bee*, 11 February 1985.

2. U.S. Bureau of the Census, *Race of the Population by States* (Washington, D.C., 1980). According to the census, 40 percent of all Chinese in America live in California, as well as 46 percent of all Filipinos, and 37 percent of all Japanese. New York ranks second for the number of Chinese residing there, and Hawaii is the second most populated state for Filipinos and Japanese.

3. Tricia Knoll, *Becoming Americans: Asian Sojourners, Immigrants, and Refugees in the Western United States* (Portland, Oreg.: Coast to Coast Books, 1982), 152.

4. U.S. Commission on Civil Rights, *Social Indicators of Equality for Minorities and Women* (Washington, D.C., 1978), 24, 50, 54, 58, 62.

5. David M. Moulton, "The Socioeconomic Status of Asian American Families in Five Major SMSAs" (Paper prepared for the Conference of Pacific and Asian American Families and HEW-related Issues, San Francisco, 1978). No comparative data were available on blacks for the fifth SMSA, Honolulu.

6. James E. Blackwell, *Mainstreaming Outsiders* (New York: General Hall, Inc., 1981), 306; and Commission on Civil Rights, *Social Indicators*, 9.

7. California Advisory Committee, "A Dream Unfulfilled: Korean and Pilipino Health Professionals in California" (Report prepared for submission to U.S. Commission on Civil Rights, May 1975), iii.

8. See Amado Y. Cabezas, "A View of Poor Linkages between Education, Occupation and Earnings for Asian Americans" (Paper presented at the Third National Forum on Education and Work, San Francisco, 1977), 17; and Census of Population, PUS, 1980.

9. Census of the Population, PUS, 1970, 1980.

10. A 1977 report on California families showed that an average of 9.3 percent of Japanese, Chinese, and Filipino families were below the poverty level, but that only 5.4 percent of these families received public assistance. The corresponding figures for Anglos were 6.3 percent and 5.9 percent. From Harold T. Yee, "The General Level of Well-Being of Asian Americans" (Paper presented to U.S. government officials in partial response to Justice Department amicus).

11. Moulton, "Socioeconomic Status," 70–71.

12. Commission on Civil Rights, *Social Indicators*, 54.

13. Bob H. Suzuki, "Education and the Socialization of Asian Americans: A Revisionist Analysis of the 'Model Minority' Thesis," *Amerasia Journal* 4:2 (1977): 43. See also Fong and Cabezas, "Economic and Employment Status," 48–49; and Commission on Civil Rights, *Social Indicators*, 97–98.

14. U.S. Commission on Civil Rights, "Education Issues" in *Civil Rights Issues of Asian and Pacific Americans: Myths and Realities* (Washington, D.C., 1979), 370–376. This material was presented by Ling-chi Wang, University of California, Berkeley.

15. Illsoo Kim, *New Urban Immigrants: The Korean Community in New York* (Princeton, N.J.: Princeton University Press, 1981).

16. For further discussion of the model minority myth and interpretation of census data, see Deborah Woo, "The Socioeconomic Status of Asian American Women in the Labor Force: An Alternative View," *Sociological Perspectives* 28:3 (July 1985): 307–338.

THE LATINO POPULATION: *The Importance of Economic Restructuring*

24

Joan Moore and Raquel Pinderhughes

. . . American minorities have been incorporated into the general social fabric in a variety of ways. Just as Chicago's black ghettos reflect a history of slavery, Jim Crow legislation, and struggles for civil and economic rights, so the nation's Latino barrios reflect a history of conquest, immigration, and a struggle to maintain cultural identity.

In 1990 there were some 22 million Latinos residing in the United States, approximately 9 percent of the total population. Of these, 61 percent were Mexican in origin, 12 percent Puerto Rican, and 5 percent Cuban. These three groups were the largest, yet 13 percent of Latinos were of Central and South American origin and another 9 percent were classified as "other Hispanic". Latinos were among the fastest-growing segments of the American population, increasing by 7.6 million, or 53 percent, between 1980 and 1990. There are predictions that Latinos will outnumber blacks by the twenty-first century. If Latino immigration and fertility continue at their current rate, there will be over 54 million Latinos in the United States by the year 2020.

From: Joan Moore and Raquel Pinderhughes (eds.), *In the Barrios: Latinos and the Underclass Debate* (New York: Russell Sage Foundation, 1993), pp. xvi–xxix. © 1993 Russell Sage Foundation. Reprinted by permission of Russell Sage Foundation.

This is an old population: as early as the sixteenth century, Spanish explorers settled what is now the American Southwest. In 1848, Spanish and Mexican settlers who lived in that region became United States citizens as a result of the Mexican-American War. Although the aftermath of conquest left a small elite population, the precarious position of the masses combined with the peculiarities of southwestern economic development to lay the foundation for poverty in the current period (see Barrera 1979; Moore and Pachon 1985).

In addition to those Mexicans who were incorporated into the United States after the Treaty of Guadalupe Hidalgo, Mexicans have continually crossed the border into the United States, where they have been used as a source of cheap labor by U.S. employers. The volume of immigration from Mexico has been highly dependent on fluctuations in certain segments of the U.S. economy. This dependence became glaringly obvious earlier in this century. During the Great Depression of the 1930s state and local governments "repatriated" hundreds of thousands of unemployed Mexicans, and just a few years later World War II labor shortages reversed the process as Mexican contract-laborers (*braceros*) were eagerly sought. A little later, in the 1950s, massive deportations recurred when "operation Wetback" repatriated hundreds of thousands of Mexicans. Once again, in the 1980s, hundreds of thousands crossed the border to work in the United States, despite increasingly restrictive legislation.

High levels of immigration and high fertility mean that the Mexican-origin population is quite young—on the average, 9.5 years younger than the non-Latino population—and the typical household is large, with 3.8 persons, as compared with 2.6 persons in non-Latino households (U.S. Bureau of the Census 1991). Heavy immigration, problems in schooling, and industrial changes in the Southwest combine to constrain advancement. The occupational structure remains relatively steady, and though there is a growing middle class, there is also a growing number of very poor people.

The incorporation of Puerto Ricans into the United States began in 1898, when the United States took possession of Puerto Rico and Cuba during the Spanish-American War. Although Cuba gained its independence in 1902, Puerto Rico became a commonwealth of the United States in 1952. Thus Puerto Rican citizens are also citizens of the United States. The colonial relationship strongly influenced the structure of the Puerto Rican economy and the migration of Puerto Ricans to the mainland. As a result of the U.S. invasion, the island's economy was transformed from a diversified, subsistence economy, which emphasized tobacco, cattle, coffee, and sugar, to a one-crop sugar economy, of which more than 60 percent was controlled by absentee U.S. owners (Steward 1956). The constriction of the sugar economy in the 1920s resulted in high unemployment and widespread poverty, and propelled the first wave of Puerto Rican migration to the United States (C. Rodriguez 1989).

Puerto Rican migration to the mainland took place in roughly three periods. The first, 1900–1945, was marked by the arrival of rural migrants forced to leave the island to find work after some of these economic transformations. Many migrants directly responded to U.S. companies who valued Puerto Rican citizenship status and experience in agriculture and recruited Puerto Rican laborers for agriculture and industry in the United States (Morales 1986; Maldonado-Denis 1972). Almost all settled in New York City, most working in low-skilled occupations.

The second period, 1946–1964, is known as the "great migration" because it was during this period that the greatest number of Puerto Ricans migrated. This movement reflected factors that included the search for work, artificially low fares between the island and New York arranged by the island government, labor recruitment, and the emergence of Puerto Rican settlements on the mainland. Though Puerto Ricans were still relegated to low-wage jobs, they were employed in large numbers.

The period after 1965 has been characterized by a fluctuating pattern of net migration as well as greater dispersion to parts of the United States away from New York City (C. Rodriguez 1989). It is known as the "revolving-door migration": during most of this period the heavy flow from the island to the mainland was balanced by equally substantial flows in the opposite direction. However, since 1980 the net outflows from Puerto Rico have rivaled those experienced in the 1950s.

Over the past three decades the economic status of Puerto Ricans dropped precipitously. By 1990, 38 percent of all Puerto Rican families were below the poverty line. A growing proportion of these families were concentrated in poor urban neighborhoods located in declining industrial centers in the Northeast and Midwest, which experienced massive economic restructuring and diminished employment opportunities for those with less education and weaker skills. The rising poverty rate has also been linked to a dramatic increase in female-headed households. Recent studies show that the majority of recent migrants were not previously employed on the island. Many were single women who migrated with their young children (Falcon and Gurak 1991). Currently, Puerto Ricans are the most economically disadvantaged group of all Latinos. As a group they are poorer than African Americans.

Unlike other Latino migrants, who entered the United States as subordinate workers and were viewed as sources of cheap labor, the first large waves of Cuban refugees were educated middle- and upper-class professionals. Arriving in large numbers after Castro's 1959 revolution, Cubans were welcomed by the federal government as bona fide political refugees fleeing communism and were assisted in ways that significantly contributed to their economic well-being. Cubans had access to job-training programs and placement services, housing subsidies, English-language programs, and small-business loans. Federal and state assistance contributed to the growth of a vigorous enclave economy (with Cubans owning many of the businesses and hiring fellow Cubans) and also to the emergence of Miami as a center for Latin American trade. Cubans have the highest family income of all Latino groups. Nevertheless, in 1990, 16.9 percent of the Cuban population lived below the poverty line.

In recent years large numbers of Salvadorans and Guatemalans have come to the United States in search of refuge from political repression. But unlike Cubans, few have been recognized by the U.S. government as bona fide refugees. Their settlement and position in the labor market have been influenced by their undocumented (illegal) status. Dominicans have also come in large numbers to East Coast cities, many also arriving as undocumented workers. Working for the lowest wages and minimum job security, undocumented workers are among the poorest in the nation.

Despite their long history and large numbers, Latinos have been an "invisible minority" in the United States. Until recently, few social scientists and policy analysts concerned with understanding stratification and social problems in the United States have noticed them. Because they were almost exclusively concerned with relations between blacks and whites, social scientists were primarily concerned with generating demographic information on the nation's black and white populations, providing almost no information on other groups. Consequently, it has been difficult, sometimes impossible, to obtain accurate data about Latinos.

Latinos began to be considered an important minority group when census figures showed a huge increase in the population. By 1980 there were significant Latino communities in almost every metropolitan area in the nation. As a group, Latinos have low education, low family incomes, and are more clustered in low-paid, less-skilled occupations. Most Latinos live in cities, and poverty has become an increasing problem. On the whole, Latinos are more likely to live in poverty than the general U.S. population: poverty is widespread for all Latino subgroups except Cubans. They were affected by structural factors that influenced the socioeconomic status of all U.S. workers. In 1990, 28 percent were poor as compared with 13 percent of African Americans (U.S. Bureau of the Census 1991). Puerto Ricans were particularly likely to be poor. . . .

. . . The earliest evidence of a new and important economic change appeared in the 1970s. Jobs seemed to be relocating: they declined massively in some formerly prosperous parts of the country and grew quickly in other, formerly peripheral regions—especially the South and West (Perry and Watkins 1977). It became obvious that the nation as a whole was losing "good" manufacturing jobs as production became internationalized (AFL-CIO Industrial Union Department 1986; Bluestone and Harrison 1982). In the 1990s, white-collar employment began to be restructured as well.

By the late 1980s there was consensus that the geographical shift in the location of job growth was a manifestation of a second and more important aspect of economic restructuring—the shift from a manufacturing to a service economy, and of the increasing globalization of the economy. This was a major transformation, and it became obvious that traditional manufacturing was not going to revive. Jobs continued to be created in the new service and information economy, but many were disproportionately at either the high or the low end of the wage and salary distributions, and many of the new firms functioned without the internal differentiation that might permit workers to move up within the company.

Rustbelt manufacturing decline and Sunbelt growth have come to epitomize what economic restructuring means. But in reality things are a lot more subtle, a lot more complex, and demand a more elaborate conceptualization, especially as these trends affect Latino poverty. Elements of a more complex model are being developed by a number of researchers, but as of this writing none is yet adequate to understand the shifts that are evident in the cities represented in this volume. Several of these deserve particular emphasis.

First, there is the "Rustbelt in the Sunbelt" phenomenon. Some researchers have argued that deindustrialization has been limited to the Rustbelt, and that the causal chain adduced by Wilson therefore does not apply outside that region. But the fact is that many Sunbelt cities developed manufacturing industries, particularly during and after World War II. Thus Rustbelt-style economic restructuring—deindustrialization, in particular—has also affected them deeply. In the late 1970s and early 1980s cities like Los Angeles experienced a major wave of plant closings that put a fair number of Latinos out of work (Morales 1985; Soja, Morales, and Wolff 1983).

Second, there has been significant reindustrialization and many new jobs in many of these cities, a trend that is easily overlooked. Most of the expanding low-wage service and manufacturing industries, like electronics and garment manufacturing, employ Latinos (McCarthy and Valdez 1986; Muller and Espenshade 1986), and some depend almost completely on immigrant labor working at minimum wage (Fernandez-Kelly and Sassen 1991). In short, neither the Rustbelt nor the Sunbelt has seen uniform economic restructuring.

Third, Latinos are affected by the "global cities" phenomenon, particularly evident in New York and Chicago. This term refers to a particular mix of new jobs and populations and an expansion of both high- and low-paid service jobs (see Sassen-Koob 1984). When large multinational corporations centralize their service functions, upper-level service jobs expand. The growing corporate elite want more restaurants, more entertainment, more clothing, and more care for their homes and children, but these new consumer services usually pay low wages and offer only temporary and part-time work. The new service workers in turn generate their own demand for low-cost goods and services. Many of them are Latino immigrants and they create what Sassen calls a "Third World city . . . located in dense groupings spread all over the city": this new "city" also provides new jobs (1989, p. 70).

Los Angeles . . . has experienced many of these patterns. The loss of manufacturing jobs has been far less visible than in New York or Chicago, for although traditional manufacturing declined, until the 1990s high-tech manufacturing did not. Moreover, Los Angeles' international financial and trade functions flourished (Soja 1987). The real difference between Los Angeles on the one hand and New York and Chicago on the other was that more poor people in Los Angeles seemed to be working. In all three cities internationalization had similar consequences for the *structure* of jobs for the poor. More of the immigrants pouring into Los Angeles were finding jobs, while the poor residents of New York and Chicago were not.

Fourth, even though the deindustrialization framework remains of overarching importance in understanding variations in the urban context of Latino poverty, we must also understand that economic restructuring shows many different faces. It is different in economically specialized cities. Houston, for example, has been called "the oil capital of the world", and most of the devastating economic shifts in that city were due to "crisis and reorganization in the world oil-gas industry" (Hill and Feagin 1987, p. 174). Miami is another special case. The economic changes that have swept Miami have little to do with deindustrialization, or with Europe or the Pacific Rim, and much to do with the overpowering influence of its Cuban population, its important "enclave economy", and its "Latino Rim" functions (see Portes and Stepick 1993).

Finally, economic change has a different effect in peripheral areas. Both Albuquerque and Tucson are regional centers in an economically peripheral area. Historically, these two cities served the ranches, farms, and mines of their desert hinterlands. Since World War II, both became military centers, with substantial high-tech defense industrialization. Both cities are accustomed to having a large, poor Latino population, whose poverty is rarely viewed as a crisis. In Tucson, for example, unemployment for Mexican Americans has been low, and there is stable year-round income. But both cities remain marginal to the national economy, and this means that the fate of their poor depends more on local factors.

Laredo has many features in common with other cities along the Texas border, with its substantial military installations, and agricultural and tourist functions. All of these cities have been affected by general swings in the American and Texan economy. These border communities have long been the poorest in the nation, and their largely Mexican American populations have suffered even more from recent economic downturns. They are peripheral to the U.S. economy, but the important point is that their economic well-being is intimately tied to the Mexican economy. They were devastated by the collapse of the peso in the 1980s. They are also more involved than most American cities in international trade in illicit goods, and poverty in Laredo has been deeply affected by smuggling. Though Texas has a long history of discrimination against Mexican Americans, race is not an issue within Laredo itself, where most of the population—elite as well as poor—is of Mexican descent. This fact is of particular importance in evaluating the underclass debate.

THE INFORMAL AND ILLICIT ECONOMIES

The growth of an informal economy is part and parcel of late twentieth-century economic restructuring. Particularly in the global cities, a variety of "informal" economic activities proliferates—activities that are small-scale, informally organized, and largely outside government regulations (cf. Portes, Castells, and Benton 1989). Some low-wage reindustrialization, for example, makes use of new arrangements in well-established industries (like home work in the garment industry, as seamstresses take their work home with them). Small-scale individual activities such as street vending and "handyman" house repairs and alterations affect communities in peripheral as well as global cities. . . .

And, finally, there are illicit activities—most notoriously, a burgeoning drug market. There is not much doubt that the new poverty in the United States has often been accompanied by a resurgence of illicit economic activities (see Fagan, forthcoming, for details on five cities). . . . Most . . . Latino communities . . . have been able to contain or encapsulate such activities so that they do not dominate neighborhood life. But in most of them there is also little doubt that illicit economic activities form an "expanded industry". They rarely provide more than a pittance for the average worker: but for a very small fraction of barrio households they are part of the battery of survival strategies. . . .

IMMIGRATION

Immigration—both international and from Puerto Rico—is of major significance for poor Latino communities in almost every city in every region of the country. Further, there is every reason to believe that immigration will continue to be important.

First, it has important economic consequences. Immigration is a central feature of the economic life of global cities: for example, Los Angeles has been called the "capital of the Third World" because of its huge Latino and Asian immigration (Rieff 1991). . . . Those cities most bound to world trends (New York, Los Angeles, Chicago, Houston, and Miami) experienced massive Latino immigration in the 1980s. In . . . Los Angeles, Houston, and Miami . . . , immigration is a major factor in the labor market, and the residents of the "second settlement" Puerto Rican communities . . . in New York and Chicago operate within a context of both racial and ethnic change and of increased Latino immigration. The restructured economy provides marginal jobs for immigrant workers, and wage scales seem to drop for native-born Latinos in areas where immigration is high. This is a more complicated scenario than the simple loss of jobs accompanying Rustbelt deindustrialization. Immigrants are ineligible for most government benefits, are usually highly motivated, and are driven to take even the poorest-paying jobs. They are also more vulnerable to labor-market swings.

These may be construed as rather negative consequences, but in addition, immigrants have been a constructive force in many cities. For example, [there is] the economic vitality of immigrant-serving businesses [and] the revival of language and of traditional social controls, the strengthening of networks, and the emergence of new community institutions. . . .

References

AFL-CIO Industrial Union Department 1986. *The Polarization of America*. Washington, DC: AFL-CIO Industrial Union Department.

Barrera, Mario 1979. *Race and Class in the Southwest*. Notre Dame, IN: University of Notre Dame Press.

Bluestone, Barry, and Bennett Harrison 1982. *The Deindustrialization of America*. New York: Basic Books.

Falcon, Luis, and Douglas Gurak 1991. "Features of the Hispanic Underclass: Puerto Ricans and Dominicans in New York." Unpublished manuscript.

Fernandez-Kelly, Patricia, and Saskia Sassen 1991. "A Collaborative Study of Hispanic Women in the Garment and Electronics Industries: Executive Summary." New York: New York University, Center for Latin American and Caribbean Studies.

Hill, Richard Child, and Joe R. Feagin 1987. "Detroit and Houston: Two Cities in Global Perspective." In Michael Peter Smith and Joe R. Feagin, eds. In *The Capitalist City*, pp. 155–177. New York: Basil Blackwell.

Maldonado-Denis, Manuel 1972. *Puerto Rico: A Sociohistoric Interpretation.* New York: Random House.

McCarthy, Kevin, and R. B. Valdez 1986. *Current and Future Effects of Mexican Immigration in California.* Santa Monica, CA: Rand Corporation.

Moore, Joan, and Harry Pachon 1985. *Hispanics in the United States.* Englewood Cliffs, NJ: Prentice Hall.

Morales, Julio 1986. *Puerto Rican Poverty and Migration: We Just Had to Try Elsewhere.* New York: Praeger.

Morales, Rebecca 1985. "Transitional Labor: Undocumented Workers in the Los Angeles Automobile Industry." *International Migration Review* 17:570–596.

Muller, Thomas, and Thomas J. Espenshade 1986. *The Fourth Wave.* Washington, DC: Urban Institute Press.

Perry, David, and Alfred Watkins 1977. *The Rise of the Sunbelt Cities.* Beverly Hills, CA: Sage.

Portes, Alejandro, Manuel Castells, and Lauren A. Benton 1989. *The Informal Economy.* Baltimore: Johns Hopkins University Press.

Portes, Alejandro, and Juan Clark 1987. "Mariel Refugees: Six Years After." *Migration World Magazine* 15:14–18.

Portes, Alejandro, and Alex Stepick 1993. *City on the Edge: The Transformation of Miami.* Berkeley: University of California Press.

Rieff, David 1987. *Going to Miami: Exiles, Tourists, and Refugees in the New America.* Boston: Little Brown.

Rodriguez, Clara 1989. *Puerto Ricans: Born in the U.S.A.* Boston: Unwin Hyman.

Sassen, Saskia 1989. "New Trends in the Sociospatial Organization of the New York City Economy." In Robert Beauregard, ed. *Economic Restructuring and Political Response.* Newbury Park, CA.

Sassen-Koob, Saskia 1984. "The New Labor Demand in Global Cities." In Michael Smith, ed. *Cities in Transformation.* Beverly Hills, CA: Sage.

Soja, Edward 1987. "Economic Restructuring and the Internationalization of the Los Angeles Region." In Michael Peter Smith and Joe R. Feagin, eds. *The Capitalist City*, pp. 178–198. New York: Basil Blackwell.

Soja, Edward W., Rebecca Morales, and G. Wolff 1983. "Urban Restructuring: An Analysis of Social and Spatial Change in Los Angeles." *Economic Geography* 59:195–230.

Stevens-Arroyo, Antonio M. 1974. *The Political Philosophy of Pedro Abizu Campos: Its Theory and Practice.* Ibero American Language and Area Center. New York: New York University Press.

Steward, Julian H. 1956. *The People of Puerto Rico.* Urbana, IL: University of Illinois Press.

U.S. Bureau of the Census 1991. *The Hispanic Population in the United States: March 1991.* Current Population Reports, Series P–20, No. 455. Washington, DC: U.S. Government Printing Office.

Families

OUR MOTHERS' GRIEF: *Racial Ethnic Women and the Maintenance of Families*

Bonnie Thornton Dill

25

REPRODUCTIVE LABOR[1] FOR WHITE WOMEN IN EARLY AMERICA

In eighteenth- and nineteenth-century America, the lives of white[2] women in the United States were circumscribed within a legal and social system based on patriarchal authority. This authority took two forms: public and private. The social, legal, and economic position of women in this society was controlled through the private aspects of patriarchy and defined in terms of their relationship to families headed by men. The society was structured to confine white wives to reproductive labor within the domestic sphere. At the same time the formation, preservation and protection of families among white settlers was seen as crucial to the growth and development of American society. Building, maintaining, and supporting families was a concern of the State and of those organizations that prefigured the State. Thus, while white women had few legal rights as women, they were protected through public forms of patriarchy that acknowledged and supported their family roles of wives, mothers, and daughters because they were vital instruments for building American society.

From: *Journal of Family History* 13 (1988): 415–431. Reprinted by permission.

The groundwork for public support of women's family roles was laid during the colonial period. As early as 1619, the London Company began planning for the importation of single women into the colonies to marry colonists, form families, and provide for a permanent settlement. The objective was to make the men "more settled and less moveable . . . instability would breed a dissolution, and so an overthrow of the Plantation" (cited in Spruill 1972, p. 8).

In accordance with this recognition of the importance of families, the London Company provided the economic basis necessary for the development of the family as a viable and essential institution within the nascent social structure of the colonies. Shares of land were allotted for both husbands and wives in recognition of the fact that "in a new plantation it is not known whether men or women be the most necessary" (cited in Spruill 1972, p. 9).

This pattern of providing an economic base designed to attract, promote and maintain families was followed in the other colonial settlements. Lord Baltimore of Maryland " . . . offered to each adventurer a hundred acres for himself, a hundred for his wife, fifty for each child, a hundred for each man servant, and sixty for a woman servant. Women heads of families were treated just as men" (Spruill 1972, p. 11).

In Georgia, which appealed to poorer classes for settlers more than did Virginia or Maryland, " . . . among the advantages they offered men to emigrate was the gainful employment of their wives and children" (Spruill 1972, p. 16).

In colonial America, white women were seen as vital contributors to the stabilization and growth of society. They were therefore accorded some legal and economic recognition through a patriarchal family structure.

> While colonial life remained hard, . . . American women married earlier [than European women], were less restricted by dowries, and often had legal protection for themselves and their children in antenuptial contracts (Kennedy 1979, p. 7).

Throughout the colonial period, women's reproductive labor in the family was an integral part of the daily operation of small-scale family farms or artisan's shops. According to Kessler-Harris (1981), a gender-based division of labor was common, but not rigid. The participation of women in work that was essential to family survival reinforced the importance of their contributions to both the protection of the family and the growth of society.

Between the end of the eighteenth and mid-nineteenth century, what is labeled the "modern American family" developed. The growth of industrialization and an urban middle class, along with the accumulation of agrarian wealth among Southern planters, had two results that are particularly pertinent to this discussion. First, class differentiation increased and sharpened, and with it, distinctions in the content and nature of women's family lives. Second, the organization of industrial labor resulted in the separation of home and family and the assignment to women of a separate sphere of activity focused on childcare and home maintenance. Whereas men's activities became increasingly focused upon the industrial competitive sphere of work, "women's activities were increasingly confined to the care of children, the nurturing of the husband, and the physical maintenance of the home" (Degler 1980, p. 26).

This separate sphere of domesticity and piety became both an ideal for all white women as well as a source of important distinctions between them. As Matthei (1982) points out, tied to the notion of wife as homemaker is a definition of masculinity in which the husband's successful role performance was measured by his ability to keep his wife in the homemaker role. The entry of white women into the labor force came to be linked with the husband's assumed inability to fulfill his provider role.

For wealthy and middle-class women, the growth of the domestic sphere offered a potential for creative development as homemakers and mothers. Given ample financial support from their husband's earnings, some of these women were able to concentrate their energies on the development and elaboration of the more intangible elements of this separate sphere. They were also able to hire other women to perform the daily tasks such as cleaning, laundry, cooking, and ironing. Kessler-Harris cautions, however, that the separation of productive labor from the home did not seriously diminish the amount of physical drudgery associated with housework, even for middle-class women.

> It did relegate the continuing hard work to second place, transforming the public image of the household by the 1820s and 1830s from a place where productive labor was performed to one whose main goals were the preservation of virtue and morality . . . Many of the "well-run" homes of the pre–Civil War period seem to have been the dwelling of overworked women. Short of household help, without modern conveniences, and frequently pregnant, these women complained bitterly of their harsh existence (Kessler-Harris 1981, p. 39).

In effect, household labor was transformed from economic productivity done by members of the family group to home maintenance; childcare and moral uplift done by an isolated woman who perhaps supervised some servants.

Working-class white women experienced this same transformation but their families' acceptance of the domestic code meant that their labor in the home intensified. Given the meager earnings of working-class men, working-class families had to develop alternative strategies to both survive and keep the wives at home. The result was that working-class women's reproductive labor increased to fill the gap between family need and family income. Women increased their own production of household goods through things such as canning and sewing; and by developing other sources of income, including boarders and homework. A final and very important source of other income was wages earned by the participation of sons and daughters in the labor force. In fact, Matthei argues that "the domestic homemaking of married women was supported by the labors of their daughters" (1982, p. 130).

The question arises: Why did white working-class families sacrifice other aspects of this nineteenth-century notion of family, such as privacy and the protection of children, to keep wives as homemakers within the home? Zaretsky (1978) provides a possible answer.

> The Victorian emphasis on the sanctity of the family and on the autonomy of women within the family marked an advance for women of all classes over the interdependent but male dominated subsistence farm of the 18th century . . . most of women's adult life was taken up with childrearing. As a result, a special respect for her place within the home, and particularly for her childrearing activities was appreciated by working class women (p. 211).

Another way in which white women's family roles were socially acknowledged and protected was through the existence of a separate sphere for women. The code of domesticity, attainable for affluent women, became an ideal toward which nonaffluent women aspired. Notwithstanding the personal constraints placed on women's development, the notion of separate spheres promoted the growth and stability of family life among the white middle class and became the basis for working-class men's efforts to achieve a family wage, so that they could keep their wives at home. Also, women gained a distinct sphere of authority and expertise that yielded them special recognition.

During the eighteenth and nineteenth centuries, American society accorded considerable importance to the development and sustenance of European immigrant families. As primary laborers in the reproduction and maintenance of family life, women were acknowledged and accorded the privileges and protections deemed socially appropriate to their family roles. This argument acknowledges the fact that the family structure denied these women many rights and privileges and seriously constrained their individual growth and development. Because women gained social recognition primarily through their membership in families, their personal rights were few and privileges were subject to the will of the male head of the household. Nevertheless, the recognition of women's reproductive labor as an essential building block of the family, combined with a view of the family as the cornerstone of the nation, distinguished the experiences of the white, dominant culture from those of racial ethnics.

Thus, in its founding, American society initiated legal, economic, and social practices designed to promote the growth of family life among European colonists. The reception colonial families found in the United States contrasts sharply with the lack of attention given to the families of racial-ethnics. Although the presence of racial-ethnics was equally as important for the growth of the nation, their political, economic, legal, and social status was quite different.

REPRODUCTIVE LABOR AMONG RACIAL-ETHNICS IN EARLY AMERICA

Unlike white women, racial-ethnic women experienced the oppressions of a patriarchal society but were denied the protections and buffering of a patriarchal family. Their families suffered as a direct result of the organization of the labor systems in which they participated.

Racial-ethnics were brought to this country to meet the need for a cheap and exploitable labor force. Little attention was given to their family and community life except as it related to their economic productivity. Labor, and not the existence or maintenance of families, was the critical aspect of their role in building the nation. Thus they were denied the social structural supports necessary to make *their* families a vital element in the social order. Family membership was not a key means of access to participation in the wider society. The lack of social, legal, and economic support for racial-ethnic families intensified and extended women's reproductive labor, created tensions and strains in family relationships, and set the stage for a variety of creative and adaptive forms of resistance.

AFRICAN-AMERICAN SLAVES

Among students of slavery, there has been considerable debate over the relative "harshness" of American slavery, and the degree to which slaves were permitted or encouraged to form families. It is generally acknowledged that many slaveowners found it economically advantageous to encourage family formation as a way of reproducing and perpetuating the slave labor force. This became increasingly true after 1807 when the importation of African slaves was explicitly prohibited. The existence of these families and many aspects of their functioning, however, were directly controlled by the master. In other words, slaves married and formed families but these groupings were completely subject to the master's decision to let them remain intact. One study has estimated that about 32% of all recorded slave marriages were disrupted by sale, about 45% by death of a spouse, about 10% by choice, with the remaining 13% not disrupted at all (Blassingame 1972, pp. 90–92). African slaves thus quickly learned that they had a limited degree of control over the formation and maintenance of their marriages and could not be assured of keeping their children with them. The threat of disruption was perhaps the most direct and pervasive cultural assault[3] on families that slaves encountered. Yet there were a number of other aspects of the slave system which reinforced the precariousness of slave family life.

In contrast to some African traditions and the Euro-American patterns of the period, slave men were not the main provider or authority figure in the family. The mother-child tie was basic and of greatest interest to the slave-owner because it was critical in the reproduction of the labor force.

In addition to the lack of authority and economic autonomy experienced by the husband-father in the slave family, use of the rape of women slaves as a weapon of terror and control further undermined the integrity of the slave family.

> It would be a mistake to regard the institutionalized pattern of rape during slavery as an expression of white men's sexual urges, otherwise stifled by the spector of the white womanhood's chastity . . . Rape was a weapon of domination, a weapon of repression, whose covert goal was to extinguish slave women's will to resist, and in the process, to demoralize their men (Davis 1981, pp. 23–24).

The slave family, therefore, was at the heart of a peculiar tension in the master-slave relationship. On the one hand, slaveowners sought to encourage familial ties among slaves because, as Matthei (1982) states: " . . . these provided the basis of the development of the slave into a self-conscious socialized human being" (p. 81). They also hoped and believed that this socialization process would help children learn to accept their place in society as slaves. Yet the master's need to control and intervene in the familial life of the slaves is indicative of the other side of this tension. Family ties had the potential for becoming a competing and more potent source of allegiance than the slave-master himself. Also, kin were as likely to socialize children in forms of resistance as in acts of compliance.

It was within this context of surveillance, assault, and ambivalence that slave women's reproductive labor took place. She and her menfolk had the task of preserving the human and family ties that could ultimately give them a reason for living. They had to socialize their children to believe in the possibility of a life in which they were not enslaved. The slave woman's labor on behalf of the family was, as Davis (1971) has pointed out, the only labor the slave engaged in that could not be directly appropriated by the slaveowner for his own profit. Yet, its indirect appropriation, as labor crucial to the reproduction of the slaveowner's labor force, was the source of strong ambivalence for many slave women. Whereas some mothers murdered their babies to keep them from being slaves, many sought within the family sphere a degree of autonomy and creativity denied them in other realms of the society. The maintenance of a distinct African-American culture is testimony to the ways in which slaves maintained a degree of cultural autonomy and resisted the creation of a slave family that only served the needs of the master.

Gutman (1976) provides evidence of the ways in which slaves expressed a unique Afro-American culture through their family practices. He provides data on naming patterns and kinship ties among slaves that flies in the face of the dominant ideology of the period. That ideology argued that slaves were immoral and had little concern for or appreciation of family life.

Yet Gutman demonstrated that within a system which denied the father authority over his family, slave boys were frequently named after their fathers, and many children were named after blood relatives as a way of maintaining family ties. Gutman also suggested that after emancipation a number of slaves took the names of former owners in order to reestablish family ties that had been disrupted earlier. On plantation after plantation, Gutman found considerable evidence of the building and maintenance of extensive kinship ties among slaves. In instances where slave families had been disrupted, slaves in new communities reconstituted the kinds of family and kin ties that came to characterize black family life throughout the South. These patterns included, but were not limited to, a belief in the importance of marriage as a long-term commitment, rules of exogamy that included marriage between first cousins, and acceptance of women who had children outside of marriage. Kinship networks were an important source of resistance to the organization of labor that treated the individual slave, and not the family, as the unit of labor (Caulfield 1974).

Another interesting indicator of the slaves' maintenance of some degree of cultural autonomy has been pointed out by Wright (1981) in her discussion of slave housing. Until the early 1800s, slaves were often permitted to build their housing according to their own design and taste. During that period, housing built in an African style was quite common in the slave quarters. By

1830, however, slaveowners had begun to control the design and arrangement of slave housing and had introduced a degree of conformity and regularity to it that left little room for the slave's personalization of the home. Nevertheless, slaves did use some of their own techniques in construction and often hid it from their masters.

> Even the floors, which usually consisted of only tamped earth, were evidence of a hidden African tradition: slaves cooked clay over a fire, mixing in ox blood or cow dung, and then poured it in place to make hard dirt floors almost like asphalt . . . In slave houses, in contrast to other crafts, these signs of skill and tradition would then be covered over (Wright 1981, p. 48).

Housing is important in discussions of family because its design reflects sociocultural attitudes about family life. The housing that slaveowners provided for their slaves reflected a view of Black family life consistent with the stereotypes of the period. While the existence of slave families was acknowledged, it certainly was not nurtured. Thus, cabins were crowded, often containing more than one family, and there were no provisions for privacy. Slaves had to create their own.

> Slave couples hung up old clothes or quilts to establish boundaries; others built more substantial partitions from scrap wood. Parents sought to establish sexual privacy from children. A few ex-slaves described modified trundle beds designed to hide parental lovemaking . . . Even in one room cabins, sexual segregation was carefully organized (Wright 1981, p. 50).

Perhaps most critical in developing an understanding of slave women's reproductive labor is the gender-based division of labor in the domestic sphere. The organization of slave labor enforced considerable equality among men and women. The ways in which equality in the labor force was translated into the family sphere is somewhat speculative. Davis (1981), for example, suggests that egalitarianism between males and females was a direct result of slavery when she says:

> Within the confines of their family and community life, therefore, Black people managed to accomplish a magnificent feat. They transformed that negative equality which emanated from the equal oppression they suffered as slaves into a positive quality: the egalitarianism characterizing their social relations (p. 18).

It is likely, however, that this transformation was far less direct than Davis implies. We know, for example, that slave women experienced what has recently been called the "double day" before most other women in this society. Slave narratives (Jones 1985; White 1985; Blassingame 1977) reveal that women had primary responsibility for their family's domestic chores. They cooked (although on some plantations meals were prepared for all of the slaves), sewed, cared for their children, and cleaned house, all after completing a full day of labor for the master. Blassingame (1972) and others have pointed out that slave men engaged in hunting, trapping, perhaps some gardening, and furniture making as ways of contributing to the maintenance of their families. Clearly, a gender-based division of labor did exist within the family and it appears that women bore the larger share of the burden for housekeeping and child care.

By contrast to white families of the period, however, the division of labor in the domestic sphere was neither reinforced in the relationship of slave women to work nor in the social institutions of the slave community. The gender-based division of labor among the slaves existed within a social system that treated men and women as almost equal, independent units of labor.[4] Thus Matthei (1982) is probably correct in concluding that:

> Whereas . . . the white homemaker interacted with the public sphere through her husband, and had her work life determined by him, the enslaved Afro-American homemaker was directly subordinated to and determined by her owner . . . The equal enslavement of husband and wife gave the slave marriage a curious kind of equality, an equality of oppression (p. 94).

Black men were denied the male resources of a patriarchal society and therefore were unable to turn gender distinctions into female subordination, even if that had been their desire. Black women, on the other hand, were denied support and protection for their roles as mothers and wives and thus had to modify and structure those roles around the demands of their labor. Thus, reproductive labor for slave women was intensified in several ways: by the demands of slave labor that forced them into the double-day of work; by the desire and need to maintain family ties in the face of a system that gave them only limited recognition; by the stresses of building a family with men who were denied the standard social privileges of manhood; and by the struggle to raise children who could survive in a hostile environment.

This intensification of reproductive labor made networks of kin and quasi-kin important instruments in carrying out the reproductive tasks of the slave community. Given an African cultural heritage where kinship ties formed the basis of social relations, it is not at all surprising that African American slaves developed an extensive system of kinship ties and obligations (Gutman 1976; Sudarkasa 1981). Research on Black families in slavery provides considerable documentation of participation of extended kin in childrearing, childbirth, and other domestic, social, and economic activities (Gutman 1976; Blassingame 1972; Genovese 1974).

After slavery, these ties continued to be an important factor linking individual household units in a variety of domestic activities. While kinship ties were also important among native-born whites and European immigrants, Gutman (1976) has suggested that these ties:

> were comparatively more important to Afro-Americans than to lower-class native white and immigrant Americans, the result of their distinctive low economic status, a condition that denied them the advantages of an extensive associational life beyond the kin group and the advantages and disadvantages resulting from mobility opportunities (p. 213).

His argument is reaffirmed by research on Afro-American families after slavery (Shimkin et al. 1978; Aschenbrenner 1975; Davis 1981; Stack 1974). Sudarkasa (1981) takes this argument one step further and links this pattern to the African cultural heritage.

> historical realities require that the derivation of this aspect of Black family organization be traced to its African antecedents. Such a view does not deny the adaptive significance of consanguineal (kin) networks. In fact, it helps to clarify why these networks had the flexibility they had and why, they, rather than conjugal relationships came to be the stabilizing factor in Black families (p. 49).

With individual households, the gender-based division of labor experienced some important shifts during emancipation. In their first real opportunity to establish family life beyond the controls and constraints imposed by a slavemaster, family life among Black sharecroppers changed radically. Most women, at least those who were wives and daughters of able-bodied men, withdrew from field labor and concentrated on their domestic duties in the home. Husbands took primary responsibility for the fieldwork and for relations with the owners, such as signing contracts on behalf of the family. Black women were severely criticized by whites for removing themselves from field labor because they were seen to be aspiring to a model of womanhood that was considered inappropriate for them. This reorganization of female labor, however, represented an attempt on the part of Blacks to protect women from some of the abuses of the slave system and to thus secure their family life. It was more likely a response to the particular set of circumstances that the newly freed slaves faced than a reaction to the lives of their former masters. Jones (1985) argues that these patterns were "particularly significant" because at a time when industrial development was introducing a labor system that divided male and female labor, the freed black family was establishing a pattern of joint work and complementary tasks between males and females that was reminiscent of the preindustrial American families. Unfortunately, these former slaves had to do this without the institutional supports that white farm families had in the midst of a sharecropping system that deprived them of economic independence.

CHINESE SOJOURNERS

An increase in the African slave population was a desired goal. Therefore, Africans were permitted and even encouraged at times to form families subject to the authority and whim of the master. By sharp contrast, Chinese people were explicitly denied the right to form families in the United States through both law and social practice. Although male laborers began coming to the United States in sizable numbers in the middle of the nineteenth century, it was more than a century before an appreciable number of children of Chinese parents were born in America. Tom, a respondent in Nee and Nee's (1973) book, *Longtime Californ'* says: "One thing about Chinese men in America was you had to be either a merchant or a big gambler, have lot of side money to have a family here. A working man, an ordinary man, just can't!" (p. 80).

Working in the United States was a means of gaining support for one's family with an end of obtaining sufficient capital to return to China and purchase land. The practice of sojourning was reinforced by laws preventing Chinese laborers from becoming citizens, and by restrictions on their entry into this country. Chinese laborers who arrived before 1882 could not bring their wives and were prevented by law from marrying whites. Thus, it is likely that the number of Chinese-American families might have been negligible had it not been for two things: the San Francisco earthquake and fire in 1906, which destroyed all municipal records; and the ingenuity and persistence of the Chinese people who used the opportunity created by the earthquake to increase their numbers in the United States. Since relatives of citizens were permitted entry, American born Chinese (real and claimed) would visit China, report the birth of a son, and thus create an entry slot. Years later the slot could be used by a relative or purchased. The purchasers were called "paper sons." Paper sons became a major mechanism for increasing the Chinese population, but it was a slow process and the sojourner community remained predominantly male for decades.

The high concentration of males in the Chinese community before 1920 resulted in a split household form of family. As Glenn observes:

> In the split household family, production is separated from other func-
> tions and is carried out by a member living far from the rest of the household.
> The rest—consumption, reproduction and socialization—are carried out by
> the wife and other relatives from the home village . . . The split household form
> makes possible maximum exploitation of the workers . . . The labor of prime-
> age male workers can be bought relatively cheaply, since the cost of repro-
> duction and family maintenance is borne partially by unpaid subsistence work
> of women and old people in the home village (Glenn 1981, pp. 14–15).

The women who were in the United States during this period consisted of a small number who were wives and daughters of merchants and a larger percentage who were prostitutes. Hirata (1979) has suggested that Chinese prostitution was an important element in helping to maintain the split-house-hold family. In conjunction with laws prohibiting intermarriage, Chinese prostitution helped men avoid long-term relationships with women in the United States and ensured that the bulk of their meager earnings would continue to support the family at home.

The reproductive labor of Chinese women, therefore, took on two di-mensions primarily because of the split-household family form. Wives who remained in China were forced to raise children and care for in-laws on the meager remittances of their sojourning husband. Although we know few details about their lives, it is clear that the everyday work of bearing and maintaining children and a household fell entirely on their shoulders. Those women who immigrated and worked as prostitutes performed the more nurturant aspects of reproductive labor, that is, providing emotional and sexual companionship for men who were far from home. Yet their role as prostitute was more likely a means of supporting their families at home in China than a chosen vocation.

The Chinese family system during the nineteenth century was a patriar-chal one wherein girls had little value. In fact, they were considered only temporary members of their father's family because when they married, they became members of their husband's families. They also had little social value: girls were sold by some poor parents to work as prostitutes, concubines, or servants. This saved the family the expense of raising them, and their earnings also became a source of family income. For most girls, however, marriages were arranged and families sought useful connections through this process.

With the development of a sojourning pattern in the United States, some Chinese women in those regions of China where this pattern was more prevalent would be sold to become prostitutes in the United States. Most, however, were married off to men whom they saw only once or twice in the 20- or 30-year period during which he was sojourning in the United States. Her status as wife ensured that a portion of the meager wages he earned would be returned to his family in China. This arrangement required considerable sacrifice and adjustment on the part of wives who remained in China and those who joined their husbands after a long separation.

Kingston (1977) tells the story of the unhappy meeting of her aunt, Moon Orchid, with her husband from whom she had been separated for 30 years.

> For thirty years she had been receiving money from him from America. But she had never told him that she wanted to come to the United States. She waited for him to suggest it, but he never did (p. 144).

His response to her when she arrived unexpectedly was to say:

> "Look at her. She'd never fit into an American household. I have important American guests who come inside my house to eat." He turned to Moon Orchid, "You can't talk to them. You can barely talk to me." Moon Orchid was so ashamed, she held her hands over her face. She wished she could also hide her dappled hands (p. 178).

Despite these handicaps, Chinese people collaborated to establish the opportunity to form families and settle in the United States. In some cases it took as long as three generations for a child to be born on United States soil.

> In one typical history, related by a 21 year old college student, great-grandfather arrived in the States in the 1890s as a "paper son" and worked for about 20 years as a laborer. He then sent for the grandfather, who worked alongside greatgrandfather in a small business for several years. Greatgrandfather subsequently returned to China, leaving grandfather to run the business and send remittance. In the 1940s, grandfather sent for father; up to this point, none of the wives had left China. Finally, in the late 1950s father returned to China and brought his wife back with him. Thus, after nearly 70 years, the first child was born in the United States (Glenn 1981, p. 14).

CHICANOS

Africans were uprooted from their native lands and encouraged to have families in order to increase the slave labor force. Chinese people were immigrant laborers whose "permanent" presence in the country was denied. By contrast, Mexican-Americans were colonized and their traditional family life was disrupted by war and the imposition of a new set of laws and conditions of labor. The hardships faced by Chicano families, therefore, were the result of the United States colonization of the indigenous Mexican population, accompanied by the beginnings of industrial development in the region. The treaty of Guadalupe Hidalgo, signed in 1848, granted American citizenship to Mexicans living in what is now called the Southwest. The American takeover, however, resulted in the gradual displacement of Mexicans from the land and their incorporation into a colonial labor force (Barrera 1979). In addition, Mexicans who immigrated into the United States after 1848 were also absorbed into the labor force.

Whether natives of Northern Mexico (which became the United States after 1848) or immigrants from Southern Mexico, Chicanos were a largely peasant population whose lives were defined by a feudal economy and a daily struggle on the land for economic survival. Patriarchal families were important instruments of community life and nuclear family units were linked together through an elaborate system of kinship and godparenting. Traditional life was characterized by hard work and a fairly distinct pattern of sex-role segregation.

> Most Mexican women were valued for their household qualities, men by their ability to work and to provide for a family. Children were taught to get up early, to contribute to the family's labor to prepare themselves for adult life . . . Such a life demanded discipline, authority, deference—values that cemented the working of a family surrounded and shaped by the requirements of Mexico's distinctive historical pattern of agricultural development, especially its pervasive debt peonage (Saragoza 1983, p. 8).

As the primary caretakers of hearth and home in a rural environment, *Las Chicanas* labor made a vital and important contribution to family survival. A description of women's reproductive labor in the early twentieth century can be used to gain insight into the work of the nineteenth-century rural women.

> For country women, work was seldom a salaried job. More often it was the work of growing and preparing food, of making adobes and plastering houses with mud, or making their children's clothes for school and teaching them the hymns and prayers of the church, or delivering babies and treating sicknesses with herbs and patience. In almost every town there were one or two women who, in addition to working in their own homes, served other families in the community as *curanderas* (healers), *parteras* (midwives), and schoolteachers (Elasser 1980, p. 10).

Although some scholars have argued that family rituals and community life showed little change before World War I (Saragoza 1983), the American conquest of Mexican lands, the introduction of a new system of labor, the loss of Mexican-owned land through the inability to document ownership, plus the transient nature of most of the jobs in which Chicanos were employed, resulted in the gradual erosion of this pastoral way of life. Families were uprooted as the economic basis for family life changed. Some immigrated from Mexico in search of a better standard of living and worked in the mines and railroads. Others who were native to the Southwest faced a job market that no longer required their skills and moved into mining, railroad, and agricultural labor in search of a means of earning a living. According to Camarillo (1979), the influx of Anglo[5] capital into the pastoral economy of Santa Barbara rendered obsolete the skills of many Chicano males who had worked as ranchhands and farmers prior to the urbanization of that economy. While some women and children accompanied their husbands to the railroad and mine camps, they often did so despite prohibitions against it. Initially many of these camps discouraged or prohibited family settlement.

The American period (post-1848) was characterized by considerable transiency for the Chicano population. Its impact on families is seen in the growth of female-headed households, which was reflected in the data as early as 1860. Griswold del Castillo (1979) found a sharp increase in female-headed households in Los Angeles, from a low of 13% in 1844 to 31% in 1880. Camarillo (1979, p. 120) documents a similar increase in Santa Barbara from 15% in 1844 to 30% by 1880. These increases appear to be due not so much to divorce, which was infrequent in this Catholic population, but to widowhood and temporary abandonment in search of work. Given the hazardous nature of work in the mines and railroad camps, the death of a husband, father or son who was laboring in these sites was not uncommon. Griswold del Castillo (1979) reports a higher death rate among men than women in Los Angeles. The rise in female-headed households, therefore, reflects the instabilities and insecurities introduced into women's lives as a result of the changing social organization of work.

One outcome, the increasing participation of women and children in the labor force was primarily a response to economic factors that required the modification of traditional values. According to Louisa Vigil, who was born in 1890:

> The women didn't work at that time. The man was supposed to marry that girl and take [care] of her . . . Your grandpa never did let me work for nobody. He always had to work, and we never did have really bad times (Elasser 1980, p. 14).

Señora Vigil's comments are reinforced in Garcia's (1980) study of El Paso. In the 393 households he examined in the 1900 census, he found 17.1% of the women to be employed. The majority of this group were daughters, mothers with no husbands, and single women. In the cases of Los Angeles and Santa Barbara, where there were even greater work opportunities for women than in El Paso, wives who were heads of household worked in seasonal and part-time jobs and lived from the earnings of children and relatives in an effort to maintain traditional female roles.

Slowly, entire families were encouraged to go to railroad workcamps and were eventually incorporated into the agricultural labor market. This was a response both to the extremely low wages paid to Chicano laborers and to the preferences of employers who saw family labor as a way of stabilizing the workforce. For Chicanos, engaging all family members in agricultural work was a means of increasing their earnings to a level close to subsistence for the entire group and of keeping the family unit together. Camarillo (1979, p. 93) provides a picture of the interplay of work, family, and migration in the Santa Barbara area in the following observation:

> The time of year when women and children were employed in the fruit cannery and participated in the almond and olive harvests coincided with the seasons when the men were most likely to be engaged in seasonal migratory work. There were seasons, however, especially in the early summer when the entire family migrated from the city to pick fruit. This type of family seasonal harvest was evident in Santa Barbara by the 1890s. As walnuts replaced almonds and as the fruit industry expanded, Chicano family labor became essential.

This arrangement, while bringing families together, did not decrease the hardships that Chicanas had to confront in raising their families. We may infer something about the rigors of that life from Jesse Lopez de la Cruz's description of the workday of migrant farm laborers in the 1940s. Work conditions in the 1890s were as difficult, if not worse.

> We always went where the women and men were going to work, because if it were just the men working it wasn't worth going out there because we wouldn't earn enough to support a family . . . We would start around 6:30 a.m. and work for four or five hours, then walk home and eat and rest until about three-thirty in the afternoon when it cooled off. We would go back and work until we couldn't see. Then I'd clean up the kitchen. I was doing the housework and working out in the fields and taking care of two children (quoted in Goldman 1981, pp. 119–120).

In the towns, women's reproductive labor was intensified by the congested and unsanitary conditions of the *barrios* in which they lived. Garcia (1980) described the following conditions in El Paso:

> Mexican women had to haul water for washing and cooking from the river or public water pipes. To feed their families, they had to spend time marketing, often in Cuidad Juarez across the border, as well as long, hot hours

> cooking meals and coping with the burden of desert sand both inside and out-
> side their homes. Besides the problem of raising children, unsanitary living
> conditions forced Mexican mothers to deal with disease and illness in their
> families. Diphtheria, tuberculosis, typhus and influenza were never too far
> away. Some diseases could be directly traced to inferior city services . . . As a
> result, Mexican mothers had to devote much energy to caring for sick chil-
> dren, many of whom died (pp. 320–321).

While the extended family has remained an important element of Chicano life, it was eroded in the American period in several ways. Griswold del Castillo (1979), for example, points out that in 1845 about 71% of Angelenos lived in extended families and that by 1880, fewer than half did. This decrease in extended families appears to be a response to the changed economic conditions and to the instabilities generated by the new sociopolitical structure. Addition-ally, the imposition of American law and custom ignored and ultimately undermined some aspects of the extended family. The extended family in traditional Mexican life consisted of an important set of familial, religious, and community obligations. Women, while valued primarily for their domesticity, had certain legal and property rights that acknowledged the importance of their work, their families of origin and their children. In California, for example:

> Equal ownership of property between husband and wife had been one
> of the mainstays of the Spanish and Mexican family systems. Community-
> property laws were written into the civil codes with the intention of strengthen-
> ing the economic controls of the wife and her relatives. The American
> government incorporated these Mexican laws into the state constitution, but
> later court decisions interpreted these statutes so as to undermine the wife's
> economic rights. In 1861, the legislature passed a law that allowed the de-
> ceased wife's property to revert to her husband. Previously it had been inher-
> ited by her children and relatives if she died without a will (Griswold del
> Castillo 1979, p. 69).

The impact of this and other similar court rulings was to "strengthen the property rights of the husband at the expense of his wife and children" (Griswold del Castillo 1979, p. 69).

In the face of the legal, social, and economic changes that occurred during the American period, Chicanas were forced to cope with a series of dislocations in traditional life. They were caught between conflicting pressures to maintain traditional women's roles and family customs and the need to participate in the economic support of their families by working outside the home. During this period the preservation of some traditional customs became an important force for resisting complete disarray.

According to Saragoza (1983), transiency, the effects of racism, and segregation, and proximity to Mexico aided in the maintenance of traditional family practices. Garcia has suggested that women were the guardians of Mexican cultural traditions within the family. He cites the work of anthropologist Manuel Gamio, who identified the retention of many Mexican customs among Chicanos in settlements around the United States in the early 1900s.

> These included folklore, songs and ballads, birthday celebrations, saints' day, baptism, weddings, and funerals in the traditional style. Because of poverty, a lack of physicians in the barrios, and adherence to traditional customs, Mexicans continued to use medicinal herbs. Gamio also identified the maintenance of a number of oral traditions, and Mexican style cooking (Garcia 1980, p. 322).

Of vital importance to the integrity of traditional culture was the perpetuation of the Spanish language. Factors that aided in the maintenance of other aspects of Mexican culture also helped in sustaining the language. However, entry into English-language public schools introduced the children and their families to systematic efforts to erase their native tongue. Griswold del Castillo reports that in the early 1880s there was considerable pressure against the speaker of Spanish in the public school. He also found that some Chicano parents responded to this kind of discrimination by helping support independent bilingual schools. These efforts, however, were short-lived.

Another key factor in conserving Chicano culture was the extended family network, particularly the system of *compadrazgo* or godparenting. Although the full extent of the impact of the American period on the Chicano extended family is not known, it is generally acknowledged that this family system, though lacking many legal and social sanctions, played an important role in the preservation of the Mexican community (Camarillo 1979, p. 13). In Mexican society, godparents were an important way of linking family and community through respected friends or authorities. Named at the important rites of passage in a child's life, such as birth, confirmation, first communion, and marriage, *compadrazgo* created a moral obligation for godparents to act as guardians, to provide financial assistance in times of need, and to substitute in case of the death of a parent. Camarillo (1979) points out that in traditional society these bonds cut across class and racial lines.

> The rites of baptism established kinship networks between rich and poor—between Spanish, mestizo and Indian—and often carried with them political loyalty and economic-occupational ties. The leading California patriarchs in the pueblo played important roles in the compadrazgo network. They sponsored dozens of children for their workers or poor relatives. The kindness

of the *padrino* and *madrina* was repaid with respect and support from the *pobladores* (pp. 12–13).

The extended family network—which included godparents—expanded the support groups for women who were widowed or temporarily abandoned and for those who were in seasonal, part- or full-time work. It suggests, therefore, the potential for an exchange of services among poor people whose income did not provide the basis for family subsistence. Griswold del Castillo (1980) argues that family organization influenced literacy rates and socioeconomic mobility among Chicanos in Los Angeles between 1850 and 1880. His data suggest that children in extended families (defined as those with at least one relative living in a nuclear family household) had higher literacy rates than those in nuclear families. He also argues that those in larger families fared better economically, and experienced less downward mobility. The data here are too limited to generalize to the Chicano experience as a whole but they do reinforce the actual and potential importance of this family form to the continued cultural autonomy of the Chicano community.

CONCLUSION: OUR MOTHERS' GRIEF

Reproductive labor for Afro-American, Chinese-American, and Mexican-American women in the nineteenth century centered on the struggle to maintain family units in the face of a variety of cultural assaults. Treated primarily as individual units of labor rather than as members of family groups, these women labored to maintain, sustain, stabilize, and reproduce their families while working in both the public (productive) and private (reproductive) spheres. Thus, the concept of reproductive labor, when applied to women of color, must be modified to account for the fact that labor in the productive sphere was required to achieve even minimal levels of family subsistence. Long after industrialization had begun to reshape family roles among middle-class white families, driving white women into a cult of domesticity, women of color were coping with an extended day. This day included subsistence labor outside the family and domestic labor within the family. For slaves, domestics, migrant farm laborers, seasonal factory-workers, and prostitutes, the distinctions between labor that reproduced family life and which economically sustained it were minimized. The expanded workday was one of the primary ways in which reproductive labor increased.

Racial-ethnic families were sustained and maintained in the face of various forms of disruption. Yet they and their families paid a high price in the process.

High rates of infant mortality, a shortened life span, the early onset of crippling and debilitating disease provided some insight into the costs of survival.

The poor quality of housing and the neglect of communities further increased reproductive labor. Not only did racial-ethnic women work hard outside the home for a mere subsistence, they worked very hard inside the home to achieve even minimal standards of privacy and cleanliness. They were continually faced with disease and illness that directly resulted from the absence of basic sanitation. The fact that some African women murdered their children to prevent them from becoming slaves is an indication of the emotional strain associated with bearing and raising children while participating in the colonial labor system.

We have uncovered little information about the use of birth control, the prevalence of infanticide, or the motivations that may have generated these or other behaviors. We can surmise, however, that no matter how much children were accepted, loved, or valued among any of these groups of people, their futures in a colonial labor system were a source of grief for their mothers. For those children who were born, the task of keeping them alive, of helping them to understand and participate in a system that exploited them, and the challenge of maintaining a measure—no matter how small—of cultural integrity, intensified reproductive labor.

Being a racial-ethnic woman in nineteenth century American society meant having extra work both inside and outside the home. It meant having a contradictory relationship to the norms and values about women that were being generated in the dominant white culture. As pointed out earlier, the notion of separate spheres of male and female labor had contradictory outcomes for the nineteenth-century whites. It was the basis for the confinement of women to the household and for much of the protective legislation that subsequently developed. At the same time, it sustained white families by providing social acknowledgment and support to women in the performance of their family roles. For racial-ethnic women, however, the notion of separate spheres served to reinforce their subordinate status and became, in effect, another assault. As they increased their work outside the home, they were forced into a productive labor sphere that was organized for men and "desperate" women who were so unfortunate or immoral that they could not confine their work to the domestic sphere. In the productive sphere, racial-ethnic women faced exploitative jobs and depressed wages. In the reproductive sphere, however, they were denied the opportunity to embrace the dominant ideological definition of "good" wife or mother. In essence, they were faced with a double-bind situation, one that required their participation in the labor force to sustain family life but damned them as women, wives, and mothers because they did not confine their labor to the home. Thus, the conflict between ideology and reality in the lives of racial-ethnic women during the nineteenth century sets the stage for stereotypes, issues of self-esteem, and conflicts around gender-role prescriptions that surface more fully in the twentieth century. Further, the tensions and conflicts that characterized their lives during this period provided the impulse for community activism to jointly address the inequities, which they and their children and families faced.

ACKNOWLEDGMENTS

The research in this study is the result of the author's participation in a larger collaborative project examining family, community, and work lives of racial-ethnic women in the United States. The author is deeply indebted to the scholarship and creativity of members of the group in the development of this study. Appreciation is extended to Elizabeth Higginbotham, Cheryl Townsend Gilkes, Evelyn Nakano Glenn, and Ruth Zambrana (members of the original working group), and to the Ford Foundation for a grant that supported in part the work of this study.

NOTES

1. The term *reproductive labor* is used to refer to all of the work of women in the home. This includes but is not limited to: the buying and preparation of food and clothing, provision of emotional support and nurturance for all family members, bearing children, and planning, organizing, and carrying out a wide variety of tasks associated with their socialization. All of these activities are necessary for the growth of patriarchal capitalism because they maintain, sustain, stabilize, and *reproduce* (both biologically and socially) the labor force.

2. The term *white* is a global construct used to characterize peoples of European descent who migrated to and helped colonize America. In the seventeenth century, most of these immigrants were from the British Isles. However, during the time period covered by this article, European immigrants became increasingly diverse. It is a limitation of this article that time and space does not permit a fuller discussion of the variations in the white European immigrant experience. For the purposes of the argument made herein and of the contrast it seeks to draw between the experiences of mainstream (European) cultural groups and that of racial/ethnic minorities, the differences among European settlers are joined and the broad similarities emphasized.

3. Cultural assaults, according to Caulfield (1974) are benign and systematic attacks on the institutions and forms of social organization that are fundamental to the maintenance and flourishing of a group's culture.

4. Recent research suggests that there were some tasks that were primarily assigned to males and some others to females. Whereas some gender-role distinctions with regard to work may have existed on some plantations, it is clear that slave women were not exempt from strenuous physical labor.

5. This term is used to refer to white Americans of European ancestry.

REFERENCES

Aschenbrenner, Joyce. 1975. *Lifelines: Black Families in Chicago.* New York, NY: Holt, Rinehart, and Winston.

Barrera, Mario. 1979. *Race and Class in the Southwest.* South Bend, IN: Notre Dame University Press.

Blassingame, John. 1972. *The Slave Community: Plantation Life in the Antebellum South.* New York: Oxford University Press.

————. 1977. *Slave Testimony: Two Centuries of Letters, Speeches, Interviews, and Auto-biographies.* Baton Rouge, LA: Louisiana State University Press.

Camarillo, Albert. 1979. *Chicanos in a Changing Society.* Cambridge, MA: Harvard University Press.

Caulfield, Mina Davis. 1974. "Imperialism, The Family, and Cultures of Resistance." *Socialist Review* 4(2)(October): 67–85.

Davis, Angela. 1971. "The Black Woman's Role in the Community of Slaves." *Black Scholar* 3(4)(December): 2–15.

————. 1981. *Women, Race and Class.* New York: Random House.

Degler, Carl. 1980. *At Odds.* New York: Oxford University Press.

Elasser, Nan Kyle MacKenzie, and Yvonne Tixier Y. Vigil. 1980. *Las Mujeres.* New York: The Feminist Press.

Garcia, Mario T. 1980. "The Chicano in American History: The Mexican Women of El Paso, 1880–1920—A Case Study." *Pacific Historical Review* 49(2)(May): 315–358.

Genovese, Eugene D. and Elinor Miller, eds. 1974. *Plantation, Town, and County: Essays on the Local History of American Slave Society.* Urbana: University of Illinois Press.

Glenn, Evelyn Nakano. 1981. "Family Strategies of Chinese-Americans: An Institutional Analysis." Paper presented at the Society for the Study of Social Problems Annual Meetings.

Goldman, Marion S. 1981. *Gold Diggers and Silver Miners*. Ann Arbor: The University of Michigan Press.

Griswold del Castillo, Richard. 1979. *The Los Angeles Barrio: 1850–1890*. Los Angeles: The University of California Press.

Gutman, Herbert. 1976. *The Black Family in Slavery and Freedom: 1750–1925*. New York: Pantheon.

Hirata, Lucie Cheng. 1979. "Free, Indentured, Enslaved: Chinese Prostitutes in Nineteenth-Century America." *Signs* 5 (Autumn): 3–29.

Jones, Jacqueline. 1985. *Labor of Love, Labor of Sorrow*. New York: Basic Books.

Kennedy, Susan Estabrook. 1979. *If All We Did Was to Weep at Home: A History of White Working-Class Women in America*. Bloomington: Indiana University Press.

Kessler-Harris, Alice. 1981. *Women Have Always Worked*. Old Westbury: The Feminist Press.

———. 1982. *Out to Work*. New York: Oxford University Press.

Kingston, Maxine Hong. 1977. *The Woman Warrior*. Vintage Books.

Matthei, Julie. 1982. *An Economic History of Women in America*. New York: Schocken Books.

Nee, Victor G., and Brett de Bary Nee. 1973. *Longtime Californ'*. New York: Pantheon Books.

Saragoza, Alex M. 1983. "The Conceptualization of the History of the Chicano Family: Work, Family, and Migration in Chicanos." Research Proceedings of the Symposium on Chicano Research and Public Policy. Stanford, CA: Stanford University, Center for Chicano Research.

Shimkin, Demetri, E. M. Shimkin, and D. A. Frate, eds. 1978. *The Extended Family in Black Societies*. The Hague: Mouton.

Spruill, Julia Cherry. 1972. *Women's Life and Work in the Southern Colonies*. New York: W. W. Norton and Company (First published in 1938, University of North Carolina Press).

Stack, Carol S. 1974. *All Our Kin: Strategies for Survival in a Black Community*. Harper and Row.

Sudarkasa, Niara. 1981. "Interpreting the African Heritage in Afro-American Family Organization." Pp. 37–53 in *Black Families*, edited by Harriette Pipes McAdoo. Beverly Hills, CA: Sage Publications.

White, Deborah Gray. 1985. *Ar'n't I a Woman?: Female Slaves in the Plantation South.* New York: W. W. Norton.

Wright, Gwendolyn. 1981. *Building the Dream: A Social History of Housing in America.* New York: Pantheon Books.

Zaretsky, Eli. 1978. "The Effects of the Economic Crisis on the Family." Pp. 209–218 in *U.S. Capitalism in Crisis,* edited by Crisis Reader Editorial Collective. New York: Union of Radical Political Economists.

PUERTO RICAN ELDERLY WOMEN:

Shared Meanings and Informal Supportive Networks

26

Melba Sánchez-Ayéndez

INTRODUCTION

Studies of older adults' support systems have seldom taken into account how values within a specific cultural context affect expectations of support and patterns of assistance in social networks. Such networks and supportive relations have a cultural dimension reflecting a system of shared meanings. These meanings affect social interaction and the expectations people have of their relationships with others.

Ethnicity and gender affect a person's adjustment to old age. Although sharing a "minority" position produces similar consequences among members of different ethnic minority groups, the groups' diversity lies in their distinctive systems of shared meanings. Studies of older adults in ethnic minority groups have rarely focused on the cultural contents of ethnicity affecting the aging process, particularly of women (Barth 1969). Cultural value orientations are central to understanding how minority elders approach growing old and how they meet the physical and emotional changes associated with aging.

This article describes the interplay between values and behavior in family and community of a group of older Puerto Rican women living on low incomes in Boston.[1] It explores how values emphasizing family interdependence and different roles of women and men shape the women's expectations, behavior, and supportive familial and community networks.

BEING A WOMAN IS DIFFERENT FROM BEING A MAN

The women interviewed believe in a dual standard of conduct for men and women. This dual standard is apparent in different attributes assigned to women and men, roles expected of them, and authority exercised by them.

The principal role of men in the family is viewed as that of provider; their main responsibility is economic in nature. Although fathers are expected to be affectionate with their children, child care is not seen to be a man's responsibility. Men are not envisioned within the domestic sphere.

The "ideal" man must be the protector of the family, able to control his emotions and be self-sufficient. Men enjoy more freedom in the public world than do women. From the women's perspective, the ideal of maleness is linked to the concept of *machismo*. This concept assumes men have a stronger sexual drive than women, a need to prove virility by the conquest of women, a dominant position in relation to females, and a belligerent attitude when confronted by male peers.

The women see themselves as subordinate to men and recognize the preeminence of male authority. They believe women ought to be patient and largely forbearing in their relations with men, particularly male family members. Patience and forbearance, however, are not confused with passivity or total submissiveness. The elderly Puerto Rican women do not conceive of themselves or other women as "resigned females" but as dynamic beings, continually devising strategies to improve everyday situations within and outside the household.

Rosa Mendoza,[2] now sixty-five, feels no regrets for having decided at thirty years of age and after nine years of marriage not to put up with her husband's heavy drinking any longer. She moved out of her house and went to live with her mother.

> I was patient for many years. I put up with his drunkenness and worked hard to earn money. One day I decided I'd be better off without him. One thing is to be patient, and another to be a complete fool. So I moved out.

Although conscious of their subordinate status to their husbands, wives are also aware of their power and the demands they can make. Ana Fuentes recalls when her husband had a mistress. Ana was thirty-eight.

> I knew he had a mistress in a nearby town. I was patient for a long time, hoping it would end. Most men, sooner or later, have a mistress somewhere. But when it didn't end after quite a time and everyone in the neighborhood knew about it, I said "I am fed up!" He came home one evening and the things I told him! I even said I'd go to that woman's house and beat her if I had to. . . . He knew I was not bluffing; that this was not just another argument. He tried to answer back and I didn't let him. He remained silent. . . . And you know what? He stopped seeing her! A woman can endure many things for a long time, but the time comes when she has to defend her rights.

These older Puerto Rican women perceive the home as the center around which the female world revolves. Home is the woman's domain; women generally make decisions about household maintenance and men seldom intervene.

Family relations are considered part of the domestic sphere and therefore a female responsibility. The women believe that success in marriage depends on the woman's ability to "make the marriage work."

> A marriage lasts as long as the woman decides it will last. It is us who make a marriage work, who put up with things, who try to make ends meet, who yield.

The norm of female subordination is evident in the view that marriage will last as long as the woman "puts up with things" and deals with marriage from her subordinate status. Good relations with affinal kin are also a woman's responsibility. They are perceived as relations between the wife's domestic unit and other women's domestic units.

Motherhood

Motherhood is seen by these older Puerto Rican women as the central role of women. Their concept of motherhood is based on the female capacity to bear children and on the notion of *marianismo*, which presents the Virgin Mary as a role model (Stevens 1973). *Marianismo* presupposes that it is through motherhood that a woman realizes herself and derives her life's greatest satisfactions.

A woman's reproductive role is viewed as leading her toward more commitment to and a better understanding of her children than is shown by the father. One of the women emphasized this view:

> It is easier for a man to leave his children and form a new home with another woman, or not to be as forgiving of children as a mother is. They will

never know what it is like to carry a child inside, feel it growing, and then bring that child into the world. This is why a mother is always willing to forgive and make sacrifices. That creature is part of you; it nourished from you and came from within you. But it is not so for men. To them, a child is a being they receive once it is born. The attachment can never be the same.

The view that childrearing is their main responsibility in life comes from this conceptualization of the mother-child bond. For the older women, raising children means more than looking after the needs of offspring. It involves being able to offer them every possible opportunity for a better life, during childhood or adulthood, even if this requires personal sacrifices.

As mother and head of the domestic domain, a woman is also responsible for establishing the bases for close and good relations among her children. From childhood through adulthood, the creation and maintenance of family unity among offspring is considered another female responsibility.

FAMILY UNITY AND INTERDEPENDENCE

Family Unity

Ideal family relations are seen as based on two interrelated themes, family unity and family interdependence. Family unity refers to the desirability of close and intimate kin ties, with members getting along well and keeping in frequent contact despite dispersal.

Celebration of holidays and special occasions are seen as opportunities for kin to be together and strengthen family ties. Family members, particularly grandparents, adult children, and grandchildren, are often reunited at Christmas, New Year's, Mother's and Father's days, Easter, and Thanksgiving. Special celebrations like weddings, baptisms, first communions, birthdays, graduations, and funerals occasion reunions with other family members. Whether to celebrate happy or sad events, the older women encourage family gatherings as a way of strengthening kinship ties and fostering family continuity.

The value the women place on family unity is also evident in their desire for frequent interaction with kin members. Visits and telephone calls demonstrate a caring attitude by family members which cements family unity.

Family unity is viewed as contributing to the strengthening of family interdependence. Many of the older women repeat a proverb when referring to family unity: *En la unión está la fuerza.* ("In union there is strength.") They believe that the greater the degree of unity in the family, the greater the emphasis family members will place on interdependence and familial obligation.

Family Interdependence

Despite adaptation to life in a culturally different society, Puerto Rican families in the United States are still defined by strong norms of reciprocity among family members, especially those in the immediate kinship group (Cantor 1979; Carrasquillo 1982; Delgado 1981; Donaldson and Martínez 1980; Sánchez-Ayéndez 1984). Interdependence within the Puerto Rican symbolic framework "fits an orientation to life that stresses that the individual is not capable of doing everything and doing it well. Therefore, he should rely on others for assistance" (Bastida 1979: 70). Individualism and self-reliance assume a different meaning from the one prevailing in the dominant U.S. cultural tradition. Individuals in Puerto Rican families will expect and ask for assistance from certain people in their social networks without any derogatory implications for self-esteem.

Family interdependence is a value to which these older Puerto Rican women strongly adhere. It influences patterns of mutual assistance with their children as well as expectations of support. The older women expect to be taken care of during old age by their adult children. The notion of filial duty ensues from the value orientation of interdependence. Adult children are understood to have a responsibility toward their aged parents in exchange for the functions that parents performed for them throughout their upbringing. Expected reciprocity from offspring is intertwined with the concept of filial love and the nature of the parent-child relationship.

Parental duties of childrearing are perceived as inherent in the "parent" role and also lay the basis for long-term reciprocity with children, particularly during old age. The centrality that motherhood has in the lives of the older women contributes to creating great expectations among them of reciprocity from children. More elderly women than men verbalize disappointment when one of their children does not participate in the expected interdependence ties. Disappointment is unlikely to arise when an adult child cannot help due to financial or personal reasons. However, it is bound to arise when a child chooses not to assist the older parent for other reasons.

These older Puerto Rican women stress that good offspring ought to help their parents, contingent upon available resources. Statements such as the following are common:

> Of course I go to my children when I have a problem! To whom would I turn? I raised them and worked very hard to give them the little I could. Now that I am old, they try to help me in whatever they can. . . . Good offspring should help their aged parents as much as they are able to.

Interdependence for Puerto Rican older parents also means helping their children and grandchildren. Many times they provide help when it is not explicitly requested. They are happy when they can perform supportive tasks for their children's families. The child who needs help, no matter how old, is not judged as dependent or a failure.

Reciprocity is not based on strictly equal exchanges. Due to the rapid pace of life, lack of financial resources, or personal problems, adult children are not always able to provide the care the elder parent needs. Many times, the older adults provide their families with more financial and instrumental assistance than their children are able to provide them. Of utmost importance to the older women is not that their children be able to help all the time, but that they visit or call frequently. They place more emphasis on emotional support from their offspring than on any other form of support.

Gloria Santos, for example, has a son and a daughter. While they do not live in the same state as their mother, they each send her fifty to seventy dollars every month. Yet, she is disappointed with her children and explains why:

> They both have good salaries but call me only once or twice a month. I hardly know my grandchildren. All I ask from them is that they be closer to me, that they visit and call me more often. They only visit me once a year and only for one or two days. I've told my daughter that instead of sending me money she could call me more often. I was a good mother and worked hard in order for them to get a good education and have everything. All I expected from them was to show me they care, that they love me.

The importance that the older women attach to family interdependence does not imply that they constantly require assistance from children or that they do not value their independence. They prefer to live in their own households rather than with their adult children. They also try to solve as many problems as possible by themselves. But when support is needed, the adult children are expected to assist the aged parent to the degree they are able. This does not engender conflict or lowered self-esteem for the aged adult. Conflict and dissatisfaction are caused when adult children do not offer any support at all.

SEX ROLES AND FAMILIAL SUPPORTIVE NETWORKS

The family is the predominant source of support for most of these older women, providing instrumental and emotional support in daily life as well as assistance during health crises or times of need. Adult children play a central role in providing familial support to old parents. For married women, husbands are also an important component of their support system. At the same time, most of the older women still perform functional roles for their families.

Support from Adult Children

The support and helpfulness expected from offspring is related to perceptions of the difference between men and women. Older women seek different types of assistance from daughters than from sons. Daughters are perceived as being inherently better able to understand their mothers due to their shared status and qualities as women; they are also considered more reliable. Sons are not expected to help as much as daughters or in the same way. When a daughter does not fulfill the obligations expected of her, complaints are more bitter than if the same were true of a son: "Men are different; they do not feel as we feel. But she is a woman; she should know better." Daughters are also expected to visit and/or call more frequently than are sons. As women are linked closely to the domestic domain, they are held responsible for the care of family relations.

Motherhood is perceived as creating an emotional bond among women. When daughters become mothers, the older women anticipate stronger ties and more support from them.

> Once a daughter experiences motherhood, she understands the suffering and hardships you underwent for her. Sons will never be able to understand this.
>
> My daughter always helped me. But when she became a mother for the first time, she grew much closer to me. It was then when she was able to understand how much a mother can love.

Most of the older women go to a daughter first when confronted by an emotional problem. Daughters are felt to be more patient and better able to understand them as women. It is not that older women never discuss their emotional problems with their sons, but they prefer to discuss them with their daughters. For example, Juana Rivera has two sons who live in the same city as she and a daughter who resides in Puerto Rico. She and her sons get along well and see each other often. The sons stop by their mother's house every day after work, talk about daily happenings, and assist her with some tasks. However, when a physical exam revealed a breast tumor thought to be malignant, it was to her daughter in Puerto Rico that the old woman expressed her worries. She recalls that time of crisis:

> Eddie was with me when the doctor told me of the possibility of a tumor. I was brave. I didn't want him to see me upset. They [sons] get nervous when I

get upset or cry. . . . That evening I called my daughter and talked to her. . . . She was very understanding and comforted me. I can always depend on her to understand me. She is the person who better understands me. My sons are also understanding, but she is a woman and understands more.

Although adult children are sources of assistance during the illnesses of their mothers, it is generally daughters from whom more is expected. Quite often daughters take their sick parents into their homes or stay overnight in the parental household in order to provide better care. Sons, as well as daughters, take the aged parent to the hospital or doctors' offices and buy medicines if necessary. However, it is more often daughters who check on their parents, provide care, and perform household chores when the parent is sick.

When the old women have been hospitalized, adult children living nearby tend to visit the hospital daily. Daughters and daughters-in-law sometimes cook special meals for the sick parent and bring the meals to the hospital. Quite often, adult children living in other states or in Puerto Rico come to help care for the aged parent or be present at the time of an operation. When Juana Rivera had exploratory surgery on her breast, her daughter came from Puerto Rico and stayed with her mother throughout the convalescence. Similarly, when Ana Toledo suffered a stroke and remained unconscious for four days, three of her six children residing in other states came to be with her and their siblings. After her release from the hospital, a daughter from New Jersey stayed at her mother's house for a week. When she left, the children who live near the old woman took turns looking after her.

Most adult children are also helpful in assisting with chores of daily living. At times, offspring take their widowed mothers grocery shopping. Other times, the older women give their children money to do the shopping for them. Daughters are more often asked to do these favors and to also buy personal care items and clothes for their mothers. Some adult offspring also assist by depositing Social Security checks, checking post office boxes, and buying money orders.

Support from Elderly Mothers

The Puerto Rican older women play an active role in providing assistance to their adult children. Gender affects the frequency of emotional support offered as well as the dynamics of the support. The older women offer advice more often to daughters than to sons on matters related to childrearing. And the approach used differs according to the children's gender. For example, one older woman stated,

> I never ask my son openly what is wrong with him. I do not want him to think that I believe he needs help to solve his problems; he is a man. . . . Yet, as a mother I worry. It is my duty to listen and offer him advice. With my daughter it is different; I can be more direct. She doesn't have to prove to me that she is self-sufficient.

Another woman expressed similar views:

> Of course I give advice to my sons! When they have had problems with their wives, their children, even among themselves, I listen to them, and tell them what I think. But with my daughters I am more open. You see, if I ask one of my sons what is wrong and he doesn't want to tell me, I don't insist too much; I'll ask later, maybe in a different way; and they will tell me sooner or later. With my daughters, if they don't want to tell me, I insist. They know I am a mother and a woman like them and that I can understand.

Older mothers perceive sons and daughters as in equal need of support. Daughters, however, are understood to face additional problems in areas such as conjugal relations, childrearing, and sexual harassment, due to their status as women.

Emotional support to daughters-in-law is also offered, particularly when they are encountering marriage or childrearing problems. Josefina Montes explains the active role she played in comforting her daughter-in-law, whose husband was having an extramarital affair:

> I told her not to give up, that she had to defend what was hers. I always listened to her and tried to offer some comfort. . . . When my son would come to my home to visit I would ask him "What is wrong with you? Don't you realize what a good mother and wife that woman is?" . . . I made it my business that he did not forget the exceptional woman she is. . . . I told him I didn't want to ever see him with the other one and not to mention her name in front of me. . . . I was on his case for almost two years. . . . All the time I told her to be patient. . . . It took time but he finally broke up with the other one.

When relations between mother and daughters-in-law are not friendly, support is not usually present. Eulalia Valle says that when her son left his wife and children to move in with another woman, there was not much she could do for her daughter-in-law.

> There was not much I could do. What could I tell him? I couldn't say she was nice to me. . . . Once I tried to make him see how much she was hurting and he replied: "Don't defend her. She has never been fond of you and you know it." What could I reply to that? All I said was, "That's true but, still, she must be very hurt." But there was nothing positive to say about her!

Monetary assistance generally flows from the older parent to the adult children, although few old people are able to offer substantial financial help. Direct monetary assistance, rarely exceeding fifty dollars, is less frequent than gift-giving. Gift-giving usually takes the form of monetary contributions for specific articles needed by their children or children's families. In this way the older people contribute indirectly to the maintenance of their children's families.

The older women also play an active role in the observance of special family occasions and holidays. On the days preceding the celebration, they are busy cooking traditional Puerto Rican foods. It is expected that those in good health will participate in the preparation of foods. This is especially true on Christmas and Easter when traditional foods are an essential component of the celebrations.

Cooking for offspring is also a part of everyday life. In many of the households, meals prepared in the Puerto Rican tradition are cooked daily "in case children or grandchildren come by." Josefina Montes, for example, cooks a large quantity of food every day for herself, her husband, and their adult children and grandchildren. Her daughters come by after work to visit and pick up their youngest children, who stay with grandparents after school. The youngest daughter eats dinner at her parents' home. The oldest takes enough food home to serve her family. Doña[3] Josefina's sons frequently drop by after work or during lunch and she always insists that they eat something.

The older women also provide assistance to their children during health crises. When Juana Rivera's son was hospitalized for a hernia operation, she visited the hospital every day, occasionally bringing food she had prepared for him. When her son was released, Doña Juana stayed in his household through-out his convalescence, caring for him while her daughter-in-law went off to work.

The aged women also assist their children by taking care of grandchil-dren. Grandchildren go to their grandmother's house after school and stay until their parents stop by after work. If the children are not old enough to walk home by themselves, the grandparent waits for them at school and brings them home. The women also take care of their grandchildren when they are not old enough to attend school or are sick. They see their role as grandmothers as a continuation or reenactment of their role as mothers and childrearers.

The women, despite old age, have a place in the functional structure of their families. The older women's assistance is an important contribution to their children's households and also helps validate the women's sense of their importance and helpfulness.

Mutual Assistance in Elderly Couples

Different conceptions of women and men influence interdependence between husband and wife as well as their daily tasks. Older married women are responsible for domestic tasks and perform household chores. They also take care of grandchildren, grocery shopping, and maintaining family relations. Older married men have among their chores depositing Social Security checks, going to the post office, and buying money orders. Although they stay in the house for long periods, the men go out into the community more often than do their wives. They usually stop at the *bodegas*,[4] which serve as a place for socializing and exchange of information, to buy items needed at home and newspapers from Puerto Rico.

Most married couples have a distinctive newspaper reading pattern. The husband comments on the news to his wife as he reads or after he has finished. Sometimes, after her husband finishes reading and commenting on the news, the older woman reads about it herself. Husbands also inform their wives of ongoing neighborhood events learned on their daily stops at the *bodegas*. Wives, on the other hand, inform husbands of familial events learned through their daily telephone conversations and visits from children and other kin members.

The older couple escort each other to service-providing agencies, even though they are usually accompanied by an adult child, adolescent grandchild, or social worker serving as translator. An older man still perceives himself in the role of "family protector" by escorting the women in his family, particularly his wife.

Older husbands and wives provide each other with emotional assistance. They are daily companions and serve as primary sources of confidence for each other, most often sharing children's and grandchildren's problems, health concerns, or financial worries. The couple do not always agree on solutions or approaches for assisting children when sharing their worries about offspring. Many times the woman serves as a mediator in communicating her husband's problems to adult children. The men tend to keep their problems, particularly financial and emotional ones, to themselves or tell their wives but not their children. This behavior rests upon the notion of men as financially responsible for the family, more self-sufficient, and less emotional than women.

Among the older couples, the husband or wife is generally the principal caregiver during the health crises of their spouse. Carmen Ruiz, for example, suffers from chronic anemia and tires easily. Her husband used to be a cook and has taken responsibility for cooking meals and looking after the household. When Providencia Cruz's husband was hospitalized she spent many hours each day at the hospital, wanting to be certain he was comfortable. She brought meals she had cooked for him, arranged his pillows, rubbed him with bay leaf rubbing alcohol, or watched him as he slept. When he was convalescing at home, she was his principal caregiver. Doña Providencia suffers from osteoarthritis and gastric acidity. When she is in pain and spends the day in bed, her husband provides most of the assistance she needs. He goes to the drugstore to buy medicine or ingredients used in folk remedies. He knows how to prepare the mint and chamomile teas she drinks when not feeling well. He also rubs her legs and hands with ointments when the arthritic pain is more intense than usual. Furthermore, during the days that Doña Providencia's ailments last, he performs most of the household chores.

While both spouses live, the couple manages many of their problems on their own. Assistance from other family members with daily chores or help during an illness is less frequent when the woman still lives with her husband than when she lives alone. However, if one or both spouses is ill, help from adult children is more common.

FRIENDS AND NEIGHBORS AS COMMUNITY SOURCES OF SUPPORT

Friends and neighbors form part of the older women's support network. However, the women differentiate between "neighbors" and "friends." Neighbors, unlike kin and friends, are not an essential component of the network which provides emotional support. They may or may not become friends. Supportive relations with friends involve being instrumental helpers, companions, and confidants. Neighbors are involved only in instrumental help.

Neighbors as Sources of Support

Contact with neighbors takes the form of greetings, occasional visits, and exchanges of food, all of which help to build the basis for reciprocity when and if the need arises. The establishment and maintenance of good relations with neighbors is considered to be important since neighbors are potentially helpful during emergencies or unexpected events. Views such as the following are common: "It is good to get acquainted with your neighbors; you never know when you might need them."

Josefina Rosario, a widow, has lived next door to an older Puerto Rican couple for three years. Exchange of food and occasional visits are part of her interaction with them. Her neighbor's husband, in his mid-sixties, occasionally runs errands for Doña Josefina, who suffers from rheumatoid arthritis and needs a walker to move around. If she runs out of a specific food item, he goes to the grocery store for her. Other times, he buys stamps, mails letters, or goes to the drugstore to pick up some medicines for her. Although Doña Josefina cannot reciprocate in the same way, she repays her neighbors by visiting every other week and exchanging food. Her neighbors tell her she is to call them day or night if she ever feels sick. Although glad to have such "good neighbors" as she call them, she stresses she does not consider them friends and therefore does not confide her personal problems to them.

Supportive Relationships Among Friends

Although friends perform instrumental tasks, the older women believe that a good friend's most important quality is being able to provide emotional support. A friend is someone willing to help during the "good" and "bad" times, and is trustworthy and reserved. Problems may be shared with a friend with the certainty that confidences will not be betrayed. A friend provides emotional support not only during a crisis or problem, but in everyday life. Friends are companions, visiting and/or calling on a regular basis.

Friendship for this group of women is determined along gender lines. They tend to be careful about men. Relationships with males outside the immediate familial group are usually kept at a formal level. Mistrust of men is based upon the women's notion of *machismo*. Since men are conceived of as having a stronger sexual drive, the women are wary of the possibility of sexual advances, either physical or verbal. None of the women regards a male as a confidant friend. Many even emphasize the word *amiga* ("female friend") instead of *amigo* ("male friend"). Remarks such as the following are common:

> I've never had an *amigo*. Men cannot be trusted too much. They might misunderstand your motives and some even try to make a pass at you.

The few times the women refer to a male as a friend they use the term *amigo de la familia* ("friend of the family"). This expression conveys that the friendly relations are not solely between the woman and the man. The expression is generally used to refer to a close friend of the husband. *Amigos de la familia* may perform instrumental tasks, be present at family gatherings and unhappy events, or drop by to chat with the respondent's husband during the day. However, relations are not based on male-female relationships.

Age similarity is another factor that seems to affect selection of friends. The friendship networks of the older women are mainly composed of people sixty years of age and older. Friends who fill the role of confidant are generally women of a similar age. The women believe that younger generations, generally, have little interest in the elders. They also state that people their own age are better able to understand their problems because they share many of the same difficulties and worries.

Friends often serve as escorts, particularly in the case of women who live alone. Those who know some English serve as translators on some occasions. Close friends also help illiterate friends by reading and writing letters.

Most of the support friends provide one another is of an emotional nature, which involves sharing personal problems. Close friends entrust one another with family and health problems. This exchange occurs when friends either visit or call each other on the telephone. A pattern commonly observed between dyads of friends is daily calls. Many women who live alone usually call the friend during the morning hours, to make sure she is all right and to find out how she is feeling.

Another aspect of the emotional support the older women provide one another is daily companionship, occurring more often among those who live alone. For example, Hilda Montes and Rosa Mendoza sit together from 1:00 to 3:00 in the afternoon to watch soap operas and talk about family events, neighborhood happenings, and household management. At 3:00 P.M., whoever is at the other's apartment leaves because their grandchildren usually arrive from school around 4:00 P.M.

Friends are also supportive during health crises. If they cannot come to visit, they inquire daily about their friend's health by telephone. When their health permits, some friends perform menial household chores and always bring food for the sick person. If the occasion requires it, they prepare and/or administer home remedies. Friends, in this sense, alleviate the stress adult children often feel in assisting their aged mothers, particularly those who live by themselves. Friends take turns among themselves or with kin in taking care of the ill during the daytime. Children generally stay throughout the night.

Exchange ties with female friends include instrumental support, companionship, and problem sharing. Friends, particularly age cohorts, play an important role in the emotional well-being of the elders.

The relevance of culture to experience of old age is seen in the influence of value orientations on the expectations these Puerto Rican women have of themselves and those in their informal supportive networks. The way a group's cultural tradition defines and interprets relationships influences how elders use their networks to secure the support needed in old age. At the same time, the extent to which reality fits culturally-based expectations will contribute, to a large extent, to elders' sense of well-being.

NOTES

1. The article is based on a nineteen-month ethnographic study. The research was supported by the Danforth Foundation; Sigma Xi; the Scientific Research Society; and the Delta Kappa Gamma Society International.

2. All names are fictitious.

3. The deference term *Doña* followed by the woman's first name is a common way by which to address elderly Puerto Rican women and the one preferred by those who participated in the study.

4. Neighborhood grocery stores, generally owned by Puerto Ricans or other Hispanics, where ethnic foods can be purchased.

REFERENCES CITED

Barth, F. 1969. Introduction to *Ethnic Groups and Boundaries*, F. Barth, ed. Boston: Little, Brown.

Bastida, E. 1979. "Family Integration and Adjustment to Aging Among Hispanic American Elderly." Ph.D. dissertation, University of Kansas.

Cantor, M. H. 1979. "The Informal Support System of New York's Inner City Elderly: Is Ethnicity a Factor?" In *Ethnicity and Aging*, D. L. Gelfand and A. J. Kutzik, eds. New York: Springer.

Carrasquillo, H. 1982. "Perceived Social Reciprocity and Self-Esteem Among Elderly Barrio Antillean Hispanics and Their Familial Informal Networks." Ph.D. dissertation, Syracuse University.

Delgado, M. 1981. "Hispanic Elderly and Natural Support Systems: A Special Focus on Puerto Ricans." Paper presented at the Scientific Meeting of the Boston Society for Gerontological Psychiatry, November, Boston, Mass.

Donaldson, E. and E. Martínez. 1980. "The Hispanic Elderly of East Harlem." *Aging* 305–306: 6–11.

Sánchez-Ayéndez, M. 1984. "Puerto Rican Elderly Women: Aging in an Ethnic Minority Group in the United States." Ph.D. dissertation, University of Massachusetts at Amherst.

Stevens, E. P. 1973. "Marianismo: The Other Face of Machismo in Latin America." In *Female and Male in Latin America*, A. Pescatello, ed. Pittsburgh: University of Pittsburgh Press.

MAN CHILD: *A Black Lesbian* Feminist's Response

27

Audre Lorde

This article is not a theoretical discussion of Lesbian Mothers and their Sons, nor a how-to article. It is an attempt to scrutinize and share some pieces of that common history belonging to my son and me. I have two children: a fifteen-and-a-half-year-old daughter Beth, and a fourteen-year-old son Jonathan. This is the way it was/is with me and Jonathan, and I leave the theory to another time and person. This is one woman's telling.

I have no golden message about the raising of sons for other lesbian mothers, no secret to transpose your questions into certain light. I have my own ways of rewording those same questions, hoping we will all come to speak those questions and pieces of our lives we need to share. We are women making contact within ourselves and with each other across the restrictions of a printed page, bent upon the use of our own/one another's knowledges.

The truest direction comes from inside. I give the most strength to my children by being willing to look within myself, and by being honest with them about what I find there, without expecting a response beyond their years. In this way they begin to learn to look beyond their own fears.

All our children are outriders for a queendom not yet assured.

My adolescent son's growing sexuality is a conscious dynamic between Jonathan and me. It would be presumptuous of me to discuss Jonathan's sexuality here, except to state my belief that whomever he chooses to explore this area with, his choices will be nonoppressive, joyful, and deeply felt from within, places of growth.

One of the difficulties in writing this piece has been temporal; this is the summer when Jonathan is becoming a man, physically. And our sons must become men—such men as we hope our daughters, born and unborn, will be pleased to live among. Our sons will not grow into women. Their way is more difficult than that of our daughters, for they must move away from us, without

From: Audre Lorde, *Sister Outsider* (Freedom, CA: Crossing Press, 1984), pp. 72–80. First published in *Conditions: Four* (1979). Reprinted by permission.

us. Hopefully, our sons have what they have learned from us, and a howness to forge it into their own image.

Our daughters have us, for measure or rebellion or outline or dream; but the sons of lesbians have to make their own definitions of self as men. This is both power and vulnerability. The sons of lesbians have the advantage of our blueprints for survival, but they must take what we know and transpose it into their own maleness. May the goddess be kind to my son, Jonathan.

Recently I have met young Black men about whom I am pleased to say that their future and their visions, as well as their concerns within the present, intersect more closely with Jonathan's than do my own. I have shared vision with these men as well as temporal strategies for our survivals and I appreciate the spaces in which we could sit down together. Some of these men I met at the First Annual Conference of Third World Lesbians and Gays held in Washington D.C. in October, 1979. I have met others in different places and do not know how they identify themselves sexually. Some of these men are raising families alone. Some have adopted sons. They are Black men who dream and who act and who own their feelings, questioning. It is heartening to know our sons do not step out alone.

When Jonathan makes me angriest, I always say he is bringing out the testosterone in me. What I mean is that he is representing some piece of myself as a woman that I am reluctant to acknowledge or explore. For instance, what does "acting like a man" mean? For me, what I reject? For Jonathan, what he is trying to redefine?

Raising Black children—female and male— in the mouth of a racist, sexist, suicidal dragon is perilous and chancy. If they cannot love and resist at the same time, they will probably not survive. And in order to survive they must let go. This is what mothers teach—love, survival—that is, self-definition and letting go. For each of these, the ability to feel strongly and to recognize those feelings is central: how to feel love, how to neither discount fear nor be overwhelmed by it, how to enjoy feeling deeply.

I wish to raise a Black man who will not be destroyed by, nor settle for, those corruptions called *power* by the white fathers who mean his destruction as surely as they mean mine. I wish to raise a Black man who will recognize that the legitimate objects of his hostility are not women, but the particulars of a structure that programs him to fear and despise women as well as his own Black self.

For me, this task begins with teaching my son that I do not exist to do his feeling for him.

Men who are afraid to feel must keep women around to do their feeling for them while dismissing us for the same supposedly "inferior" capacity to feel deeply. But in this way also, men deny themselves their own essential humanity, becoming trapped in dependency and fear.

As a Black woman committed to a liveable future, and as a mother loving and raising a boy who will become a man, I must examine all my possibilities of being within such a destructive system.

Jonathan was three-and-one-half when Frances, my lover, and I met; he was seven when we all began to live together permanently. From the start, Frances' and my insistence that there be no secrets in our household about the fact that we were lesbians has been the source of problems and strengths for both children. In the beginning, this insistence grew out of the knowledge, on both our parts, that whatever was hidden out of fear could always be used either against the children or ourselves—one imperfect but useful argument for honesty. The knowledge of fear can help make us free.

> for the embattled
> there is no place
> that cannot be
> home
> nor is.*

For survival, Black children in america must be raised to be warriors. For survival, they must also be raised to recognize the enemy's many faces. Black children of lesbian couples have an advantage because they learn, very early, that oppression comes in many different forms, none of which have anything to do with their own worth.

To help give me perspective, I remember that for years, in the namecalling at school, boys shouted at Jonathan not—"your mother's a lesbian"—but rather—"your mother's a nigger."

When Jonathan was eight years old and in the third grade we moved, and he went to a new school where his life was hellish as a new boy on the block. He did not like to play rough games. He did not like to fight. He did not like to stone dogs. And all this marked him early on as an easy target.

When he came in crying one afternoon, I heard from Beth how the corner bullies were making Jonathan wipe their shoes on the way home whenever Beth wasn't there to fight them off. And when I heard that the ringleader was a little boy in Jonathan's class his own size, an interesting and very disturbing thing happened to me.

My fury at my own long-ago impotence, and my present pain at his suffering, made me start to forget all that I knew about violence and fear, and

---

*From "School Note" in *The Black Unicorn* (W.W. Norton and Company, New York, 1978), p. 55.

blaming the victim, I started to hiss at the weeping child. "The next time you come in here crying . . . ," and I suddenly caught myself in horror.

This is the way we allow the destruction of our sons to begin—in the name of protection and to ease our own pain. *My* son get beaten up? I was about to demand that he buy that first lesson in the corruption of power, that might makes right. I could hear myself beginning to perpetuate the age-old distortions about what strength and bravery really are.

And no, Jonathan didn't have to fight if he didn't want to, but somehow he did have to feel better about not fighting. An old horror rolled over me of being the fat kid who ran away, terrified of getting her glasses broken.

About that time a very wise woman said to me, "Have you ever told Jonathan that once you used to be afraid, too?"

The idea seemed far-out to me at the time, but the next time he came in crying and sweaty from having run away again, I could see that he felt shamed at having failed me, or some image he and I had created in his head of mother/woman. This image of woman being able to handle it all was bolstered by the fact that he lived in a household with three strong women, his lesbian parents and his forthright older sister. At home, for Jonathan, power was clearly female.

And because our society teaches us to think in an either/or mode—kill or be killed, dominate or be dominated—this meant that he must either surpass or be lacking. I could see the implications of this line of thought. Consider the two western classic myth/models of mother/son relationships: Jocasta/Oedipus, the son who fucks his mother, and Clytemnestra/Orestes, the son who kills his mother.

It all felt connected to me.

I sat down on the hallway steps and took Jonathan on my lap and wiped his tears. "Did I ever tell you about how I used to be afraid when I was your age?"

I will never forget the look on that little boy's face as I told him the tale of my glasses and my after-school fights. It was a look of relief and total disbelief, all rolled into one.

It is as hard for our children to believe that we are not omnipotent as it is for us to know it, as parents. But that knowledge is necessary as the first step in the reassessment of power as something other than might, age, privilege, or the lack of fear. It is an important step for a boy, whose societal destruction begins when he is forced to believe that he can only be strong if he doesn't feel, or if he wins.

I thought about all this one year later when Beth and Jonathan, ten and nine, were asked by an interviewer how they thought they had been affected by being children of a feminist.

Jonathan said that he didn't think there was too much in feminism for boys, although it certainly was good to be able to cry if he felt like it and not to have to play football if he didn't want to. I think of this sometimes now when I see him practicing for his Brown Belt in Tai Kwon Do.

The strongest lesson I can teach my son is the same lesson I teach my daughter: how to be who he wishes to be for himself. And the best way I can do this is to be who I am and hope that he will learn from this not how to be me, which is not possible, but how to be himself. And this means how to move to that voice from within himself, rather than to those raucous, persuasive, or threatening voices from outside, pressuring him to be what the world wants him to be.

And that is hard enough.

Jonathan is learning to find within himself some of the different faces of courage and strength, whatever he chooses to call them. Two years ago, when Jonathan was twelve and in the seventh grade, one of his friends at school who had been to the house persisted in calling Frances "the maid." When Jonathan corrected him, the boy then referred to her as "the cleaning woman." Finally Jonathan said, simply, "Frances is not the cleaning women, she's my mother's lover." Interestingly enough, it is the teachers at this school who still have not recovered from his openness.

Frances and I were considering attending a Lesbian/Feminist conference this summer, when we were notified that no boys over ten were allowed. This presented logistic as well as philosophical problems for us, and we sent the following letter:

> Sisters:
> Ten years as an interracial lesbian couple has taught us both the dangers of an oversimplified approach to the nature and solutions of any oppression, as well as the danger inherent in an incomplete vision.
> Our thirteen-year-old son represents as much hope for our future world as does our fifteen-year-old daughter, and we are not willing to abandon him to the killing streets of New York City while we journey west to help form a Lesbian-Feminist vision of the future world in which we can all survive and flourish. I hope we can continue this dialogue in the near future, as I feel it is important to our vision and our survival.

The question of separatism is by no means simple. I am thankful that one of my children is male, since that helps to keep me honest. Every line I write shrieks there are no easy solutions.

I grew up in largely female environments, and I know how crucial that has been to my own development. I feel the want and need often for the society of women, exclusively. I recognize that our own spaces are essential for developing and recharging.

As a Black woman, I find it necessary to withdraw into all-Black groups at times for exactly the same reasons—differences in stages of development and differences in levels of interaction. Frequently, when speaking with men and white women, I am reminded of how difficult and time-consuming it is to have to reinvent the pencil every time you want to send a message.

But this does not mean that my responsibility for my son's education stops at age ten, any more than it does for my daughter's. However, for each of them, that responsibility does grow less and less as they become more woman and man.

Both Beth and Jonathan need to know what they can share and what they cannot, how they are joined and how they are not. And Frances and I, as grown women and lesbians coming more and more into our power, need to relearn the experience that difference does not have to be threatening.

When I envision the future, I think of the world I crave for my daughters and my sons. It is thinking for survival of the species—thinking for life.

Most likely there will always be women who move with women, women who live with men, men who choose men. I work for a time when women with women, women with men, men with men, all share the work of a world that does not barter bread or self for obedience, nor beauty, nor love. And in that world we will raise our children free to choose how best to fulfill themselves. For we are jointly responsible for the care and raising of the young, since *that* they be raised is a function, ultimately, of the species.

Within that tripartite pattern of relating/existence, the raising of the young will be the joint responsibility of all adults who choose to be associated with children. Obviously, the children raised within each of these three relationships will be different, lending a special savor to that eternal inquiry into how best can we live our lives.

Jonathan was three-and-a-half when Frances and I met. He is now fourteen years old. I feel the living perspective that having lesbian parents has brought to Jonathan is a valuable addition to his human sensitivity.

Jonathan has had the advantage of growing up within a nonsexist relationship, one in which this society's pseudonatural assumptions of ruler/ruled are being challenged. And this is not only because Frances and I are lesbians, for unfortunately there are some lesbians who are still locked into patriarchal patterns of unequal power relationships.

These assumptions of power relationships are being questioned because Frances and I, often painfully and with varying degrees of success, attempt to evaluate and measure over and over again our feelings concerning power, our own and others'. And we explore with care those areas concerning how it is used and expressed between us and between us and the children, openly and otherwise. A good part of our biweekly family meetings are devoted to this exploration.

As parents, Frances and I have given Jonathan our love, our openness, and our dreams to help form his visions. Most importantly, as the son of lesbians, he has had an invaluable model—not only of a relationship—but of relating.

Jonathan is fourteen now. In talking over this paper with him and asking his permission to share some pieces of his life, I asked Jonathan what he felt were the strongest negative and the strongest positive aspects for him in having grown up with lesbian parents.

He said the strongest benefit he felt he had gained was that he knew a lot more about people than most other kids his age that he knew, and that he did not have a lot of the hang-ups that some other boys did about men and women.

And the most negative aspect he felt, Jonathan said, was the ridicule he got from some kids with straight parents.

"You mean, from your peers?" I said.

"Oh no," he answered promptly. "My peers know better. I mean other kids."

REPORTS FROM THE FRONT:

Welfare Mothers up in Arms

28

Diane Dujon, Judy Gradford, and Dottie Stevens

Women on welfare know they are in constant battle to provide for themselves and their children. Their "enemies" are multiple, often including husbands, the welfare bureaucracy, and the attitudes of the general public. In this essay representatives from a group of welfare recipients in the Boston area who have organized into a welfare-rights group called ARMS (Advocacy for Resources for Modern Survival) describe some of the skirmishes they face every day.*

---

*In addition to the three authors, several other members of the ARMS collective should be mentioned for their contributions to the larger unpublished paper from which this essay is taken, "Welfare Mothers up in Arms." They are Angela Hannon, Marion Graham, Jeannie MacKenzie, Carolyn Turner, and Hope Habtemarian. The cited quotations in this essay are taken from interviews with various ARMS members.

From: Rochelle Lefkowitz and Ann Withorn (eds.), *For Crying Out Loud: Women and Poverty in the United States* (New York: Pilgrim Press, 1986), pp. 211–219. Reprinted by permission of the Pilgrim Press, Cleveland, Ohio.

OUR LIVES NO LONGER BELONG TO US

One of my sons was diagnosed as having a high lead level in his blood. The Welfare Department placed my son under protective services and told me that I would have to find another place to live or they would put my son into a foster home. With six children on a welfare budget, it's not easy to find an apartment. And I had to find one within thirty days! To keep the state from taking my son, I was forced to move into the first available housing I could find.

Since I was an emergency case and eligible for a housing subsidy, my name was placed at the top of the list. I had to take the first available unit offered by the Housing Authority. The offer: a brand new town-house-type apartment *fifty miles away* in a white, middle-class suburb!

I knew this move would devastate my family because we would be so far away from our relatives and friends. When you're poor, you have to depend on your family and friends to help you through when you don't have the money to help yourself. At least two or three times every month I take my children to my mother's house to eat. How would we ever be able to get to her house from fifty miles away?

I also knew that my neighbors would not welcome me and my children: a black single woman with six children. I imagined the sneers of the merchants as I paid for my groceries with food stamps and the grunts of the doctors as I pulled out my Medicaid card.

I thought about the problems of transportation that were sure to crop up. How would I get my children to school? What if they got sick; how far was the nearest hospital? I envisioned the seven of us walking for miles with grocery bags. In short, I felt no relief at having found a nice clean apartment within the allotted time, but my back was against the wall. I could not refuse or my son would be put into foster care.

We now live in a totally hostile environment severed from our family and friends. And although we live in a physically beautiful development, life for us is hard. A poor family with no transportation is lost in the suburbs. We are as isolated as if we lived on a remote island in the Pacific.

Situations like the one described by this woman show how our lives no longer belong to us. We have, in effect, married the state. To comply with the conditions of our recipient status, we cannot make any personal decisions ourselves. We must consult the Welfare Department first, and the final decision is theirs. The state is a domineering, chauvinistic spouse.

Politicians often boast or complain about the many services and benefits welfare recipients receive. For us, these services and benefits are the bait that the predatory department uses to entrap us and our families. Like the wiley fox, we are driven by hunger to the trap. We must carefully trip the trap, retrieve the bait, and escape, hopefully unscathed. Also similar to the fox, our incompetence can lead to starvation, disease, and death for ourselves and our families.

This may seem an unlikely analogy to some; but the Welfare Department, in its *eagerness* to help us constantly adopts policies that put us in catch-22 situations. We are continuously in a dilemma over whether we should seek the help we desperately need or not. The purportedly "free" social services that are available to us extort a usurer's fee in mental and physical anguish. So, although there are several services available, we are often unable or reluctant to receive them. Below we describe some of the catch-22's that constitute mental cruelty for us.

CATCH-22: A LOW BUDGET

The first catch-22 we encounter is living under the conditions set up by the state for recipients. Under penalty of law we are required adequately to house, clothe, feed, and otherwise care for our children on a budget that is two thirds of the amount considered to be at the poverty line. If we fail to fulfill our obligation in the opinion of friends, strangers, neighbors, relatives, enemies, or representatives of the Welfare Department, the state can and *will* take our children from our homes. Anyone can call the department anonymously and report that we are neglecting or abusing our children. With no further questions, the state initiates an investigation, which further jeopardizes our family stability. While it is necessary for the state to protect children from abuse and neglect, a large portion of the investigations are based on unfounded allegations for which no one can be held accountable. It is no easy task to fulfill the basic obligation of surviving on welfare, because we often pay as much as 85 to 95 percent of our income for rent and utilities.

On the other hand, there is no reward for a job well done. If we manage to clothe, feed, and house our children, the risk is the same: at the least, biting remarks from people in public, and at the most, an investigation of fraud.

We are constantly under public scrutiny. We are made to feel uncomfortable if we wear jewelry, or buy a nice blouse, or own a warm coat. It's as if we're not supposed to have families or friends who love us and might give us a birthday or Christmas present. Absolutely no thought is given to the fact that we may have had a life before welfare! Heaven forbid that anyone should honor the great job we must do as shoppers!

There are a few legal ways in which welfare mothers can supplement their monthly grant. Some of these supports are available upon eligibility for Aid to Families with Dependent Children (AFDC); others, termed "social services," have additional individual eligibility requirements. Each has its quota of catch-22's:

CATCH-22: EMERGENCY ASSISTANCE

I was $300 in arrears with my electric bill. The electric company sent me several reminders, but I didn't have the money to pay my bill. It made me very nervous. It was winter, and although I had oil, if my electricity were turned off, my pilot light would go out and my children would be cold.

I took my bill and the warning notices to my social worker at the Welfare Department. She told me that I could receive up to $500 of Emergency Assistance per year, but that only one such grant could be made in any twelve-month period. However, I could not receive any Emergency Assistance until I received a 'shut-off' notice. She further explained that I should wait because if I received the $300 EA grant, I could not get the additional $200 to which I was entitled that year. I would have to wait twelve months before I could be eligible for another grant. I really wasn't interested in getting all I could, I just wanted to be able to sleep at night; but since I didn't have a shut-off notice, I had no choice but to wait.

I received a shut-off notice when my bill was about $400. I applied for, and received, the EA grant.

The very next year, I was in a similar situation. I again attempted to wait for the shut-off notice. One day I received a notice from the electric company stating that I was scheduled for a "field collection." My social worker reminded me that EA can only be awarded upon my receipt of a shut-off notice. Even though the notice stated that my service would be "interrupted" if I failed to honor the collector, it did not have the specific words *shut off* and, therefore, I was ineligible for EA. I couldn't believe what she was telling me! To become eligible, I had to go to the electric company to ask them to stamp "shut off" on my bill. I was angry that I was being forced to humiliate myself by revealing my personal business to the electric company representative.

CATCH-22: MEDICAID

Medicaid is the most treasured benefit to families who must depend on AFDC, but it, too, falls short of the expectations of beneficiaries. Doctors, hospitals, druggists, and other health providers often refuse to accept Medicaid patients. The amount of paper work that is required for each and every patient is tedious and time-consuming. Medicaid also sets limits on the type and amount of treatments it will cover.

Every trip to the doctor is a grueling, costly, and time-consuming event. First, we must search for a doctor or medical facility that will accept Medicaid. If we are lucky enough to have a neighborhood health center nearby, we will often go to the clinic. In either situation, we usually have to wait for hours to receive the medical attention we need. Doctors often remark about the amount of paperwork that is required by the state and the fact that they often have to wait six to eight months to obtain their fees from the state. It is exceedingly distressing to be sick and to have to hear about the doctor's problems.

> I had periodontal disease once. The dentist explained that an infection had settled under my gums and he would have to cut my gums and scrape the infection away. Since I was on Medicaid, I was required to wait until Medicaid approved the dental procedure.
>
> After several weeks, the approval arrived. My dentist informed me that although Medicaid approved the procedure, the amount approved was too low to allow him to use gas as was customary. I had a choice: either I could pay him the difference, and *enjoy* a painless procedure; or I could have the procedure done for the cost Medicaid allotted and he could use novocaine, which would be at least moderately painful.
>
> I didn't have the money to pay the difference, but I knew that if I delayed the operation I stood a good chance of losing my teeth. I decided to brave the novocaine.
>
> The dentist had to cut deep into my gums and the novocaine did nothing for the pain below the surface. I tried hard to be still and keep my mouth open wide, but I was in agony with the pain. The procedure took four hours and required sixty-four stitches and forty-seven injections of novocaine!

CATCH-22: FOOD STAMPS

Food Stamps are a symbol of the government's benevolence. Rather than increase the amount of the welfare budget so that we can afford to buy more food, the government *supplements* our budgets with food stamps. As the name suggests, food is all that can be purchased with them. And not much food at that. Households receiving the maximum amount of food stamps receive an average of forty cents per meal per person. All other commodities, such as soap, detergent, toilet paper, diapers, etc., must be separated at the time of purchase. Food, by anyone's definition, is a necessity for sustaining life. Why, then, do we feel as if food is a luxury?

Contrary to public opinion, we pay for these food stamps at a cost significantly higher than a cup of coffee per meal. We pay with the anguish of wondering how we are going to maintain healthy children on $1.20 per day. Food stamps last an average of ten days, depending on the supply of staples (flour, sugar, cereal, salt, spaghetti, etc.) we have on hand; the rest of the month we struggle to keep up with the milk, eggs, juice, fruits, vegetables, and bread so vital to good health. The last two weeks we are challenged to use our imaginations to ensure that our children receive the best nutrition possible. For those of us who are lucky enough to be able to commit "fraud" through friends and relatives, it's a little easier; but for many of us it's often an impossible task!

> The way the food stamp budget is calculated, it's as if our diet is supposed to shrink in the summer. Our fuel costs are counted as an expense in the winter, so we receive more food stamps in the winter and less in the summer. This budget policy is ludicrous because most of our fuel costs in the winter are paid with Fuel Assistance.
>
> I had trouble one time receiving my food stamps. Every month I would have to commute to the next town where my welfare office was located to report that I had not received my food stamps. Each time I was interrogated by the food stamps worker about whether I had cashed my food stamps and was trying to get some more under false pretenses. I had to sign a sworn statement to the effect that I had not received my food stamps before they could issue replacement stamps.
>
> I finally decided that rather than go through the hassle and expense of picking up my food stamps from the welfare office each month, I would purchase a post office box. When I went to inform my worker of my box number, I was told that food stamps could not be sent to a post office box. This policy was supposed to deter fraud. I was forced to continue to pick up my stamps from the welfare office each month.

Food stamp redemption centers are generally located in areas that are virtually inaccessible to those of us without cars. Much of the money we are supposed to be saving with food stamps is spent on transportation to the centers.

> My food stamps usually come on the due date, but my welfare check is often late. This creates a problem for me because I always need the food, but with no cash money it is almost impossible to do the shopping I need to do the first time. I have to make at least two trips to the supermarket: the first trip for food only with the food stamps and the second, when my check comes, to buy soap, cleaning products, etc. Two trips to the store doubles the amount of transportation expenses, too.

CATCH-22: WELFARE FRAUD

Fraud within the welfare system might also be called "devising a way to survive." The federal and state laws call it fraud if a welfare recipient uses up her allotment for any reason and seeks assistance from a friend, relative, or acquaintance (be it ten cents or $10). If she does not report this money to the welfare office, she is technically considered to be defrauding the Welfare Department. Such unrealistic definitions leave all of us vulnerable to fraud and make it difficult to separate honest need from intentional deception.

According to the narrow, unrealistic guidelines of the state and federal government, most or all welfare recipients could be accused of having committed fraud at some time during their ordeal with the welfare system, even though they would not have meant to defraud anyone. We are stuck in a system that inadequately provides for us and that even the social workers know, depends upon our having a "little help from our friends." However, if we are caught doing what we all have to do to survive, we may even be made into an "example" and used to discredit the difficulties faced by women on welfare.

We are poor because we do not receive enough money from the Welfare Department to live decently. If we lack family or friends, we may feel forced to find ways to get a little extra money for our families. Many who have been discovered working to buy Christmas presents, for example, have been brought to court by the Welfare Department and either fined, jailed, or made to pay back all monies received while working.

Although there seems to be large-scale vendor fraud among those who supply Medicaid services, it is seldom investigated or taken to court. While we are hounded for minor infractions, little is done to the unscrupulous doctors, dentists, druggists, nursing-home operators, and others in the health field who blatantly commit welfare fraud as a regular practice. Because of their "respectability" in the community, these providers are in a position to bill Medicaid for services never rendered, and they do so with some regularity. When these abuses are publicized, public outcry is minimal and fleeting at best. It is so much easier to blame "those people."

Even in the event of discovery by a Medicaid "fraud squad," the welfare recipient, rather than the health professional suspected of illegally using the system, often becomes the target of their investigation.

> I was ordered by our Medicaid fraud squad to appear at my local welfare office within the week. A dentist who had done surgery on my mouth for a periodontal disease was under suspicion of committing fraud.
>
> I arrived at the office not knowing what to expect. Two men who resembled G-Men arrived and hustled me into a cubicle and began interrogating me. Not being satisfied with my answers, they proceeded to look into my mouth at every tooth in my head to prove I had fillings where they were not supposed to be, according to the computer printout they were studying. They also wanted me to account for every filling, extraction, check-up, and cleaning of my three children, which had been done by the same dentist. This took me another week.
>
> I was treated as if I was guilty of something. When you are on welfare, no one cares about your feelings—they don't count.

The welfare system, in reality, has been set up to promote fraud as a means of survival. They *know* we can't live on budgets "below the poverty line." With a more reasonable system of providing financial assistance, there would be less "fraud" because people on welfare would be less desperate. But as it is, women are punished for being on welfare and pushed into impossible binds.

ENOUGH ALREADY

Families who must rely on the welfare system to survive are constantly torn to pieces by the bureaucratic policies that supply the services they need. Each benefit comes with its own rules and regulations, which must be followed if recipients are to remain eligible. The policies are designed separately, with little regard to policies of other agencies, so we are continuously in compromising positions.

Welfare policies are written in "still life." Like the prepackaged vegetables in the supermarkets, they look fine on the surface: but turn them over and look beneath the surface and you may see rotten spots. Living, breathing people need policies that allow for individuality and flexibility. No two families are alike or have the same needs. We should not be lumped together and threatened with extinction if we complain. Women who are already in crises do not need the added stress of conflicting policies among the services that are ours by right. We are strong, capable, and often wise beyond our years. We demand that we be allowed to have some control over our own lives. No governor, president, general, or legislator should be able to dictate to us where we live, what we eat, or where, or even, whether we should work outside the home when we are already taking care of our children.

Education

EDUCATION AND THE STRUGGLE AGAINST RACE, CLASS, AND GENDER INEQUALITY

29

Roslyn Arlin Mickelson and Stephen Samuel Smith

INTRODUCTION

For the past thirty-five years, policymakers have claimed that federal education policies seek, among other things, to further equality among the races, between the sexes, and, to a much lesser extent, among social classes.[1] In this respect, educational policymakers have shared one of the assumptions that has long been part of the putative dominant ideology: a "good education" is *the* meal ticket. It will unlock the door to economic opportunity and thus enable disadvantaged groups or individuals to improve their lot dramatically. According to the dominant ideology, the United States is basically a meritocracy in which hard work and individual effort are rewarded, especially in financial terms. Related to this central belief are a series of culturally enshrined misconceptions about poverty and wealth. The central one is that poverty and wealth are the result of individual inadequacies or strengths rather than the results of the distributive mechanisms of the capitalist economy. A second misconception is the belief that everyone is the master of her

From: Berch Berberoglu (ed.), *Critical Perspectives In Sociology: A Reader* (Dubuque, Iowa: Kendall/Hunt Publishing, 1991). Copyright 1991 by Kendall/Hunt Publishing Company. Reprinted by permission of the author and the publisher.

or his own fate.[2] The dominant ideology assumes that American society is open and competitive, a place where an individual's status depends on talent and motivation, not inherited position. To compete, everyone must have access to education free of the fetters of family background or ascriptive factors like gender and race.[3] Since the middle of this century the reform policies of the federal government have been designed, at least officially, to enhance individuals' opportunities to acquire education.

We begin this essay by discussing the major educational doctrines and policies of the past thirty-five years that claim to have been aimed at promoting equality through greater equality of educational opportunity. This discussion includes the success and failures of programs such as school desegregation, compensatory education, Title IX, and job training. We then focus on the barriers such programs face in actually promoting equality. Here our point is that inequality is so deeply rooted in the structure and operation of the U.S. political economy that, at best, educational reforms can play only a limited role in ameliorating such inequality. Considerable evidence indicates that the educational system helps legitimate, if not actually reproduce, significant aspects of social inequality.

First, let us distinguish among equality, equality of opportunity, and equality of educational opportunity. The term *equality* has been the subject of extensive scholarly and political debate, much of which is beyond the scope of this essay. Most Americans reject equality of life conditions as a goal, because it would require a fundamental transformation of our basic economic and political institutions, a scenario most are unwilling to accept.[4] In the words of one observer, "So long as we live in a democratic capitalist society—that is, so long as we maintain the formal promise of political and social equality while encouraging the practice of economic inequality—we need the idea of equal opportunity to bridge that otherwise unacceptable contradiction."[5] The distinction between equality of opportunity and equality of outcome is important. Through this country's history, equality has most typically been understood in the former way. Rather than a call for the equal distribution of money, property, or many other social goods, the concern over equality has been with equal opportunity in pursuit of these goods. As Ralph Waldo Emerson put it: "The genius of our country has worked out our true policy—opportunity. Opportunity of civil rights, of education, of personal power, and not less of wealth; doors wide open."[6] To use a current metaphor: If life is a game, the playing field must be level; if life is a race, the starting line must be in the same place for everyone. For the playing field to be level, many believe education is crucial, giving individuals the wherewithal to compete in the allegedly meritocratic system. Thus equality of opportunity hinges on equality of educational opportunity.[7]

THE SPOTTY RECORD OF FEDERAL EDUCATIONAL REFORMS

In the past thirty-five years a series of educational reforms initiated at the national level has been introduced into local school systems. All of the reforms aimed to move education closer to the ideal of equality of educational opportunity. Here we discuss several of these reforms and how the evolution of the concept of equality of educational opportunity, spurred on by the Coleman Report, shaped many of these reforms during the past two decades. Given the importance of race and racism in American social history, many of the federal education policies during this period attempted to redress the most egregious forms of inequality based on race.

School Desegregation

Although American society has long claimed to be based on freedom, justice, and equality of opportunity, the history of race relations has long suggested the opposite. Perhaps the most influential early discussion of this disparity was Gunnar Myrdal's *An American Dilemma*, published in 1944, which vividly exposed the contradictions between the ethos of freedom, justice, equality of opportunity and the actual experiences of African Americans in the United States.[8] Segregated schools presented observers like Myrdal with direct evidence of the shallowness of American claims to equality for all.

The school desegregation movement, whose first phase culminated in the 1954 *Brown* decision outlawing de jure segregation in schools, was the first orchestrated attempt in U.S. history to directly address inequality of educational opportunity.[9] The links among desegregation, equality of educational opportunity, and the larger issue of equality of opportunity are very clear from the history of the desegregation movement. The NAACP strategically chose school segregation to be the camel's nose under the tent of Jim Crow society.[10] That one of the nation's foremost civil rights organizations saw the attack on segregated schools as the opening salvo in the battle against society-wide inequality indicates the pivotal role of education in the American belief system in promoting equality of opportunity.

Has desegregation succeeded? This is really two questions: First, have desegregation efforts integrated public schools? Second, have desegregation efforts enhanced students' educational opportunities? Since 1954 progress toward integrated education has been limited at best. As Hochschild notes, racial isolation has only diminished; it has not gone away. . . . Furthermore, integrated schools are often resegregated at the classroom level by tracking or ability grouping. Since 1954 racial isolation has declined everywhere but the Northeast. Ironically, today the greatest degree of racial segregation occurs in the Northeast and the least is in the South.[11] The answer to the first question, then, is that desegregation policies have had only a limited effect on overall school integration.

Has desegregation helped to equalize educational outcomes nationwide? As Hochschild points out, a better question might be which desegregation programs under which circumstances accomplish which goals.[12] Where desegregation has succeeded, its effects have been for the most part positive.[13] Evidence from desegregation research suggests that, overall, minority and majority children benefit academically and socially from well-run programs. The city in which we live, Charlotte, North Carolina, is an example of one such success story.[14] Despite these limited but positive outcomes, in the last decade of the twentieth century, most American children attend schools segregated by race, ethnicity, and class. Consequently, thirty-five years of official federal interventions aimed at achieving equality of educational opportunity through school desegregation have not achieved that goal; children from different race and class backgrounds continue to receive significantly segregated and largely unequal educations.[15]

The Coleman Report

Based largely on the massive data introduced in the 1954 *Brown* desegregation case, which showed that resources in black and white schools were grossly unequal, Congress mandated in 1964 a national study of the "lack of availability of equality of educational opportunity for individuals due to race, color, religious, or national origin in public schools." The authors of the study, James Coleman and his associates, expected to find glaring disparities in educational resources available to African-American and white students and that these differences would explain the substantial achievement differences between white and minority students.[16]

Instead, the Coleman Report, released in 1966, produced some very unexpected findings. The researchers found that twelve years after *Brown*, most Americans still attended segregated schools but that the characteristics of black and white schools (such as facilities, books, labs, teacher experience, and expenditures) were surprisingly similar. Apparently, segregated Southern districts upgraded black educational facilities in the wake of the *Brown* decisions. More importantly, Coleman and his colleagues found that school resources had relatively little to do with variations in students' school performance. Instead, they found that family background influenced school achievement more than any other factor, including school characteristics.[17]

The effects of the last finding were dramatic. It deflected attention from how schools operate and instead focused public policy upon poor and minority children and their families as the ultimate sources of unequal school outcomes. Numerous observers concluded that schools were not primarily to blame for black-white educational differences, overlooking another Coleman Report finding that could implicate schools in inequality of educational outcomes. The overlooked finding showed that African-American and white achievement differences increased with every year of schooling. That is, the achievement gap between black and white first graders was much smaller than the gap between twelfth graders. This finding suggests that at best schools reinforce the disadvantages of race and class and at worst are themselves a major source of educational inequality.

Although published a quarter of a century ago, the Coleman Report remains one of the most important and controversial pieces of social research ever completed. One of its many lasting results was a redefinition of the concept of equality of educational opportunity because it made clear that greater inputs into schools were not associated with greater student achievement. No longer were financial inputs a satisfactory measure of equality of opportunity. Only to the extent that academic outcomes of achievement (how well a student performs in school) and attainment (how many years of education a student acquires) are equal can claims be advanced about the putative extent of equality of educational opportunity.[18]

Compensatory Education

A second lasting outcome of the Coleman Report was widespread support for compensatory education. Policymakers interpreted the finding that family background explains more of the variance in students' achievement than any other factor as evidence of "cultural deprivation" among poor and minority families. This notion was consistent with Oscar Lewis's then-popular thesis on the culture of poverty.[19] Such an interpretation of the Coleman Report gave impetus to an education movement to compensate for the alleged cultural deficiencies of non–middle class, nonwhite families so that when so-called disadvantaged children came to school they could compete without the handicap of their background.

Compensatory education refers to the many programs that began with the passage of the Elementary and Secondary Education Act in 1965. These programs target children who are both poor and underachieving and provide them with developmental preschool or a variety of individualized programs in math, reading, and language arts once they are in elementary school. The following are included under the compensatory education rubric:

- Early childhood education such as Head Start
- Follow Through, where Head Start children, now in elementary school, continue to receive special programs
- Bilingual education
- Chapter I (formerly called Title I), which provides language arts and math programs plus food, medicine, and clothing to needy children in primary schools
- Guidance and counseling in secondary schools
- Higher education programs designed to identify potential college students in high schools and special admissions, transition, and retention programs for qualified students going to college[20]

Compensatory education has a controversial history. Initially, critics from the left charged that the underlying premise of compensatory education, that poor and minority families were deficient relative to middle-class white families, was racist and elitist. Compensatory education's most famous critic on the right was Arthur Jensen. His 1969 article on IQ and scholastic achievement argued that compensatory education was a waste of time and money because the lower African-American achievement scores indicated that blacks were less intelligent than whites.[21] This criticism of compensatory education miscast the debate over poverty and education into one about race and education because it ignored the fact that many compensatory education students were white and most African-American children at the time did not participate in the programs.

Despite attacks like Jensen's and an initial absence of evidence that the programs accomplished any of their goals, the compensatory education movement survived the past twenty years. Recent research has begun to demonstrate both the cognitive and social benefits from compensatory education,[22] although the achievement gaps between minority and white and between working- and middle-class children remain. No doubt this is true in part because today only 18 percent of income-eligible students participate in the most famous and successful program, Head Start.[23] While the recent evidence regarding the effects of these programs continues to be positive, we must conclude that compensatory education, like integrated education, has not brought about equality of school outcomes.

Human Capital Theory and Job Training Programs

The notion that "good" education is *the* meal ticket has received theoretical exposition in human capital theory, which views education as a capital investment in human beings. Theodore Schultz, a University of Chicago economist, originally put forth this theory. He argued that "earnings, especially those of minority groups, reflect . . . inadequate investment in their health and education."[24] The social policies developed to remedy shortages in human capital among minority and working-class youth included job training programs like the Comprehensive Education and Training Act (CETA) and its successor, the Job Training and Partnership Act (JTPA). A central problem with both programs is that graduates have few if any jobs once the training is over.[25] The failure of both programs to eradicate inequality stems from the faulty premise that individual, not structural, factors are at the heart of poverty. The poor often have a great deal of skills and education. What they lack is available, well-paying jobs in which to invest their skills.[26] We elaborate on this crucial point later in this article.

These criticisms apply as well to more conventional high school vocational education programs. Vocational students tend to be disproportionately working class and minority, and the courses are highly segregated by gender. Vocational education classes in high school and community colleges are frequently tailored to local businesses' labor force requirements. Too often they train students in obsolete, narrow skills that are virtually useless once the student graduates from the educational program. Much of the recent corporate interest in school reform stems from the results of these processes.[27] The track record of job training programs (CETA and JTPA) and secondary school vocational education programs indicate human capital–based educational reforms do not, and cannot, narrow race, class, and gender differences in equality of educational opportunity in this society.

Title IX

Title IX is the primary federal law prohibiting sex discrimination in education. It states, "No person in the United States shall, on the basis of sex, be excluded from participation in, be denied benefit of, or be subjected to discrimination under any program or activity receiving Federal financial assistance." Until its passage in 1972, gender inequality in educational opportunity received minimal legislative attention. Title IX covers admissions quotas by sex, different courses, and athletic programs. It requires existing school programs be examined for gender-based discrimination and mandates equal treatment of all students in courses, financial aid, counseling services, and employment.[28]

Although Title IX was passed in 1972, it lacked implementing regulations until 1975. Nor was it ever accompanied by substantial federal or state financial assistance. Title IX was potentially damaged in February 1984, when the

Supreme Court ruled in *Grove City College* v. *Bell* that coverage of Title IX was limited to programs and activities, rather than entire institutions, receiving federal money.[29] The effects of this ruling remain unclear.[30]

Sexism in education persists despite laws prohibiting it. Even though female achievement and attainment levels are comparable to those of males, many sexist practices remain an integral part of schooling at all levels. For example, curricular materials from kindergarten to college reveal a preponderance of male characters. In addition, male and female characters usually reflect traditional gender roles and behaviors. Although the gap is narrowing, women who graduate from high school are frequently less well prepared in college-level math and science than men are and consequently cannot enroll in math, science, engineering, or computer science courses in the same frequencies as men.[31] Vocational education at the high school and college level remains highly sex segregated. Administrators are overwhelmingly male although most teachers are female. Women in academia face barriers to promotion and continue to earn less than their male colleagues.[32]

Recent civil rights legislation fails to address any of these problems. Like the laws and policies aimed at eliminating race differences in school processes and outcomes, those aiming to eliminate gender differences in educational opportunities have, at best, only narrowed differences. Educational opportunity in the United States remains highly unequal for people of different gender, race, ethnic, and class backgrounds.

EQUALITY OF EDUCATIONAL OPPORTUNITY AND EQUALITY OF INCOME

The educational reforms intended to lessen inequality were not designed primarily to give all students access to better playgrounds, computers, books, teachers, and athletic facilities. Greater equality in the distribution of these resources was seen as a means to better equip minorities, women, and to a lesser extent, working-class whites to improve their chances in life once they become adults. Equality of educational opportunity, as noted earlier, was viewed as means of promoting equality of opportunity.

Whatever increase in the equality of educational opportunity has resulted from the reforms discussed earlier, it has not led to a significantly greater equality of life-chances, at least as measured by income, one of the most telling and broadest gauges of equality in the United States. Much of a person's social standing depends on income because in a capitalist society income is related to all other forms of inequality. Although this measure does not address the social class basis of inequality, income is one useful and clear measure of inequality in this society.[33] In addition, the disjunction between the greater equality of educational opportunity and lack of a corresponding increase in the equality of incomes can be explained by the nature of the U.S. political economy. The main cause of income inequality is the structure and operation of U.S. capitalism, which have scarcely been affected by the educational reforms discussed earlier.

Several years after the publication of the landmark Coleman Report, Jencks and his associates reanalyzed the data on which it was based to explore the conventional wisdom that a "good education" was how disadvantaged individuals and groups could improve their economic situation. Their findings were published in a now famous book, *Inequality* (1972). Among its conclusions was the following:

> The evidence suggests that equalizing educational opportunity [achievement and attainment] would do very little to make adults more equal. . . . The experience of the past 25 years suggests that even fairly substantial reductions in the range of educational attainments do not appreciably reduce economic inequality among adults.[34]

Almost twenty years of educational reforms have elapsed since those remarks were made. Sadly, the evidence in support of that conclusion is stronger now than it was in 1972. Current statistics on educational and income attainment show that race and gender differences in educational achievement continue to narrow and differences in attainment have all but disappeared.[35] Nevertheless, race and gender differences in income remain stable over time. . . .

While race differences in educational attainment have virtually disappeared, and women attain slightly more education than men, both minorities and women remain "less equal" than men—that is, on average they earn significantly less income than white males with comparable educational credentials. . . . Equality of educational opportunity, indicated by years of attainment, simply has not produced anything resembling equality of income.

The greater equality of educational opportunity has not led to a corresponding increase in the equality of incomes because educational reforms do not create more good-paying jobs, affect sex-segregated and racially segmented occupational structures, or limit the mobility of capital either between regions of the country or between the United States and other countries. For example, no matter how good an education white working-class or minority youth may receive, it does nothing to alter the fact that thousands of relatively good paying manufacturing jobs have left Northern inner cities for Northern suburbs, the Sunbelt, or foreign countries.[36] Perhaps service jobs are left in the wake of such capital flight or have been created in its place, but they pay less than manufacturing jobs. An economy in which McDonald's employs more people than USX Corporation (the former industrial giant U.S. Steel Corporation) is one in which the economic opportunities for minority and working-class youth are limited. Without changes in the structure and operation of the capitalist economy, educational reforms may enable some members of disadvantaged groups to improve their situation, but they cannot markedly improve the social and economic position of these groups in their entirety. This is the primary reason that educational reforms do little to affect the gross social inequalities that inspired them in the first place.

BEYOND ATTAINMENT: THE PERSISTENCE OF EDUCATIONAL INEQUALITY

Educational reforms have not led to greater equality for several additional reasons. Many aspects of school processes and curricular content are deeply connected to race, class, and gender inequality. But gross measures of educational outputs, such as median years of schooling completed, mask these indicators of inequality. The data in Figure 1 indicate that today African Americans as a group have attained essentially as much education as have whites as a group; the same is true for women compared to men. But aggregate data like these obscure gross differences in attainment within race and gender groups. These within-group differences are primarily linked to class and region. . . . Similar differences by class exist for whites. Within-group differences demonstrate the extent to which educational opportunity remains inequitable. Despite significant within-group differences, the comparable median levels of educational attainment by race and gender can illustrate our theoretical point regarding the structural inability of increases in education to reduce income inequality. Were we to compare income by race and by gender for different levels of educational attainment, our argument would be even stronger. For any level of educational attainment, blacks earn less than whites with comparable levels of education and women earn less than men with similar credentials. Furthermore, the disparities in returns to education increase as levels of educational attainment increase.[37]

Another example of persistent inequality of educational opportunity is credential inflation. Even though women, minorities, and members of the working class now obtain higher levels of education than they did before, the most privileged in society gain even higher levels of education. At the same time, the educational requirements for the best jobs (those with the highest salaries, benefits, agreeable working conditions, autonomy, responsibility) are growing so that only those with the most education from the best schools are eligible for the best jobs. Because the more privileged are almost always in a better position to gain these desirable educational credentials, members of the working class, women, and minorities are still at a competitive disadvantage. Due to credential inflation, previous educational requirements for good jobs, now within the reach of many dispossessed groups, are inadequate and insufficient in today's labor market.[38]

Although gaps in educational attainment between males and females and between minorities and whites have narrowed (see Figure 1), not all educational experiences are alike. Four years of public high school in Beverly Hills are quite different from four years in an inner-city school.[39] Dreeben and Gamoran examined race differences in learning to read among first-grade children in the Chicago area. They concluded that the explanation for why nonblacks learned more than blacks rests, in part, in features of teacher instructional practices, which differ by school. Black children in their sample who attended minority schools were provided with less time in basal instruction—the core activity of primary school reading—and covered fewer new vocabulary words than did black and white children in either integrated or all white schools. The authors concluded, "Blacks learned less than nonblacks . . . mainly because of the instructional inadequacies prevailing in all-black schools."[40]

One additional aspect of these differences is what sociologists of education call the hidden curriculum, which refers to two separate but related processes. The first is that the content and process of education differ for children according to their race, gender, and class. The second is that these differences help reproduce the inequalities based on race, gender, and class that characterize U.S. society.

One aspect of the hidden curriculum is the formal curriculum's ideological content. Anyon's work on U.S. history texts demonstrates that children from more privileged backgrounds are more likely to be exposed to rich, sophisticated, and complex materials than are their working-class counterparts.[41] Another aspect of the hidden curriculum concerns the social organization of the school and the classroom. Some hidden curriculum theorists suggest that tracking, ability grouping, and conventional teacher-centered classroom interactions contribute to the reproduction of the social relations of production at the workplace. This perspective derives from the work of Bowles and Gintis; their seminal work, *Schooling in Capitalist America*, proposes that there is a correspondence among the social relations of the workplace, the home, and the school.[42] Lower-track classrooms are disproportionately filled with working-class and minority students. Students in lower tracks are more likely than those in higher tracks to be assigned repetitive exercises at a very low level of cognitive challenge. Lower-track students are likely to work individually and to lack classroom experience with problem solving or other independent, creative activities.[43] Such activities are more conducive to preparing students for working-class jobs than for professional and managerial positions. Many proponents of the correspondence principle argue that educational experiences from preschool to high school differentially prepare students for their ultimate positions in the work force, and a student's placement in various school programs is based primarily on her or his race and class origin.[44] The correspondence principle has sometimes been applied in too deterministic and mechanical a fashion.[45] Nonetheless, in certain cases, it is a compelling contribution to explanations of how and why school processes and outcomes are so markedly affected by the race, gender, and class of students. . . .

CONCLUSION

In this chapter we have argued that educational reforms cannot eliminate inequality. But education remains important to any struggle to reduce inequality. Moreover, it is more than a meal ticket; it is intrinsically worthwhile. Desegregation, compensatory education, Title IX, and bilingual education all facilitate cognitive growth and promote important nonsexist, nonracist attitudes and practices; they make schools a more humane place for adults and children. Furthermore, education, even reformist liberal education, contains the seeds of individual and social transformation. Those of us committed to the struggle against inequality cannot be paralyzed by the structural barriers that make it impossible for education to eliminate inequality. We must look upon the schools as arenas of struggle. To do otherwise would be to concede what must be contested in every way.

ENDNOTES

1. This essay draws on an article by Roslyn Arlin Mickelson that appeared as "Education and the Struggle Against Race, Class, and Gender Inequality," *Humanity and Society* 11, no. 4 (1987), pp. 440–464. The authors thank Kevin Hawk for his technical assistance in the preparation of the graphs.

2. R. H. deLone, *Small Futures* (New York: Harcourt Brace Jovanovich, 1979). Ascertaining whether a set of beliefs constitutes the dominant ideology in a particular society or social formation involves a host of difficult theoretical and empirical questions. For this reason we use the term *putative dominant ideology*. For discussions of these questions, see Nicholas Abercrombie et al., *The Dominant Ideology Thesis*

(London: George Allen & Unwin, 1980); James C. Scott, *Weapons of the Weak* (New Haven: Yale University Press, 1985); Stephen Samuel Smith, "Political Acquiescence and Beliefs about State Coercion" (unpublished Ph.D. dissertation, Stanford University, 1990).

3. Whether racism and sexism are epiphenomena is too large an issue to address in this article. We take the position that the logic of capitalism breeds sexism and racism, but that both have a quasi-independent cultural history and are therefore important social phenomena in their own right (see C. A. Mackinnon, "Feminism, Marxism, Method, and the State: An Agenda for Theory," in N. O. Keohane et al. (eds.), *Feminist Theory* (Chicago: University of Chicago Press, 1981).

4. Kevin Dougherty, "After the Fall: Research on School Effects Since the Coleman Report," *Harvard Educational Review* 51, no. 2 (1981), pp. 301–308.

5. Jennifer L. Hochschild, "The Double-Edged Sword of Equal Educational Opportunity." Paper presented at the meeting of the American Education Research Association, Washington, D.C., April 22, 1987.

6. Douglas Rae, *Equalities* (Cambridge, Mass.: Harvard University Press, 1981), p. 64.

7. C. J. Hurn, *The Limits and Possibilities of Schooling* (Boston: Allyn and Bacon, 1978).

8. G. Myrdal, *An American Dilemma: The Negro Problem and Modern Democracy* (New York: Harper & Row, 1944).

9. R. Kluger, *Simple Justice* (New York: Knopf, 1975).

10. Ibid.

11. J. L. Hochschild, *The New American Dilemma* (New Haven: Yale University Press, 1984).

12. Ibid.

13. Jomills H. Braddock II, Robert L. Crain, and James M. McPartland, "A Long-Term View of School Desegregation: Some Recent Studies of Graduates as Adults," in Jeanne H. Ballantine (ed.), *Schools and Society* (Mountain View, Calif.: Mayfield, 1989), pp. 303–313; Robert L. Crain et al., *Making Desegregation Work* (Cambridge, Mass.: Ballinger, 1982); Hochschild, *New American Dilemma*; Charles V. Willie, *Desegregation Plans That Work* (Westport, Conn.: Greenwood Press, 1984).

14. Frye Gaillard, *The Dream Long Deferred* (Chapel Hill: University of North Carolina Press, 1989).

15. Fiske, "Integration Lags."

16. J. Karabel and A. H. Halsey, *Power and Ideology in Education* (New York: Oxford University Press, 1977), p. 20.

17. J. S. Coleman et al., *Equality of Educational Opportunity* (Washington, D.C.: Government Printing Office, 1966); Karabel and Halsey, *Power and Ideology.*

18. B. Heyns, "Educational Effects: Issues in Conceptualization and Measurement," in J. C. Richardson (ed.), *Handbook of Theory and Research for the Sociology of Education* (New York: Greenwood Press, 1986).

19. O. Lewis, "The Culture of Poverty," *Scientific American* (October 1966), pp. 19–25.

20. J. H. Ballantine, *The Sociology of Education* (Englewood Cliffs, N.J.: Prentice-Hall, 1983).

21. A. F. Jensen, "How Much Can We Boost IQ and Scholastic Achievement?" *Harvard Educational Review* 39 (Winter 1969), pp. 1–123. The genetic model of human intelligence Jensen used to substantiate his claims of African Americans' alleged lower levels of intelligence was the work of the late Sir Cyril Burt. In the mid-1970s, however, scientists exposed Burt's work as a major scientific fraud because he had fabricated his data on the IQs of separated identical twins as well as the existence of the two women who supposedly assisted him during his research. Stephen J. Gould's *The Mismeasure of Man* (New York: Norton, 1981) is a masterful exposé of Burt's fakery as well as the scientific and moral bankruptcy of the sexist, classist, and racist underpinnings of much of the intelligence testing movement.

22. J. F. Berrueta-Clement et al., *Changed Lives: The Effects of the Perry Preschool Project on Youths Through Age 19* (Ypsilanti, Mich.: The High/Scope Press, 1985); L. F. Carter, "The Sustaining Effects of Compensatory and Elementary Education," *Educational Researcher* 13, no. 7 (1984), pp. 3–11.

23. Ronald Henkoff, "Now Everyone Loves Head Start," *Fortune*, Special Issue on Saving Our Schools (Spring 1990), pp. 35–43.

24. T. Schultz, "Investment in Human Capital," *American Economic Review* (March 1961), pp. 1–17; Karabel and Halsey, *Power and Ideology*, pp. 313–324.

25. Wilford Wilms, personal communication (1984); Paul Weckstein, personal interview (Washington, D.C., March 8, 1990).

26. B. Bluestone, "Economic Policy and the Fate of the Poor," *Social Policy*, no. 2 (1972): 30–31.

27. Carol A. Ray and Roslyn A. Mickelson, "Corporate Leaders, Resistant Youth, and School Reform in Sunbelt City: The Political Economy of Education," *Social Problems* 37, no. 2 (1990), pp. 178–190.

28. Ballantine, *Sociology of Education*; S. S. Klein, *Handbook for Achieving Sex Equity in Education* (Baltimore: Johns Hopkins University Press, 1985).

29. Klein, *Handbook*, pp. 94–95.

30. Title IX is not the only federal law attempting to achieve gender equity in education through legislative reforms. Other federal programs that target sexism in education include Title IV of the 1964 Civil Rights Act, the Women's Educational Equity Act of 1974 and 1978, an amendment to the 1976 Vocational Education Act, the law authorizing the National Institute of Education, and the laws creating the National Science Foundation and the U.S. Commission on Civil Rights. Klein, *Handbook*; R. Salomone, *Equality of Education Under the Law* (New York: St. Martin's Press, 1986).

31. Elizabeth Fennema and Gilah C. Leder, *Mathematics and Gender* (New York: Teachers College Press, 1990); Klein, *Handbook*; Cornelius Riordan, *Girls and Boys in School: Together or Separate* (New York: Teachers College Press, 1990).

32. L. McMillen, "Women Professors Pressing to Close Salary Gap: Some Colleges Adjust Pay, Others Face Lawsuits," *Chronicle of Higher Education* 33, (1987), p. 1.

33. To be sure, income does not measure class-based inequality, but there is a positive correlation between income and class. Income has the additional advantage of being easily quantifiable. Were we to use another measure of inequality—wealth—the disjuncture between it and increases in educational attainment would appear even larger. The distribution of wealth in U.S. society has remained fairly stable since the Depression, with the richest 10 percent of the population owning about 65 percent of the total wealth. Since the Reagan administration's regressive fiscal and monetary policies have been in effect, the wealth of this country has been further redistributed upward at the expense of the working and middle classes.

34. C. Jencks et al., *Inequality* (New York: Harper & Row, 1972), p. 11.

35. Ira Shor, *Culture Wars* (Boston: Routledge & Kegan Paul, 1986).

36. Lois Weis, *Working Class Without Work: High School Students in a Deindustrialized Economy* (New York: Routledge, 1990); William J. Wilson, *The Truly Disadvantaged* (Chicago: University of Chicago Press, 1988).

37. Reynolds Farley, *Blacks and Whites: Narrowing the Gap?* (Cambridge, Mass.: Harvard University Press, 1984).

38. R. Collins, *Credential Society* (New York: Academic Press, 1979); R. B. Freeman, *The Over-educated American* (New York: Academic Press, 1976).

39. Jean Anyon, "Social Class and the Hidden Curriculum of Work," *Journal of Education* 162, no. 1 (1980), pp. 67–92; Jean Anyon, "Social Class and School Knowledge," *Curriculum Inquiry*, no. 10 (1981), pp. 3–42; Roslyn Mickelson, "The Secondary School's Role in Social Stratification: A Comparison of Beverly Hills High School and Morningside High School," *Journal of Education* 162, no. 4 (1980), pp. 83–112.

40. Robert Dreeben and Adam Gamoran, "Race, Instruction, and Learning," *American Sociological Review* 51, no. 5 (1986), pp. 660–669.

41. Anyon, op. cit.

42. S. Bowles and H. Gintis, *Schooling in Capitalist America: Educational Reform and the Contradictions of Economic Life* (New York: Basic Books, 1976).

43. Jeannie Oakes, *Keeping Track* (New Haven, Conn.: Yale University Press, 1985).

44. Ibid.; S. Lubeck, *Sandbox Society* (Philadelphia: Falmer Press, 1985); Mickelson, "Secondary School's Role"; R. A. Mickelson, "The Case of the Missing Brackets: Teachers and Social Reproduction," *Journal of Education* 169, no. 2 (1987), pp. 78–88.

45. M. Apple and L. Weis (eds.), *Ideology and Practice in Schooling* (Philadelphia: Temple University Press, 1983); H. Giroux, *Theory and Resistance in Education* (Boston: Bergin and Garvey, 1983).

CANTO, LOCURA Y POESIA

30

Olivia Castellano

I am a walking contradiction. I have no Ph.D. yet I'm a full professor of English at a state university. By all definitions and designs I should not even have made it to college. I am the second of five children of a Southern Pacific Railroad worker with a fifth-grade education and a woman who dropped out of the second grade to help raise ten siblings—while her mother worked ten hours a day cleaning houses and doing laundry for rich Texan ranchers.

In Comstock, the Tex-Mex border town about fifteen miles from the Rio Grande where I spent the first twelve years of my life, I saw the despair that poverty and hopelessness had etched in the faces of young Chicano men who, like my father, walked back and forth on the dusty path between Comstock and the Southern Pacific Railroad station. They would set out every day on rail carts to repair the railroad. The women of Comstock fared no better. Most married early. I had seen them in their kitchens toiling at a stove, with one

From: *Women's Review of Books* 7, no. 5 (Feb. 1990). Reprinted by permission of the author.

baby propped on one hip and two toddlers tugging at their skirts. Or they followed their working mothers' route, cleaning houses and doing laundry for rich Texan ranchers who paid them a pittance. I decided very early that this was not the future I wanted.

In 1958 my father, tired of seeing his days fade into each other without promise, moved us to California where we became farmworkers in the San Jose area (then a major agricultural center). I saw the same futile look in the faces of young Chicanos and Chicanas working beside my family. Those faces already lined so young with sadness made me deadly serious about my books and my education.

At a young age—between eleven and fourteen—I began my intellectual and spiritual rebellion against my parents and society. I fell in love with books and created space of my own where I could dare to dream. Yet in school I remained shy and introverted, terrified of my white, male professors. In my adolescence I rebelled against my mother's insistence that Mexican girls should marry young, as she did at eighteen. I told her that I didn't care if my cousins Alicia and Anita were getting married and having babies early. "I was put on this earth to make books, not babies!" I announced and ran into my room.

Books were my obsession. I wanted to read everything that I was not supposed to. By fourteen I was already getting to know the Marquis de Sade, Rimbaud, Lautréamont, Whitman, Dostoyevsky, Marx. I came by these writers serendipitously. To get from home to Sacramento High School I had to walk through one of the toughest neighborhoods in the city, Oak Park. There were men hanging out with liquor in brown paper bags, playing dice, shooting craps and calling from cars: "Hey, baby, get in here with me!" I'd run into a little library called Oak Park Library, which turned out to have a little bit of everything. I would walk around and stare at the shelves, killing time till the shifty-eyed men would go away.

The librarians knew and tolerated me with skepticism: "Are you sure you're going to read the Marquis de Sade? Do your parents know you're checking out this material? What are you doing with the *Communist Manifesto?*" One librarian even forbade me to check the books out, so I'd sit reading in the library for hours on end. Later, at sixteen or seventeen, I was allowed to check anything and everything out.

So it was that I came to grapple with tough language and ideas. These books were hot! Yet I also was obsessed with wanting to be pretty, mysterious, silent and sexy. I wanted to have long curly hair, red lips and long red nails; to wear black tight dresses and high heels. I wanted desperately to look like the sensuous femmes fatales of the Mexican cinema—María Féliz, one of the most beautiful and famous of Mexico's screen goddesses, and Libertad Lamarque, the smoky-voiced, green-eyed Argentinian singer. These were the women I admired when my mother and I went to the movies together. So these were my "outward" models. My "inward" models, the voices of the intellect that spoke to me when I shut the door to my room, were, as you have gathered, a writer of erotica, two mad surrealists, a crazy Romantic, an epileptic literary genius and a radical socialist.

I needed to sabotage society in a major, intellectually radical way. I needed to be a warrior who would catch everyone off guard. But to be a warrior, you must never let your opponent figure you out. When the bullets of racism and sexism are flying at you, you must be very clever in deciding how you want to live. I knew that everything around me—school, teachers, television, friends, men, even my own parents, who in their internalized racism and self-hatred didn't really believe I'd amount to much though they hoped like hell that life would prove them wrong—everything was against me, and I understood this fully.

To protect myself I fell in love with language—all of it, poems, stories, novels, plays, songs, biographies, "cuentos" or little vignettes, movies—all manifestations of spoken and written language. I fell in love with ideas, with essays by writers like Bacon or Montaigne. I began my serious reading crusade around age eleven, when I was already convinced that books were central to my life. Only through them and through songs, I felt, would I be free to structure some kind of future for myself.

I wanted to prove to anyone who cared to ask (though by now I was convinced no one gave a damn) that I, the daughter of a laborer-farmworker, could dare to be somebody. Try to imagine what it is like to be always full of rage—rage at everything: at white teachers who could never even pronounce my name (I was often called anything from "Odilia" to "Otilia" to "Estela"); rage at those teachers who asked me point-blank, "But how did you get to be so smart? You are Mexican, aren't you?"; rage at my eleventh-grade English teacher who said to me in front of the class, "You stick to essay writing; never try to write a poem again because a poet you are not!" (This, after I had worked for two diligent weeks on an imitation of "La Belle Dame Sans Merci"! Now I can laugh. Then it was pitiful.)

From age thirteen I was also angry at boys who hounded me for dates. When I'd reject them they'd yell, "So what do you plan to do for the rest of your life, fuck a book?" I was angry at my Chicana classmates in high school who, perhaps jealous of my high grades, would say, "What are you trying to do, be like the whites?" I regret to say that I was also angry at my parents, exasperated by their docility, their limited expectations of me. Oh, I knew they were proud; but sometimes, in their own misdirected rage (maybe afraid of my little successes), they would make painful comments. "Te vas a volver loca con esos jodidos libros" ("You'll go nuts with those fucking books") was my mother's frequent statement. Or the even more sickening, "Esta nunca se va a casar." ("Give up on this one; she'll never get married.") This was the tenor of my adolescent years. When nothing on either side of the two cultures, Mexican or Anglo-American, affirms your existence, that is how rage is shaped.

While I managed to escape at least from the obvious entrapments—a teen pregnancy, a destructive early marriage—I did not escape years of being told I wasn't quite right, that because of my ethnicity and gender I was somehow defective, incomplete. Those years left wounds on my self-esteem, wounds so deep that even armed with my books and stolen knowledge I could not entirely escape deep feelings of unworthiness.

By the time I graduated from high school and managed to get a little scholarship to California State University in Sacramento, where I now teach (in 1962 it was called Sacramento State College), I had become very unassertive, immensely shy. I was afraid to look unfeminine if I raised my hand in class, afraid to seem ridiculous if I asked a "bad" question and all eyes turned on me. A deeper part of me was afraid that my rage might rear its ugly head and I would be considered "an angry Mexican accusing everybody of racism." I was painfully concerned with my physical appearance: wasn't I supposed to look beautiful like Féliz and Lamarque? Yet while I wanted to look pretty for the boys, the thought of having sex terrified me. What if I got pregnant, had to quit college and couldn't read my books any more? The more I feared boys, the more I made myself attractive for them and the more they would make advances, the more I rejected them.

The constant tension sapped my energy and distracted me from my creative journeys into language. Oh, I would write little things (poems, sketches for stories, journal entries), but I was afraid to show them to anyone. Besides, no one knew I was writing them. I was so frightened by my white, male professors, especially in the English department—they looked so arrogant and were so ungiving of their knowledge—that I didn't have the nerve to major in English, though it was the major I really wanted.

Instead, I chose to major in French. The "Parisiens" and "Québecois" in the French department faculty admired my French accent: "Mademoiselle, êtes-vous certaine que vous n'êtes pas parisienne?" they would ask. In short, they cared. They engaged me in dialogue, asked why I preferred to study French instead of Spanish. ("I already know Spanish," I'd say.) French became my adopted language. I could play with it, sing songs in it and sound exotic. It complemented my Spanish; besides, I didn't have to worry about speaking English with my heavy Spanish accent and risk being ridiculed. At one point, my spoken French was better than my oral Spanish; my written French has remained better than my written Spanish.

At 23, armed with a secondary school teaching credential and B.A. in French with an English minor, I became a high school teacher of French and English. Soon after that I began to work for a school district where the majority of the students were Chicanos and Blacks from families on welfare and/or from households run by women.

After two years of high school teaching, I returned to Cal State at Sacramento for the Master's degree. Professionally and artistically, it was the best decision I have ever made. The Master's program to which I applied was a pilot program in its second year at CSUS. Called the Mexican American Experienced Teachers' Fellowship, it was run by a team of anthropology professors, central among whom was Professor Steven Arvizu. The program was designed to turn us into "agents of cultural change." It was 1969 and the program was one of the first federally funded (Title V) ones to address Mexican American students' needs by re-educating their teachers.

My interests were literary, but all twenty of us "fellows" had to get an M.A. in social anthropology, since this experiment took the "anthropologizing education" approach. We studied social dynamics, psycholinguistics, history of Mexico, history of the American Southwest, community activism and confrontational strategies and the nature of the Chicano movement. The courses were eye-openers. I had never heard the terms Chicano, biculturalism, marginality, assimilation, Chicanismo, protest art. I had never heard of Cesar Chavez and the farmworkers nor of Luis Valdez and the Teatro Campesino. I had never studied the nature of racism and identity. The theme of the program was that culture is a powerful tool for learning, self-expression, solidarity and positive change. Exploring it can help Chicano students understand their bicultural circumstances.

The program brought me face to face with nineteen other Chicano men and women, all experienced public school teachers like myself, with backgrounds like mine. The program challenged every aspect of my life. Through group counseling, group encounter, classroom interaction, course content and community involvement I was allowed to express my rage and to examine it in the company of peers who had a similar anger. Most of our instructors, moreover, were Chicano or white professors sensitive to Chicanos. For the first time, at 25, I had found my role models. I vowed to do for other students what these people had done for me.

Eighteen years of teaching primarily white women students, Chicanos and Blacks at California State University, Sacramento, have led me to see myself less as a teacher and more as a cultural worker, struggling against society to undo the damage of years of abuse. I continue to see myself as a warrior empowered by my rage. Racism and sexism leave two clear-cut scars on my students; internalized self-hatred and fear of their own creative passion, in my view the two most serious obstacles in the classroom. Confronting this two-headed monster has made me razor-sharp. Given their tragic personal stories, the hope in my students' eyes reconfirms daily the incredible beauty, the tenacity of the human spirit.

Teaching white women students (ages 30–45) is no different from working with Chicano and Black students (both men and women): you have to bring about changes in the way they view themselves, their abilities, their right to get educated and their relation to a world that has systematically oppressed them simply for being who they are. You have to help them channel and understand the seething rage they carry deep inside, a rage which, left unexpressed, can make them turn against each other and, more sadly, against themselves.

I teach four courses per semester: English 109G, Writing for Proficiency for Bilingual/Bidialectal Students (a course taken mainly by Chicano and Black students, ages 19–24); English 115A, Pedagogy/Language Arts for Prospective Elementary School Teachers (a course taken mainly by women aged 25–45, 50 percent white, 50 percent Chicano); English 180G, Chicano Literature, an advanced studies General Education course for non-English majors (taken by excellent students, aged 24–45, about 40 percent white, 40 percent Chicano, 20 percent Black/Vietnamese/Filipino/South American). The fourth course is English 1, Basic Language Skills, a pre-freshman composition course taken primarily by Black and Chicano freshmen, male and female, aged 18–22, who score too low on the English Placement Test to be placed in "regular" Freshman Composition.

Mine is a teaching load that, in my younger days at CSUS, used to drive me close to insanity from physical, mental and spiritual exhaustion—spiritual from having internalized my students' pain. Perhaps not fully empowered myself, not fully emplumed in the feathers of my own creativity (to borrow the wonderful "emplumada" metaphor coined by Lorna Dee Cervantes, the brilliant Chicana poet), I allowed their rage to become part of mine. This kind of rage can kill you. And so through years of working with these kinds of students I have learned to make my spirit strong with "canto, locura y poesia" (song, madness, and poetry). Judging from my students' progress, the songs have worked.

Truly, it takes a conjurer, a magus with all her teaching cards up her sleeve, to deal with the fragmented souls that show up in my classes. Among the Chicanos and Blacks I get ex-offenders (mostly men but occasionally a woman who has done time), orphans, single women heads of household, high school dropouts who took years to complete their Graduation Equivalency Diploma.

I get women who have been raped and/or who have been sexually abused either by a father figure or by male relatives—Sylvia Tracey, for example, a 30-year-old Chicana feminist, mother of two, whose parents pressured her to marry her (white) rapist and who is going through divorce after ten years of marriage. I get women who have been battered. And, of course, I get the young Chicano and Black little yuppies who don't believe the world existed before 1970, who know nothing about the sixties' history of struggle and student protest, who—in the case of the Chicanos—feel ashamed that their parents speak English with an accent or were once farmworkers. I get Chicanos, Blacks and white women, especially, who are ashamed of their writing skills, who have never once been told that they could succeed in school.

Annetta Jones is typical. A 45-year-old Black woman, who single-hand-edly raised three children, all college-educated and successful, she is still married to a man who served ten years in prison for being a "hit man." She visited him faithfully in prison and underwent all kinds of humiliation at the hands of correctional officers—even granting them sexual favors just to be allowed to have conjugal visits. When her husband completed his time he fell in love with a young woman from Chicago, where he now lives.

Among my white women students (ranging in age from 25 to 40, though occasionally I get a 45-year-old woman who wants to be an elementary or high school teacher and "help out young kids so they won't have to go through what I went through"—their exact words) I get women who are either divorced or divorcing; rarely do I get a "happily" married woman. This is especially true of the white women who take my Chicano literature and my credential-ped-agogy classes. Take Lynne Trebeck, for instance, a white woman about 40 years old who runs a farm. When she entered the university her husband objected, so she divorced him! They continue to live in the same house (he refused to leave), "but now he has no control over me," she told me trium-phantly midway through the semester. She has two sons, fifteen and eighteen years old; as a young woman she did jail time as the accomplice of a convicted drug dealer.

Every semester I get two or three white lesbian feminists. This semester there was Vivianne Rose, about 40, in my Chicano literature class. Apparently sensing too much conservatism in the students, and knowing that she wanted to be an elementary school teacher, she chose to conceal her sexual orientation. On the first day of class she wore Levi pants, a baggy sweat shirt, white tennis shoes and a beige baseball cap. By the end of the first week she had switched to ultrafeminine dresses and skirts, brightly colored blouses, nylons and medium-heeled black shoes, not to mention lipstick and eye makeup. When she spoke in class she occasionally made references to "my husband who is Native American." She and Sylvia Tracey became very close friends. Halfway through the course they informed me that "Shit, it's about time we tell her." (This, from Sylvia.) "Oh hell, why not," Vivianne said; "my 'husband' is a woman." The woman *is* Native American; Vivianne Rose lived on a reservation for years and taught young Native American children to read and write. She speaks "Res" talk (reservation speech) and has adopted her "husband's" last name.

Among my white women students there are also divorced women who are raising two to four children, usually between the ages of eight and seventeen. Sometimes I get older widowed white women who are taking classes for their own enjoyment, not for a degree. These women also tell stories of torment: rapes, beatings, verbal and emotional harassment from their men. On occasion I get women who have done jail time, usually for taking the rap for drug-connected boyfriends. I rarely get a married woman, but when I do there is pain: "My husband doesn't really want me in school." "My husband doesn't really care what I do in college as long as I take care of his needs and the kids' needs." "My husband doesn't really know what I'm studying—he has never asked and I've never told him."

Most of the white women as well as the minority students come to the university under special programs. There is the "Educational Opportunity Program" for students who do not meet all university entrance requirements or whose grade point average is simply too low for regular admission. There is the "Student Affirmative Action Program" for students who need special counseling and tutoring to bring their academic skills up to par or deal with emotional trauma. There is the "College Assistance Migrant Program" for students whose parents are migrant farmworkers in the agricultural areas surrounding Sacramento. There is a wonderful program called PASAR for older women students entering the university for the first time or returning after a multiple-year absence. The Women's Resource Center also provides small grants and scholarships for re-entry women. A large number of my students (both white and minority women) come severely handicapped in their basic language, math and science skills; many have never used a computer. It is not uncommon (especially among Chicanos and Blacks) to get an incoming student who scores at the fifth- and sixth-grade reading levels.

The task is herculean, the rewards spiritually fulfilling. I would not have it any other way. Every day is a lesson in humility and audacity. That my students have endured nothing but obstacles and putdowns, yet still have the courage and strength to seek a college education, humbles me. They are, like me, walking paradoxes. They have won against all the odds (their very presence on campus attests to that). Yet really they haven't won: they carry a deeply ingrained sense of inferiority, a firm conviction that they are not worthy of success.

This is my challenge: I embrace it wholeheartedly. There is no place I'd rather be, no profession more noble. Sure, I sometimes have doubts: every day something new, sad, even tragic comes up. Just as I was typing this article, for instance, Vicky, one of the white students in my Chicano literature class, called in tears, barely able to talk. "Professor, I can't possibly turn in my paper to your mailbox by four o'clock," she cried. "Everything in my house is falling apart! My husband just fought with my oldest daughter [from a previous marriage], has thrown her out of the house. He's running up and down the street, yelling and threatening to leave us. And I'm sitting here trying to write your paper! I'm going crazy. I feel like walking away from it all!" I took an hour from writing this article to help her contain herself. By the end of our conversation, I had her laughing. I also put her in touch with a counselor friend of mine and gave her a two-day extension for her final paper. And naturally I was one more hour late with my own writing!

I teach in a totally non-traditional way. I use every trick in the book: lots of positive reinforcement, both oral and written; lots of one-on-one conferences. I network women with each other, refer them to professor friends who can help them; connect them to graduate students and/or former students who are already pursuing careers. In the classroom I force my students to come up in front of their classmates, explain concepts or read their essays aloud. I create panels representing opposing viewpoints and hold debates—lots of oral participation, role-playing, reading their own texts. Their own writing and opinions become part of the course. On exams I ask them questions about their classmates' presentations. I meet with individual students in local coffeehouses or taverns: it's much easier to talk about personal pain over coffee or a beer or a glass of wine than in my office. My students, for the most part, do not have a network of support away from the university. There are no supportive husbands, lovers (except on rare occasions, as with my lesbian students), no relatives saying, "Yes, you can do it."

Is it any wonder that when these students come to me they have a deep sense of personal shame about everything—poor skills, being older students? They are also very angry, not only at themselves but at the schools for having victimized them; at poor, uninspired teaching; at their parents for not having had high enough expectations of them or (in the case of the women) for having allowed them to marry so young. Sylvia, my Chicana feminist student, put it best when I was pointing out incomplete sentences in her essay: "Where the hell was I when all this was being taught in high school? And why didn't anybody give a damn that I wasn't learning it?"

I never teach content for the first two weeks of any of my courses. I talk about anger, sexism, racism and the sixties—a time when people believed in something larger than themselves. We dialogue—about prisons and why so many Chicano and Black young men are behind bars in California; why people fear differences; why they are so homophobic. I give my students a chance to talk about their anger ("coraje" in Spanish). I often read them the poem by my friend and colleague Jose Montoya, called "Eslipping and Esliding," where he talks about "locura" (craziness) and says that with a little locura, a little eslipping and esliding, we can survive the madness that surrounds us. We laugh at ourselves, sharing our tragic, tattered pasts; we undo everything and let the anger out. "I know why so many of you are afraid of doing well," I say. "You've been told you can't do it, and you're so pissed off about it, you can't concentrate." Courage takes pure concentration. By the end of these initial two or three weeks we have become friends and defined our mutual respect. Only then do we enter the course content.

I am not good at endings; I prefer to celebrate beginnings. The struggle continues and the success stories abound. Students come back, year after year, to say "Thank you." Usually I pull these visitors into the classroom: "Tell my class that they can do it. Tell them how you did it!" They start talking and can't stop. "Look, Olivia, when I first came into your class," said Sylvia, "I couldn't even put a fucking sentence together. And now look at me, three years later I'm even writing poetry!"

REMINISCENCE OF A POST-INTEGRATION KID: *Or, Where Have We Come Since Then?*

31

Gaye Williams

I want to dedicate this to the young woman who sat in the University of Tennessee dining hall while a table of seven white kids—who would share her microbiology class an hour later—stood to leave as she sat down. They stood to prove that they were too good to sit next to a Black woman. However, one of them (her lab partner) could later laugh and joke with her by the safety of the microscope where no one could see them, as if nothing had happened.

From: *Sage: A Scholarly Journal on Black Women* 1 (Spring, 1984):20–21. Reprinted by permission.

Sharing space were the young man and the woman, who was smarter but darker, and who knew what it was to be hurt.

Post-segregation. I asked my mother why, when I got around to attending schools, everything had calmed down. "After the whites realized they had lost," she said, "they mostly gave up fighting desegregation." So, no mobs lined up to keep me out of North Roebuck School's third grade class, or my Black teacher from her job teaching all those white kids. I even won Top Scholar of my class.

Someone asked me when I first began to identify as a feminist. I realized then that I was outraged about the unequal treatment of women in relation to men long before I understood what racism meant. Before being asked the question, I had not thought about the fact that I lacked a racial consciousness growing up in Birmingham, Alabama, in the sixties. I thought some more.

The absence of a racial consciousness was a result of my being a post-integration kid. I came up through a conspiracy to silence the Black movement, which was not a topic in the white schools I attended. Even though one of my babysitters was among the girls killed in the 16th Street Baptist Church bombing, everything was fine, or so the silence suggested. There were no bombs, no riots, no state troopers taking me to school. There was also no one to reinforce the radio's message that Black was beautiful, just James Brown shouting as I got ready for school "Say It Loud, I'm Black and I'm Proud!"

But JB's message did not sink in. In elementary school, I thought the white kids would not know I was Black if I did not tell them, although I am not at all light enough to pass. I made myself invisible and pretended I was someone else, for protection from the self-hatred that I had learned from somewhere. The fact that my imaginary self was always male, and more often white than Black, says something about the effects of curriculum, climate, and the availability of suitable role models to my sense of self-validation. Very little in my pre-college schooling gave me an alternative to hating my Blackness and femaleness, although at the time I am sure I would not have spoken of it that way. I retreated into identities I felt were safer, more attractive than my own.

Was that all the fault of curriculum, climate, lack of suitable role models? Maybe not, but I spent the great bulk of my pre-college time either in school or studying, with very little free time left over. When I was not studying, I was acting, and the classes and plays were little different in content from my school lessons.

Very early in my education, the lines were drawn between the kids singled out as smart and the others. I think about the church schools I attended—no rods spared nor children spoiled there. Being "good" and "smart" were my best defenses. From there I went to public elementary school for second grade, where I wandered outside of class to the school library and the reading specialist's lab. I was a favored child, along with my friend Eric. We were taught at home by our school teacher/administrator families. We both advanced beyond the other kids and were treated something like a prince and princess—"golden" children.

I thought I was rich, with my light skin and proper talk, because I was set apart by the teacher, attractive to boys, and picked on by at least one girl. I knew, even while she made me cry, that Debra was as much attracted to me as she was envious or curious. That was the year when the teacher performed a wedding ceremony for Eric and me in front of all the others.

The chance to develop my talent also set me apart from other Black kids. I was not a singer, coming along in the bosom of a Black church choir, but acting in plays that had Black characters only because their actors happened to be Black. After a successful audition, I went to the Alabama School of the Fine Arts in drama. The number of Black kids in the school slowly grew, but most of them were musicians. Even among the Black actors, I was still set apart by my lack of a Black or southern accent.

The other side of being a golden child marked for success was the feeling of being separated from my peers. I was held up for praise and vulnerable to disdain. One problem of post-integration kids is that the divisions between us of class, sex, and color have become more virulent. These differences have always been present in the Black community, and their resulting hierarchy a divisive problem. But this distancing effect of the combination of light skin, brains and opportunity became much greater post-integration because the possibility of admission to white society became much more probable. With post-integration came more options.

I came from a family that sent its children, daughters included, to the best schools the Black community had to offer, at whatever sacrifice. They considered education the necessary ticket to advancement. By the time I was ready for college, even the Ivy League was within reach. My admission was bought by sit-ins, marches, riots, freedom rides, and legislation, but I had only the vaguest understanding of the importance of these occurrences.

My mother once said that she wondered whether she had done the best thing sending me to mostly white private schools. Which is more important—that kids have the best opportunity to develop their minds in language, art and science, or that their souls and spirits benefit from the best the Black tradition has to offer, so that they face the world's hostility feeling part of a strong community and heritage? She could not ask nor answer then the question of why we should have to choose between the two.

It was in college that I began to gain a full sense of what it meant to me to be born Black and female. Knowing that the Ivy League probably had little idea of what a young Black girl/woman from the South needed or wanted so far away from home, I had the sense to build a community of friends around me very soon. Convinced that I was not prepared to be in this prestigious university and that I was going to have a difficult time academically (even though I never before had trouble in school), I surrounded myself with people and activities to help see me through. First, I found upperclass women students who were intrigued, I think, by my boldness and always ready with advice and comfort. They were my survival, along with special teachers, the Boston women's community, and the Black children and their families I met in the Intergenerational Outreach Saturday Educational Program.

I met older Black kids who had a strong racial and feminist consciousness. But they had also felt separated from most Blacks for the same reasons I had. These folks were discovering together how to bridge gaps, knowing that these were the most important lessons to learn. I grew from a base of self-love to alliances and friendships with students who were Chicano, Puerto Rican, African, Asian and American Indian, and the gay folks of all ethnicities.

The most important thing about going to college when I did and where I did was that I began to broaden my idea of education. Thinking about what education has meant to me seems almost the same as thinking about what being young means. There is the education that happens inside classrooms or other clearly designated learning places. But it is difficult, and perhaps undesirable, to separate the learning I did there from the constant process of figuring out what I was doing in this world. The point to consider is how the learning in everyday life and that of school fit together. Then I can begin to answer questions of how my racial/feminist consciousness was shaped.

I went to a lot of conferences and events in the Boston area while in school. The most significant one was the 1981 Women and Law Conference. Meeting feminist women of color there had a catalytic effect on my education, helping to bring everything together. Afterwards, I began demanding that my course work in school complement the questions I was framing and hearing outside the classroom. I insisted that this expensive schooling give me something that would help me understand the meaning of my life as a Black woman. From the example of the women I met there and at subsequent events, I gained the inspiration and courage necessary to run successfully for president of the Harvard-Radcliffe Black Students Association, and learned to bring the force that women's work embodied onto the campus and into the lives of the other students.

I have not liked feeling cut off from people, especially from Black folks. It has been my good fortune to make friendships that showed me how silly and immaterial were the barriers which I thought kept me apart. With the help of my friends of all races I have come to insist that the post-integration decades be just that: a time that allows me to "integrate" and to use all the parts of myself. I insist on it being a time that does not demand that I be either Black **or** female, but recognizes that I am both, **all** of the time. From there I can go on to the business of helping make this world fit for all of us to live in.

Of course, my education does not end with the close of classes. The beat, I know, will keep going on.

INTEGRATING THE AMERICAN MIND　　32

Henry Louis Gates, Jr.

When I'm asked to talk about the opening of the American mind, or the decentering of the humanities, or the new multiculturalism—or any number of such putative "developments"—I have to say my reaction is pretty much Mahatma Gandhi's when they asked him what he thought about Western civilization. He said he thought it would be a very good idea. My sentiments exactly.

This decade has, to be sure, witnessed an interesting coupling of trends. On the one hand, we've seen calls from on high to reclaim a legacy, to fend off the barbarians at the gates and return to some prelapsarian state of grace. On the other hand (or is it the same hand?), we've seen a disturbing recrudescence of campus racism sweeping the nation. Many of you will have seen the articles about this in the recent media, as the topic has been in the news for quite some time. For people who agitated in the civil rights era and saw real gains in the college curriculum in the 1970s, the new conservatism seems to have succeeded in their own efforts rather as the Redemption politicians followed the Reconstruction, threatening to undo what progress had been made.

One thing is clear. Education in a democratic society (or in one that aspires to that ideal) has particular burdens placed upon it: few theorists of American education, in this century or the preceding one, have separated pedagogy from the needs of citizenship. The usual term, here, is often given a sinister intonation: *social reproduction*. Yet this country has always had an evolutionary view of what reproduction entails: we've never been content with more replication, we've sought improvement. We want our kids to be better than we are.

So it's discouraging, even painful, to look about our colleges, bastions of liberal education, and find that people are now beginning to talk—and with justice, it seems—about the "new racism." I don't want to offer a simple diagnosis, but perhaps the phenomenon isn't completely unconnected to larger political trends. It's been pointed out that today's freshmen were ten years old when the Reagan era began; presumably the public discourse of the 1980s had something to do with the forming of political sensibilities.

But whatever the causes, the climate on campus has been worsening: according to one monitoring group, racial incidents have been reported at over 175 colleges since the 1986–87 school year. And that's just counting the cases that made the papers.

At the same time, there's been, since 1977, a marked decline in overall black enrollment in colleges. The evidence suggests the decline is connected to a slipping economic situation, and to cuts in available federal aid. In the decade since 1977, federal grants and scholarships have fallen 62 percent, and that, of course, disproportionately affects minority students. Almost half of all black children (46.7 percent) live under the poverty line, according to the Congressional Research Service. Indeed, if you look at students at traditionally black colleges, you find that 42 percent of them come from families with income below the poverty line; a third of these students come from families with a total family income less than $6,000 a year. So when it comes to larger economic trends, blacks are like the canaries in the coal mine: the first to go when things are going wrong.

But there's an even bigger problem than getting these students, and that's keeping them. The attrition rate is depressing. At Berkeley, one in four black students will graduate. The fact is, according to the National Center of Education Statistics, that of freshmen blacks in 1980, only 31 percent had graduated by 1986. And while financial pressures explain some of it, they don't explain all of it.

Down the educational pike, things get worse. Just 4.0 percent of our full-time college professors are black, and the number is said to be decreasing. In 1986, only 820 of the 32,000 Ph.D.'s awarded went to blacks; less than half those 820 planned a college career (that's 0.015 percent of our new Ph.D.'s).

In short, it's a bad situation. But it's not a conspiracy; nobody wants it to be the way it is. In general, our colleges really are devoted to diversity: people are genuinely upset when they fail to incorporate diversity among their students and faculty. I said before that the peculiar charge of our education system is the shaping of a democratic polity. It's a reflection of the public consensus on this matter that one of the few bipartisan issues in the last presidential campaign had to do with equitable access to higher education. Pollsters on both sides found that this was an issue that made the American heart skip a beat. Equal opportunity in education is an idea with very broad appeal in this country. And that has something to do with what education means to us. So one thing I want to bring out is that the schools are a site where real contradictions and ambivalences are played out.

I would like to think about institutions for higher learning in terms of the larger objectives of what we call a liberal education; as unfashionable as it is among many of my fellow theorists, I do believe in the humanities, very broadly conceived. But it's that breadth of conception I want to address. We hear the complaints. Allan Bloom, for example, laments that "just at the moment when everyone else has become 'a person,' blacks have become blacks." (Needless to say, "everyone else" can become a person precisely when the category *person* comes to be defined in contradistinction to *black*.) Many thoughtful educators are dismayed, even bewildered, when minority students—at Berkeley or Stanford or Texas or Oberlin, the sentiment's widespread—say that they feel like visitors, like guests, like foreign or colonized citizens in relation to a traditional canon that fails to represent their cultural identities. I'm not interested in simply endorsing that sentiment; it's not a reasoned argument, this reaction, but it is a playing out—a logical extension— of an ideology resident in the traditional rhetoric about Western civilization. And I want to consider it in that light.

Once upon a time, there was a race of men who could claim all of knowledge as their purview. Someone like Francis Bacon really did try to organize all of knowledge into a single capacious but coherent structure. And even into the nineteenth century, the creed of universal knowledge—*mathesis universalis*—still reigned. There's a wonderful piece of nineteenth-century student doggerel about Jowett, the Victorian classicist and master of Balliol College, Oxford, which rather sums up the philosophy:

> *Here stand I, my name is Jowett,*
> *If there's knowledge, then I know it;*
> *I am the master of this college,*
> *What I know not, is not knowledge.*

The question this raises for us, of course, is: How does something get to count as knowledge? Intellectuals, Gramsci famously observes, can be defined as experts in legitimation. And the academy, today, is an institution of legitimation—establishing what counts as knowledge, what counts as culture. In the most spirited attacks on the movement toward multiculturalism in the academy today, there's a whiff of this: We are the masters of this college—What we know not, is not knowledge. So that in the wake of Bacon's epistemic megalomania, there's been a contrary movement, a constriction of what counts as even worth knowing. We've got our culture, what more do we need? Besides, there was Heidegger on stage right, assuring us that "philosophy speaks Greek." And beyond the cartography of Western culture? A cryptic warning: Here Be Monsters.

I got mine: The rhetoric of liberal education remains suffused with the imagery of possession, patrimony, legacy, lineage, inheritance—call it cultural geneticism (in the broadest sense of that term). At the same moment, the rhetoric of possession and lineage subsists upon, and perpetuates, a division: between us and them, we the heirs of our tradition, and you, the Others, whose difference defines our identity. (In the French colonies, in Africa and the Caribbean, a classroom of African students would dutifully read from their textbook, "Our ancestors, the Gauls . . . " Well, you could see that wasn't going to last.)

What happens, though, if you buy into that rhetoric—if you accept its terms and presuppositions about cultural geneticism? Then you will say: Yes, I am Other, and if the aim of education is to reinforce an individual's rightful cultural legacy, then I don't belong here—I am a guest at someone else's banquet. Foucault called this kind of contestation that of "reverse discourse": it remains entrapped within the presuppositions of the discourse it means to oppose, enacts a conflict internal to that "master discourse"; but when the terms of argument have already been defined, it may look like the only form of contestation possible. . . .

. . . Broadening our educational vistas is not only sweet, but useful. As most of you will know, a panel of U.S. governors recently concluded that America's economic and cultural dominance has been endangered by our vast ignorance of the languages and cultures of other nations. In a report made public recently, the National Governors Association argued for broad changes in the way we teach foreign languages and basic geography.

Among the report's more startling findings are the following:

- A recent Gallup poll revealed that young American adults know less about geography than their peers in six developed countries.
- One in seven American adults cannot locate the United States on a world map.

- Twenty-five percent of a sample of high school seniors in Dallas did not know that Mexico was the country bordering the United States to the south.
- Only 20 percent of American high school graduates receive more than two years' instruction in a foreign language.

In response to this report, Ernest L. Boyer, the president of the Carnegie Foundation for the Advancement of Teaching, remarked that "a curriculum with international perspective" is "critically important to the future of our nation."

Well, Americans know so very little about world history and culture in part because high school and college core curricula, in this country, center upon European and American societies, with America represented as the logical conclusion or summary of civilization since the Greeks, in the same way that Christians believe that Christ is the culmination of the Old Testament prophesies. Our ignorance of physical geography is a symptom of a much broader ignorance of the world's cultural geography. Since the trivium and quadrivium of the Latin Middle Ages, "the humanities" has *not* meant the best that has been thought by all human beings; rather, "the humanities" has meant the best that has been thought by white males in the Greco-Roman, Judeo-Christian traditions. A tyrannical pun obtains between the words *humanity*, on the one hand, and *humanities*, on the other.

We need to reform our entire notion of core curricula to account for the comparable eloquence of the African, the Asian, the Latin American, and the Middle Eastern traditions, to prepare our students for their roles in the twenty-first century as citizens of a world culture, educated through a truly human notion of the humanities.

Now, I talked earlier about the long-dead ideal of universal knowledge. Today, you look back to C. P. Snow's complaint about the gulf between the "two cultures," and you think, *two?* Keep counting, C. P. The familiar buzz words here are the "fragmentation of humanistic knowledge." And there are people who think that the decentering of the humanities that I advocate just makes a bad situation worse: Bring on the Ivory Towers of Babel. So I want to say a few words about this.

There are, certainly, different kinds of fragmentation. One kind of fragmentation is just the inevitable result of the knowledge explosion; specialized fields produce specialized knowledge, and there's too much to keep up with. But there's another kind of fragmentation which does deserve scrutiny: the ways in which knowledge produced in one discipline setting is inaccessible to scholars in another discipline, even when it would be useful to them in solving their problems. And here, what I call the decentering of the humanities can help us rethink some of the ways traditional subjects are constituted and can allow us a critical purchase helpful in cultural studies quite generally. Indeed, far from being inimical to traditional Western scholarship, humanistic scholarship in Asian and African cultures can be mutually enriching to it, to the humanities in general.

And that's as you'd expect: The study of the humanities is the study of the possibilities of human life in culture. It thrives on diversity. And when you get down to cases, it's hard to deny that what you could call the new scholarship has invigorated the traditional disciplines. Historians of black America, for example, have pioneered work in oral history that's had a significant effect on the way nineteenth-century social history is done. . . .

In a more practical vein: It turns out that the affirmative action programs for recruiting minority faculty have been successful only at institutions where strong ethnic studies programs exist. Many ambitious "minority" scholars of my generation, feeling secure in their academic credentials and their ethnic identities, have tried to fill a lacuna they perceived in their own education by producing scholarship about, well, "their own people." A lot of the social commitment that emerged during the 1960s has been redirected to the scholarly arena: continents of ignorance have been explored and charted. At the same time, "minority studies" (so-called) are not "for" minorities, any more than "majority studies" (let's say) are for majorities. And it is wrong simply to conflate affirmative action objectives in employment with the teaching of such subjects.

I respect what Robert Nisbet calls the Academic Dogma: knowledge for its own sake (I suspect it doesn't quite exist; but that's another matter). At the same time, I believe that truly humane learning can't help but expand the constricted boundaries of human sympathy, of social tolerance. Maybe the truest thing to be said about racism is that it represents a profound failure of imagination. I've talked a good deal about cultural pluralism as a good in itself, as the natural shape of scholarship untrammeled by narrow ethnocentrism. And the best ethnic studies departments have made a real contribution to this ideal of scholarly diversity. As I said, I respect the ideal of the disinterested pursuit of knowledge, however unattainable, and I don't think classes should be converted into consciousness-raising sessions, Lord knows; at the same time, anyone who's not a positivist realizes that "moral education" is a pleonasm: in the humanities, facts and values don't exist in neatly disjunct regimes of knowledge. Allan Bloom is right to ask about the effect of higher education on our kids' moral development, even though that's probably the only thing he is right about.

Amy Gutmann said something important in her recent book *Democratic Education:* "In a democracy, political disagreement is not something that we should generally seek to avoid. Political controversies over our educational problems are a particularly important source of social progress because they have the potential for educating so many citizens." I think that's true; I think a lot of us feel that any clamor or conflict over the curriculum is just a bad thing in itself, that it somehow undermines the legitimacy of the institutions of knowledge—a sort of no-news-is-good-news attitude on the subject of education; they think if you even look at a university cross-eyed, it'll dry up and blow away and then where will you be? In contrast, I think one of the most renewing activities we can do is to rethink the institutions where we teach people to think; we invest in myths of continuity, but universities have constantly been molting and creating themselves anew for the last millennium, and there's no reason to think that'll change in the next. Gerald Graff has been saying, where there's no consensus—and there's no consensus—teach the conflicts. In fact, I think at the better colleges, we do. We don't seem to be able not to. And that's nothing to be embarrassed about: College isn't kindergarten, and our job isn't to present a seemly, dignified, unified front to the students. College students are too old to *form*—we shouldn't delude ourselves—but they're not too old to challenge. I'm reminded of something that the college president and educator Robert Maynard Hutchins wrote, in a book he published during the height of the McCarthy era. He recounted a conversation he'd had with a distinguished doctor about the attempt of the Board of Regents of the University of California to extort (as Hutchins put it) "an illegal and unconstitutional oath of loyalty from the faculty of that great institution." "Yes, but" the doctor said, "if we are going to hire these people to look after our children we are entitled to know what their opinions are." And Hutchins grandly remarked, "I think it is clear that the collapse of liberal education in the United States is related as cause or effect or both to the notion that professors are people who are hired to look after children." Wise words, those.

In all events, the sort of pluralism I've been recommending has one evolutionary advantage over its opponents. If you ask, How do we form a consensus around such a "de-centering" proposal? the answer is that it doesn't exactly require a consensus. Which is why, in the words of John Dewey, "pluralism is the greatest philosophical idea of our times." Not that this puts us home free. As Dewey also said,

What philosophers have got to do is to work out a fresh analysis of the relations between the one and the many. Our shrinking world presents that issue today in a thousand different forms. . . . How are we going to make the most of

The State and Social Policy

BLACK MALES AND SOCIAL POLICY: **33**
Breaking the Cycle of Disadvantage

Ronald L. Taylor

Federal social policy, whether in the area of employment, welfare, education, social insurance, poverty, or criminal justice, has often had a disproportionate impact on the lives of Black Americans. Their historically disadvantaged position in American society has magnified the impact of federal programs, whether their effect was to provide opportunities or to create obstacles to the achievement of racial equality. For example, the civil-rights legislation of the 1960s inspired an array of federally sponsored programs that contributed to significant improvements in the socioeconomic status and general well-being of Black Americans. In contrast, federal retreat from such programs and policies in the 1970s contributed to a deepening of Black social problems in the 1980s, especially in employment, education, health, and family stability.

From: Ronald L. Taylor, "Black Males and Social Policy: Breaking the Cycle of Disadvantage," in *The American Black Male: His Present Status and His Future*, ed. Richard G. Majors and Jacob U. Gordon (Chicago: Nelson-Hall Publishers, 1994), pp. 148–166. Reprinted by permission.

Such reversals in federal initiatives have had a particularly adverse effect on the quality of life among young Black men, as evidenced by a number of social indicators. . . .

Here we will examine how the shift in emphasis of federal policy during the past two decades—from a focus on the structural sources of Black inequality and social problems to one emphasizing the behavioral and value deficiencies of the Black "underclass"—contributed to the current crisis among young Black males in the inner cities and elsewhere in the country. . . .

FROM BLAMING THE SYSTEM TO BLAMING THE VICTIMS

Federal initiatives designed to combat such national social problems as poverty, unemployment, ill health, and the shortcomings of the educational system are relatively recent developments in the United States. While the 1930s saw the beginnings of programmatic structures and social policy innovations designed to ameliorate the economic conditions of some segments of the population (e.g., public assistance for the needy over sixty-five, the blind, and dependent children under the Social Security Act of 1935), it was not until the 1960s that a more comprehensive system of income supports and other social provisions was established (Weir et al. 1988). One of the central objectives of federal programs and policies inaugurated during this period was the reduction of poverty and the myriad of social problems associated with it, especially among Blacks and other minority populations. Under the banner of the Great Society, the Johnson Administration declared a War on Poverty that included an expansion of existing entitlement programs and the establishment of such new training programs as food stamps, Medicare, subsidized housing, and job training to enable the poor to break the cycle of poverty and become productive members of society. . . .

. . . By the late 1970s, a new analysis of poverty and other social problems besetting Black communities gained ascendancy among intellectuals and policymakers, to wit, that poverty, unemployment, welfare dependency, and female-headed households, among other problems, were less the result of such structural factors as racial discrimination and social isolation than of excessive welfare programs and policies. Such initiatives had destroyed the incentive of the poor to be productive and exacerbated their behavioral deficiencies. Indeed, this was the major theme of presidential candidate Ronald Reagan, who vowed to dismantle much of the Great Society's social legislation and entitlement programs were he to be elected. With his election came drastic cuts in social programs targeted on the poor and the most sweeping policy reversals in the thirty-five years since the New Deal (Orfield 1988).

With the election of Ronald Reagan to the presidency in 1980, federal policy had come full circle, from an emphasis on the structural sources of poverty, unemployment, and related social problems in the 1960s to an emphasis on the behavioral characteristics, norms, and values of the poor, which in the 1980s were seen to produce such problems. The latter focus implies that social programs can do little to correct the behavioral deficiencies of the Black underclass or promote improvements in their social condition. But such a conclusion is based on the faulty assumption that, despite enormous expenditures on social programs targeted on the poor, little was accomplished. . . .

. . . As the erosion of public support, cutbacks and/or dismantling of federal programs targeted on the poor continued, social conditions in many inner cities, and among Black men in particular, grew progressively worse. The deterioration was due in large measure to shifting policy objectives and the failure of such policies to address the underlying structural conditions that promoted the persistence of a variety of social problems in Black communities.

BLACK MALES: ISSUES, PERSPECTIVES, AND SOCIAL POLICY

. . . The lack of attention to Black social problems has been especially costly to young Black males, given their high-risk status. On nearly every socioeconomic measure, from infant mortality to life expectancy, Black males are worse off today than they were some two decades ago (Gibbs 1988; Taylor 1987). As Marshall (1988) has noted, because of a complex set of mutually reinforcing factors, Black males start life with serious disadvantages. They are, for instance, more likely to be born to unwed teenage mothers who are poorly educated and more likely to neglect or abuse their children; the children of these mothers are also more likely to be born underweight and to experience injuries or neurological defects that require long-term care. Moreover, such children are more likely to be labelled "slow learners" or "educable mentally retarded," to have learning difficulties in school, to lag behind their peers in basic educational competences or skills, and to drop out of school at an early age. Black boys are also more likely to be institutionalized or placed in foster care (Gibbs 1988).

To be sure, the sources of the problems many Black males experience in education, employment, the criminal justice system, and in other areas of social life can be traced, not only to public and scholarly indifference to such problems, but to recent structural changes in the economy, perverse demographic trends, and intractably high levels of urban poverty, which converged to exacerbate conditions in inner cities across the country and the predicament of Black males in particular (Wacquant and Wilson 1989). With respect to demographic changes, for example, the proportion of Black youth in central cities increased by more than 70 percent during the past two decades. The result may have been a "critical mass" of Black youth which, it has been suggested, is a condition that can set in motion "a self-sustaining chain reaction" that contributes to "an explosive increase in the amount of crime, addiction, and welfare dependency" in the inner cities (Wilson 1977, 20). Under these conditions, community institutions and local labor markets have proved inadequate to meet the needs and cope with the consequences of a substantially enlarged Black youth population. As William Wilson (1983) has observed, "on the basis of these demographic changes alone, one would expect Black youth to account disproportionately for the increasing social problems of the central city" (p. 83).

To such demographic changes must be added continuing problems in education, employment, poverty, and the criminal justice system which diminish the life chances of Black males and imperil the future of Black communities.

Black Males and Education

Data on school enrollment and educational attainment at the precollege level for Black youth and young adults show marked improvements during the past two decades. . . . The gap in median years of schooling between Blacks and whites declined from 2.7 years in 1960 to less than one-half year in 1980: 12.6 years for Blacks and 13.0 years for whites (National Research Council 1989).

These data, however, mask significant differences between Black and white youths in rate of delayed education (i.e., being behind in school) and in nonattendance. For example, Black youths are much more likely to be enrolled below the modal grade for their age group than are white youth, a differential that increases in higher grades. . . . There remains a significant racial difference in the completion of high school: 21 percent of all Blacks eighteen and nineteen years old, and 25 percent of these twenty and twenty-one years old had neither completed nor were enrolled in high school in 1980, compared to 14.9 percent and 14.5 percent, respectively, of white youths in these age groups. The dropout rates for Black males in major cities like New York, Chicago, Philadelphia, and Detroit are considerably higher than the rates for Black youths as a whole, ranging from 40 to 70 percent in these cities. In general, Black males in major cities are about two and one-half times more likely to drop out of school than Black females and have the highest school dropout rate of all gender groups, leaving nearly a quarter of them ill-equipped to enter the job market, military service, or post secondary education (Gibbs 1988). . . .

In general, research concerning race and gender differences in the quality of school experiences suggests that the relatively poor scholastic performance, negative attitudes, and high attrition rates of Black males are at least in part a function of the nature of their interactions and treatment in public schools. The implications of such a finding become apparent when viewed in the context of a finding reported in other studies: that limited or negative academic feedback to Black males is related to the tendency among these youth to lower their expectations, degrade their scholastic abilities, and substitute compensatory terms for positive self-regard and sense of achievement (Hunt and Hunt, 1977; Hare and Castenell, 1985; Irvine, 1990).

Whatever the mix and contribution of these factors to the underachievement and failure of Black males in public schools, the consequences are fewer employment possibilities, high rates of joblessness, and long-term limitations on economic and social mobility.

Black Males and Employment

Despite evidence of significant gains in their educational and occupational status during the past twenty years, the labor market position and employment problems of young Black males grew worse during this period, reaching what some analysts describe as "catastrophic" proportions in the 1980s. As recently as 1954 employment rates for young Black and white males were identical (52 percent in each case). By the early 1960s, 59 percent of Black males ages sixteen to twenty-four were employed, compared to 68 percent of white males in these age groups. [There has been] a startling decline in the employment condition of young Black males between 1970 and 1985. . . .

Various analyses of the Black youth unemployment crisis identify the sources of the problem in deteriorating local economies and functional transformations in urban structures, increased job competition between Black youth and older women in the labor force, increases in minimum wage, discriminatory employer behavior, increased opportunities for criminal activities as alternative sources of employment and income, inadequate education, lack of marketable skills, and low motivation and/or aspiration on the part of Black youth (Kasarda 1989; Thomas and Scott 1979; Freeman 1986). Based on their analysis of extensive surveys and interviews of Black males, ages sixteen to nineteen, in the inner-city poverty areas of Boston, Philadelphia, and Chicago, Freeman and Holzer (1986) found "no single cause" of the decline in employment among these men. Rather, Freeman (1986) observes, "joblessness among young Black men is part-and-parcel of other social pathologies that go beyond the labor market, including youth crime, and drug and alcohol abuse, residual employer discrimination and performance on the job, particularly absenteeism" (p. 2F).

Yet, as Kasarda (1989) has shown, fundamental changes in the structure of city economies (from centers of goods processing to centers of information processing) during the past two decades have severely affected the employment prospects of disadvantaged urban Blacks, particularly males, with less than a high school education. According to Kasarda, "these structural changes led to a substantial reduction of lower-skilled jobs in traditional employing institutions that attracted and economically upgraded previous generations of urban Blacks" (p. 27). Such losses in employment opportunities, Kasarda concludes, have had "devastating effects on Black families, which further exacerbated the problems of the economically displaced" (p. 27). . . . In short, conditions that prevent or make it difficult for young men to find work discourage them from trying to obtain economic independence and promote growing disaffection or alienation from society, contributing to a host of social problems and self-destructive behaviors (Taylor 1991).

Crime, Delinquency, and Substance Abuse

Blacks in general and Black males in particular are arrested, convicted, and incarcerated for criminal offenses at rates considerably higher than are whites (National Research Council 1989). Although Black men represent approximately 6 percent of the total population, they constitute nearly 50 percent of the inmates in local, state, and federal jails. The rate of incarceration of Black males was 1,581 per 100,000 population in 1985, compared with 252 per 100,000 for white males. In short, Black males were being imprisoned for criminal offenses at a rate six times the rate for white males. Among youth, Black males are disproportionately represented in arrest statistics on crime and delinquency. While Black youth constituted approximately 14 percent of all youths aged fifteen to nineteen in 1984, they were involved in 69 percent of the robbery arrests, 54 percent of the rape arrests, and 39 percent of the aggravated assaults attributable to persons under eighteen years of age (Federal Bureau of Investigation 1986). In fact, more than half of all arrests for violent crimes and a quarter of all property crimes reported in 1984 involved Black youths. By the time they reach age nineteen, one in six Black males will have been arrested.

Black males are not only more likely to be arrested for violent crimes but are more likely to be the victims of crimes as well. According to a recent report by the U.S. Department of Justice (1985), young Black males have the highest violence/victimization rate of all groups in the United States, including murder, robbery, and aggravated assault. Blacks in general are five times more likely than whites to be homicide victims, and the lifetime risk of being a homicide victim is far greater for Black males than for any other group: only 1 out of 179 white males is likely to be a homicide victim in his lifetime, compared to 1 in 30 Black males, 1 in 495 white females, and 1 in 132 Black females. In almost all cases of Black male homicide, the offender is another Black male, a fact which in part accounts for the overrepresentation of Black males in correctional institutions. This pattern is true for most crimes of violence and for many property offenses as well (O'Brien 1987).

The overrepresentation of Black youths, especially males, in crime and delinquency has been attributed to a variety of factors, including bias on the part of law enforcement agencies in the treatment and/or disposition of cases involving young Blacks; high rates of joblessness, poverty, and welfare dependency among these youths and their families in inner cities; and a "subculture of violence" created by these conditions, which fosters high levels of interpersonal violence, including homicide (Freeman and Holzer 1986; National Research Council 1989). In addition, the sharp rise in "crack," a form of cocaine, and other drugs in Black communities has contributed to a marked increase in individual and gang violence and crime among Black males during the past decade, with the wider distribution of handguns among Black youth identified as a major contributing factor in the rise of mortality from homicide (National Center for Health Statistics 1989). . . .

Poverty

The growth and concentration of poverty in inner-city communities during the past two decades are perhaps key to understanding some of the changes in other measures of socioeconomic well-being among Black males, since poverty has long been shown to have adverse affects on educational attainment (Rumberger 1983) and employment opportunities (Freeman 1978), and to be positively associated with crime and delinquency (Freeman and Holzer 1985), family disruption (Wilson 1987), and teenage pregnancy (Hogan and Kitagawa, 1985).

In their assessment of poverty in the inner cities, Wacquant and Wilson (1989) conclude that "the urban black poor of today differ both from their counterparts of earlier years and from the white poor in that they are becoming increasingly concentrated in dilapidated territorial enclaves that epitomize acute social and economic marginalization" (p. 9). . . .

In recent decades, many inner-city communities have experienced sharp increases in the percentage of poor families, rapid exodus in record numbers of working and middle-income Black families, near collapse of local economies and community institutions, and high levels of unemployment. Moreover, the majority of households in many inner-city neighborhoods are headed by women who are economically worse off than other poor families and are more dependent on public assistance to support their children. In short, the increasing social and spatial concentration of poverty in the inner cities of this country, and the growing predominance of social ills long associated with poverty—family dissolution, school failure, drugs, violent crimes, housing deterioration—have qualitatively and quantitatively altered the psychosocial as well as the material foundations of life for most of their inhabitants, accelerating the process of economic marginality and social isolation (Wilson 1987).

These developments have affected young Black males in the inner cities in a variety of ways. Widespread joblessness, welfare dependency, and social isolation from mainstream institutions have not only deprived many of these young men of opportunities to acquire essential work experience for gainful employment and eventual self-sufficiency, but have discouraged them from marrying and forming families (Testa et al. 1989). . . .

The existence of large enclaves of young male adults in the inner cities, mired in poverty and a predatory subculture, without adequate education, marketable skills, or opportunities for earning legitimate incomes, constitutes a genuine crisis in Black communities and a major challenge (and threat) for the larger society.

BREAKING THE CYCLE OF DISADVANTAGE

. . . Although a public policy aimed at reinforcing poor Black families is an essential step toward improvement in the status of Black males, so too are efforts to enhance their employment prospects. While many federally sponsored employment and training programs launched during the 1960s and 1970s were plagued by a variety of administrative, political, and organizational problems, the effectiveness of some of these programs in improving the long-term employment prospects and life chances of disadvantaged Black males has been well documented (Hahn and Lerman 1985; Taylor 1990).

Under these programs, thousands of young Black men received remedial assistance and work experience which enabled them to earn their first steady incomes and to become acquainted with the world of work. As the employment crisis among inner-city Black male youths deepens across the country, such programs may play a decidedly more important role in helping to reduce high levels of joblessness, crime, school truancy, and early parenthood among these men, provided such programs effectively address their personal and educational, as well as their employment needs. Since most of these problems are concentrated among a relatively small subset of young Black males residing in the fifteen to twenty largest U.S. cities (Smith et al. 1987), "a major infusion of resources should be directed to a target group of cities that house these large pools of needy youths . . . the intent would be to build a full program infrastructure at a level of services far closer to the level of need than available resources now permit" (p. 44). In the absence of such a concentrated programming strategy, it is unlikely that the employment or earning prospects of Black males will substantially improve.

In the final analysis, a service strategy must be supplemented with or replaced by an employment strategy designed to create plentiful job opportunities for all if Black males are to find useful and productive roles in a society whose economic institutions are undergoing rapid change. For without the latter, no amount of remediation, employment training, or other social services is likely to improve the job prospects and life chances of Black males. Hence, the key to breaking the cycle of disadvantage among Black males is a federal social policy that seeks to address the employment and familial needs of these men. Indeed, an emphasis on the former may make it less necessary to deal with the problems associated with the latter.

References

Federal Bureau of Investigation. 1986. *Crime in the United States, 1985, Uniform Crime Reports.* Washington, DC: U.S. Dept. of Justice.

Freeman, R. 1978. "Black Economic Progress Since 1964." *The Public Interest* 52: 52–68.

———. 1986. "Cutting Back Youth Unemployment." *New York Times,* July 20.

Freeman, R., and N. Holzer. 1986. *The Black Youth Employment Crisis.* Chicago: University of Chicago Press.

Gibbs, J. T., ed. 1988. *Young, Black, and Male in America.* Dover, MA: Auburn.

Hahn, A., and R. Lerman. 1985. *What Works in Youth Employment Policy.* Washington, DC: Committee on New American Realities.

Hare, B., and L. Castenell. 1985. "No Place to Run, No Place to Hide: Comparative Status and Future Prospects of Black Boys." In M. Spencer; G. Brookins; and W. Allen (eds.), *Beginnings: The Social and Affective Development of Black Children*. Hillsdale, NJ: Erlbaum.

Hogan, D., and E. Kitagawa. 1985. "The Impact of Social Status, Family Structure, and Neighborhood on the Fertility of Black Adolescents." *American Journal of Sociology* 90: 825–55.

Hunt, J., and L. Hunt. 1977. "Racial Inequality and Self-Image: Identity Maintenance as Identity Diffusion." *Sociology and Social Research* 61: 539–59.

Irvine, J. 1990. *Black Students and School Failure*. New York: Greenwood Press.

Kasarda, J. 1989. "Urban Industrial Transition and the Underclass." *The Annals* 501: 26–47.

Marshall, R. 1988. "Foreword." In J. Gibbs (ed.), *Young, Black, and Male in America*. Dover, MA: Auburn House.

National Center for Health Statistics. 1989. "Firearm Mortality among Children and Youth" (advance data). Hyattsville, MD: U.S. Department of Health and Human Services.

National Research Council. 1989. *A Common Destiny: Blacks and American Society*. Washington, DC: National Academy Press.

O'Brien, R. 1987. "The Interracial Nature of Violent Crimes: A Reexamination." *American Journal of Sociology* 92: 817–35.

Orfield, G. 1988. "Race and the Liberal Agenda: The Loss of the Integrationist Dream, 1965–1974." In M. Weir, A. Orloff, and T. Skocpol (eds.), *The Politics of Social Policy in the United States*. Princeton, NJ: Princeton University Press.

Rumberger, R. 1983. "Dropping Out of High School: The Influence of Race, Sex, and Family Background." *American Educational Research Journal* 20: 199–220.

Smith, T.; C. Walker; and R. Baker. 1987. *Youth and the Work-Place: Second-Chance Programs and the Hard-to-Serve*. New York: Wm. T. Grant Foundation.

Taylor, R. 1987. "Black Youth in Crisis." *Humboldt Journal of Social Relations* 14: 106–33.

———. 1990. "Improving the Status of Black Youth: Some Lessons from Recent National Experiments." *Youth and Society* 22: 85–107.

———. 1991. "Poverty and Adolescent Black Males." In P. Edelman and J. Ladner (eds.), *Adolescence and Poverty: Challenge for the 1990s*. Washington, DC: Center for National Policy.

Testa, M.; N. Astone; M. Krogh; and K. Neckerman. 1989. "Employment and Marriage among Inner-City Fathers." *The Annals* 501: 79–91.

Thomas, G., and W. Scott. 1979. "Black Youth and the Labor Market: The Unemployment Dilemma." *Youth and Society* 11: 163–89.

U.S. Department of Justice. 1985. *Criminal Victimization in the United States, 1983.* Washington, DC: U.S. Government Printing Office.

Wacquant, L., and W. Wilson. 1989. The Cost of Racial and Class Exclusion in the Inner City. *The Annals* 501: 8–25.

Weir, M.; A. Orloff; and T. Skocpol, eds. 1988. *The Politics of Social Policy in the United States.* Princeton, NJ: Princeton University Press.

Wilson, J. 1977. *Thinking about Crime.* New York: Vintage.

Wilson, W. 1983. Inner-City Dislocations. *Society* 21: 80–86.

———. 1987. *The Truly Disadvantaged: The Inner City, the Underclass and Public Policy.* Chicago: University of Chicago Press.

THOUGHTS ON CLASS, RACE, AND PRISON

34

Alan Berkman and Tim Blunk

Having been locked up as political prisoners for the past five years, we have found a lot in several recent articles which was familiar and thought provoking. Certain descriptions of ping-ponging from jail to jail could be a log of some of our own trips conducted by the Bureau of Prisons (BoP), and we can verify accounts of how prisoners typically support one another. Perhaps most importantly, we strongly agree with the view—best expressed in an article by Sam Day published in the October 13, 1989 issue of *Isthmus*—that the threat of prison should not deter political activists from acting. We are concerned,

From: Alan Berkman and Tim Blunk, "Thoughts on Class, Race, and Prison," in *Cages of Steel: The Politics of Imprisonment in the United States*, ed. Ward Churchill and J. J. Vanderwall (Washington, DC: Maisonnueve Press, 1992), pp. 190–193. Reprinted by permission.

though, that in so vividly portraying his own experience and utilizing it to demystify prison life, he may have played into some common misrepresentations of life in U.S. prisons.

First, a word about our own experiences. We've each spent time in prison for a variety of politically motivated acts. One of us (Alan Berkman) spent two years in a number of county jails in the Philadelphia area. The other (Tim Blunk) spent a year in New York's Riker's Island. We've both been in a number of federal prisons, including the BoP's most repressive facility at Marion, Illinois. For the past eighteen months, we've been together in the equivalent of a county jail in Washington, D.C. where we and four women activists face federal charges of resisting U.S. war crimes through "violent and illegal means" (we call this the Resistance Conspiracy Case). Tim is presently serving a fifty-eight year sentence; Alan, twelve years. Having spent considerable time in the whole range of penal institutions, from county jails where most prisoners are pre-trial to Marion, where the average sentence is forty years, we've been struck by how directly prisons reflect the social and economic realities of society as a whole.

We live in a country where large numbers of people, particularly young Afro-American, Latino and Native American women and men, have been written off by society. Government leaders look at the human devastation caused by their policies and declare war on the victims (the "war on drugs"). The racism and malignant neglect that permeate the schools, the labor market, the welfare system and social services of the Third World and poor white neighborhoods of our cities bear their inevitable fruit in the prisons. In prison, though, even the usual facade of "fairness" and "justice" by which such inequity is usually accompanied is dropped.

There are almost a million people in U.S. jails, state and federal prisons. More than ninety percent of these prisoners are in county jails and state prisons. D.C. jail, for example, is filled to capacity with poor, mis-educated, profoundly alienated young Afroamerican men and women. It's a warehouse. There are no programs, no contact visiting, no privacy, and most importantly, no justice.

Particularly for political activists who are white and middle class, prison can be one of the few places where we look at America from the bottom up. We can understand poverty differently when we live with young people who've never had $20 in their pocket and who are constantly bombarded with the Gucci ads, the BMWs on *Miami Vice*, and the idea of free trips to Rio on *The Price is Right*. In a consumer society, if you've got nothing, you're considered to be nothing, and the frustration of this reality leads people into crime and drugs. You can see the deadness and despair in the eyes of too many of the people around you as they realize the American Dream is not for them now, and probably wasn't from the day they were born into the ghetto, the barrio or the reservation. And you'll get a feeling in your gut about the rage racism and poverty generate in the hearts of the oppressed. In fact, if you allow yourself, you'll learn quite a lot about the feelings that accompany that word, "oppressed."

What we've come to understand ever more deeply from our experience as prisoners is that the essence of oppression lies in dehumanization and disrespect. Being oppressed isn't just being a bit poorer or having a rougher life than your oppressor; it means being treated as if you're less of a human being than they are. For instance, we've consistently experienced the fact that you can be getting along okay with the guards and prison officials when they suddenly do something which is not only totally outrageous, but often a complete violation of their own supposed rule structure. You call them on it, and their response is to look at you—*through* you would be more accurate—as if you just weren't there. It doesn't matter a bit whether you're right or wrong. They deny that you have the right even to hold an opinion on the matter. It cuts at the very core of your sense of humanity and self-worth. It is infuriating and degrading.

Self-worth. Dignity. Self-respect. These are the feelings prison con-sciously or unconsciously is designed to destroy. These are the same feelings that racism and economic inequality in society work to destroy in the poor and people of color. This is why we believe Sam and others err when they reduce prison life to "three hots [meals] and a cot" and other minor indignities. It's like reducing chattel slavery to hard work, bad pay and poor living conditions. It leaves out the truth of the human dimension of oppression. And it leaves out the human reaction to prison conditions: rebellion. Prison "riots" are fundamentally slave rebellions.

Over the past few weeks, there have been one major and two minor prison uprisings in the Pennsylvania prison system. At Camp Hill, where prisoners destroyed much of the prison, the precipitating event seems trivial: the prison cancelled the semi-annual family day picnics where prisoners' relatives can bring home-cooked food and eat with their loved ones out on the yard. What's the big deal about two afternoons per year? Well, at one level, when you don't have much, even a *little* more deprivation hurts rather badly, especially when that deprivation concerns those precious bits of time you can spend with those closest to you and when you manage to feel most fully human. At a deeper level, the arbitrariness of the prison officials' decision in Pennsylvania gave the prisoners a clear message that they had nothing at all coming to them: *everything* is a "privilege," *nothing* is a right. The oppressor giveth, the oppressor taketh away. The oppressed have nothing to say in the matter, one way or the other. The oppressed—the prisoner in this instance—is a cipher, a statistic, nothing more than an animal whose basic needs of calories, shelter and waste disposal must (usually) be met. No more.

The "senseless violence" of burning buildings and smashing equipment which marked the Camp Hill rebellion—and which has marked prison uprisings from Attica to Atlanta, New Mexico, Los Angeles, and elsewhere—is, perhaps paradoxically, the most basic assertion of human dignity and self-respect. "We do have a say," the prisoners are insisting. "We will exercise our rights as human beings one way or another, either as reasonable human beings in a reasonable situation, or this way if you insist the situation remain unreasonable. But either way you can ultimately do nothing to provide our demonstration of the fact that *we are people*." It is a very human response to a very inhuman situation, the same impulse which led to the great ghetto uprisings of Watts, Detroit and Newark during the '60s, and which has led to similar phenomena in places like Miami during the '80s.

Political activists who come to jail and are able to keep their eyes and hearts open will learn an enormous amount, both about themselves and about the world they live in. How do we deal with our own race/class biases to build principled relationships with those we live with? If we're white, how do we respond when the Aryan Brotherhood tells us they'll kick our ass if we eat with Afro-American, Latino or American Indian prisoners in the *de facto* segregated mess hall? How do we struggle against the pervasive and intense sexism in men's prisons? How do we remain caring and giving people without allowing ourselves to be taken advantage of and disrespected? How do we decide when to directly confront the power of the officials, and when not to? When do we get involved in other people's beefs, whether among prisoners or between prisoners and guards?

There's a lot to learn in prison, although it's certainly not the only, or even necessarily the best, place to learn it. Whatever else may be said, time spent behind bars—no more or less than any other sort of time—need never be "wasted." We agree with Sam and others when they say that fear of prison should never be paralyzing. While locked up, you'll more than likely get in shape, have the opportunity to read many of the books you always meant to plow through but never managed to get around to, learn to write letters which say more than just "howdy-do." Most importantly, you'll finally have the chance to identify with the oppressed on a gut level which is typically absent from the life experiences of most Euroamericans. You'll come to understand power relationships in a new and far more meaningful way. And we're willing to bet that you'll come out more committed than ever to the need for fundamental social change.

THE VANISHED NATIVE AMERICANS

35

Cynthia Brown

ADRIAN: Mr. President, I'm here today as a Lumbee Indian of North Carolina, yet under the law I'm not an Indian. What are you going to do to resolve this problem?
CLINTON: Why is that? I don't understand it. . . . I didn't know that there were Native American tribes that hadn't been formally recognized.
ADRIAN: Yes, there's lots.

This exchange, which aired this past February on the ABC News special *President Clinton Answering Children's Questions*, probably surprised the audience as much as the President. Apart from a few household names, the more than 500 Native American tribes currently recognized as such by the Bureau of Indian Affairs (B.I.A.) are virtually unknown to most Americans. Equally unknown is the fact that a complex process must be undergone to acquire legal status as an Indian tribe. At present more than a hundred groups have asked to be recognized as Indian nations; they are willing to undergo a process that can last nearly a decade and cost hundreds of thousands of dollars.

The federal government does not have figures on how many people are represented by these would-be tribes, but partial figures suggest a total of 80,000 to 100,000. The petitioning groups are scattered across thirty states; in Vermont are the St. Francis/Sokoki Band of Abenakis; in Louisiana the Jena Band of Choctaws; in Washington the Duwamish Indian Tribe; in Nevada the Pahrump Band of Paiutes. Some represent split-offs from the larger tribes—Cherokee, Apache, Chippewa—while others are the remnants of nations forced to move and re-establish themselves one or more times over 300 years.

If acknowledged by the federal government, a tribe gains access to the aid provided by the B.I.A., which includes some start-up funds for the tribal government and a needs-based program of health, housing and educational assistance, in addition to the right to manage natural resources on their lands. This aid often sounds more impressive than it is, but hard-pressed groups consider it far better than nothing. A recognized Indian nation also has its own government and some degree of cultural autonomy; as a byproduct, it can participate fully in the lobbying efforts of the U.S.-wide Native American

From: Cynthia Brown, "The Vanished Native Americans," *The Nation* 257 (October 11, 1993): 384–389. Copyright © The Nation Company, Inc. Reprinted by permission of the publisher.

community, which has coalitions dealing with everything from federal management of tribal lands held in trust to reform of the B.I.A., which is part of the Interior Department. By contrast, as Bret Bradigan, representing the Kern Valley Sun Indian Council in California, told a Congressional committee in 1989, unacknowledged Native American communities experience "denial on issues of education, housing, economic development . . . sacred grounds, preservation of our ancestral grounds, and freedom to practice our religious, cultural and traditional ways." John Echohawk, executive director of the Native American Rights Fund, calls federal acknowledgment literally a "matter of life and death."

Ron Goode, chairman of California's North Fork Mono Tribe, a nonrecognized group, told the above-mentioned Congressional committee, "We are constantly threatened by loss of our health services, our young are without grant funds to attend college, our adults are denied career training and educational services, our elders' homes are in need of repair and are denied housing services, and our sacred sites are unprotected and their use constantly threatened. Even our cultural museum property has been attacked by the county for condemnation for eminent domain purposes for a road realignment."

Yet fewer than half of petitions for acknowledgment have been successful. Petitioners first face the need for years of research, most of which is beyond the abilities of the layperson and so must be contracted for by groups of people who are economically desperate. Under Presidents Reagan and Bush, petitioners also faced an obvious bias against acknowledging "new" tribes. And some believe they face a lack of sympathy from recognized tribes.

The regulations that currently govern federal recognition of Indian tribes went into effect in 1978, but the status difficulties of many Native American groups go back well over a century. In California, a state with one of the highest concentrations of groups currently seeking federal acknowledgment, treaties of the mid-1880s granted Native Americans huge tracts of land in the Sierras, but those treaties were never ratified by Congress because of the pressure from white settlers swamping the state during the Gold Rush. The tribal populations dispersed, or clustered in small settlements called *rancherías*, as a result of the land loss. Now groups like the Wukchumni Council and the Choinumni Council are seeking recognition. In Indiana, President Andrew Jackson's removal policies led to U.S. purchase of Miami tribal land and, in 1846, to the relocation of half the Miami tribe, which settled in Oklahoma. The two halves were recognized in 1854 as separate nations, but through a series of acts, revisions and reversals of policy over the years, the Indiana Miamis were "terminated" as a tribe in 1897 and have been unable to obtain acknowledgment since. The 1887 General Allotment Act, which divided tribal lands into parcels for individual farming, also resulted in fragmentation of communities in many parts of the country. This "integration" policy may have been well intentioned, but it was devastating to cultural and political cohesion.

The Indian Reorganization Act of 1934 tried to "rationalize" Indian tribal structures. The B.I.A. worked with some tribes (among them the Oklahoma Miamis but not the Indiana nation) and had them vote on constitutions. After World War II, faced with civil suits for breaches of trust responsibility, the federal government created the Indian Claims Commission to pay off the claims and "resolve" longstanding disputes over land ownership. Then, continuing the policy of assimilation to its logical extreme, Congress "terminated" dozens of tribes during the 1950s and early 1960s. The struggle to restore federal recognition to tribes terminated during this period has been largely won. And Ada Deer, the Clinton Administration's choice to head the B.I.A., who first came to Washington to lobby for restoration of her Menominee tribe of Wisconsin, is known to be sympathetic to restoration of the handful of terminated tribes that remain from that era.

But with regard to first-time acknowledgment of "new" tribes, the situation is more complicated. Some activists claim that established tribes don't want to share the federal pie with newcomers; the B.I.A. aid works out to only a few thousand dollars a person as it is, and in areas where recognition would allow new groups to exploit natural resources on trust lands—like fishing rights in the Northwest—the established tribes don't look forward to competition. But in other cases, such as that of the Lumbee of North Carolina, many tribes have supported recognition.

The B.I.A.'s specialized branch on acknowledgment is known for having better qualified staff than any other section of the controversial agency. Most have graduate degrees in anthropology and genealogy because the 1978 regulations require an investigation of the history of each tribe seeking recognition, to establish that it has existed continuously, with members living near one another, and has functioned over time as a cohesive political organization. And though the regulations were supposed to be relatively simple to follow, their creator, former acknowledgment branch chief John (Bud) Shapard, contends that in practice they have turned out to be too burdensome for most Indian groups to handle. Investigators must trace kinship lines for cultural groups that did not keep written records for most of their history; continuity is often difficult to document, as nearly every group has suffered disruption, reversals of federal policy and economic pressures to leave poor land. Groups now regularly turn to a program in the Health and Human Services Department that makes grants to pay for expert help in researching and filing B.I.A. petitions for acknowledgment. But these do not cover all costs. Once the petition is filed, it may be several years before the acknowledgment branch offers its initial critique; more time passes as the petitioners remedy deficiencies, and while the B.I.A. reaches its initial and final opinions.

In addition to the financial burden involved, groups seeking acknowledgment have faced two further disadvantages. First, there is the effect of past dispersal. As current branch chief Holly Record notes, referring to the seven criteria outlined in the regulations, "Fairness is not our eighth criterion." She explains that although a petitioning group may once have existed as a tribe and been dispersed or abused by federal policy, the acknowledgment process does not make up for that; under the regulations, a tribe is a tribe only if it can prove its continuous and cohesive existence since at least the early nineteenth century. Second, the Reagan and Bush administrations were unenthusiastic about taking more Indians onto the federal rolls and possibly having to augment the grossly inadequate B.I.A. budget. The bureau currently receives around $1.6 billion in program funds to administer the agency and service some 1 million members of recognized tribes who live on reservations or other tribal land. With the Departments of Health and Human Services and Housing and Urban Development giving some aid to unrecognized groups, the Reagan and Bush administrations strongly favored the status quo, even if that means one-third of Native Americans living below the poverty line and more than 50 percent unemployment on most reservations.

Since 1978, 143 groups residing in thirty states have sought recognition; most of these claims are still pending, proceeding at a snail's pace. Of the twenty-three cases resolved, only eight petitioners have been recognized. Holly Record, in defending the B.I.A.'s acknowledgment branch, notes that of the thirteen groups turned down, seven were found to have no Indian ancestry. Other cases have been more debatable; after the branch turned down the Miamis of Indiana in 1992, the group, represented by the Native American Rights Fund, took its case to court.

Perhaps more than the regulations themselves, which offer some room for interpretation, the attitude of the Interior Department under the Republican administrations was definitive. The case of the North Carolina Lumbee—a group of about 40,000, including the child who questioned President Clinton—is currently the most controversial. A crucial fact is that the Lumbee are an unusually large group. Although the best-known tribes are larger (the Navajo number about 165,000, the Cherokee 95,000, according to B.I.A.), most tribes, and certainly all groups recognized as tribes through the regulation process, are considerably smaller, on the order of 500. Even the Miamis of Indiana, with about 4,500 members, are uncommonly numerous, and their supporters claim that this may explain federal reluctance to recognize them while recently acknowledging the Snoqualmies of Washington, a group of fewer than 1,000 members with a history very similar to that of the Miamis. In the Lumbee's case, the criticisms are especially pointed, both because the group's petition was reportedly one of the most convincing yet presented and because of the way the Bush Administration's Interior Department responded to it.

Ninety percent of the Lumbee reside, as they have traditionally, in Robeson County and adjoining counties. This area was never heavily settled by non-Indians; the tribe lived without much outside contact until the 1830s. The tribal language Cheraw, is dead, but membership in the tribe has always depended on maintenance of tribal relationships, so the community is tight-knit; according to the group's lawyer, Arlinda Locklear, herself a descendant of one of the principal family lines of the Lumbee, 95 percent of Lumbees marry within the tribe. Since 1885 the group has been recognized as a tribe by the State of North Carolina, but it is governed traditionally, along family lines; it has no constitution, having been neglected during the Indian Reorganization Act process after 1934. Nonetheless, state recognition meant that, until desegregation became law in the 1960s, the Lumbee operated their own school system for some eighty years and kept records of the ancestry of every child admitted to that system.

The Lumbee's search for recognition has foundered repeatedly on economic grounds. In 1890, for example, Interior rejected the group, saying that the Treasury was already overburdened and the Lumbee were too "civilized" to need aid. Similarly, under the 1956 Lumbee Act, Congress recognized the tribe for all purposes *except* federal funding. The 1978 acknowledgment regulations contain language excluding tribes terminated or subject to statute regarding federal benefits, but during the Carter Administration the B.I.A. suggested the Lumbee seek recognition through the acknowledgment process anyway. According to Locklear, it took eight years and well over $500,000 to prepare the petition, which was submitted in 1987.

The following year, the bureau turned down the Ysleta del Sur, a Texas group with a similar statutory history, and advised the Lumbee that they would likely be rejected as well. In 1989, according to Locklear, the B.I.A.'s solicitor issued a formal opinion on the 1956 act, to the effect that the Lumbee could not be eligible for recognition through the acknowledgment process. This might be justifiable as a technical, close reading of the law. But when the Lumbee pursued recognition through special legislation, in 1988, the Interior Department testified against them as well. Yet since then, says Locklear, the B.I.A. has proposed repeal of the 1956 act. So now the Lumbee may lose what federal status they do have, with no guarantees that the acknowledgment process will work for them. A few months ago, the Lumbee withdrew their B.I.A. petition in disgust. It is unclear what the tribe's next step will be.

The Lumbee have unusually broad support, but the B.I.A. is not alone in its restrictive approach to recognition. George Waters, a lobbyist representing a number of established tribes, including the Yakima of Washington, does not think much of the Lumbee's claim, arguing that the tribe no longer uses its original language or observes religious ceremonies (although these are not specifically required under the regulations). He thinks tribal sovereignty, the principle behind recognition, should not be belittled by awarding it without careful review. And he believes that looser regulations would lead to frivolous petitions. It was opinions like this—as well as Administration resistance, and, in one case, a threatened Bush veto—that blocked several bills during the 1980s to review the acknowledgment process or bypass it with individual tribal recognitions.

The Clinton campaign received near-total support from Native Americans, and the Administration's choice of Ada Deer as B.I.A. director is a popular one. In addition, according to bureau branch chief Record, two changes are in the offing. After twelve years of Republican understaffing that delayed review of petitions, the acknowledgment branch will now run on full staff. And proposed amendments to the 1978 regulations would ease the requirements somewhat for those groups with previous federal relationships.

But those expecting much more liberal policies may be disappointed. The B.I.A.'s budget request for fiscal year 1994 is $100 million higher than the budget request for this year, but for those tribes who do achieve federal recognition, there still won't be much to go around.

THE BRUTALITY OF THE BUREAUCRACY

36

Theresa Funiciello

Poverty is the worst form of violence.

—Mahatma Gandhi

From: Theresa Funiciello, "The Brutality of the Bureaucracy," in *The Tyranny of Kindness* (New York: Atlantic Monthly Press, 1993), pp. 24–29. Copyright © 1993 by Theresa Funiciello. Reprinted by permission of Grove/Atlantic, Inc.

In our own time, attention to experience may signal that the greatest threat to due process principles is formal reason severed from the insights of passion. . . . In the bureaucratic welfare state of the late twentieth century, it may be that what we need most acutely is the passion that understands the pulse of life beneath the official version of events . . . the characteristic complaint of our time seems to be not that government provides no reasons, but that its reasons often seem remote from the human beings who must live with their consequences.

—Justice William J. Brennan, Jr.

Think of the worst experience you've ever had with a clerk in some government service job—motor vehicles, hospital, whatever—and add the life-threatening condition of impending starvation or homelessness to the waiting line, multiply the anxiety by an exponent of ten, and you have some idea of what it's like in a welfare center. You wait and wait, shuttling back and forth in various lines like cattle to the slaughter. You want to wring the workers' necks, but you don't dare talk back. The slightest remark can set your case back hours, days, weeks, or forever. Occasionally someone loses it and starts cursing at the top of her lungs. Then she's carted away by security guards. In the early days, I thought this meant she was getting served faster for having had the nerve to lay it all out. It wasn't long before it became evident that these were among the ones who would be arrested and sometimes beaten up. It's truly amazing that more welfare workers aren't killed; the torment so many of them inflict would break the patience of anyone whose life wasn't on the line. But that's always their ace in the hole. No check, no life.

Babies and other small children are squirming all over the place, and always at least one worker shoots verbal bullets if they cry or run around. Half the time they're hungry as well as bored, but the mother has run out of the food she brought, had none to begin with, or can't leave to get some for fear her number will be called while she's out tracking down an apple, candy bar, or quart of milk. If she yells at the kid, a worker yells at her. If she doesn't yell at the kid, a worker yells at her.

You run the same risk taking one of the kids or yourself to the bathroom, so "accidents" are common. In any event, the bathrooms in welfare centers are not to be believed. The stalls have no doors, and there's rarely any toilet paper. If they are ever cleaned, it must be on holidays; they are putrid with the stench of weeks-old rags piled up anywhere because there are usually no bins to put them in. You could yearn for the good life in a prison after walking into a welfare center bathroom.

Not all workers were ghouls, but they all had to contend with their own predicaments, too. What should have happened and didn't was in many ways the consequence of the draw—what worker you got on what day of the week.

Most workers know little about the actual law governing eligibility for public assistance, and even those who start out with good intentions often get blown away by the ever-changing legal script. Each month, new regulations by the dozens are distributed. Most of the workers are so overwhelmed with the sheer volume of clients that only the truly stalwart keep up with the changes.

On the one side, the department is constantly trying to figure itself out, so as "to minimize waste and abuse." On the other, slews of lawsuits are constantly changing policies out of compliance with state or federal laws. And, because most lawmakers are oblivious of the Constitution as it pertains to poor people, it is not uncommon for laws that are unconstitutional to be passed, put into operation, challenged, and overturned. A welfare worker in Kingston, New York, once told one of our members, "When you sign up for welfare, that's when you sign all your rights away." Though this attitude is pervasive throughout the system, it is simply not true.

If you have the uncanny luck of getting a worker who knows the law, you also have to get that person on a good day—which makes all the difference between being treated decently and being treated inhumanely. Inhumanity, by the nature of the job, is the more common treatment. God forbid your worker had a fight with a family member that morning. If the worker has previously been on welfare herself, she might be more helpful, but it's more likely that she'll be more bitter about having to "work" at this low-paying, essentially dead-end job while someone else makes off with a welfare check. Still other workers have this attitude that the welfare is coming right out of their pockets—an outlook the hierarchy likes to cultivate. Some welfare centers even had a system of rewards for intake workers who denied the most applications or cut off the most recipients in a given period.

When you walk into a center, you have the right to get and file an application. To do that, you must first be deemed worthy of one by the persons whose job it is to hand them out and initiate the herding process. Often people never get past point one, because the worker *who is obligated by law* to give you that application, not to question or determine your eligibility, often refuses to do so. When that happens, people who are rejected at the door, who have no resources to turn to for advice, just fall through the cracks. They are included in no statistics, anywhere. For all practical intents and purposes, from the state's point of view, they do not exist unless they have submitted an application. At DWAC we used to get many people with this problem. There had to be thousands of others, because we were such a small, virtually unknown operation at the time.

Getting the application into the hands of an "intake worker" who will interview you and check that you have proper documentation—like birth certificates, leases, and social security numbers—is the next hurdle. We were convinced that some of the workers were sadists and would find any excuse to humiliate an applicant, like the one who raged that an applicant was using a verboten red pen. The applicant had made a quarter-inch mark on the form, not even completing the first letter of her name. "You can't use red ink," bellowed the application approver nearby. Okay. The applicant switched to black ink and continued. But over an hour later, as the applicant handed the worker the completed form in black ink, she was told that the entire application was no good because of the tiny red mark on it.

This worker, whose job was to look at applications to determine if they were completely filled out, simply refused to pass the form on to the intake worker. She insisted the applicant do the whole thing over again. She could have told her to get a new form the moment she spotted the red mark. She could have simply handed her another. She could have ignored the obviously irrelevant red mark. But this may have been the only part of her life in which she held the reins of power. She obviously liked it.

There is no rational, uniform method for qualifying for welfare. The attitude of the worker prevails on the one end and that of the state on the other, with rules of the federal government (subject to the politics of the moment) having some jurisdiction over all. At a minimum, there are fifty administrative systems for welfare, one in each state. That's then subdivided into whatever local categories the state chooses—like by county. In New York at the time, to qualify for welfare under the best of circumstances, you could have zero dollars in your possession, and if you had outside income (from a job or child support) it had to be less than the maximum you could have gotten on welfare if you had nothing. As a practical matter, if you walked in and said you had fifty dollars to your name, which the worker would know could not possibly last until your case was accepted, you could not qualify for welfare. If you had any asset whatsoever that the worker deemed sellable—like an old wedding ring—it had to be sold. *All* your resources had to be expended before you were eligible. One of the few things that improved for New Yorkers at least, during the Reagan administration, was the establishment of a one-thousand-dollar "resource limit," which meant an applicant could hold on to any asset deemed worth less than a thousand dollars and still qualify as long as all the other income requirements were met.

In states like Mississippi, income of more than $48 a month for a three-person family would, as a practical matter, disqualify them for welfare eligibility. In New York City, the maximum possible benefit for a three-person family at that time was $332. Job or other outside net income would have to be less than that.

When the intake worker gets through with you, the case supervisor normally has the final thumbs up or down. So all the variables that affected your outcome with the intake worker are once again operative with the supervisor. The case supervisor can demand that you see still other workers to approve this or that required document or answer on your seventeen-page application form. You can be sent to a dozen places before a determination is made. Pregnant women virtually ready to give birth must supply a doctor's verification of their pregnancies. Processing the application takes weeks, sometimes even months (in spite of the law, which says a determination must be made within thirty days). Even after you are determined eligible, you can wait days for the money to come through. Whether you are in heartrending, desperate, obvious need with a sick baby on your lap does not matter. People can fake heartrending, desperate, obvious need. That possibility guides the entire department as well as the local, state, and federal policies. For the poor there is no honor system like the one that guides paying taxes. No innocent until proven guilty like in the criminal justice system. No siree, Betsy.

More often than not, it was quite the opposite. There was a woman who had been raped and was being denied welfare for "refusing to comply" by naming the father. Father? Who the hell were they talking about? This young woman was leaving her infant home alone in a makeshift apartment in someone's attic while she worked a few hours a night in an all-night diner. She was taking home $41 a week. She called in tears. She simply couldn't make it on $41 a week.

If the welfare system worked the way it should, it would still not give people enough to live on. But securing and keeping what little it is possible to get would be difficult for an FBI agent. For instance, since the rape victim had only $41 a week for herself and her child, she was eligible for the difference between what she was paid on her job and the amount set for a welfare grant for two people. However, when asked who the father of her child was, she was in an impossible bind. He had told her neither his name nor his address during the rape, and she'd never seen him before in her life. Her inability to supply his name drove the case worker into a paroxysm of doubt. If she had known who the father was and he was not providing for the child, even if they were married she would still have been eligible for welfare.

Once the worker was dissatisfied with the answer to the name-that-father question, whole avenues of questions followed. Among them: had she reported the rape to the police? She hadn't. Now the worker felt justified in rejecting the application or postponing action until the woman went to the police to report a rape that had occurred over a year ago, securing proof of the report in the process. As you might imagine, this intrusion on their day-to-day work does not make the police happy and can and usually does evolve into another hassle.

Though the welfare is required to assist the applicant in obtaining documents that might be difficult for her to obtain on her own, this assistance is rarely given. Reluctantly, she went to the police, who resisted the paperwork. She called me from the police station, frustrated by another bureaucracy. I had to pressure them to give her a statement verifying that she made the report, which they thought was crazy, given the time lapse. Then it was back to the welfare department for the next round. We hoped she had satisfied them, but they wouldn't say. She was told that she would hear by mail.

This time they rejected her for lack of an address at which to contact a man whose name she did not know. An so on and on. In cases like these, the only way an applicant usually gets through the maze is by getting the aggressive intervention of an advocate or friend who knows the rules and how to enforce them or by going to court, which, given the dearth of lawyers who can or will take such a case, is rare.

Analyzing Social Issues

Public debates about issues such as multicultural education, abortion, school prayer, and welfare reform reflect race, class, and gender arrangements in society. Race, class, and gender are systemic features of society and do not change rapidly; however, the issues they generate change over time. How does thinking inclusively about race, class, and gender help us analyze the social issues that concern us today?

Using a race-, class-, and gender-inclusive framework to analyze social issues is significant in at least three ways. First, themes that emerge as social issues (for example, concern with the environment, homelessness, AIDS, sexuality, and immigration) are products of social institutions and institutional policies reflecting race, class, and gender arrangements. Since social institutions themselves are framed by race, class, and gender politics, what counts as a social issue reflects these institutional politics.

The organization of health care in the United States offers one compelling case of how the politics of race, class, and gender operate. Depending on their placement in the race, class, and gender system, different groups have widely varying patterns of health and illness. Something as fundamental as who gets to live and who is left to die reflects the politics of race, class, and gender. Death rates for diabetes, for example—a disease exacerbated by poor nutrition and health care—are noticeably higher among African American, Latino, and Native American women than among Asian or White women. Access to health care also reflects the politics of race, class, and gender. People of color are much less likely than Whites to have health insurance. About 10 percent of Whites report they have no health coverage, compared to 17 percent of Asians, 20 percent of African Americans and Native Americans, and 33 percent of Latinos (O'Hare 1992: 17). Race, class, and gender also shape the contours of health research and health care delivery. For example, differences in funding for breast cancer research and heart disease research speak to the differential power of men and women in the health care system. Similarly, funding for infertility treatments for middle-class White couples is quite different than ongoing efforts to control the fertility of poor women via funding for sterilization.

The ways in which social issues are presented and understood also operate within a race-, class-, gender-inclusive framework; thus, ideologies of race, class, and gender frame social issues. Ironically, despite the centrality of race, class, and gender in shaping social issues, the importance of these systems of oppression typically remains hidden. This is because the systems obscure existing systems of inequality that produce social issues.

Media representations of homelessness provide an example of how dominant ideologies frame our understanding of an important social issue. Increasing levels of underemployment coupled with a housing shortage of affordable, low-income housing pushed many people into poverty in the 1980s. The result was a growing level of homelessness in the United States. But while the social issue of homelessness affected men and women, Whites and people of color alike, not all segments of the homeless population were seen as equally worthy. Building on long-standing notions of the deserving and undeserving poor, newly homeless families, especially women with dependent children, garnered more sympathy than groups long seen as part of the undeserving poor— namely, single men, persons with addictions, veterans, and the mentally ill.

Moreover, racial politics intersected with class and gender to further subdivide the homeless—White families became seen as more deserving than people of color.

Since current arrangements of race, class, and gender create social issues, an equally inclusive framework will be required to effectively address such issues. It is not enough to critique existing social structures; how we use our understanding of race, class, and gender to analyze existing social issues matters. For example, how can we hope to redesign health care so that it is equitable and responsible to all citizens? How might we provide safe, afford-able housing for all members of our respective communities so that everyone has a home? Social institutions are sites that not only structure and legitimate existing race, class, and gender power relations, but individuals and groups use these same social institutions as sites of struggle. Thinking inclusively allows us to create alternative analyses of social issues and more effectively use our social locations to effect change.

As an example, think about the environment. Thinking narrowly about the environment leaves many people of color with the perception that the environmental movement is more concerned with protecting animals and other wildlife than it is with protecting people. In which types of communities are polluting industries located? What is the least desirable land and who is forced to live there? Some respond that communities with high concentrations of poor people and/or people of color incur more of the negative costs of industrial development than more affluent communities. A more broadly defined movement recognizes that patterns of environmental use of natural resources reflect institutional decision making. A more broadly defined envi-ronmental movement, such as the environmental justice movement—a na-tional coalition of grass-roots organizations from all types of communities—takes into account all people and their relationship to the environment.

In Part Four, we include readings on three significant social issues that are currently contested in American public discourse. We examine American national identity, sexuality, and violence—social issues that we think can only be understood if they are located within a race, class, and gender analysis. While these three issues are not the only ones meriting attention, they currently are important to the public agenda. Contemporary discussions about what constitutes American identity and culture, about sexual politics, and about the causes and consequences of violence are framed by implicit and explicit assumptions that stem from competing ideologies of race, class, and gender. How do race, class, and gender help our analysis of these important social issues? Conversely, how do these issues aid our understanding of the politics of race, class, and gender?

AMERICAN IDENTITY AND CULTURE

Despite the ideology of the melting pot, American national identity has been closely linked to a history of racial privilege. The term *American* is assumed to mean White; other types of "Americans," such as African Americans and Asian Americans, become distinguished from the "real" Americans by virtue of their race. "To be an 'all-American' is, by definition, *not* to be an Asian American, Pacific American, American Indian, Latino, Arab American, or African American," observes political theorist Manning Marable in "Beyond Racial Identity Politics." More importantly, certain benefits are reserved for those deemed deserving "Americans."

This long-standing view of American national identity and of the social rewards that should accrue to American citizens is being challenged by the changing contours of the population in the United States. The overall percentage of the American population comprised of people of color is changing. Between 1900 and 1960, the percentage of the population comprised of people of color increased slightly, from 13.1 to 14.9 percent. In contrast, between 1960 and 1990, the share of the U.S. population of people of color tripled in size, to approximately 25 percent. In 1992, the combined population of Native

Americans, African Americans, Latinos, and Asian Americans was estimated at 64.3 million; moreover, the composition of people of color in the United States is also undergoing substantial change. Just a few generations ago, African Americans comprised the largest "minority" group. Although still the most numerous, African Americans are now less than half of all people of color, and their share is declining. Latino and Asian American populations are growing more rapidly than African American or Native American populations (O'Hare 1992: 9).

Changes of this magnitude raise fundamental questions about American national identity. Who are we as a nation? How does the increased visibility and vocalism of people of color shape who we become as a nation? Who will control the social institutions reproducing American culture, such as the schools and the media? What will the content of American culture produced by such a diverse population be? How do we craft one country from so much difference created by the interlocking structures of race, class, and gender? How do we reconcile different perspectives on these questions that result from our varying placement in interlocking systems of race, class, and gender?

Several themes emerge in rethinking American identity in this context of the politics of race, class, and gender. One is the issue of Americans who never "melted" into the melting pot. Many people think that race is like ethnicity and that the failure of people of color to shed their cultures and assimilate into the mainstream as White ethnic groups have purportedly done represents an unwillingness to embrace American national identity. This view represents a serious misreading of the meaning of race in shaping American national identity. Race is not like ethnicity because of the centrality of whiteness within the dominant national identity.

For Native Americans and African Americans, the issue of the failure to assimilate into American national culture takes on special meaning. Ward Churchill in "Crimes Against Humanity" points out that, while Native American symbolism and imagery is ubiquitous, the use of Native American mascots for sports teams actually operates to keep Native Americans subordinate. Ironically, while the actual original "Americans" experienced and continue to experience oppression, Native Americans are simultaneously used as symbols for what makes America "American."

Current trends in Black culture can also be seen as a response to these larger themes of American national identity. Dominant ideologies about race, class, and gender and ideologies of resistance developed by subordinated groups gain institutional expression in language, the media, humor, and music. In "Rap, Race, and Politics" Clarence Lusane discusses rap as a culture of resistance and offers a complex view of how the politics of race, class, and gender work together in shaping African American music, and how African American artists themselves both reproduce these structures and resist them. On the one hand, institutional structures of racism have restricted Black access to wealth, power, and prestige; African Americans have lacked opportunities to assimilate, to "melt" into the American identity. On the other hand, the segregation of African Americans simultaneously has provided community structures with their own outcomes. Young Black people in particular see bleak futures for themselves, are warehoused in deteriorating neighborhoods and schools, and, in response, have created hip hop as a culture of resistance. The refusal to "melt" or assimilate is often accomplished by holding on to culture.

Ironically, while Native Americans and African Americans struggle to broaden definitions of American to include their experiences, Asians and Latinos migrate to the United States, in part also in search of an American identity. This leads to a second theme regarding American national identity— namely, the impact that changing population patterns are having on American culture. Many people of color see the United States as a land of opportunity, and their arrival and struggles fit into existing structures of race, class, and gender to reveal the tenacity of these structures and how these structures are being changed. For example, in "You're Short, Besides!" Sucheng Chan describes her experiences of immigration and of "becoming an American," revealing how American identity looks different to those choosing to migrate and to those lacking these choices.

Recent migrations by people of color have fostered a corresponding discussion of the rights that accrue to citizens. What happens if citizens are defined as "White" and if English is the only "American" language? Which benefits should be reserved for American citizens and which citizens should receive these benefits? In the United States, this process is profoundly affected by race, class, gender, and sexuality. Often this process fosters conflict among people of color, as Manning Marable reminds us in "Beyond Racial Identity Politics." Sometimes such conflict can have tragic consequences, as was the case of the Los Angeles riot in 1992.

Changing gender relations in the United States present yet another challenge to the process of rethinking American national identity. While the term *American* historically has meant "White," it has simultaneously been cast in gendered terms as well. Efforts to come to terms with contested American racial identity are occurring in conjunction with a renegotiated gender identity. People with the power to define reality create dominant ideologies reflecting their own interests, as Gloria Steinem reminds us in "If Men Could Menstruate." These ideologies shape all social institutions, especially those central to reinforcing beliefs. Changing gender politics resulting from feminist activism in the 1970s and 1980s have also fostered an ideological backlash against feminism, as Susan Faludi shows in "Blame It on Feminism." What is being contested is what it means to be an "American woman" or an "American man."

With so much at stake, it should not surprise us if the inability to think inclusively about American national identity also permeates forms of resistance to dominant ideology and the practices they defend. Clarence Lusane's discussion of "Rap, Race, and Politics" demonstrates how access to a culture of resistance allows African Americans to resist powerful dominant ideologies concerning race and class, but Lusane also points to the contradictory nature of hip hop culture. While it may challenge race and class politics in the United States, it simultaneously falls short in analyzing gender politics necessary for a more democratic, humane society.

While many social issues intersect with this overarching theme of crafting an American national identity and culture that is inclusive, two social issues present powerful challenges to this process. The challenges are those raised by the changing nature of sexuality and violence as part of American national identity and culture.

SEXUALITY

Sexuality in American culture has emerged as a central social issue in contemporary American life. Sexuality is a critical part of the structures of race, class, and gender. While we often think of sexuality as a social issue, it has two meanings. Sexuality links systems of race, class, and gender, and it can be seen as a dimension of each. For example, constructing images about Black sexuality was central to maintaining institutional racism. Similarly, beliefs about women's sexuality operate in structuring gender oppression. At the same time, sexuality operates as a system of oppression comparable to the systems of race, class, and gender. *Compulsory heterosexuality*, the institutionalized structures and beliefs that define and enforce heterosexual behavior as the only natural and permissible form of sexual expression, can be seen as a system of oppression that affects everyone.

Currently, both meanings of sexuality—namely, sexuality as a dimension of race, class, and gender oppression and sexuality as a system of oppression—are being contested in American culture. Some challenge how sexuality is linked to the social class system. Women's sexuality clearly is manipulated by business in order to sell products. As Naomi Wolf points out in "The Beauty Myth," women's sexuality is used to sell products and their sexuality is itself turned into a commodity for sale on the open market.

This commodification of sexuality operates in a race-specific fashion. Different groups encounter different treatment based on their placement in systems of race, class, and gender. In "Cultural and Historical Influences on Sexuality in Hispanic/Latin Women," Oliva Espín demonstrates how systems of sexuality are racialized. This manipulation of sexuality in structuring systems of race, class, and gender is itself based on assumptions of heterosexism and definitions of so-called normal sexuality. In "A New Politics of Sexuality" June Jordan refers to this situation as the "politics of sexuality" and posits that it "subsumes all of the different ways in which some of us seek to dictate to others of us what we should do, what we should desire, what we should dream about, and how we should behave ourselves, generally." Pervasive heterosexism works to create a basic anxiety about sexuality that permeates a whole series of human relationships.

Since sexuality has been constructed in the context of social institutions affected by systems of race, class, and gender, the system of compulsory heterosexuality itself reflects these same systems. Barbara Smith's article, "Homophobia" points out several misconceptions and attitudes especially troubling to gays, lesbians, and bisexuals, particularly those of color. She takes issue with beliefs that being gay, lesbian, or bisexual are private issues and not political matters, that homosexuality is a "White problem" that does not affect people of color, and that hate speech and other forms of violence against gays, lesbians, and bisexuals can be tolerated.

Pervasive heterosexism and homophobia shape current social issues. Evelynn Hammonds in "Race, Sex, AIDS" examines the tragic consequences of not respecting sexual differences in responding to a pressing public health issue. Hammonds not only points to the differential health care based on group membership but also explores how very definitions of health and disease themselves vary by race, class, gender, and sexuality.

VIOLENCE AND SOCIAL CONTROL

The escalating level of violence permeating American society has emerged as another social issue in American national culture that challenges the very terms of that culture. Rather than seeing violence as the result of individual acts, violence represents a device of social control used to maintain systems of race, class, and gender inequality. Within social institutions organized around the politics of race, class, and gender, actual violence or the threat of violence has long been essential in maintaining oppression. When people resist, violence is often used to keep them in their place.

Many of the current forms of violence are directly linked to inequality. For example, Sumi Cho's article "Korean Americans vs. African Americans: Conflict and Construction" examines some the tensions among Black people and Koreans. Many also argue that pornography is a form of violence against women. It represents the link between commercialized sexuality and violence because both sell products. Joining sexuality and violence in this particular product fuels the social class system's ongoing search for new consumer markets. The fact that women and children are used in this industry reflects inequalities of gender and age.

The arming of America via gun sales and promotion represents another example of the marketing of violence within American culture and illustrates the consequences that this practice can have within American society. Unfortunately, violence often hurts the most vulnerable segments of society. For example, violence against children has reached alarming proportions. In 1991, homicide was the fourth leading cause of death for children one to four years old, and the third leading cause of death for those five to fourteen years old. These data are heightened for African American children. During 1991, homicide was the second leading cause of death for Black children ages one to fourteen; a Black child died from a gunshot wound every six hours (Robinson and Briggs 1993: 47).

It is important to see violence not as acts of individual social deviance but as one outcome of how the politics of race, class, and gender permeate a range of social institutions. Violent acts, the threat of violence, and more generalized policies based on the use of force find organizational homes in designated institutions of social control—primarily the police, the military, and the criminal justice system. But violence simultaneously permeates a range of other social institutions. Some social institutions are known for their use of force, while others are less peaceful than they appear. Male violence against women and children, whether physical, emotional, or sexual, in families of all social classes and racial/ethnic compositions contradicts the dominant belief that the home is a place of tranquility and love. Sexual harassment on the job as a dimension of social control belies the belief that men and women encounter a similar work environment. In the media, depicting rape as pleasurable to women and portraying violence against Native Americans, Asian Americans, and African Americans in numerous movies as justified contributes to a generalized belief system condoning violence; thus, violence is simultaneously concentrated and diffuse.

One prominent theme in the readings in "Violence and Social Control" concerns the links between individual acts of violence and more routinized, systemic violence. Rape, for example, appears to be a private act, but as Patricia Martin and Robert Hummer compellingly explore in "Fraternities and Rape on Campus," rapes occur as part of a generalized climate condoning violence against women. Lynching African American men can be seen as the random acts of unruly mobs; yet, as Jacquelyn Hall argues in her "The Mind That Burns in Each Body," the lack of arrest, prosecution, and convictions of those who lynched human beings is powerful testament to public endorsement of these seemingly individual acts of violence. Although violence may be experienced individually, it occurs in specific organizational and institutional contexts.

Race, class, and gender are fundamental determinants of organizational policies that foster social control. As Elijah Anderson points out in "The Police and the Black Male," African American men's experiences with the police exemplify the differential treatment that individuals receive based on race, class, and gender. Similarly, Jacquelyn Hall's historical analysis of the specific forms of assault against African American men and women—lynching and institutionalized rape—explores the complex interconnections among race, class, and gender in making African American women and men vulnerable to racial violence in gender-specific ways.

Investigating violence and social control also reveals the interconnected nature of dominant ideologies of masculinity, the individualized and systemic forms of violence they justify, and the mechanisms of social control that support masculinity. "More Power Than We Want" by Bruce Kokopeli and George Lakey investigates these links and elaborates on Martin and Hummer's discussion of the connection between masculine identity and campus rape and Hall's analysis of lynching and white masculinity. Although we include no articles on violence against gays and lesbians, homophobic violence and violence against women are both ways of enforcing interlocking belief systems concerning heterosexism and masculinity. Violence based on race, gender, and heterosexism is part of one system of social control: All of these types of violence aim to reinforce systems of privilege.

In a sense, the escalating violence within American society is a curious outcome of these same conditions of inequality. Trying to renegotiate a new American national identity and culture that no longer has the certainty of fixed race, class, and gender categories creates change, confusion, and often violence. When people are wronged, they can use violence as a weapon; thus, the degree of violence can be linked to the issue of a changing American national identity.

Sexuality and violence are but two social issues that intertwine with the need to rethink American national identity. A call for the need to build a more inclusive, less hierarchical, multicultural American national community resonates throughout Manning Marable's discussion of racial identity politics.

How can we build an American national community that acknowledges the differences among us created by politics of race, class, and gender? As Marable points out, "the challenge begins by constructing new cultural and political 'identities,' based on the realities of America's changing multicultural, democratic milieu."

This final and, in some ways, most central theme to rethinking American national identity requires avoiding two problematic stances to social change. On the one hand, structures where whiteness, masculinity, and wealth set the standards by which everyone else is judged cannot continue. While this way of building community—namely, placing one group in the center and requiring everyone else to try and be like them—can foster a certain sense of certainty within the American national community, it is unjust. Social issues such as oppressive sexualities and societal violence emerge in this context. On the other hand, a politics of self-interest where each group lobbies only for itself and cares little about the American national community as a whole is equally undesirable. Crafting new views of community that take histories of race, class, and gender relations into account yet aim to construct one American community where equity becomes a reality is a vital task that promises to engage us all for some time.

References

O'Hare, William P. 1992. "America's Minorities—The Demographics of Diversity." *Population Bulletin* 47, no. 4 (December). Washington, DC: Population Reference Bureau.

Robinson, Lori S., and Jimmie Briggs. 1993. "Kids and Violence." *Emerge* 5, no. 11 (November): 44–50.

American Identity and Culture

BEYOND RACIAL IDENTITY POLITICS: **37**
Towards a Liberation Theory for Multicultural Democracy

Manning Marable

Americans are arguably the most 'race-conscious' people on earth. Even in South Africa, the masters of apartheid recognized the necessity to distinguish between 'coloureds' and 'black Africans'. Under the bizarre regulations of apartheid, a visiting delegation of Japanese corporate executives or the diplomatic corps of a client African regime such as Malawi could be classified as 'honorary whites'. But in the US, 'nationality' has been closely linked historically to the categories and hierarchy of national racial identity. Despite the orthodox cultural ideology of the so-called 'melting pot', power, privilege and the ownership of productive resources and property has always been unequally

From: Manning Marable, "Beyond Racial Identity Politics: Towards a Liberation Theory for Multicultural Democracy," *Race & Class*, Vol. 35, No. 1 (July/September 1993): 113–130, Institute of Race Relations, London. Reprinted by permission.

allocated in a social hierarchy stratified by class, gender and race. Those who benefit directly from these institutional arrangements have historically been defined as 'white', overwhelmingly upper class and male. And it is precisely here, within this structure of power and privilege, that 'national identity' in the context of mass political culture is located. To be an 'all-American' is, by definition, *not* to be an Asian-American, Pacific-American, American Indian, Latino, Arab-American or African-American. Or, viewed another way, the hegemonic ideology of 'whiteness' is absolutely central in rationalising and justifying the gross inequalities of race, gender, and class experienced by millions of Americans relegated to the politically peripheral status of 'Others'. As marxist cultural critic E. San Juan has observed: 'Whenever the question of the national identity is at stake, boundaries in space and time are drawn . . . A decision is made to represent the Others—people of color—as missing, absent, or supplemental.'[1] 'Whiteness' becomes the very 'centre' for the dominant criteria for national prestige, decision-making, authority and intellectual leadership.

Ironically, because of the centrality of 'whiteness' within the dominant national identity, Americans generally make few distinctions between 'ethnicity' and 'race', and the two concepts are usually used interchangeably. Both the oppressors and those who are oppressed are, therefore, imprisoned by the closed dialectic of race. 'Black' and 'white' are usually viewed as fixed, permanent and often antagonistic social categories. Yet, in reality, 'race' should be understood not as an entity, within the histories of all human societies, or grounded to some inescapable or permanent biological or genetic differences between human beings. 'Race' is, first and foremost, an unequal relationship between social aggregates, characterised by dominant and subordinate forms of social interaction, and reinforced by the intricate patterns of public discourse, power, ownership and privilege within the economic, social and political institutions of society.

Race only becomes 'real' as a social force when individuals or groups behave towards each other in ways which either reflect or perpetuate the hegemonic ideology of subordination and the patterns of inequality in daily life. These are, in turn, justified and explained by assumed differences in physical and biological characteristics, or in theories of cultural deprivation or intellectual inferiority. Thus, far from being static or fixed, race as an oppressive concept within social relations is fluid and ever changing. What is an oppressed 'racial group' changes over time, geographical space and historical conjuncture. What is termed 'black', 'Hispanic' or 'Oriental' by those in power to describe one human being's 'racial background' in a particular setting can have little historical or practical meaning within another social formation which is also racially stratified, but in a different manner.

Since so many Americans view the world through the prism of permanent racial categories, it is difficult to convey the idea that radically different ethnic groups may have a roughly identical 'racial identity' imposed on them. For example, although native-born African-Americans, Trinidadians, Haitians, Nigerians and Afro-Brazilians would all be termed 'black' on the streets of New York City, they have remarkably little in common in terms of language, culture, ethnic traditions, rituals and religious affiliations. Yet they are all 'black' racially, in the sense that they will share many of the pitfalls and prejudices built into the institutional arrangements of the established social order for those defined as 'black'. Similarly, an even wider spectrum of divergent ethnic groups—from Japanese-Americans, Chinese-Americans, Filipino-Americans and Korean-Americans to Hawaiians, Pakistanis, Vietnamese, Arabs and Uzbekis—are described and defined by the dominant society as 'Asians' or, worse yet, as 'Orientals'. In the rigid, racially stratified American social order, the specific nationality, ethnicity and culture of a person of colour has traditionally been secondary to an individual's 'racial category', a label of inequality which is imposed from without rather than constructed by the group from within. Yet as Michael Omi, Asian-American Studies Professor at the University of California at Berkeley, has observed, we are also 'in a period in which our conception of racial categories is being radially transformed'. The waves of recent immigrants create new concepts of what the older ethnic communities have been. The observations and generalisations we imparted 'to racial identities' in the past no longer make that much sense.

In the United States, 'race' for the oppressed has also come to mean an identity of survival, victimisation and opposition to those racial groups or elites which exercise power and privilege. What we are looking at here is *not* an *ethnic* identification or culture, but an awareness of shared experience, suffering and struggles against the barriers of racial division. These collective experiences, survival tales and grievances form the basis of an historical consciousness, a group's recognition of what it has witnessed and what it can anticipate in the near future. This second distinct sense of racial identity is both imposed on the oppressed and yet represents a reconstructed critical memory of the character of the group's collective ordeals. Both definitions of 'race' and 'racial identity' give character and substance to the movements for power and influence among people of colour. . . .

NOTE

1. E. San Juan, 'Racism, ideology and resistance', *Forward Motion* (Vol. 10, no. 3, September 1991), pp. 35–42.

CRIMES AGAINST HUMANITY

38

Ward Churchill

If nifty little "pep" gestures like the "Indian Chant" and the "Tomahawk Chop" are just good clean fun, then let's spread the fun around, shall we?

During the past couple of seasons, there has been an increasing wave of controversy regarding the names of professional sports teams like the Atlanta "Braves," Cleveland "Indians," Washington "Redskins," and Kansas City "Chiefs." The issue extends to the names of college teams like Florida State University "Seminoles," University of Illinois "Fighting Illini," and so on, right on down to high school outfits like the Lamar (Colorado) "Savages." Also involved have been team adoption of "mascots," replete with feathers, buckskins, beads, spears, and "warpaint" (some fans have opted to adorn themselves in the same fashion), and nifty little "pep" gestures like the "Indian Chant" and "Tomahawk Chop."

A substantial number of American Indians have protested that use of native names, images and symbols as sports team mascots and the like is, by definition, a virulently racist practice. Given the historical relationship between Indians and non-Indians during what has been called the "Conquest of America," American Indian Movement leader (and American Indian Anti-Defamation Council founder) Russell Means has compared the practice to contemporary Germans naming their soccer teams the "Jews," "Hebrews," and "Yids," while adorning their uniforms with grotesque caricatures of Jewish faces taken from the Nazis' anti-Semitic propaganda of the 1930s. Numerous demonstrations have occurred in conjunction with games—most notably during the November 15, 1992 match-up between the Chiefs and Redskins in Kansas City—by angry Indians and their supporters.

In response, a number of players—especially African Americans and other minority athletes—have been trotted out by professional team owners like Ted Turner, as well as university and public school officials, to announce that they mean not to insult but to honor native people. They have been

From: Ward Churchill, "Crimes Against Humanity," *Z Magazine* 6 (March 1993): 43–47. Reprinted by permission of the author.

joined by the television networks and most major newspapers, all of which have editorialized that Indian discomfort with the situation is "no big deal," insisting that the whole thing is just "good, clean fun." The country needs more such fun, they've argued, and "a few disgruntled Native Americans" have no right to undermine the nation's enjoyment of its leisure time by complaining. This is especially the case, some have argued, "in hard times like these." It has even been contended that Indian outrage at being systematically degraded—rather than the degradation itself—creates "a serious barrier to the sort of intergroup communication so necessary in a multicultural society such as ours."

Okay, let's communicate. We are frankly dubious that those advancing such positions really believe their own rhetoric, but, just for the sake of argument, let's accept the premise that they are sincere. If what they say is true, then isn't it time we spread such "inoffensiveness" and "good cheer" around among *all* groups so that *everybody* can participate *equally* in fostering the round of national laughs they call for? Sure it is—the country can't have too much fun or "intergroup involvement"—so the more, the merrier. Simple consistency demands that anyone who thinks the Tomahawk Chop is a swell pastime must be just as hearty in their endorsement of the following ideas—by the logic used to defend the defamation of American Indians—should help us all really start yukking it up.

First, as a counterpart to the Redskins, we need an NFL team called "Niggers" to honor Afro-Americans. Half-time festivities for fans might include a simulated stewing of the opposing coach in a large pot while players and cheerleaders dance around it, garbed in leopard skins and wearing fake bones in their noses. This concept obviously goes along with the kind of gaiety attending the Chop, but also with the actions of the Kansas City Chiefs, whose team members—prominently including black team members—lately appeared on a poster looking "fierce" and "savage" by way of wearing Indian regalia. Just a bit of harmless "morale boosting," says the Chiefs' front office. You bet.

So that the newly-formed Niggers sports club won't end up too out of sync while expressing the "spirit" and "identity" of Afro-Americans in the above fashion, a baseball franchise—let's call this one the "Sambos"—should be formed. How about a basketball team called the "Spearchuckers?" A hockey team called the "Jungle Bunnies?" Maybe the "essence" of these teams could be depicted by images of tiny black faces adorned with huge pairs of lips. The players could appear on TV every week or so gnawing on chicken legs and spitting watermelon seeds at one another. Catchy, eh? Well, there's "nothing to be upset about," according to those who love wearing "war bonnets" to the

Super Bowl or having "Chief Illiniwik" dance around the sports arenas of Urbana, Illinois.

And why stop there? There are plenty of other groups to include. "Hispanics?" They can be "represented" by the Galveston "Greasers" and San Diego "Spics," at least until the Wisconsin "Wetbacks" and Baltimore "Beaners" get off the ground. Asian Americans? How about the "Slopes," "Dinks," "Gooks," and "Zipperheads?" Owners of the latter teams might get their logo ideas from editorial page cartoons printed in the nation's newspapers during World War II: slant-eyes, buck teeth, big glasses, but nothing racially insulting or derogatory, according to the editors and artists involved at the time. Indeed, this Second World War-vintage stuff can be seen as just another barrel of laughs, at least by what current editors say are their "local standards" concerning American Indians.

Let's see. Who's been left out? Teams like the Kansas City "Kikes," Hanover "Honkies," San Leandro "Shylocks," Daytona "Dagos," and Pittsburgh "Polacks" will fill a certain social void among white folk. Have a religious belief? Let's all go for the gusto and gear up the Milwaukee "Mackerel Snappers" and Hollywood "Holy Rollers." The Fighting Irish of Notre Dame can be rechristened the "Drunken Irish" or "Papist Pigs." Issues of gender and sexual preferences can be addressed through creation of teams like the St. Louis "Sluts," Boston "Bimbos," Detroit "Dykes," and the Fresno "Fags." How about the Gainesville "Gimps" and Richmond "Retards," so the physically and mentally impaired won't be excluded from our fun and games?

Now, don't go getting "overly sensitive" out there. None of this is demeaning or insulting, at least not when it's being done to Indians. Just ask the folks who are doing it, or their apologists like Andy Rooney in the national media. They'll tell you—as in fact they *have* been telling you—that there's been no harm done, regardless of what their victims think, feel, or say. The situation is exactly the same as when those with precisely the same mentality used to insist that Step 'n' Fetchit was okay, or Rochester on the Jack Benny Show, or Amos and Andy, Charlie Chan, the Frito Bandito, or any of the other cutsey symbols making up the lexicon of American racism. Have we communicated yet?

Let's get just a little bit real here. The notion of "fun" embodied in rituals like the Tomahawk Chop must be understood for what it is. There's not a single non-Indian example used above which can be considered socially acceptable in even the most marginal sense. The reasons are obvious enough. So why is it different where American Indians are concerned? One can only conclude that, in contrast to the other groups at issue, Indians are (falsely) perceived as being too few, and therefore too weak, to defend themselves effectively against racist and otherwise offensive behavior.

Fortunately, there are some glimmers of hope. A few teams and their fans have gotten the message and have responded appropriately. Stanford University, which opted to drop the name "Indians" from Stanford, has experienced no resulting drop-off in attendance. Meanwhile, the local newspaper in Portland, Oregon recently decided its long-standing editorial policy prohibiting use of racial epithets should include derogatory team names. The Redskins, for instance, are now referred to as "the Washington team," and will continue to be described in this way until the franchise adopts an inoffensive moniker (newspaper sales in Portland have suffered no decline as a result).

Such examples are to be applauded and encouraged. They stand as figurative beacons in the night, proving beyond all doubt that it is quite possible to indulge in the pleasure of athletics without accepting blatant racism into the bargain.

NUREMBERG PRECEDENTS

On October 16, 1946, a man named Julius Streicher mounted the steps of a gallows. Moments later he was dead, the sentence of an international tribunal composed of representatives of the United States, France, Great Britain, and the Soviet Union having been imposed. Streicher's body was then cremated, and—so horrendous were his crimes thought to have been—his ashes dumped into an unspecified German river so that "no one should ever know a particular place to go for reasons of mourning his memory."

Julius Streicher had been convicted at Nuremberg, Germany of what were termed "Crimes Against Humanity." The lead prosecutor in his case—Justice Robert Jackson of the United States Supreme Court—had not argued that the defendant had killed anyone, nor that he had personally committed any especially violent act. Nor was it contended that Streicher had held any particularly important position in the German government during the period in which the so-called Third Reich had exterminated some 6,000,000 Jews, as well as several million Gypsies, Poles, Slavs, homosexuals, and other *untermenschen* (subhumans).

The sole offense for which the accused was ordered put to death was in having served as publisher/editor of a Bavarian tabloid entitled *Der Stürmer* during the early-to-mid 1930s, years before the Nazi genocide actually began. In this capacity, he had penned a long series of virulently anti-Semitic editorials and "news" stories, usually accompanied by cartoons and other images graphically depicting Jews in extraordinarily derogatory fashion. This, the prosecution asserted, had done much to "dehumanize" the targets of his distortion in the mind of the German public. In turn, such dehumanization had made it possible—or at least easier—for average Germans to later indulge in the outright liquidation of Jewish "vermin." The tribunal agreed, holding that Streicher was therefore complicit in genocide and deserving of death by hanging.

During his remarks to the Nuremberg tribunal, Justice Jackson observed that, in implementing its sentences, the participating powers were morally and legally binding themselves to adhere forever after to the same standards of conduct that were being applied to Streicher and the other Nazi leaders. In the alternative, he said, the victorious allies would have committed "pure murder" at Nuremberg—no different in substance from that carried out by those they presumed to judge—rather than establishing the "permanent benchmark for justice" which was intended.

Yet in the United States of Robert Jackson, the indigenous American Indian population had already been reduced, in a process which is ongoing to this day, from perhaps 12.5 million in the year 1500 to fewer than 250,000 by the beginning of the 20th century. This was accomplished, according to official sources, "largely through the cruelty of [EuroAmerican] settlers," and an informal but clear governmental policy which had made it an articulated goal to "exterminate these red vermin," or at least whole segments of them.

Bounties had been placed on the scalps of Indians—any Indians—in places as diverse as Georgia, Kentucky, Texas, the Dakotas, Oregon, and California, and had been maintained until resident Indian populations were decimated or disappeared altogether. Entire peoples such as the Cherokee had been reduced to half their size through a policy of forced removal from their homelands east of the Mississippi River to what were then considered less preferable areas in the West.

Others, such as the Navajo, suffered the same fate while under military guard for years on end. The United States Army had also perpetrated a long series of wholesale massacres of Indians at places like Horseshoe Bend, Bear River, Sand Creek, the Washita River, the Marias River, Camp Robinson, and Wounded Knee.

Through it all, hundreds of popular novels—each competing with the next to make Indians appear more grotesque, menacing, and inhuman—were sold in the tens of millions of copies in the U.S. Plainly, the Euro-American public was being conditioned to see Indians in such a way as to allow their eradication to continue. And continue it did until the Manifest Destiny of the U.S.—a direct precursor to what Hitler would subsequently call Lebensraumpolitik (the politics of living space)—was consummated.

By 1900, the national project of "clearing" Native Americans from their land and replacing them with "superior" Anglo-American settlers was complete: the indigenous population had been reduced by as much as 98 percent while approximately 97.5 percent of their original territory had "passed" to the invaders. The survivors had been concentrated, out of sight and mind of the public, on scattered "reservations," all of them under the self-assigned "plenary" (full) power of the federal government. There was, of course, no Nuremberg-style tribunal passing judgment on those who had fostered such circumstances in North America. No U.S. official or private citizen was ever imprisoned—never mind hanged—for implementing or propagandizing what had been done. Nor had the process of genocide afflicting Indians been completed. Instead, it merely changed form.

Between the 1880s and the 1980s, nearly half of all Native American children were coercively transferred from their own families, communities, and cultures to those of the conquering society. This was done through compulsory attendance at remote boarding schools, often hundreds of miles from their homes, where native children were kept for years on end while being systematically "deculturated" (indoctrinated to think and act in the manner of Euro Americans rather than Indians). It was also accomplished through a pervasive foster home and adoption program—including "blind" adoptions, where children would be permanently denied information as to who they were/are and where they'd come from—placing native youths in non-Indian homes.

The express purpose of all this was to facilitate a U.S. governmental policy to bring about the "assimilation" (dissolution) of indigenous societies. In other words, Indian cultures as such were to be caused to disappear. Such policy objectives are directly contrary to the United Nations 1948 Convention on Punishment and Prevention of the Crime of Genocide, an element of international law arising from the Nuremberg proceedings. The forced "transfer of the children" of a targeted "racial, ethnical, or religious group" is explicitly prohibited as a genocidal activity under the Convention's second article.

Article II of the Genocide Convention also expressly prohibits involuntary sterilization as a means of "preventing births among" a targeted population. Yet, in 1975, it was conceded by the U.S. government that its Indian Health Service (IHS) then a subpart of the Bureau of Indian Affairs (BIA), was even then conducting a secret program of involuntary sterilization that had affected approximately 40 percent of all Indian women. The program was allegedly discontinued and the IHS was transferred to the Public Health Service, but no one was punished. In 1990, it came out that the IHS was inoculating Inuit children in Alaska with Hepatitis-B vaccine. The vaccine had already been banned by the World Health Organization as having a demonstrated correlation with the HIV-Syndrome which is itself correlated to AIDS. As this is written, a "field test" of Hepatitis-A vaccine, also HIV-correlated, is being conducted on Indian reservations in the northern plains region.

The Genocide Convention makes it a "crime against humanity" to create conditions leading to the destruction of an identifiable human group, as such. Yet the BIA has utilized the government's plenary prerogatives to negotiate mineral leases "on behalf of" Indian peoples paying a fraction of standard royalty rates. The result has been "super profits" for a number of preferred U.S. corporations. Meanwhile, Indians, whose reservations ironically turned out to be in some of the most mineral-rich areas of North America, which makes us, the nominally wealthiest segment of the continent's population, live in dire poverty.

By the government's own data in the mid-1980s, Indians received the lowest annual and lifetime per capita incomes of any aggregate population group in the United States. Concomitantly, we suffer the highest rate of infant mortality, death by exposure and malnutrition, disease, and the like. Under such circumstances, alcoholism and other escapist forms of substance abuse are endemic in the Indian community, a situation which leads both to a general physical debilitation of the population and a catastrophic accident rate. Teen suicide among Indians is several times the national average.

The average life expectancy of a reservation-based Native American man is barely 45 years; women can expect to live less than three years longer.

Such itemizations could be continued at great length, including matters like the radioactive contamination of large portions of contemporary Indian Country, the forced relocation of traditional Navajos, and so on. But the point should be made: Genocide, as defined in international law, is a continuing fact of day-to-day life (and death) for North America's native peoples. Yet there has been—and is—only the barest flicker of public concern about, or even consciousness of, this reality. Absent any serious expression of public outrage, no one is punished and the process continues.

A salient reason for public acquiescence before the ongoing holocaust in Native North America has been a continuation of the popular legacy, often through more effective media. Since 1925, Hollywood has released more than 2,000 films, many of them rerun frequently on television, portraying Indians as strange, perverted, ridiculous, and often dangerous things of the past. Moreover, we are habitually presented to mass audiences one-dimensionally, devoid of recognizable human motivations and emotions; Indians thus serve as props, little more. We have thus been thoroughly and systematically dehumanized.

Nor is this the extent of it. Everywhere, we are used as logos, as mascots, as jokes: "Big-Chief" writing tablets, "Red Man" chewing tobacco, "Winnebago" campers, "Navajo" and "Cherokee" and "Pontiac" and "Cadillac" pickups and automobiles. There are the Cleveland "Indians," the Kansas City "Chiefs," the Atlanta "Braves" and the Washington "Redskins" professional sports teams—not to mention those in thousands of colleges, high schools, and elementary schools across the country—each with their own degrading caricatures and parodies of Indians and/or things Indian. Pop fiction continues in the same vein, including an unending stream of New Age manuals purporting to expose the inner works of indigenous spirituality in everything from pseudo-philosophical to do-it-yourself styles. Blond yuppies from Beverly Hills amble about the country claiming to be reincarnated 17th century Cheyenne Ushamans ready to perform previously secret ceremonies.

In effect, a concerted, sustained, and in some ways accelerating effort has gone into making Indians unreal. It is thus of obvious importance that the American public begin to think about the implications of such things the next time they witness a gaggle of face-painted and war-bonneted buffoons doing the "Tomahawk Chop" at a baseball or football game. It is necessary that they think about the implications of the grade-school teacher adorning their child in turkey feathers to commemorate Thanksgiving. Think about the significance of John Wayne or Charleton Heston killing a dozen "savages" with a single bullet the next time a western comes on TV. Think about why Land-o-Lakes finds it appropriate to market its butter with the stereotyped image of an "Indian princess" on the wrapper. Think about what it means when non-Indian academics profess—as they often do—to "know more about Indians than Indians do themselves." Think about the significance of charlatans like Carlos Castaneda and Jamake Highwater and Mary Summer Rain and Lynn Andrews churning out "Indian" bestsellers, one after the other, while Indians typically can't get into print.

Think about the real situation of American Indians. Think about Julius Streicher. Remember Justice Jackson's admonition. Understand that the treatment of Indians in American popular culture is not "cute" or "amusing" or just "good, clean fun."

Know that it causes real pain and real suffering to real people. Know that it threatens our very survival. And know that this is just as much a crime against humanity as anything the Nazis ever did. It is likely that the indigenous people of the United States will never demand that those guilty of such criminal activity be punished for their deeds. But the least we have the right to expect—indeed, to demand—is that such practices finally be brought to a halt.

YOU'RE SHORT, BESIDES!   **39**

Sucheng Chan

When asked to write about being a physically handicapped Asian American woman, I considered it an insult. After all, my accomplishments are many, yet I was not asked to write about any of them. Is being handicapped the most salient feature about me? The fact that it might be in the eyes of others made me decide to write the essay as requested. I realized that the way I think about myself may differ considerably from the way others perceive me. And maybe that's what being physically handicapped is all about.

I was stricken simultaneously with pneumonia and polio at the age of four. Uncertain whether I had polio of the lungs, seven of the eight doctors who attended me—all practitioners of Western medicine—told my parents they should not feel optimistic about my survival. A Chinese fortune teller my mother consulted also gave a grim prognosis, but for an entirely different reason: I had been stricken because my name was offensive to the gods. My grandmother had named me "grandchild of wisdom," a name that the fortune teller said was too presumptuous for a girl. So he advised my parents to change my name to "chaste virgin." All these pessimistic predictions notwithstanding, I hung onto life, if only by a thread. For three years, my body was periodically pierced with electric shocks as the muscles of my legs atrophied. Before my illness, I had been an active, rambunctious, precocious, and very curious child. Being confined to bed was thus a mental agony as great as my physical pain. Living in war-torn China, I received little medical attention; physical therapy was unheard of. But I was determined to walk. So one day, when I was six or seven, I instructed my mother to set up two rows of chairs to face each other so that I could use them as I would parallel bars. I attempted to walk by holding my body up and moving it forward with my arms while dragging my legs along behind. Each time I fell, my mother gasped, but I badgered her until she let me try again. After four nonambulatory years, I finally walked once more by pressing my hands against my thighs so my knees wouldn't buckle.

From: Asian Women United of California (eds.), *Making Waves: An Anthology of Writings By and About Asian American Women* (Boston: Beacon Press, 1989), pp. 265–272. Copyright © 1989 by Asian Women United. Reprinted by permission of Beacon Press.

My father had been away from home during most of those years because of the war. When he returned, I had to confront the guilt he felt about my condition. In many East Asian cultures, there is a strong folk belief that a person's physical state in this life is a reflection of how morally or sinfully he or she lived in previous lives. Furthermore, because of the tendency to view the family as a single unit, it is believed that the fate of one member can be caused by the behavior of another. Some of my father's relatives told him that my illness had doubtless been caused by the wild carousing he did in his youth. A well-meaning but somewhat simple man, my father believed them.

Throughout my childhood, he sometimes apologized to me for having to suffer retribution for his former bad behavior. This upset me; it was bad enough that I had to deal with the anguish of not being able to walk, but to have to assuage his guilt as well was a real burden! In other ways, my father was very good to me. He took me out often, carrying me on his shoulders or back, to give me fresh air and sunshine. He did this until I was too large and heavy for him to carry. And ever since I can remember, he has told me that I am pretty.

After getting over her anxieties about my constant falls, my mother decided to send me to school. I had already learned to read some words of Chinese at the age of three by asking my parents to teach me the sounds and meaning of various characters in the daily newspaper. But between the ages of four and eight, I received no education since just staying alive was a full-time job. Much to her chagrin, my mother found no school in Shanghai, where we lived at the time, which would accept me as a student. Finally, as a last resort, she approached the American School which agreed to enroll me only if my family kept an *amah* (a servant who takes care of children) by my side at all times. The tuition at the school was twenty U.S. dollars per month—a huge sum of money during those years of runaway inflation in China—and payable only in U.S. dollars. My family afforded the high cost of tuition and the expense of employing a full-time *amah* for less than a year.

We left China as the Communist forces swept across the country in victory. We found an apartment in Hong Kong across the street from a school run by Seventh-Day Adventists. By that time I could walk a little, so the principal was persuaded to accept me. An *amah* now had to take care of me only during recess when my classmates might easily knock me over as they ran about the playground.

After a year and a half in Hong Kong, we moved to Malaysia, where my father's family had lived for four generations. There I learned to swim in the lovely warm waters of the tropics and fell in love with the sea. On land I was a cripple; in the ocean I could move with the grace of a fish. I liked the freedom of being in the water so much that many years later, when I was a graduate student in Hawaii, I became greatly enamored with a man just because he called me a "Polynesian water nymph."

As my overall health improved, my mother became less anxious about all aspects of my life. She did everything possible to enable me to lead as normal a life as possible. I remember how once some of her colleagues in the high school where she taught criticized her for letting me wear short skirts. They felt my legs should not be exposed to public view. My mother's response was, "All girls her age wear short skirts, so why shouldn't she?"

The years in Malaysia were the happiest of my childhood, even though I was constantly fending off children who ran after me calling, "*Baikah! Baikah!*" ("Cripple! Cripple!" in the Hokkien dialect commonly spoken in Malaysia). The taunts of children mattered little because I was a star pupil. I won one award after another for general scholarship as well as for art and public speaking. Whenever the school had important visitors my teacher always called on me to recite in front of the class.

A significant event that marked me indelibly occurred when I was twelve. That year my school held a music recital and I was one of the students chosen to play the piano. I managed to get up the steps to the stage without any problem, but as I walked across the stage, I fell. Out of the audience, a voice said loudly and clearly, "Ayah! A *baikah* shouldn't be allowed to perform in public." I got up before anyone could get on stage to help me and, with tears streaming uncontrollably down my face, I rushed to the piano and began to play. Beethoven's "Für Elise" had never been played so fiendishly fast before or since, but I managed to finish the whole piece. That I managed to do so made me feel really strong. I never again feared ridicule.

In later years I was reminded of this experience from time to time. During my fourth year as an assistant professor at the University of California at Berkeley, I won a distinguished teaching award. Some weeks later I ran into a former professor who congratulated me enthusiastically. But I said to him, "You know what? I became a distinguished teacher by *limping* across the stage of Dwinelle 155!" (Dwinelle 155 is a large, cold classroom that most colleagues of mine hate to teach in.) I was rude not because I lacked graciousness but because this man, who had told me that my dissertation was the finest piece of work he had read in fifteen years, had nevertheless advised me to eschew a teaching career.

"Why?" I asked.

"Your leg . . . " he responded.

"What about my leg?" I said, puzzled.

"Well, how would you feel standing in front of a large lecture class?"

"If it makes any difference, I want you to know I've won a number of speech contests in my life, and I am not the least bit self-conscious about speaking in front of large audiences. . . . Look, why don't you write me a letter of recommendation to tell people how brilliant I am, and let *me* worry about my leg!"

This incident is worth recounting only because it illustrates a dilemma that handicapped persons face frequently: those who care about us sometimes get so protective that they unwittingly limit our growth. This former professor of mine had been one of my greatest supporters for two decades. Time after time, he had written glowing letters of recommendation on my behalf. He had spoken as he did because he thought he had my best interests at heart; he thought that if I got a desk job rather than one that required me to be a visible, public person, I would be spared the misery of being stared at.

Americans, for the most part, do not believe as Asians do that physically handicapped persons are morally flawed. But they are equally inept at interacting with those of us who are not able-bodied. Cultural differences in the perception and treatment of handicapped people are most clearly expressed by adults. Children, regardless of where they are, tend to be openly curious about people who do not look "normal." Adults in Asia have no hesitation in asking visibly handicapped people what is wrong with them, often expressing their sympathy with looks of pity, whereas adults in the United States try desperately to be polite by pretending not to notice.

One interesting response I often elicited from people in Asia but have never encountered in America is the attempt to link my physical condition to the state of my soul. Many a time while living and traveling in Asia people would ask me what religion I belonged to. I would tell them that my mother is a devout Buddhist, that my father was baptized a Catholic but has never practiced Catholicism, and that I am an agnostic. Upon hearing this, people would try strenuously to convert me to their religion so that whichever God they believed in could bless me. If I would only attend this church or that temple regularly, they urged, I would surely get cured. Catholics and Buddhists alike have pressed religious medallions into my palm, telling me if I would wear these, the relevant deity or saint would make me well. Once while visiting the tomb of Muhammad Ali Jinnah in Karachi, Pakistan, an old Muslim, after finishing his evening prayers, spotted me, gestured toward my legs, raised his arms heavenward, and began a new round of prayers, apparently on my behalf.

In the United States adults who try to act "civilized" towards handicapped people by pretending they don't notice anything unusual sometimes end up ignoring handicapped people completely. In the first few months I lived in this country, I was struck by the fact that whenever children asked me what was the matter with my leg, their adult companions would hurriedly shush them up, furtively look at me, mumble apologies, and rush their children away. After a few months of such encounters, I decided it was my responsibility to educate these people. So I would say to the flustered adults, "It's okay, let the kid ask." Turning to the child, I would say, "When I was a little girl, no bigger than you are, I became sick with something called polio. The muscles in my leg shrank up and I couldn't walk very well. You're much luckier than I am because now you can get a vaccine to make sure you never get my disease. So don't cry when your mommy takes you to get a polio vaccine, okay?" Some adults and their little companions I talked to this way were glad to be rescued from embarrassment; others thought I was strange.

Americans have another way of covering up their uneasiness: they become jovially patronizing. Sometimes when people spot my crutch, they ask if I've had a skiing accident. When I answer that unfortunately it is something less glamorous than that, they say, "I bet you *could* ski if you put your mind to it!" Alternately, at parties where people dance, men who ask me to dance with them get almost belligerent when I decline their invitation. They say, "Of course you can dance if you *want* to!" Some have given me pep talks about how if I would only develop the right mental attitude, I would have more fun in life.

Different cultural attitudes toward handicapped persons came out clearly during my wedding. My father-in-law, as solid a representative of middle America as could be found, had no qualms about objecting to the marriage on racial grounds, but he could bring himself to comment on my handicap only indirectly. He wondered why his son, who had dated numerous high school and college beauty queens, couldn't marry one of them instead of me. My mother-in-law, a devout Christian, did not share her husband's prejudices, but she worried aloud about whether I could have children. Some Chinese friends of my parents, on the other hand, said that I was lucky to have found such a noble man, one who would marry me despite my handicap. I, for my part, appeared in church in a white lace wedding dress I had designed and made myself—a miniskirt!

How Asian Americans treat me with respect to my handicap tells me a great deal about their degree of acculturation. Recent immigrants behave just like Asians in Asia; those who have been here longer or who grew up in the United States behave more like their white counterparts. I have not encountered any distinctly Asian American pattern of response. What makes the experience of Asian American handicapped people unique is the duality of responses we elicit.

Regardless of racial or cultural background, most handicapped people have to learn to find a balance between the desire to attain physical independence and the need to take care of ourselves by not overtaxing our bodies. In my case, I've had to learn to accept the fact that leading an active life has its price. Between the ages of eight and eighteen, I walked without using crutches or braces but the effort caused my right leg to become badly misaligned. Soon after I came to the United States, I had a series of operations to straighten out the bones of my right leg; afterwards though my leg looked straighter and presumably better, I could no longer walk on my own. Initially my doctors fitted me with a brace, but I found wearing one cumbersome and soon gave it up. I could move around much more easily—and more important, faster—by using one crutch. One orthopedist after another warned me that using a single crutch was a bad practice. They were right. Over the years my spine developed a double-S curve and for the last twenty years I have suffered from severe, chronic back pains, which neither conventional physical therapy nor a lighter work load can eliminate.

The only thing that helps my backaches is a good massage, but the soothing effect lasts no more than a day or two. Massages are expensive, especially when one needs them three times a week. So I found a job that pays better, but at which I have to work longer hours, consequently increasing the physical strain on my body—a sort of vicious circle. When I was in my thirties, my doctors told me that if I kept leading the strenuous life I did, I would be in a wheelchair by the time I was forty. They were right on target: I bought myself a wheelchair when I was forty-one. But being the incorrigible character that I am, I use it only when I am *not* in a hurry!

It is a good thing, however, that I am too busy to think much about my handicap or my backaches because pain can physically debilitate as well as cause depression. And there are days when my spirits get rather low. What has helped me is realizing that being handicapped is akin to growing old at an accelerated rate. The contradiction I experience is that often my mind races along as though I'm only twenty while my body feels about sixty. But fifteen or twenty years hence, unlike my peers who will have to cope with aging for the first time, I shall be full of cheer because I will have already fought, and I hope won, that battle long ago.

Beyond learning how to be physically independent and, for some of us, living with chronic pain or other kinds of discomfort, the most difficult thing a handicapped person has to deal with, especially during puberty and early adulthood, is relating to potential sexual partners. Because American culture places so much emphasis on physical attractiveness, a person with a shriveled limb, or a tilt to the head, or the inability to speak clearly, experiences great uncertainty—indeed trauma—when interacting with someone to whom he or she is attracted. My problem was that I was not only physically handicapped, small, and short, but worse, I also wore glasses and was smarter than all the boys I knew! Alas, an insurmountable combination. Yet somehow I have managed to have intimate relationships, all of them with extraordinary men. Not surprisingly, there have also been countless men who broke my heart—men who enjoyed my company "as a friend," but who never found the courage to date or make love with me, although I am sure my experience in this regard is no different from that of many able-bodied persons.

The day came when my backaches got in the way of having an active sex life. Surprisingly that development was liberating because I stopped worrying about being attractive to men. No matter how headstrong I had been, I, like most women of my generation, had had the desire to be alluring to men ingrained into me. And that longing had always worked like a brake on my behavior. When what men think of me ceased to be compelling, I gained greater freedom to be myself.

I've often wondered if I would have been a different person had I not been physically handicapped. I really don't know, though there is no question that being handicapped has marked me. But at the same time I usually do not *feel* handicapped—and consequently, I do not *act* handicapped. People are therefore less likely to treat me as a handicapped person. There is no doubt, however, that the lives of my parents, sister, husband, other family members, and some close friends have been affected by my physical condition. They have had to learn not to hide me away at home, not to feel embarrassed by how I look or react to people who say silly things to me, and not to resent me for the extra demands my condition makes on them. Perhaps the hardest thing for those who live with handicapped people is to know when and how to offer help. There are no guidelines applicable to all situations. My advice is, when in doubt, ask, but ask, in a way that does not smack of pity or embarrassment. Most important, please don't talk to us as though we are children.

So, has being physically handicapped been a handicap? It all depends on one's attitude. Some years ago, I told a friend that I had once said to an affirmative action compliance officer (somewhat sardonically since I do not believe in the head count approach to affirmative action) that the institution which employs me is triply lucky because it can count me as nonwhite, female and handicapped. He responded, "Why don't you tell them to count you four times? . . . Remember, you're short, besides!"

RAP, RACE AND POLITICS

40

Clarence Lusane

'Whatever may be the conditions of a people's political and social factors . . . it is generally within the culture that we find the seed of opposition, which leads to the structuring and development of the liberation movement.'

—Amilcar Cabral[1]

For many black youths in the United States, in the words of the classic song by War, the world is a ghetto. Trapped in and witness to cycles of violence, destitution and lives of desperation, their aspirations and views find expression in political behavior, social practice, economic activities and cultural outlets. These streams came together and informed a culture of resistance that has been termed Hip Hop whose most dynamic expression is in the form of rap music. On the one hand, rap is the voice of alienated, frustrated and rebellious black youth who recognise their vulnerability and marginality in post-industrial America. On the other hand, rap is the packaging and marketing of social discontent by some of the most skilled ad agencies and largest record producers in the world. It's the duality that has made rap and rappers an explosive issue in the politics of power that shaped the 1992 US elections and beyond. It's also this duality that has given rap its many dimensions and flavours; its spiralling matrix of empowerment and reaction.

Influenced by a tradition of oral leaders and artists, from Malcolm X, Martin Luther King and Nikki Giovanni to Gil Scott Heron and the Last Poets, young black cultural activists evolved from the urban cosmos of the early 1980s ready for rap. Denied opportunity for more formal music training and access to instruments due to Reagan-era budget cuts in education and school music programmes, turntables became instruments and lyrical acrobatics became a cultural outlet. Initially underground, by the late 1980s, rap and the broad spectrum of Hip Hop had become the dominant cultural environment of young African-Americans, particularly males.

From: Clarence Lusane, "Rap, Race, and Politics," *Race & Class*, Vol. 35, No. 1 (July/September 1993): 41–55, Institute of Race Relations, London. Reprinted by permission.

Following the historic example of the cultural modes of the civil rights and Black Power movements, rap has had an international impact. Just as the Vietnamese sang civil rights freedom songs, so have the political imperatives of rap traversed the globe and found expression in venues from Mexico to India. In Czechoslovakia, local rappers rap about the struggle of being young and penniless. Wearing baseball caps and half-laced sneakers, youth in the Ivory Coast have found a bond in the music. Australian rappers kick it about the mistreatment of the Aborigine people. Tributes to the victims of US atomic bombs form the substance of local rappers in Japan.[2]

The cultural power of rap as global protest music is undeniable. To understand the genesis of this power, however, requires a return to the source. It is rap's impact on the economics, politics and gender issues in the African-American community that must be examined, if only briefly, to sense its significance, possibilities and contradictions. The enemies of rap, as one observer noted, have gone after 'the message, the messenger and the medium'.[3] And it is not just whites who have dismissed and criticized rap. As Salim Muwakkil wrote in *In These Times*, 'for many middle-class black Americans, rap is . . . a soundtrack for sociopaths'.[4]

HIP HOP CAPITALISM

From slave town to Motown, from Bebop to Hip Hop, black music has been shaped by the material conditions of black life. Contextually, today's black youth culture flows out of the changes that affected the political economy of US capitalism over the last two decades. Incremental economic and social gains made in the late 1960s and the 1970s were destroyed with a vengeance in the Reagan and Bush years. Many observers of black politics saw the handwriting on the wall when Reagan came into power. In 1982, political writer Manning Marable prophesied:

> The acceleration of black unemployment and underemployment, the capitula-
> tion of many civil rights and Black Power leaders to the Right, the demise of
> militant black working-class institutions and caucuses, and the growing de-
> pendency of broad segments of the black community upon public assistance
> programs and transfer payments of various kinds; these interdependent reali-
> ties within the contemporary black political economy are the beginning of a
> new and profound crisis for black labor in America.[5]

Thus, in the late 1980s and early 1990s, the material basis for the production and reproduction of black youth alienation is the growing immiseration of millions of African-American working-class families. Between 1986 and 1992, according to the Census Bureau, an additional 1.2 million African-Americans fell below the poverty line.[6] As stunning as that may be, the Bush administration achieved the same result in half the time. A report issued by the Children's Defense Fund documents that 841,000 youth fell into poverty in the first two years of the Bush administration, affecting, in some cities, as many as two-thirds of minority children.[7] The official poverty rate for blacks is 32.7 per cent, 10.2 million people, which is higher than for Hispanics (28.7 per cent), Asians (13.8 per cent), or whites (11.3 per cent).[8]

Most critical, however, has been the unemployment situation of African-American youth and what has happened to black youth economically over the last three decades. Since 1960, black youth suffered the largest decline in employment of all component groups of all races. In 1986, in the middle of the Republican years, black teenage unemployment was officially as high as 43.7 per cent. In October 1992, six years later, the numbers remained virtually unchanged, with black youth unemployment officially at 42.5 per cent.[9] One does not have to agree with the rantings and rage of Ice T, Sister Souljah or other rappers to unite with their sense of isolation, anger and refusal to go down quietly. Ignored and 'dissed' by both major political parties and much of what passes for national black leadership, is it any wonder that Ice Cube reflects the views of so many youth when he sings:

> Do I have to sell me a whole lot of crack
> For decent shelter and clothes on my back?
> Or should I just wait for President Bush
> Or Jesse Jackson and Operation PUSH?[10]

It was, then, perfectly logical that Hip Hop culture should initially emerge most strongly in those cities hardest hit by Reaganomics with large minority youth populations—New York, Los Angeles, Houston and Oakland. For many of these youth, rap became not only an outlet for social and political discourse, but also an economic opportunity that required little investment other than boldness and a competitive edge. In a period when black labour was in low demand, if one could not shoot a basketball like Michael Jordan, then the entertainment industry was one of the few legal avenues available for the get-rich consciousness that dominated the social ethos of the 1980s.

Rap music is big business. According to the *Los Angeles Times*, in 1990 rap brought in $600 million; . . . in 1991, sales rose to about $700 million.[11] 2 Live Crew's *As Nasty as They Wanna Be*, the subject of law suits and arrests, sold more than two million copies. In their debut album, *Straight Outta Compton*,

NWA sold over a million copies and followed that up in 1992 by breaking all sales records with their *Efil4zaggin* (Niggaz 4 life spelled backwards) album. The album sold an unprecedented 900,000 copies in its first week of release and later went on to sell millions.[12]

Rap is attractive because it requires generally low-investment costs for the corporations. According to one producer, a rap album can be produced for less than $50,000, while an equivalent album for an established rock group or popular R&B group can cost $100,000–300,000. And while rap artists are signed with a bewildering frenzy, they are also dropped more rapidly than musicians from other music forms. If an artist or group doesn't do well within the first six to eight weeks of their release, they are often sent packing.

Annually, young black consumers, aged 15–24, spend about $23 billion a year in the United States, of which about $100 million is spent on records and tapes.[13] African-Americans, however, are not the main purchasers of rap as, increasingly, rap is being bought by non-blacks. A survey taken in mid-1992 found that 74 per cent of rap sold in the first six months of that year was bought by whites.[14] This is one reason why every major record company and communications conglomerate, from Sony to Atlantic, has made significant investments in rap music.

For many rappers, Hip Hop capitalism promises both riches and racial integrity. Rappers found that they could yell at the system and be paid (highly) by it at the same time. A legitimate desire and need for economic empowerment could be turned into profit with only minor ideological adjustments and rationalisations about 'free speech' by capital. Some of those who have been the target of censorship, such as rapper Ice T, would argue that free enterprise will only let free speech go so far. As he says in his song 'Freedom of Speech':

> *Freedom of Speech*
> *That's some mutherfuckin' bullshit*
> *You say the wrong thing*
> *They'll lock your ass up quick*

In the laissez-faire capitalist atmosphere that dominated the early years of modern rap, a number of black entrepreneurs were able to enter the business and become highly successful. Queen Latifah's Flavor Unit Management and Records is home to popular groups such as Nikki D, Black Sheep, D Nice, Pete Rock & CL Smooth and Naughty By Nature.

No one better symbolises the contradictory aspirations of the rappers than Russell Simmons and his phenomenal achievements with Rush Communications and its rap label, Def Jam. By any estimation, Rush Communications is huge. Home to top rap groups such as Public Enemy, Run DMC and Big Daddy Kane and producer of the highly-rated, hip hopish cable comedy series Def Comedy Jam, Simmons has transformed what was essentially a small basement operation into a $34 million conglomerate. Rap artists at Rush have earned ten gold records, six platinum records, and two multiplatinum records. Plans are afoot to expand the conglomerate into film production and even sell public stock. Rush Communications is the thirty-second largest black-owned business in the United States and the second largest black-owned entertainment company.[15]

As CEO, Simmons earns an estimated $5 million annually. Usually attired in sneakers, sweatsuits and baseball caps, Simmons is a major driving force behind the music and in attacking the racist structures of the popular music business that have historically reduced the role of African-American to that of powerless entertainer.

Simmons' success and efforts are as laudable as they are remarkable. They do not represent, however, a break from the economic system that is responsible for the misery that forms the substance of the music that Rush produces. The commodification of black resistance is not the same as resistance to a society built upon commodification. Rap artists, even those who obtained some level of economic power and independence, are still slaves to a market system that requires an economic elite and mass deprivation.

It's critical to note that it has been more than just the multinational recording industry that has benefited from the reduction of black culture to the circumscribed limits of Hip Hop. The alcohol, tennis shoe, clothing, hat and film industries have boomed as a result of the new markets that have opened up or expanded, based on the spread of Hip Hop and the often exploitative use of rap artists in advertising.

Alcohol companies, already complicit in the disproportionate targeting of the black community for liquor sales, were quick to front rap stars to sell their product. . . . Ice Cube, Eric B. & Rakim, EPMD, the Geto Boys, Compton's Most Wanted, Yo! MTV Rap's Fab Freddie and Yo-Yo—who was not even drinking age at the time—were all used to sell highly potent malt liquor. Sexually-suggestive scripts also attempted to convince consumers that malt liquors are aphrodisiacs. Yo-Yo would moan that St Ides Malt Liquor 'puts you in the mood [and] makes you wanna go oooh'. Ice Cube claimed that with St Ides you could 'get your girl in the mood quicker' and that the beverage would make your 'jimmy thicker'. The alcohol content of St Ides, Elephant, Magnum, Crazy Horse, Olde English 800, Red Bull Malt Liquor, PowerMaster and other malt beers is greater than regular beer—nearly twice as great in some cases. Malt beer accounts for only about 3 per cent of all beer sold, yet more than 30 per cent of its sales are in the black community.

This exploitation of these rappers' popularity was denounced by community activists and black health advocates around the country. Makani Themba of the Marin Institute in California pointed out astutely that the beer companies were 'appropriating a very important part of our culture to sell what is a dangerous product for many of these kids'.[16] . . .

The political economy of Hip Hop, i.e., its capacity to open markets, maximise profits, and commodify legitimate grief and unrest, is the material basis that drives it forward. However, the nature of Hip Hop, its political soul, is to provoke and agitate.

FEAR OF A BLACK PLANET: THE POLITICS OF PROVOCATION

> *Nightmare. That's what I am*
> *America's nightmare*
> *I am what you made me*
> *The hate and evil that you gave me . . .*
> *America, reap what you sow.*
> *—2 PAC*

The dominant ideological trend of the rappers is black nationalism. Universally wedded to the notion that black leadership, for the most part, has sold out, the black nationalist rhetoric of Hip Hop becomes a challenging and liberating political paradigm in the face of surrender on the part of many political forces in the black community. While there are leftist rappers, such as the Disposal Heroes of Hiphoprisy and KRS-One, who to some extent embody Jesse Jackson's Rainbow Coalition notion of politics, most range from the soft-core nationalism of Arrested Development to the hard-core nationalist influenced raps of the political and gangsta rappers.

In particular, minister Louis Farrakhan and his Nation of Islam have had tremendous influence on the political views of black youth, in general, and of rappers, more specifically. Ice Cube, for example, joined the organisation, stating that, 'To me, the best organisation around for black people is the Nation of Islam'.[17] And a whole set of Muslim rappers has come on the scene. This includes groups such as Brand Nubian, Poor Righteous Teachers, King Sun, Movement Ex and Paris, who created his own mini-controversy when he wrote the song 'Bush Killer'—the title of which should be explanation enough. . . .

GANGSTA RAP

Much of rap's political pedagogy comes from the so-called gangsta rappers. Dismissed by many as vulgar, profane, misogynist, racist, anti-Semitic and juvenile—accusations that carry a great deal of validity—gangsta rap, at the same time, reflects and projects what scholar Robin D.G. Kelley calls 'the lessons of lived experiences'.[18] In a sense, Cube, NWA, Too Short, the Geto Boys and others are the 'organic intellectuals' of the inner-city black poor, documenting as they do their generally hidden conditions and life-style choices. Naughty by Nature's 'Ghetto Bastard' is a captivating and engrossing piece of verbal literature and sociology. This autobiographical tale of a black male teenager's urban experience is a brilliant exposition of what Foucault has termed the 'insurrection of subjugated knowledges'. The song is rich in a wide array of themes—absent ghetto fathers, the attractiveness of lumpen activities, the racist assumptions of the education system, the vicissitudes of consumer culture, dilapidated housing and the ever-present threat of violence—that reflect an experience that is collectively endured, daily, by millions of African-Americans. It's social anthropology with rhythm. Naughty by Nature's Treach projects the anger and frustration of many when he raps,

> *Say somethin' positive, well positive ain't where I live*
> *I live right around the corner from West Hell*
> *Two blocks from South Shit and once in a jail cell*[19]

An examination of today's rap songs quickly demonstrates that the principal topic of the music is the social crisis engulfing working-class black America. Unlike the moralistic preaching, escapism or sentimentality that defines most popular music, including the moderated rap of Hammer, hardcore rappers detail the unemployment, miseducation, discrimination, homicides, gang life, class oppression, police brutality and regressive gender politics that dominate the lives of many black youth. The macho boasting, misogyny, violent fantasies and false consciousness exist side by side with an immature, but clear, critique of authority, a loathing of the oppressive character of wage labour, a hatred of racism and an exposé of Reaganism.

Many rappers, for example, address the racist character of the nation's war on drugs. Although blacks make up only about 15 per cent of the nation's drug users, they are close to 50 per cent of those arrested on drug charges, mainly for possession. The drug war's collateral damage continues to grow in what one senate committee calls a '$32 billion failure'.[20] Raps like NWA's 'Dope Man', Ice T's 'New Jack Hustler', Ice Cube's 'The Product' and CPO's 'The

Wall' all expose the bankruptcy of the war on drugs and its deadly impact on the black community. . . .

Debates over police brutality were also forced into the public arena as a result of rap songs. While the video of the Rodney King beating introduced many in the United States to the reality of police brutality in the black community, the rap community noted that it had been discoursing about the issue for years. . . .

Not surprisingly, there have been moves to censor rap. One organisation active in this is the Parents Music Resource Center (PMRC) founded by Tipper Gore (wife of vice-president Al Gore) and Susan Baker (wife of Bush's campaign manager, James Baker). In 1985, PMRC led the movement that forced congressional hearings on record labelling. Record companies suc-cumbed to the pressure and began to put warning labels on rap and rock music felt to be obscene and too explicit. The first album to have a warning label placed on it was Ice T's *Rhyme Pays*. Evidence indicates that most of the groups targeted for labelling are black. In a 1989 newsletter put out by PMRC, every song listed as having warning labels was done by a black artist. Other groups calling for censorship of rap records have been more explicitly racist. Missouri Project Rock passed out information packets that criticised 'race-mixing' and called Martin Luther King 'Martin Lucifer King'.[21]

Tipper Gore battled Ice T on the Oprah Winfrey show and wrote about it in the *Washington Post*. In an article titled 'Hate, rape and rap', she justifiably criticised some of the vile sexist statements made by rappers, particularly Ice T. In words that specifically seem to be addressed to Ice T, she said, 'We must raise our voices in protest and put pressure on those who not only reflect this hatred but also package, polish, promote and market it; those who would make words like "nigger" acceptable.'[22] In highly moral tones, she then unconvinc-ingly attempted to make a link between rap music and rape.

One of Ice T's responses was to write a song, 'Freedom of Speech'. In the song, he attacks Gore personally. He says:

> *Think I give a fuck about a silly bitch named Gore?*
> *Yo, PMRC, here we go, war!*[23]

THE EVIL THAT MEN DO: RAP'S PHALLO-CENTRIC MUSINGS

Gender issues in rap remain controversial. From its earliest days to the present, women and more than a few men have rightfully condemned much of rap music as misogynist and degrading to women. National Council of Negro

Women president Dorothy Height states: 'This music is damaging because it is degrading to women to have it suggested in our popular music that [women] are to be abused.'[24] Former head of the NAACP, Benjamin Hooks, echoes that sentiment. He says, in reference to the music, 'our [black] cultural experience does not include debasing women'.[25]

Scholar Marilyn Lashley is uncompromising in her denunciation of the portrayal of women in rap music. It is 'explicitly and gratuitously sexual, occasionally bestial and frequently violent. These images, in the guise of "art and music", exploit, degrade and denigrate African-American women as well as the race. They encourage sexual harassment, exploitation and misogyny at their best and sexual abuse at their worst', she states.[26]

Others, mainly men, have defended these projections as part of a continuity in black culture that is not as harmful as it appears. No less than Harvard scholar and cultural critic Henry Louis Gates walks softly on this turf. He pooh-poohs the uproar by stating that the male rappers are playing out the old black tradition of 'signifying',[27] a practice that is relatively harmless and culturally important. Some have attempted to justify the degradation of women by arguing for the singular uplifting of black males. Ice Cube, for example, argued in an interview with Angela Davis that black women have to wait for black men to be uplifted first.[28]

Rap has become a forum for debating the nature of gender relations among black youth. The name-calling, descriptions of graphic rapes and other negative encounters between young black women and men dominate the music's gender politics. For many of the rap groups, their songs are one long extended sex party. This aspect of the music has also drawn fire, though many of the male rappers have argued that they are engaging in meaningless fantasies. However, as one feminist correctly pointed out, it's 'not so much the issue of sex as an obsession, but that of sex as a violent weapon against women'. She goes on to say that songs like 'Treat Her Like a Prostitute', 'One Less Bitch', 'Pop that Coochie', 'Baby Got Back', 'Me So Horny', 'That Bitch Betta Have My Money' and 'She Swallowed It' 'not only desensitise their audiences to violence against women, they also help rationalise and reinforce a nihilistic mentality among those who already suffer from the effects of ghetto reality'.[29] The escalating incidence of rape and sexual harassment against women in general, and black women in particular, underscores her concerns.

A number of positive female rappers have emerged to challenge the musical and ideological dominance of the male rappers. Strong women, such as Queen Latifah, Monie Love, Queen Mother Rage, Isis and MC Lyte, have produced popular songs that have advocated positive relations among men and women, called for sisterhood and projected what scholar Patricia Hill Collins calls an 'Afrocentric feminist epistemology'[30]. . . .

In addition, some rap male and gender-mixed groups, such as the Disposal Heroes of Hiphoprisy and Arrested Development, have shown that a positive perspective on black female and male relations is possible.

But, for every (social) action, there is an opposite and equal (social) reaction. Hard-core female rappers, such as Bytches With Problems, Nikki D, Hoes With an Attitude and LA Starr, have come on the scene and demonstrated that they can be as vulgar, blasphemous and homicidal as the men. BWP's Lyndah and Tanisha, whose records are distributed by mega-corp Columbia Records, have been called the 'Thelma and Louise' of rap.[31] In song after song, they gun down men, cops and anyone else who crosses or is perceived to have crossed their path. When they are not committing homicide, they are busy either screwing men to death or ripping them off. All the while, they hold high the banner of women's liberation. . . .

It is ironic that BWP, like many of the hard-core women's groups, are produced by men who also write many of the lyrics. That women producers and managers are far and few between is one of the main reasons why the music remains so misogynist. Progressive women rappers complain incessantly about how difficult it is for their music to be produced. Even some of the songs produced by the political, usually black nationalist, rappers run counter to women's liberation. While eschewing the violence and sexual exploitation of the hard-core gangsta rappers, groups such as Public Enemy and X-Clan will often project a romanticised notion of black womanhood that does not funda-mentally challenge male domination. More critically, they will also use language that fundamentally reinforces the power relations of gender oppression.

As one scholar noted, for many, the rappers are 'urban griots dispensing social and cultural critiques'.[32] While this may be true, the nature of those critiques is simultaneously both painfully naive and incredibly insightful; abjectly dehumanising and rich in human spirit. Rap's pedagogy, like the initial stages of all pedagogies or oppressed people, emerges incomplete, contradic-tory and struggling for coherence. If we look closely, the birth and evolution of rap tells us as much about the current state of black America as rap's content and form.

At the same time, we must be careful not to reduce African-American culture to the commodities and political ambiguities of Hip Hop. Music forms, including jazz and blues and other non-Hip Hop cultural expression deserve criticism, reaffirmation and validation. In the end, Hip Hop is neither the cultural beast that will destroy black America nor the political panacea that will save it, but is a part of the ongoing African-American struggle constantly reaching for higher and higher modes of liberation.

References

1. Amilcar Cabral, 'National liberation and culture', in *Return to the Source* (New York, Africa Information Service, 1973), p. 43.

2. See articles on rap worldwide by B. Bollag, K. Noble, J. Bernard and S. Weisman in *New York Times* (23 August 1992).

3. Robin Givhan, 'Of rap, racism and fear: why does this message music seem so menacing?', *San Francisco Chronicle* (6 August 1992).

4. Quoted in Kathleen M. Sullivan, '2 Live Crew and the cultural contradictions of Miller', *Reconstruction* (Vol. 1, no. 2, 1990), p. 23.

5. Manning Marable, 'The crisis of the black working-class: an economic and historical analysis', *Science & Society* (Summer 1982), p. 156.

6. National Urban League, *The State of Black America 1993* (New York, National Urban League, 1993), p. 168.

7. Barbara Vobejda, 'Children's poverty rose in '80s; suburbs, rural areas also show increase', *Washington Post* (12 August 1992).

8. Press release, US Department of Commerce, Bureau of Census, 3 September 1992.

9. See *The State of Black America*, op. cit. and K. Jennings, 'Understanding the persisting crisis of black youth unemployment' in J. Jennings (ed.), *Race, Politics and Economic Development* (New York, 1992).

10. Ice Cube, 'A bird in the hand', *Death Certificate* (Priority Records, 1991).

11. Paul Grein, 'It was feast or famine in '90 cents; platinum ranks thin, but smashes soar', *Billboard* (12 January 1991), p. 9.

12. See John Leland, 'Rap and race', *Newsweek* (29 June 1992), p. 49, and James T. Jones IV, 'NWA's career gets a jolt from lyric's shock value', *USA Today* (21 June 1991).

13. Bruce Horovitz, 'Quincy Jones, Time Warner launch rap lovers magazine', *Los Angeles Times* (15 September 1992).

14. Chuck Philips, 'The uncivil war: the battle between the establishment and the supporters of rap opens old wounds of race and class', *Los Angeles Times* (19 July 1992).

15. Christopher Vaughn, 'Simmons' rush for profits', *Black Enterprise* (December 1992), p. 67.

16. Media Action Alert issued by the Marin Institute, 23 July 1991.

17. Ice Cube and Angela Davis, 'Nappy happy', *Transition* (no. 58, 1992), p. 191.

18. Robin D.G. Kelly, 'Straight from underground', *The Nation*, p. 796.

19. Naughty by Nature, 'Ghetto Bastard,' *Naughty by Nature* (Tommy Boy Records).

20. *The President's Drug Strategy: has it worked*, Majority Staff of the Senate Judiciary Committee and the International Narcotics Control Caucus, September 1992.

21. Sullivan, op. cit.

22. Tipper Gore, 'Hate, rape and rap', *Washington Post* (8 January 1990).

23. Ice T, 'Freedom of Speech', (Rhyme Syndicate, 1990).

24. 'Other rappers accused of "nasty" influences', *Washington Times* (16 June 1992).

25. Ibid.

26. Marilyn Lashley, 'Bad rap', *Washington Post* (25 September 1992).

27. 'Other rappers accused of "nasty" influences', op. cit.

28. Cube and Davis, op. cit., p. 186.

29. Sonja Peterson-Lewis, 'A feminist analysis of the defenses of obscene rap lyrics', *Black Sacred Music* (Summer 1991), p. 78.

30. P.H. Collins, *Black Feminist Thought: knowledge, consciousness, and the politics of empowerment* (Boston, 1990).

31. K. Carroll, 'Word on Bytches with problems', *Black Arts Bulletin* (Vol. 1, no. 8), p. 1.

32. M.E. Dyson, 'Performance, protest and prophecy in the culture of Hip Hop', *Black Sacred Music* (Summer 1991), p. 22.

IF MEN COULD MENSTRUATE—   **41**

Gloria Steinem

A white minority of the world has spent centuries conning us into thinking that a white skin makes people superior—even though the only thing it really does is make them more subject to ultraviolet rays and to wrinkles. Male human beings have built whole cultures around the idea that penis-envy is "natural" to women—though having such an unprotected organ might be said to make men vulnerable, and the power to give birth makes womb-envy at least as logical.

From: *Ms.* VII (October 1978): 110. © Gloria Steinem. Reprinted by permission.

In short, the characteristics of the powerful, whatever they may be, are thought to be better than the characteristics of the powerless—and logic has nothing to do with it.

What would happen, for instance, if suddenly, magically, men could menstruate and women could not?

The answer is clear—menstruation would become an enviable, boast-worthy, masculine event:

Men would brag about how long and how much.

Boys would mark the onset of menses, that longed-for proof of manhood, with religious ritual and stag parties.

Congress would fund a National Institute of Dysmenorrhea to help stamp out monthly discomforts.

Sanitary supplies would be federally funded and free. (Of course, some men would still pay for the prestige of commercial brands such as John Wayne Tampons, Muhammad Ali's Rope-a-dope Pads, Joe Namath Jock Shields—"For Those Light Bachelor Days," and Robert "Baretta" Blake Maxi-Pads.)

Military men, right-wing politicians, and religious fundamentalists would cite menstruation ("*men*-struation") as proof that only men could serve in the Army ("you have to give blood to take blood"), occupy political office ("can women be aggressive without that steadfast cycle governed by the planet Mars?"), be priests and ministers ("how could a woman give her blood for our sins?"), or rabbis ("without the monthly loss of impurities, women remain unclean").

Male radicals, left-wing politicians, and mystics, however, would insist that women are equal, just different; and that any woman could enter their ranks if only she were willing to self-inflict a major wound every month ("you *must* give blood for the revolution"), recognize the preeminence of menstrual issues, or subordinate her selfness to all men in the Cycle of Enlightenment.

Street guys would brag ("I'm a three-pad man") or answer praise from a buddy ("Man, you lookin' *good!*") by giving fives and saying, "Yeah, man, I'm on the rag!"

TV shows would treat the subject at length. ("Happy Days": Richie and Potsie try to convince Fonzie that he is still "The Fonz," though he has missed two periods in a row.) So would newspapers. (SHARK SCARE THREAT-ENS MENSTRUATING MEN. JUDGE CITES MONTHLY STRESS IN PARDONING RAPIST.) And movies. (Newman and Redford in "Blood Brothers"!)

Men would convince women that intercourse was *more* pleasurable at "that time of the month." Lesbians would be said to fear blood and therefore life itself—though probably only because they needed a good menstruating man.

Of course, male intellectuals would offer the most moral and logical arguments. How could a woman master any discipline that demanded a sense of time, space, mathematics, or measurement, for instance, without that in-built gift for measuring the cycles of the moon and planets—and thus for measuring anything at all? In the rarefied fields of philosophy and religion, could women compensate for missing the rhythm of the universe? Or for their lack of symbolic death-and-resurrection every month?

Liberal males in every field would try to be kind: the fact that "these people" have no gift of measuring life or connecting to the universe, the liberals would explain, should be punishment enough.

And how would women be trained to react? One can imagine traditional women agreeing to all these arguments with a staunch and smiling masochism. ("The ERA would force housewives to wound themselves every month": Phyllis Shlafley. "Your husband's blood is as sacred as that of Jesus—and so sexy, too!": Marabel Morgan.) Reformers and Queen Bees would try to imitate men, and *pretend* to have a monthly cycle. All feminists would explain endlessly that men, too, needed to be liberated from the false idea of Martian aggressiveness, just as women needed to escape the bonds of menses-envy. Radical feminists would add that the oppression of the nonmenstrual was the pattern for all other oppressions. ("Vampires were our first freedom fighters!") Cultural feminists would develop a bloodless imagery in art and literature. Socialist feminists would insist that only under capitalism would men be able to monopolize menstrual blood. . . .

In fact, if men could menstruate, the power justifications could probably go on forever.

If we let them.

BLAME IT ON FEMINISM   42

Susan Faludi

⌐

To be a woman in America at the close of the 20th century—what good fortune. That's what we keep hearing, anyway. The barricades have fallen, politicians assure us. Women have "made it," Madison Avenue cheers.

From: Susan Faludi, "Blame It on Feminism," in *Backlash: The Undeclared War Against Women* (New York: Crown, 1991), pp. ix–xxiii. Copyright © 1991 by Susan Faludi. Reprinted by permission of Crown Publishers, Inc.

Women's fight for equality has "largely been won," *Time* magazine announces. Enroll at any university, join any law firm, apply for credit at any bank. Women have so many opportunities now, corporate leaders say, that we don't really need equal opportunity policies. Women are so equal now, lawmakers say, that we no longer need an Equal Rights Amendment. Women have "so much," former President Ronald Reagan says, that the White House no longer needs to appoint them to higher office. Even American Express ads are saluting a woman's freedom to charge it. At last, women have received their full citizenship papers.

And yet . . .

Behind this celebration of the American woman's victory, behind the news, cheerfully and endlessly repeated, that the struggle for women's rights is won, another message flashes. You may be free and equal now, it says to women, but you have never been more miserable.

This bulletin of despair is posted everywhere—at the newsstand, on the TV set, at the movies, in advertisements and doctors' offices and academic journals. Professional women are suffering "burnout" and succumbing to an "infertility epidemic." Single women are grieving from a "man shortage." The *New York Times* reports: Childless women are "depressed and confused" and their ranks are swelling. *Newsweek* says: Unwed women are "hysterical" and crumbling under a "profound crisis of confidence." The health advice manuals inform: High-powered career women are stricken with unprecedented outbreaks of "stress-induced disorders," hair loss, bad nerves, alcoholism, and even heart attacks. The psychology books advise: Independent women's loneliness represents "a major mental health problem today." Even founding feminist Betty Friedan has been spreading the word: she warns that women now suffer from a new identity crisis and "new 'problems that have no name.' "

How can American women be in so much trouble at the same time that they are supposed to be so blessed? If the status of women has never been higher, why is their emotional state so low? If women got what they asked for, what could possibly be the matter now?

The prevailing wisdom of the past decade has supported one, and only one, answer to this riddle: it must be all that equality that's causing all that pain. Women are unhappy precisely *because* they are free. Women are enslaved by their own liberation. They have grabbed at the gold ring of independence, only to miss the one ring that really matters. They have gained control of their fertility, only to destroy it. They have pursued their own professional dreams—and lost out on the greatest female adventure. The women's movement, as we are told time and again, has proved women's own worst enemy.

"In dispensing its spoils, women's liberation has given my generation high incomes, our own cigarette, the option of single parenthood, rape crisis centers, personal lines of credit, free love, and female gynecologists," Mona Charen, a young law student, writes in the *National Review*, in an article titled "The Feminist Mistake." "In return it has effectively robbed us of one thing upon which the happiness of most women rests—men." The *National Review* is a conservative publication, but such charges against the women's movement are not confined to its pages. "Our generation was the human sacrifice" to the women's movement, *Los Angeles Times* feature writer Elizabeth Mehren contends in a *Time* cover story. Baby-boom women like her, she says, have been duped by feminism: "We believed the rhetoric." In *Newsweek*, writer Kay Ebeling dubs feminism "the Great Experiment That Failed" and asserts "women in my generation, its perpetrators, are the casualties." Even the beauty magazines are saying it: *Harper's Bazaar* accuses the women's movement of having "lost us [women] ground instead of gaining it."

In the last decade, publications from the *New York Times* to *Vanity Fair* to the *Nation* have issued a steady stream of indictments against the women's movement, with such headlines as WHEN FEMINISM FAILED or THE AWFUL TRUTH ABOUT WOMEN'S LIB. They hold the campaign for women's equality responsible for nearly every woe besetting women, from mental depression to meager savings accounts, from teenage suicides to eating disorders to bad complexions. The "Today" show says women's liberation is to blame for bag ladies. A guest columnist in the *Baltimore Sun* even proposes that feminists produced the rise in slasher movies. By making the "violence" of abortion more acceptable, the author reasons, women's rights activists made it all right to show graphic murders on screen. . . .

Popular psychology manuals peddle the same diagnosis for contemporary female distress. "Feminism, having promised her a stronger sense of her own identity, has given her little more than an identity *crisis*," the best-selling advice manual *Being a Woman* asserts. The authors of the era's self-help classic *Smart Women/Foolish Choices* proclaim that women's distress was "an unfortunate consequence of feminism," because "it created a myth among women that the apex of self-realization could be achieved only through autonomy, independence, and career."

In the Reagan and Bush years, government officials have needed no prompting to endorse this thesis. Reagan spokeswoman Faith Whittlesey declared feminism a "straitjacket" for women, in the White House's only policy speech on the status of the American female population—entitled "Radical Feminism in Retreat." Law enforcement officers and judges, too, have pointed a damning finger at feminism, claiming that they can chart a path from rising female independence to rising female pathology. As a California sheriff explained it to the press, "Women are enjoying a lot more freedom now, and as a result, they are committing more crimes." The U.S. Attorney General's Commission on Pornography even proposed that women's professional advancement might be responsible for rising rape rates. With more women in college and at work now, the commission members reasoned in their report, women just have more opportunities to be raped.

Some academics have signed on to the consensus, too—and they are the "experts" who have enjoyed the highest profiles on the media circuit. On network news and talk shows, they have advised millions of women that feminism has condemned them to "a lesser life." Legal scholars have railed against "the equality trap." Sociologists have claimed that "feminist-inspired" legislative reforms have stripped women of special "protections." Economists have argued that well-paid working women have created "a less stable American family." And demographers, with greatest fanfare, have legitimated the prevailing wisdom with so-called neutral data on sex ratios and fertility trends; they say they actually have the numbers to prove that equality doesn't mix with marriage and motherhood. . . .

But what "equality" are all these authorities talking about?

If American women are so equal, why do they represent two-thirds of all poor adults? Why are more than 80 percent of full-time working women making less than $20,000 a year, nearly double the male rate? Why are they still far more likely than men to live in poor housing and receive no health insurance, and twice as likely to draw no pension? Why does the average working woman's salary still lag as far behind the average man's as it did twenty years ago? Why does the average female college graduate today earn less than a man with no more than a high school diploma (just as she did in the '50s)—and why does the average female high school graduate today earn less than a male high school dropout? Why do American women, in fact, face the worst gender-based pay gap in the developed world?

If women have "made it," then why are nearly 80 percent of working women still stuck in traditional "female" jobs—as secretaries, administrative "support" workers and salesclerks? And, conversely, why are they less than 8 percent of all federal and state judges, less than 6 percent of all law partners, and less than one half of 1 percent of top corporate managers? Why are there only three female state governors, two female U.S. senators, and two Fortune 500 chief executives? Why are only nineteen of the four thousand corporate officers and directors women—and why do more than half the boards of Fortune companies still lack even one female member?

If women "have it all," then why don't they have the most basic requirements to achieve equality in the work force? Unlike virtually all other industrialized nations, the U.S. government still has no family-leave and child care programs—and more than 99 percent of American private employers don't offer child care either. Though business leaders say they are aware of and deplore sex discrimination, corporate America has yet to make an honest effort toward eradicating it. In a 1990 national poll of chief executives at Fortune 1000 companies, more than 80 percent acknowledged that discrimination impedes female employees' progress—yet, less than 1 percent of these same companies regarded *remedying* sex discrimination as a goal that their personnel departments should pursue. In fact, when the companies' human resource officers were asked to rate their department's priorities, women's advancement ranked last.

If women are so "free," why are their reproductive freedoms in greater jeopardy today than a decade earlier? Why do women who want to postpone childbearing now have fewer options than ten years ago? The availability of different forms of contraception has declined, research for new birth control has virtually halted, new laws restricting abortion—or even information about abortion—for young and poor women have been passed, and the U.S. Supreme Court has shown little ardor in defending the right it granted in 1973.

Nor is women's struggle for equal education over; as a 1989 study found, three-fourths of all high schools still violate the federal law banning sex discrimination in education. In colleges, undergraduate women receive only 70 percent of the aid undergraduate men get in grants and work-study jobs—and women's sports programs receive a pittance compared with men's. A review of state equal-education laws in the late '80s found that only thirteen states had adopted the minimum provisions required by the federal Title IX law—and only seven states had anti-discrimination regulations that covered all education levels.

Nor do women enjoy equality in their own homes, where they still shoulder 70 percent of the household duties—and the only major change in the last fifteen years is that now middle-class men *think* they do more around the house. (In fact, a national poll finds the ranks of women saying their husbands share equally in child care shrunk to 31 percent in 1987 from 40 percent three years earlier.) Furthermore, in thirty states, it is still generally legal for husbands to rape their wives; and only ten states have laws mandating arrest for domestic violence—even though battering was the leading cause of injury of women in the late '80s. Women who have no other option but to flee find that isn't much of an alternative either. Federal funding for battered women's shelters has been withheld and one third of the 1 million battered women who seek emergency shelter each year can find none. Blows from men contributed far more to the rising numbers of "bag ladies" than the ill effects of feminism. In the '80s, almost half of all homeless women (the fastest growing segment of the homeless) were refugees of domestic violence.

The word may be that women have been "liberated," but women themselves seem to feel otherwise. Repeatedly in national surveys, majorities of women say they are still far from equality. Nearly 70 percent of women polled by the *New York Times* in 1989 said the movement for women's rights had only just begun. Most women in the 1990 Virginia Slims opinion poll agreed with the statement that conditions for their sex in American society had improved "a little, not a lot." In poll after poll in the decade, overwhelming majorities of women said they needed equal pay and equal job opportunities, they needed an Equal Rights Amendment, they needed the right to an abortion without government interference, they needed a federal law guaranteeing maternity leave, they needed decent child care services. They have none of these. So how exactly have we "won" the war for women's rights?

Seen against this background, the much ballyhooed claim that feminism is responsible for making women miserable becomes absurd—and irrelevant. . . . The afflictions ascribed to feminism are all myths. From "the man shortage" to "the infertility epidemic" to "female burnout" to "toxic day care," these so-called female crises have had their origins not in the actual conditions of women's lives but rather in a closed system that starts and ends in the media, popular culture, and advertising—an endless feedback loop that perpetuates and exaggerates its own false images of womanhood.

Women themselves don't single out the women's movement as the source of their misery. To the contrary, in national surveys 75 to 95 percent of women credit the feminist campaign with *improving* their lives, and a similar proportion say that the women's movement should keep pushing for change. Less than 8 percent think the women's movement might have actually made their lot worse.

What actually is troubling the American female population, then? If the many ponderers of the Woman Question really wanted to know, they might have asked their subjects. In public opinion surveys, women consistently rank their own *inequality*, at work and at home, among their most urgent concerns. Over and over, women complain to pollsters about a lack of economic, not marital, opportunities; they protest that working men, not working women, fail to spend time in the nursery and the kitchen. . . .

The truth is that the last decade has seen a powerful counterassault on women's rights, a backlash, an attempt to retract the handful of small and hard-won victories that the feminist movement did manage to win for women. This counterassault is largely insidious: in a kind of pop-culture version of the Big Lie, it stands the truth boldly on its head and proclaims that the very steps that have elevated women's position have actually led to their downfall.

The backlash is at once sophisticated and banal, deceptively "progressive" and proudly backward. It deploys both the "new" findings of "scientific research" and the dime-store moralism of yesteryear; it turns into media sound bites both the glib pronouncements of pop-psych trend-watchers and the frenzied rhetoric of New Right preachers. The backlash has succeeded in framing virtually the whole issue of women's rights in its own language. Just as Reaganism shifted political discourse far to the right and demonized liberalism, so the backlash convinced the public that women's "liberation" was the true contemporary American scourge—the source of an endless laundry list of personal, social, and economic problems. . . .

As the backlash has gathered force, it has cut off the few from the many—and the few women who have advanced seek to prove, as a social survival tactic, that they aren't so interested in advancement after all. Some of them parade their defection from the women's movement, while their working-class peers founder and cling to the splintered remains of the feminist cause. While a very few affluent and celebrity women who are showcased in news articles boast about having "found my niche as Mrs. Andy Mill" and going home to "bake bread," the many working-class women appeal for their economic rights—flocking to unions in record numbers, striking on their own for pay equity and establishing their own fledgling groups for working women's rights. . . .

Although the backlash is not an organized movement, that doesn't make it any less destructive. In fact, the lack of orchestration, the absence of a single string-puller, only makes it harder to see—and perhaps more effective. A backlash against women's rights succeeds to the degree that it appears *not* to be political, that it appears not to be a struggle at all. It is most powerful when it goes private, when it lodges inside a woman's mind and turns her vision inward, until she imagines the pressure is all in her head, until she begins to enforce the backlash, too—on herself. . . .

Backlash happens to be the title of a 1947 Hollywood movie in which a man frames his wife for a murder he's committed. The backlash against women's rights works in much the same way: its rhetoric charges feminists with all the crimes it perpetrates. The backlash line blames the women's movement for the "feminization of poverty"—while the backlash's own instigators in Washington pushed through the budget cuts that helped impoverish millions of women, fought pay equity proposals, and undermined equal opportunity laws. The backlash line claims the women's movement cares nothing for children's rights—while its own representatives in the capital and state legislatures have blocked one bill after another to improve child care, slashed billions of dollars in federal aid for children, and relaxed state licensing standards for day care centers. The backlash line accuses the women's movement of creating a generation of unhappy single and childless women—but its purveyors in the media are the ones guilty of making single and childless women feel like circus freaks.

To blame feminism for women's "lesser life" is to miss entirely the point of feminism, which is to win women a wider range of experience. Feminism remains a pretty simple concept, despite repeated—and enormously effective—efforts to dress it up in greasepaint and turn its proponents into gargoyles. As Rebecca West wrote sardonically in 1913, "I myself have never been able to find out precisely what feminism is: I only know that people call me a feminist whenever I express sentiments that differentiate me from a doormat."

The meaning of the word "feminist" has not really changed since it first appeared in a book review in the *Athenaeum* of April 27, 1895, describing a woman who "has in her the capacity of fighting her way back to independence." It is the basic proposition that, as Nora put it in Ibsen's *A Doll's House* a century ago, "Before everything else I'm a human being." It is the simply worded sign hoisted by a little girl in the 1970 Women's Strike for Equality: I AM NOT A BARBIE DOLL. Feminism asks the world to recognize at long last that women aren't decorative ornaments, worthy vessels, members of a "special-interest group." They are half (in fact, now more than half) of the national population, and just as deserving of rights and opportunities, just as capable of participating in the world's events, as the other half. Feminism's agenda is basic: It asks that women not be forced to "choose" between public justice and private happiness. It asks that women be free to define themselves—instead of having their identity defined for them, time and again, by their culture and their men.

The fact that these are still such incendiary notions should tell us that American women have a way to go before they enter the promised land of equality.

Sexuality

RACE, SEX, AIDS: *The Construction of "Other"*   **43**

Evelynn Hammonds

In March of this year when Richard Goldstein's article, "AIDS and Race—the Hidden Epidemic" appeared in the *Village Voice*, the following statement in the lead paragraph jumped out at me: "a black woman is thirteen times more likely than a white woman to contract AIDS, says the Centers for Disease Control; a Hispanic woman is at eleven times the risk. Ninety-one percent of infants with AIDS are non-white." My first reaction was shock. I was stunned to discover the extent and rate of spread of AIDS in the black community, especially given the lack of public mobilization either inside or outside the community. My second reaction was anger. AIDS is a disease that for the time being signals a death notice. I am angry because too many people have died and are going to die of this disease. The gay male community over these last several years has been transformed and mobilized to halt transmission and gay men (at least white gay men) with AIDS have been able to live and die with some dignity and self-esteem. People of color need the opportunity to estab-lish programs and interventions to provide education so that the spread of this disease in our communities can be halted, and to provide care so that people of color with AIDS will not live and die as pariahs.

My final reaction was despair. Of course I *knew* why information about AIDS and the black community had been buried—by both the black and white media. The white media, like the dominant power structure, have moved into

From: *Radical America* Vol. 20 No. 6 (1987), "Facing AIDS: A Special Issue": 28–36. Reprinted by permission.

their phase of "color-blindness" as a mark of progress. This ideology buries racism along with race. In the case of AIDS and race, the problem with "color-blindness" becomes clear. Race remains a reality in this society, including a reality about how perception is structured. On the one hand, race blindness means a failure to develop educational programs and materials that speak in the language of our communities and recognize the position of people of color in relation to the dominant institutions of society: medical, legal, etc. Additionally, we must ask why the vast disproportion of people of color in the AIDS statistics hasn't been seen as a remarkable fact, or as worthy of comment. By their silence, the white media fail to challenge the age-old American myth of blacks as carriers of disease, especially sexually transmitted disease. This association has quietly become incorporated into the image of AIDS.

The black community's relative silence about AIDS is in part also a response to this historical association of blacks, disease, and deviance in American society. Revealing that AIDS is prevalent in the black community raises the spectre of blacks being associated with two kinds of deviance: sexually transmitted disease and homosexuality.

As I began to make connections between AIDS and race I slowly began to pull together pieces of information and images of AIDS that I had seen in the media. Immediately I began to think about the forty year-long Tuskegee syphilis experiment on black men. I thought about the innuendoes in media reports about AIDS in Africa and Haiti that hinted at bizarre sexual practices among black people in those countries; I remembered how a black gay man had been portrayed as sexually irresponsible in a PBS documentary on AIDS; I thought about how little I had seen in the black press about AIDS and black gay men; I began to notice the thinly veiled hostility toward the increasing number of i.v. drug users with AIDS. Goldstein's article revealed dramatically, the deafening silence about who was now actually contracting and dying from AIDS—gay/bisexual black and Hispanic men (now about 50% of black and Hispanic men with AIDS); many black and Hispanic i.v. drug users; black and Hispanic women and black and Hispanic babies born to these women.

In this culture, how we think about disease determines who lives and who dies. The history of black people in this country is riddled with episodes displaying how concepts of sickness, disease, health, behavior and sexuality, and race have been entwined in the definition of normalcy and deviance. The power to define disease and normality makes AIDS a political issue.

The average black person on the street may not know the specifics of concepts of disease and race but our legacy as victims of this construction means that we know what it means to have a disease cast as the result of the immoral behavior of a group of people. Black people and other people of color notice, pay attention to what diseases are cast upon us and why. As the saying goes—"when white people get a cold, black people get pneumonia."

In this article I want to address the issues raised by the white media's silence on the connections between AIDS and race; the black media's silence on the connections between AIDS and sexuality/sexual politics, the failure of white gay men's AIDS organizations to reach the communities of people of color, and finally the implications for gay activists, progressives and feminists.

It is very important to outline the historical context in which the AIDS epidemic occurs in regards to race. The dominant media portrayals of AIDS and scientists' assertions about its origins and modes of transmission have everything to do with the history of racial groups and sexually transmitted diseases.

THE SOCIAL CONSTRUCTION OF DISEASE

> A standard feature of the vast majority of medical articles on the health of blacks was a sociomedical profile of a race whose members were rapidly becoming diseased, debilitated, and debauched and had only themselves to blame.[1]

One of the first things that white southern doctors noted about blacks imported from Africa as slaves, was that they seemed to respond differently than whites to certain diseases. Primarily they observed that some of the diseases that were epidemic in the south seemed to affect blacks less severely than whites—specifically, fevers (e.g. yellow fever). Since in the eighteenth and nineteenth centuries there was little agreement about the nature of various illnesses and the causes of many common diseases were unknown, physicians tended to attribute the differences they noted simply to race.

In the 19th century when challenges were made to the institution of slavery, white southern physicians were all too willing to provide medical evidence to justify slavery.

> They justified slavery and, after its abolition, second-class citizenship, by insisting that blacks were incapable of assuming any higher station in life. . . . Thus, medical discourses on the peculiarities of blacks offered, among other things, a pseudoscientific rationale for keeping blacks in their places.[2]

If as these physicians maintained, blacks were less susceptible to fevers than whites, then it seemed fitting that they and not whites should provide most of the labor in the hot, swampy lowlands where southern agriculture was centered. Southern physicians marshalled other "scientific" evidence, such as measurement of brain sizes and other body organs to prove that blacks constituted an inferior race. For many whites these arguments were persuasive because "objective" science offered validity to their personal "observations," prejudices and fears.

The history of sexually transmitted diseases, in particular syphilis, indicates the pervasiveness of racial/sexual stereotyping. The history of syphilis in America is complex, as Allan Brandt discloses in his book *No Magic Bullet*. According to Brandt, "venereal disease has historically been assumed to be the disease of the 'other'." Obviously the complicated interaction of sexuality and disease has deep implications for the current portrayal of AIDS.

Like AIDS, the prevailing nineteenth century view of syphilis was characterized early-on in moral terms—and when it became apparent that a high rate of syphilis occurred among blacks in the South, the morality issue heightened considerably. Diseases that are acquired through immoral behavior were considered in many parts of the culture as punishment from God, the wages of sin. Anyone with such a disease was stigmatized. A white person could avoid this sin by a change in behavior. But for blacks it was different. It was noted that one of the primary differences that separated the races was that blacks were more flagrant and loose in their sexual behavior—behaviors they could not control.

> Moreover, personal restraints on self-indulgence did not exist, physicians insisted, because the smaller brain of the Negro had failed to develop a center for inhibiting sexual behavior.[3]

Therefore blacks deserved to have syphilis, since they couldn't control their behavior and as the Tuskegee experiment carried that logic to extreme—blacks also deserved to die from syphilis.

> [B]lacks suffered from venereal diseases because they would not, or could not, refrain from sexual promiscuity. Social hygiene for whites rested on the assumption that attitudinal changes could produce behavioral changes. A single standard of high moral behavior could be produced by molding sexual attitudes through moral education. For blacks, however, a change in their very *nature* seemed to be required.[4]

If in the above quotation, you change blacks to homosexuals and whites to heterosexuals then the parallel to the media portrayal of people with AIDS is obvious.

The black community's response to the historical construction of sexually transmitted diseases as the result of bad, inherently uncontrollable behavior of blacks—is sexual conservatism. To avoid the stigma of being cast with diseases of the "other," the black media, as well as other institutions in the community, avoid public discussion of sexual behavior and other "deviant" behavior like drug use. The white media on the other hand is often quick to cast blacks and people of color as "other" either overtly or covertly.

BLACK COMMUNITY RESPONSE TO AIDS

... Black and Hispanic people make up 39% of all cases even though they account for only 17 percent of the adult population.[5] Eighty per cent of the pediatric cases are black and Hispanic. The average life expectancy after diagnosis of a white person with AIDS in the US is two years; of a person of color, nineteen weeks.[6]

When I examined the few articles that have been written about AIDS in the national black press, several themes emerged. Almost all the articles I saw tried to indicate that the black people are at risk while simultaneously trying to avoid any implication that AIDS is a "black" disease. The black media has underemphasized, though recognized, that there are significant socioeconomic cofactors in terms of the impact of AIDS in the black community. The high rate of drug use and abuse in the black community is in part a result of many other social factors—high unemployment, poor schools, inadequate housing and limited access to health care, all factors in the spread of AIDS. These affect specifically the fact that people of color with AIDS are diagnosed at more advanced stages of the disease and are dying faster. The national black media have so far also failed to deal with any larger public policy issue that the AIDS crisis will precipitate for the community; and most importantly homosexuality and bisexuality were dealt with in a very conservative and problematic fashion.

TESTING

In terms of testing *Ebony* encourages more opportunity for people to be tested anonymously; *Essence* recommends testing for women thinking of getting pregnant. Both articles mention that exposure of test results could result in discrimination in housing and employment but neither publication discusses the issue at any length. There is no mention of testing that is going on in the military and how those results are being used nor is there mention of testing in prison. It is clear from the sketchy discussion of testing that the political issues around testing are not being faced.

SEXUALITY

The most disappointing aspect of these articles is that by focusing on individual behavior as the cause of AIDS and by setting up bisexuals, homosexuals, and drug users as "other" in the black community, and as "bad," the
national black media falls into the trap of reproducing exactly how white
society has defined the issue. But unlike the situation for whites, what happens to these groups within the black community will affect the community
as a whole. Repressive practices around AIDS in prisons will affect all black
men in prison with or without AIDS and their families outside and any other
black person facing the criminal justice system; the identification of significant numbers of people of color in the military with AIDS will affect all
people of color in the military. Quarantine, suspension of civil liberties for
drug users in the black community with AIDS will affect everyone in the
community. Healthcare and housing access will be restricted for all of us. If
people with AIDS are set-off as "bad" or "other"—no change in individual
behavior in relation to them will save any of us. There can be no "us" or
"them" in our communities.

The *Ebony* article entitled: "The Truth about AIDS: Dread Disease is
Spreading Rapidly through Heterosexual Population," while highlighting the
increase of AIDS among heterosexuals in the black community, makes several
comments about black homosexuals. The author notes that there is generally
a negative attitude towards homosexuals in the community and quotes several
physicians who emphasize that the reticence on this issue is a hindrance to
AIDS education efforts in the community. It does not emphasize that, because
of this "reticence," only now as AIDS is being recognized as striking heterosexuals, is it beginning to be talked about in the black community.

> One of the greatest problems in the black community, other than igno-
> rance about the disease, is the large number of black men who engage in sex
> acts with other men but who don't consider themselves homosexuals.[7]

The point is then that since AIDS was initially characterized as a "gay disease"
and many black men don't consider themselves gay in spite of their sexual
practices, the black community did not acknowledge the presence of AIDS.

The association of AIDS with "bad" behavior is prominent in this article.
Homosexuals and drug users are described as a "physiologically and economically depressed subgroup of the black community."[8]

The message is that to deal with this disease the individual behavior of a
deviant subgroup must be changed. Additionally, the recommendation to
heterosexuals is to "not have sex" with bisexuals and drug users. There are no

recommendations about how the community can find a way to deal with the silence around the issues of homosexuality/bisexuality, sexual practices in general and drug use. The article fails to say what the implications of the sexual practices of black men are for the community.

The *Essence* article, entitled *Nobody's Safe*, avoids the issue as well.[9] The authors describe a scenario of a 38 year-old middle-class professional woman who is suddenly found to have AIDS. Her husband had died two years earlier due to a rare form of pneumonia. After testing positive for AIDS she is told by one of her husband's relatives that he had been bisexual. The text following this scenario goes on to describe how most women contract AIDS; it gives a general sketch of the origins of the disease and discusses the latency period and defines asymptomatic carriers of the virus. There is no mention of bisexuality or homosexuality. The implication is again—just don't have sex with those people if you want to avoid AIDS. It avoids discussion of the prevalence of bisexuality among black men, and consequently the way that AIDS will ultimately change sexual relationships in the black community.

EDUCATION EFFORTS AND SEXUAL BEHAVIOR IN THE AGE OF AIDS

The implications of this silence on sexuality are obvious when education efforts for black people are being discussed. But there is more at stake here than simply an acknowledgment. Both articles note the desperate need for education and material that speaks directly to the black community, so that black people can recognize that they too are at risk. But the other part of the message one gets from these articles is that black children must be taught the "facts about sex, AIDS and drug use and abuse" not about sexuality. *Essence* reports that a new group has formed in Atlanta which sponsors "Play Safe Parties" to teach women how to practice safer sex. In effect AIDS is described in terms of individual behavior. There are no specific guidelines about what safer sex is—that it is about a community response as much as it is about individual behavior; instead, there is a push for people to return to monogamous, traditional relationships without analysis as to what that means for heterosexuals in a community where women far outnumber men in the population; where traditional patriarchal relationships are not easily accepted anymore. What about discussions about "safer sex" for men? What about sexual pleasure for women and who negotiates it? These articles do not recognize that you can't simply separate sex from AIDS, nor can you respond to it by a call to a return to traditional values while not exploring the implications of that move.

What white gay men have been able to do in the face of the AIDS crisis is to use the connection between sex and community. They succeeded in validating and mobilizing the gay community to the deadly implications of AIDS while preserving their right to define sexual expression and therefore challenge the conception of homosexuality as bad. For the black community, however, "the fear of a racial backlash against minorities as they become more identified with AIDS is one of the reasons the black community has been slow to address this issue, to put it on our agenda."[10] What's at issue here is how to break the dominant culture's association of blacks with disease and immorality. The response so far has been to appeal to blacks to demonstrate our "traditions of respectability," e.g., to embrace monogamy in the face of the dominant culture's association of black people with promiscuity, and to deny the existence of homosexuality in the black community. But such a response means that the racist ideology that gives white culture the power to define morality and immorality remains intact. Black gays are rendered invisible and efforts at educating the community and providing care for people with AIDS are hampered by the need to preserve the notion that gaining respectability involves gaining authority.

Sexuality and sexual politics never came to the forefront of the civil rights agenda because of the reaction of the black community to the way in which race and sex had historically been used against the black community. What the AIDS epidemic raises is that the black political agenda has not been able to dethrone the power of that ideology.

THE MAINSTREAM (WHITE) PRESS

In general the mainstream media has been silent on the rise of AIDS in the black and Hispanic communities. Until very recently, with the exception of a few special reports, such as a quite excellent one on the PBS' MacNeil-Lehrer Report, most media reports on AIDS continue to speak of the disease without mention of its effects on people of color. In recent months specific attention has been paid to the "new" phenomenon of heterosexuals with AIDS or "heterosexual AIDS." This terminology is used without the slightest mention that among Haitians and extensively in Africa, AIDS was never a disease confined to homosexuals.

The assumption in reports about the spread of AIDS to heterosexuals is that these heterosexuals are white—read that as white, middle-class, non-drug-using, sexually-active people. The facts are that there are very few cases of AIDS among this group. As many as 90 percent of the cases of AIDS among heterosexuals are black and Hispanic. In many media reports blacks and

Hispanics with AIDS are lumped in the i.v. drug users group. What the media has picked up on is that heterosexual transmission in the US now endangers middle class whites.

A good example of the mainstream media approach is an article by Kate Leishman in the February, 1987 issue of *Atlantic Monthly*. She writes that most Americans, even liberals, have the attitude that AIDS is the result of immoral behavior. Leishman lists the statistics on heterosexual transmission of AIDS at the beginning of her article. Fifteen pages later the following information appears:

> In the case of sexually active gay men [AIDS] is a tragedy—as it is for poor black and Hispanic youths, among whom there is a nationwide epidemic of venereal disease, which is a certain cofactor in facilitating transmission of HIV. This combination with the pervasive use of drugs among blacks and Hispanics ensures that the epidemic will hit them hardest next.[11]

Her first explicit mention of people of color describes them as a group that uses drugs extensively, and as also riddled with venereal disease (a fact she does not support with any data). The image is one of the "unregenerate young street tough" that causes all the trouble in our cities, in short the conventional racist stereotype of black and Hispanic youth displayed in the press almost every day. Her use of the word tragedy because of the risk to blacks, Hispanics and gays is gratuitous at best. The main focus of the article is the risk of AIDS to white heterosexuals and the need for them to face their fears of AIDS so they can effectively change their behavior.

In a passage reminiscent of 19th century physicians' moral advice she notes the problems associated with changing people's behavior and promoting safe sex, and wonders if one can draw any lessons for heterosexual behavior from the gay male experience.

> Many people believe that the intensity or quality of homosexual drives is unique, while others argue that the ability to control sexual impulses varies extraordinarily within groups of any sexual preference.[12]

What I find striking in this passage is that there is still debate over whether certain "groups" of people have the same ability to exercise control over their sexual behavior and drives as "normal" white heterosexuals do. The passage also suggests that white heterosexuals are still the only group who have the strength, the moral fortitude, the inherent ability if educated, to control their sexual and other behavior. After all, is this a disease about behavior and not viruses, right? Leishman doesn't interview any blacks or Hispanics about their fears of AIDS, or how they want to deal with it with respect to sexual practice or other behavior.

Two months later in May several letters to the editors of *Atlantic Monthly* appeared in response to Leishman's article. In particular one reader observed her omission of statistics about the risk of AIDS to blacks and Hispanics. She responded in a fairly defensive manner:

> My article and many others have commented on the high risk of expo-
> sure to AIDS among blacks and Hispanics. Mr. Patrick's observations that
> blacks and Hispanics already account for ninety per cent of the case load
> seems oddly to suggest that AIDS is on its way to becoming a disease of mi-
> norities. But the Centers for Disease Control has stressed that the overrepre-
> sentation of blacks and Hispanics in AIDS statistics is related not to race per
> se but to underlying risk factors.[13]

The risk factor she mentions is intravenous drug use. Leishman fails to deal with the "overrepresentation" of blacks and Hispanics in AIDS statistics. To mention our higher risk only implies that AIDS is a disease of minorities if you believe minorities are inherently different or behave differently in the face of the disease or if you believe that the disease will be confined to the minority community.

So pervasive is the association of race and i.v. drug use, that the fact that a majority of black and Hispanic men who have AIDS are gay or bisexual, and *non* i.v. drug users, has remained buried in statistics.[14] In the face of the statistics, *The New York Times* continues to identify i.v. drug use as the distinguishing mode of transmission among black and Hispanic men, by focusing not on the percentage of black and Hispanic AIDS cases that are drug related, but on the percentage of drug related AIDS cases that are black or Hispanic, which is 94%. This framework, besides blocking information that the black and Hispanic communities need, also functions to keep the white community's image "clean."

CONCLUSION

As this article goes to press, media coverage of the extent of AIDS in the black and Hispanic communities is increasing daily. These latest articles are cover-ing the efforts in the black and Hispanic communities both to raise conscious-ness in these communities with respect to AIDS and to increase government funding to support culturally specific educational programs. Within the black community, the traditional source of leadership, black ministers, are now publicly expressing the reasons for their previous reluctance to speak out about AIDS. The reasons expressed tend to fall into the areas I have tried to discuss in this article, as indicated by the following comments that recently appeared in the *Boston Globe:*

> Although some black ministers described gays as the children of God and AIDS as just another virus, many more talked about homosexuality as sinful, including some who referred to AIDS as a God-sent plague to punish the sexually deviant.[15]
>
> There's a lot of fear of stigmatization when you stand up. . . . How does this label your church or the people who go to your church? said Rev. Bruce Wall, assistant pastor of Twelfth Baptist Church in Roxbury. Rev. Wall said ministers may also fear that an activist role on AIDS could prompt another question: 'Maybe that pastor is gay.'[16]

The arguments I have made as to the background of these kinds of comments continue to come out in the public discourse on AIDS and race in the national media. As the public discussion and press coverage has increased, one shift is apparent. The media is now focussing on why the black and Hispanic communities have not responded to AIDS before as a "problem" specific to these communities, while there is no acknowledgment that part of the problem is the way the media, the CDC, and the Public Health Service prevented race-specific information about AIDS from being widely disseminated. Or, to say it differently, there is no recognition of how the medical and media construction of AIDS as a "gay disease," or a disease of Haitians has affected the black and Hispanic communities.

Finally, as the black and Hispanic communities mobilize against AIDS, coalitions with established gay groups will be critical. To date, some in the black community have noted the lack of culturally specific educational material produced by these groups. Some gay groups are responding to that criticism. For progressives, feminists and gay activists, the AIDS crisis represents a crucial time when the work we have done on sexuality and sexual politics will be most needed to frame the fight against AIDS in political terms that move the politics of sexuality out of the background and challenge the repressive policies and morality that threaten not only the people with this disease but all of us.

FOOTNOTES

1. James H. Jones, *Bad Blood: The Tuskegee Syphilis Experiment* (New York: Free Press, 1981), p. 21.

2. *Ibid.*, p. 17.

3. *Ibid.*, p. 23.

4. *Ibid.*, p. 48.

5. "High AIDS Rate Spurring Efforts for Minorities," *New York Times*, Sunday, August 2, 1987.

6. *Mother Jones*, Vol. 12, May 1987.

7. *Ebony*, April, 1987, p. 128, quoting a Los Angeles AIDS expert.

8. *Ibid.*, p. 130.

9. *Essence*, June 1987.

10. John Jacob, President, National Urban League, *New York Times*, Sunday, August 2, 1987.

11. *Atlantic Monthly*, February 1987, p. 54.

12. *Ibid.*, p. 40.

13. *Atlantic Monthly*, May 1987, p. 13.

14. *New York Times*, Sunday, August 2, 1987.

15. *Boston Globe*, Sunday, August 9, 1987, p. 1.

16. *Ibid.*, p. 12.

HOMOPHOBIA: *Why Bring It Up?*   **44**

Barbara Smith

In 1977, the Combahee River Collective, a Black feminist organization in Boston of which I was a member, wrote:

> The most general statement of our politics at the present time would be that we are actively committed to struggling against racial, sexual, heterosexual, and class oppression and see as our particular task the development of inte-

From: Barbara Smith, "Homophobia: Why Bring It Up?" *Interracial Books for Children Bulletin* 14(1983): 112–113. Reprinted by permission of The Council on Interracial Books for Children, P.O. Box 1263, New York, NY 10023.

grated analysis and practice based upon the fact that the major systems of
oppression are interlocking. . . . We . . .often find it difficult to separate race
from class from sex oppression because in our lives they are most often experienced simultaneously.*

Despite the logic and clarity of Third World women's analysis of the
simultaneity of oppression, people of all colors, progressive ones included,
seem peculiarly reluctant to grasp these basic truths, especially when it comes
to incorporating an active resistance to homophobia in their every day lives.
Homophobia is usually the last oppression to be mentioned, the last to be taken
seriously, the last to go. But it is extremely serious, sometimes to the point of
being fatal.

Consider that on the night of September 29, 1982, 20–30 New York City
policemen rushed without warning into Blues, a Times Square bar. They
harassed and severely beat the patrons, vandalized the premises, emptied the
cash register and left without making a single arrest. What motivated such
brutal behavior? The answer is simple. The cops were inspired by three
cherished tenets of our society: racism, classism and homophobia: the bar's
clientele is Black, working class and gay. As the police cracked heads, they
yelled racist and homophobic epithets familiar to every school child. The
attackers' hatred of both the queer and the colored, far from making them
exceptional, put them squarely on the mainstream. If their actions were more
extreme than most, their attitudes certainly were not.

The Blues bar happens to be across the street from the offices of *The New
York Times*. The white, upper middle-class, presumably heterosexual staff of
the nation's premier newspaper, regularly calls in complaints about the bar to
the police. Not surprisingly, none of the New York daily papers, including the
Times, bothered to report the incident. A coalition of Third World and white
lesbians and gay men organized a large protest demonstration soon after the
attack occurred. Both moderate and militant civil rights and anti-racist organizations were notably absent, and they have yet to express public outrage about
a verifiable incident of police brutality, undoubtedly because the Black people
involved were not straight.

INTERTWINING "ISMS"

What happened at Blues perfectly illustrates the ways in which the major
"isms" *including* homophobia are intimately and violently intertwined. As a

* The Combahee River Collective, "A Black Feminist Statement" in *All the Women Are White, All
the Blacks Are Men, But Some of Us Are Brave: Black Women's Studies*, pp. 13, 16 (The Feminist
Press), 1982.

Black woman, a lesbian, a feminist and an activist, I have little difficulty seeing how the systems of oppression interconnect, if for no other reason than that their meshings so frequently affect my life. During the 70s and 80s political lesbians of color have often been the most astute about the necessity for developing understandings of the connections between oppressions. They have also opposed the building of hierarchies and challenged the "easy way out" of choosing a "primary oppression"and downplaying those messy inconsistencies that occur whenever race, class and sexual identity actually mix. Ironically, for the forces on the right, hating lesbians and gay men, people of color, Jews and women go hand in hand. *They* make connections between oppressions in the most negative ways with horrifying results. Supposedly progressive people, on the other hand, who oppose oppression on every other level, balk at acknowledging the societally sanctioned abuse of lesbians and gay men as a serious problem. Their tacit attitude is "Homophobia, why bring it up?"

There are numerous reasons for otherwise sensitive people's reluctance to confront homophobia in themselves and others. A major one is that people are generally threatened about issues of sexuality, and for some the mere existence of homosexuality calls their sexuality/heterosexuality into question. Unlike many other oppressed groups, homosexuals are not a group whose identity is clear from birth. Through the process of coming out, a person might indeed acquire this identity at any point in life. One way to protect one's heterosexual credentials and privilege is to put down lesbians and gay men at every turn, to make as large a gulf as possible between "we" and "they."

There are several misconceptions and attitudes which I find particularly destructive because of the way they work to isolate the concerns of lesbians and gay men:

1. **Lesbian and gay male oppression is not as serious as other oppressions. It is not a political matter, but a private concern.** The life-destroying impact of lost jobs, children, friendships and family; the demoralizing toll of living in constant fear of being discovered by the wrong person which pervades all lesbians and gay men's lives whether closeted or out; and the actual physical violence and deaths that gay men and lesbians suffer at the hands of homophobes can be, if one subscribes to this myth, completely ignored.

2. **"Gay" means gay white men with large discretionary incomes, period.** Perceiving gay people in this way allows one to ignore that some of us are women *and* people of color *and* working class *and* poor *and* disabled *and* old. Thinking narrowly of gay people as white, middle-class, and male, which is just what the establishment media want people to think, undermines consciousness of how identities and issues overlap. It is essential, however, in

making connections between homophobia and other oppressions, not to fall prey to the distorted reasoning that the justification for taking homophobia seriously is that it affects some groups who are "verifiably" oppressed, for example, people of color, women or disabled people. Homophobia is in and of itself a verifiable oppression and in a heterosexist system, all nonheterosexuals are viewed as "deviants" and are oppressed.

3. **Homosexuality is a white problem or even a "white disease."** This attitude is much too prevalent among people of color. Individuals who are militantly opposed to racism in all its forms still find lesbianism and male homosexuality something to snicker about or, worse, to despise. Homophobic people of color are oppressive not just to white people, but to members of their own groups—at least ten percent of their own groups.

4. **Expressions of homophobia are legitimate and acceptable in contexts where other kinds of verbalized bigotry would be prohibited.** Put-downs and jokes about "dykes" and "faggots" can be made without the slightest criticism in circles where "nigger" and "chink" jokes, for instance, would bring instant censure or even ostracism. One night of television viewing indicates how very acceptable public expressions of homophobia are.

How can such deeply entrenched attitudes and behavior be confronted and changed? Certainly gay and lesbian/feminist activism has made significant inroads since the late 60s, both in the public sphere and upon the awareness of individuals. These movements have served a highly educational function, but they have not had nearly enough impact upon the educational system itself. Curriculum that focuses in a positive way upon the issues of sexual identity, sexuality and sexism is still rare, particularly in primary and secondary grades. Yet schools are virtual cauldrons of homophobic sentiment, as witnessed by everything from the graffiti in the bathrooms and the put-downs yelled on the playground, to the heterosexual bias of most texts and the firing of teachers on no other basis than that they are not heterosexual.

In the current political climate schools are constantly under hostile scrutiny from well-organized conservative forces. More than a little courage is required to challenge students' negative attitudes about what it means to be homosexual, female, Third World, etc. but these attitudes *must* be challenged if pervasive taken-for-granted homophobia is ever to cease. I have found both in teaching and in speaking to a wide variety of audiences that making connections between oppressions is an excellent way to introduce the subjects of lesbian and gay male identity and homophobia, because it offers people a frame of reference to build upon. This is especially true if efforts have already been made in the classroom to teach about racism and sexism. It is factually inaccurate and strategically mistaken to present gay materials as if all gay people were white and male. Fortunately, there is an increasing body of work available, usually written by Third World feminists, that provides an integrated approach to the intersection of/and multiplicity of identities and issues.

Perhaps some readers are still wondering, "Homophobia, why bring it up?" One reason to bring it up is that at least ten percent of your students will be or already are lesbians and gay males. Ten percent of your colleagues are as well. Homophobia may well be the last oppression to go, but it will go. It will go a lot faster if people who are opposed to every form of subjugation work in coalition and make it happen.

THE BEAUTY MYTH

45

Naomi Wolf

... Beauty pornography looks like this: The perfected woman lies prone, pressing down her pelvis. Her back arches, her mouth is open, her eyes shut, her nipples erect; there is a fine spray of moisture over her golden skin. The position is female superior; the stage of arousal, the plateau phase just preceding orgasm. On the next page, a version of her, mouth open, eyes shut, is about to tongue the pink tip of a lipstick cylinder. On the page after, another version kneels in the sand on all fours, her buttocks in the air, her face pressed into a towel, mouth open, eyes shut. The reader is looking through an ordinary women's magazine. In an ad for Reebok shoes, the woman sees a naked female torso, eyes averted. In an ad for Lily of France lingerie, she sees a naked female torso, eyes shut; for Opium perfume, a naked woman, back and buttocks bare, falls facedown from the edge of a bed; for Triton showers, a naked woman, back arched, flings her arms upward; for Jogbra sports bras, a naked female torso is cut off at the neck. In these images, where the face is visible, it is expressionless in a rictus of ecstasy. The reader understands from them that she will have to look like that if she wants to feel like that.

Beauty sadomasochism is different: In an ad for Obsession perfume, a well-muscled man drapes the naked, lifeless body of a woman over his shoulder. In an ad for Hermès perfume, a blond woman trussed in black leather is hanging upside down, screaming, her wrists looped in chains, mouth bound. In an ad for Fuji cassettes, a female robot with a playmate's body, but made of steel, floats with her genitals exposed, her ankles bolted and her face a steel mask with slits for the eyes and mouth. In an ad for Erno Laszlo skin care products, a woman sits up and begs, her wrists clasped together with a leather leash that is also tied to her dog, who is sitting up in the same posture and begging. In an American ad for Newport cigarettes, two men tackle one woman and pull another by the hair; both women are screaming. In another Newport ad, a man forces a woman's head down to get her distended mouth around a length of spurting hose gripped in his fist; her eyes are terrified. In an ad for Saab automobiles, a shot up a fashion model's thighs is captioned, "Don't worry. It's ugly underneath." In a fashion layout in *The Observer* (London), five men in black menace a model, whose face is in shock, with scissors and hot iron rods. In *Tatler* and *Harper's and Queen*, "designer rape sequences (women beaten, bound and abducted, but immaculately turned out and artistically photographed)" appear. In Chris von Wangenheim's *Vogue* layout, Doberman pinschers attack a model. Geoffrey Beene's metallic sandals are displayed against a background of S and M accessories. The woman learns from these images that no matter how assertive she may be in the world, her private submission to control is what makes her desirable. . . .

Sexual "explicitness" is not the issue. We could use a lot more of that, if explicit meant honest and revealing; if there were a full spectrum of erotic images of uncoerced real women and real men in contexts of sexual trust, beauty pornography could theoretically hurt no one. Defenders of pornography base their position on the idea of freedom of speech, casting pornographic imagery as language. Using their own argument, something striking emerges about the representation of women's bodies: The representation is heavily censored. Because we see many versions of the naked Iron Maiden, we are asked to believe that our culture promotes the display of female sexuality. It actually shows almost none. It censors representations of women's bodies, so that only the official versions are visible. Rather than seeing images *of* female desire or that cater *to* female desire, we see mock-ups of living mannequins, made to contort and grimace, immobilized and uncomfortable under hot lights, professional set-pieces that reveal little about female sexuality. In the United States and Great Britain, which have no tradition of public nakedness, women rarely—and almost never outside a competitive context—see what other *women* look like naked; we see only identical humanoid products based loosely on women's bodies. . . .

. . . Leaving aside the issue of what violent sexual imagery does, it is still apparent that there is an officially enforced double standard for men's and women's nakedness in mainstream culture that bolsters power inequities.

The practice of displaying breasts, for example, in contexts in which the display of penises would be unthinkable, is portrayed as trivial because breasts are not "as naked" as penises or vaginas; and the idea of half exposing men in a similar way is moot because men don't have body parts comparable to breasts. But if we think about how women's genitals are physically concealed, unlike men's, and how women's breasts are physically exposed, unlike men's, it can be seen differently: women's breasts, then correspond to men's penises as the vulnerable "sexual flower" on the body, so that to display the former and conceal the latter makes women's bodies vulnerable while men's are protected. Cross-culturally, unequal nakedness almost always expresses power relations: In modern jails, male prisoners are stripped in front of clothed prison guards; in the antebellum South, young black male slaves were naked while serving the clothed white masters at table. To live in a culture in which women are routinely naked where men aren't is to learn inequality in little ways all day long. So even if we agree that sexual imagery is in fact a language, it is clearly one that is already heavily edited to protect men's sexual—and hence social—confidence while undermining that of women. . . .

Why this flood of images now? They do not arise simply as a market response to deep-seated, innate desires already in place. They arise also—and primarily—to set a sexual agenda and to *create* their versions of desire. The way to instill social values, writes historian Susan G. Cole, is to eroticize them. Images that turn women into objects or eroticize the degradation of women have arisen to counterbalance women's recent self-assertion. They are welcome and necessary because the sexes have come too close for the comfort of the powerful; they act to keep men and women apart, wherever the restraints of religion, law, and economics have grown too weak to continue their work of sustaining the sex war.

Heterosexual love, before the women's movement, was undermined by women's economic dependence on men. Love freely given between equals is the child of the women's movement, and a very recent historical possibility, and as such very fragile. It is also the enemy of some of the most powerful interests of this society.

If women and men in great numbers were to form bonds that were equal, nonviolent, and sexual, honoring the female principle no less or more than the male, the result would be more radical than the establishment's worst nightmares of homosexual "conversions." A mass heterosexual deviation into tenderness and mutual respect would mean real trouble for the status quo since heterosexuals are the most powerful sexual majority. The power structure would face a massive shift of allegiances: From each relationship might emerge a doubled commitment to transform society into one based publicly on what have traditionally been women's values, demonstrating all too well the appeal for both sexes of a world rescued from male dominance. The good news would get out on the street: Free women have more fun; worse, so do free men.

Male-dominated institutions—particularly corporate interests—recognize the dangers posed to them by love's escape. Women who love themselves are threatening; but men who love real women, more so. Women who have broken out of gender roles have proved manageable: Those few with power are being retrained as men. But with the apparition of numbers of men moving into passionate, sexual love of real women, serious money and authority could defect to join forces with the opposition. Such love would be a political upheaval more radical than the Russian Revolution and more destabilizing to the balance of world power than the end of the nuclear age. It would be the downfall of civilization as we know it—that is, of male dominance; and for heterosexual love, the beginning of the beginning.

Images that flatten sex into "beauty," and flatten the beauty into something inhuman, or subject her to eroticized torment, are politically and socioeconomically welcome, subverting female sexual pride and ensuring that men and women are unlikely to form common cause against the social order that feeds on their mutual antagonism, their separate versions of loneliness.

Barbara Ehrenreich, Elizabeth Hess, and Gloria Jacobs, in *Re-Making Love*, point out that the new market of sexual products demands quick-turnover sexual consumerism. That point applies beyond the sexual accessories market to the entire economy of consumption. The last thing the consumer index wants men and women to do is to figure out how to love one another: The $1.5-trillion retail-sales industry depends on sexual estrangement between men and women, and is fueled by sexual dissatisfaction. Ads do not sell sex—that would be counterproductive, if it meant that heterosexual women and men turned to one another and were gratified. What they sell is sexual discontent.

Though the survival of the planet depends on women's values balancing men's, consumer culture depends on maintaining a broken line of communication between the sexes and promoting matching sexual insecurities. Harley-Davidsons and Cuisinarts stand in for maleness and femaleness. But sexual satisfaction eases the stranglehold of materialism, since status symbols no longer look sexual, but irrelevant. Product lust weakens where emotional and sexual lust intensifies. The price we pay for artificially buoying up this market is our heart's desire. The beauty myth keeps a gap of fantasy between men and women. That gap is made with mirrors; no law of nature supports it. It keeps us spending vast sums of money and looking distractedly around us, but its smoke and reflection interfere with our freedom to be sexually ourselves.

Consumer culture is best supported by markets made up of sexual clones, men who want objects and women who want to be objects, and the object desired ever-changing, disposable, and dictated by the market. The beautiful object of consumer pornography has a built-in obsolescence, to ensure that as few men as possible will form a bond with one woman for years or for a lifetime, and to ensure that women's dissatisfaction with themselves will grow rather than diminish over time. Emotionally unstable relationships, high divorce rates, and a large population cast out into the sexual marketplace are good for business in a consumer economy. Beauty pornography is intent on making modern sex brutal and boring and only as deep as a mirror's mercury, antierotic for both men and women.

But even more powerful interests than the consumer index depend on heterosexual estrangement and are threatened by heterosexual accord. The military is supported by nearly one third of the United States government's budget; militarism depends on men choosing the bond with one another over the bond with women and children. Men who loved women would shift loyalties back to the family and community from which becoming a man is one long exile. Serious lovers and fathers would be unwilling to believe the standard propaganda of militarism: that their wives and children would benefit from their heroic death. Mothers don't fear mothers; if men's love for women and for their own children led them to define themselves first as fathers and lovers, the propaganda of war would fall on deaf ears: The enemy would be a father and partner too. This percentage of the economy is at risk from heterosexual love. Peace and trust between men and women who are lovers would be as bad for the consumer economy and the power structure as peace on earth for the military-industrial complex.

Heterosexual love threatens to lead to political change: An erotic life based on nonviolent mutuality rather than domination and pain teaches firsthand its appeal beyond the bedroom. A consequence of female self-love is that the woman grows convinced of social worth. Her love for her body will be unqualified, which is the basis of female identification. If a woman loves her own body, she doesn't grudge what other women do with theirs; if she loves femaleness, she champions its rights. It's true what they say about women: Women *are* insatiable. We *are* greedy. Our appetites do need to be controlled if things are to stay in place. If the world were ours too, if we believed we could get away with it, we *would* ask for more love, more sex, more money, more commitment to children, more food, more care. These sexual, emotional, and physical demands *would* begin to extend to social demands: payment for care of the elderly, parental leave, child-care, etc. The force of female desire would be so great that society would truly have to reckon with what women want, in bed and in the world.

The economy also depends on a male work structure that denies the family. Men police one another's sexuality, forbidding each other to put sexual love and family at the center of their lives; women define themselves as successful according to their ability to sustain sexually loving relationships. If too many men and women formed common cause, that definition of success would make its appeal to men, liberating them from the echoing wind-tunnel of competitive masculinity. Beauty pornography is useful in preventing that eventuality: When aimed at men, its effect is to keep them from finding peace in sexual love. The fleeting chimera of the airbrushed centerfold, always receding before him, keeps the man destabilized in pursuit, unable to focus on the beauty of the woman—known, marked, lined, familiar—who hands him the paper every morning.

The myth freezes the sexual revolution to bring us full circle, evading sexual love with its expensive economic price tag. The nineteenth century constrained heterosexuality in arranged marriages; today's urban overachievers sign over their sexual fate to dating services, and their libido to work: One survey found that many yuppie couples share mutual impotence. The last century kept men and women apart in rigid gender stereotypes, as they are now estranged through rigid physical stereotypes. In the Victorian marriage market, men judged and chose; in the stakes of the beauty market, men judge and choose. It is hard to love a jailer, women knew when they had no legal rights. But it is not much easier to love a judge. Beauty pornography is a war-keeping force to stabilize the institutions of a society under threat from an outbreak of heterosexual love. . . .

When they discuss [their bodies], women lean forward, their voices lower. They tell their terrible secret. It's my breasts, they say. My hips. It's my thighs. I hate my stomach. This is not aesthetic distaste, but deep sexual shame. The parts of the body vary. But what each woman who describes it shares is the conviction that *that* is what the pornography of beauty most fetishizes. Breasts, thighs, buttocks, bellies; the most sexually central parts of women, whose "ugliness" therefore becomes an obsession. Those are the parts most often battered by abusive men. The parts that sex murderers most often mutilate. The parts most often defiled by violent pornography. The parts that beauty surgeons most often cut open. The parts that bear and nurse children and feel sexual. A misogynist culture has succeeded in making women hate what misogynists hate. . . .

CULTURAL AND HISTORICAL INFLUENCES ON SEXUALITY IN HISPANIC/LATIN WOMEN:

Implications for Psychotherapy

46

Oliva M. Espín

CONTEMPORARY SEXUALITY AND THE HISPANIC WOMAN

If the role of women is currently beset with contradictions in the mainstream of American society,[1] this is probably still more true for women in Hispanic groups. The honor of Latin families is strongly tied to the sexual purity of women. And the concept of honor and dignity is one of the essential distinctive marks of Hispanic culture. For example, classical Hispanic literature gives us a clue to the importance attributed to honor and to female sexual purity in the culture. La Celestina, the protagonist of an early Spanish medieval novel, illustrates the value attached to virginity and its preservation. Celestina was an old woman who earned her living in two ways: by putting young men in touch with young maidens so they could have the sexual contact that parents would never allow, and by "sewing up" ex-virgins, so that they would be considered virgins at marriage. Celestina thus made her living out of making and unmaking virgins. The fact that she ends by being punished with death further emphasizes the gravity of what she does. In the words of a famous Spanish playwright of the seventeenth century, "al Rey la hacienda y la vida se han de dar, mas no el honor; porque el honor es patrimonio del alma y el alma solo es de Dios."[2] This quotation translates literally, "to the king you give money and life, but not your honor, because honor is part of the soul, and your soul belongs only to God."

Different penalties and sanctions for the violation of cultural norms related to female sexuality are very much associated with social class. The upper classes or those seeking an improved social status tend to be more rigid

---

Abridged from: Carole Vance (ed.), *Pleasure and Danger* (Boston: Routledge & Kegan Paul, 1984), pp. 149–164. Reprinted by permission.

about sexuality. This of course is related to the transmission of property. In the upper classes, a man needs to know that his children are in fact his before they inherit his property. The only guarantee of his paternity is that his wife does not have sexual contact with any other man. Virginity is tremendously important in that context. However, even when property is not an issue, the only thing left to a family may be the honor of its women and as such it may be guarded jealously by both males and females. Although Hispanics in the twentieth century may not hold the same strict values—and many of them certainly cannot afford the luxury to do so—women's sexual behavior is still the expression of the family's honor. The tradition of maintaining virginity until marriage that had been emphasized among women continues to be a cultural imperative. The Virgin Mary—who was a virgin and a mother, but never a sexual being—is presented as an important role model for all Hispanic women, although Hispanic unwed mothers, who have clearly overstepped the boundaries of culturally-prescribed virginity for women, usually are accepted by their families. Married women or those living in common-law marriages are supposed to accept a double standard for sexual behavior, by which their husbands may have affairs with other women, while they themselves are expected to remain faithful to one man all of their lives. However, it is not uncommon for a Hispanic woman to have the power to decide whether or not a man is going to live with her, and she may also choose to put him out if he drinks too much or is not a good provider.[3]

In fact, Latin women experience a unique combination of power and powerlessness which is characteristic of the culture. The idea that personal problems are best discussed with women is very much part of the Hispanic culture. Women in Hispanic neighborhoods and families tend to rely on other women for their important personal and practical needs. There is a widespread belief among Latin women of all social classes that most men are undependable and are not to be trusted. At the same time, many of these women will put up with a man's abuses because having a man around is an important source of a woman's sense of self-worth. Middle-aged and elderly Hispanic women retain important roles in their families even after their sons and daughters are married. Grandmothers are ever present and highly vocal in family affairs. Older women have much more status and power than their white American counterparts, who at this age may be suffering from depression due to what has been called the "empty-nest syndrome." Many Hispanic women are providers of mental health services (which sometimes include advice about sexual problems) in an unofficial way as "curanderas," "espiritistas," or "santeras," for those people who believe in these alternative approaches to health care.[4] Some of these women play a powerful role in their communities, thanks to their reputation for being able to heal mind and body.

However, at the same time that Latin women have the opportunity to exercise their power in the areas mentioned above, they also receive constant cultural messages that they should be submissive and subservient to males in order to be seen as "good women." To suffer and be a martyr is also a characteristic of a "good woman." This emphasis on self-renunciation, combined with the importance given to sexual purity for women, has a direct bearing on the development of sexuality in Latin women. To enjoy sexual pleasure, even in marriage, may indicate lack of virtue. To shun sexual pleasure and to regard sexual behavior exclusively as an unwelcome obligation toward her husband and a necessary evil in order to have children may be seen as a manifestation of virtue. In fact, some women even express pride at their own lack of sexual pleasure or desire. Their negative attitudes toward sex are frequently reinforced by the inconsiderate behavior and demands of men.

Body image and related issues are deeply connected with sexuality for all women. Even when body-related problems may not have direct implications for sexuality, the body remains for women the main vehicle for expressing their needs. The high incidence of somatic complaints presented by low-income Hispanic women in psychotherapy might be a consequence of the emphasis on "martyrdom" and self-sacrifice, or it might be a somatic expression of needs and anxieties. More directly related to sexuality are issues of birth control, pregnancy, abortion, menopause, hysterectomy and other gynecological problems. Many of these have traditionally been discussed among women only. To be brought to the attention of a male doctor may be enormously embarrassing and distressing for some of these women. Younger Hispanic women may find themselves challenging traditional sexual mores while struggling with their own conflicts about beauty and their own embarrassment about visiting male doctors.

One of the most common and pervasive stereotypes held about Hispanics is the image of the "macho" man—an image which generally conjures up the rough, tough, swaggering men who are abusive and oppressive towards women, who in turn are seen as being exclusively submissive and long-suffering.[5]

Some authors[6] recognize that "machismo"—which is nothing but the Hispanic version of the myth of male superiority supported by most cultures—is still in existence in the Latin culture, especially among those individuals who subscribe more strongly to traditional Hispanic values. Following this tradition, Latin females are expected to be subordinated to males and to the family. Males are expected to show their manhood by behaving in a strong fashion, by demonstrating sexual prowess and by asserting their authority over women. In many cases, these traditional values may not be enacted behaviorally, but are still supported as valued assumptions concerning male and female "good" behavior. According to Aramoni,[7] himself a Mexican psychologist, "machismo" may be a reaction of Latin males to a series of social conditions, including the effort to exercise control over their ever-present, powerfully demanding, and suffering mothers and to identify with their absent fathers. Adult males continue to respect and revere their mothers, even when they may not show much respect for their wives or other women. As adolescents they may have protected their mothers from fathers' abuse or indifference. As adults they accord their mother a respect that no other woman deserves, thus following their fathers' steps. The mother herself teaches her sons to be dominant and independent in relations with other women. Other psychological and social factors may be influential in the development of "machismo." It is important to remember that not all Latin males exhibit the negative behaviors implied in the "macho" stereotype, and that even when certain individuals do, these behaviors might be a reaction to oppressive social conditions by which Hispanic men too are victimized.

Sexually, "machismo" is expressed through an emphasis on multiple, uncommitted sexual contacts which start in adolescence. In a study of adolescent rituals in Latin America, Espín[8] found that many males celebrated their adolescence by visiting prostitutes. The money to pay for this sexual initiation was usually provided by fathers, uncles or older brothers. Adolescent females, on the other hand, were offered coming-out parties, the rituals of which emphasize their virginal qualities. Somehow, a man is more "macho" if he manages to have sexual relations with a virgin; thus, fathers and brothers watch over young women for fear that other men may make them their sexual prey. These same men, however, will not hesitate to take advantage of the young women in other families. Women, in turn, are seen as capable of surrendering to men's advances, without much awareness of their own decisions on the matter. "Good women" should always say no to a sexual advance. Those who say yes are automatically assumed to be less virtuous by everyone, including the same man with whom they consent to have sex.

Needless to say, sexual understanding and communication between the sexes is practically rendered impossible by these attitudes generated by "machismo." However, not all Hispanics subscribe to this perspective and some reject it outright. In a review of the literature on studies of decision-making patterns in Mexican and Chicano families the authors concluded that "Hispanic males may behave differently from non-Hispanic men in their family and marital lives, but not in the inappropriate fashion suggested by the myth with its strong connotations of social deviance."[9] This article reviews only research on the decision-making process in married couples and, thus, other aspects of male-female relationships in the Hispanic culture are not discussed.

In the context of culturally appropriate sex-roles, mothers train their daughters to remain virgins at all cost, to cater to men's sexual needs and to play "little wives" to their father and brothers from a very early age. If a mother is sick or working outside the home and there are no adult females around, the oldest daughter, no matter how young, will be in charge of caring not only for the younger siblings, but also for the father, who would continue to expect his meals to be cooked and his clothes to be washed.

Training for appropriate heterosexuality, however, is not always assimilated by all Latin women. A seldom-mentioned fact is that, as in all cultures, there are lesbians among Hispanic women. Although emotional and physical closeness among women is encouraged by the culture, overt acknowledgment of lesbianism is even more restricted than in mainstream American society. In a study about lesbians in the Puerto Rican community, Hidalgo and Hidalgo-Christensen found that "rejection of homosexuals appears to be the dominant attitude in the Puerto Rican community."[10] Although this attitude may not seem different from that of the dominant culture, there are some important differences experienced by Latin lesbian women which are directly related to Hispanic cultural patterns. Frequent contact and a strong interdependence among family members, even in adulthood, are essential features of Hispanic family life. Leading a double life becomes more of a strain in this context. "Coming out" may jeopardize not only these strong family ties, but also the possibility of serving the Hispanic community in which the talents of all members are such an important asset. Because most lesbian women are single and self-supporting, and not encumbered by the demands of husbands and children, it can be assumed that the professional experience and educational level of Hispanic lesbians will tend to be relatively high. If this is true, professional experience and education will frequently place Hispanic lesbian women in positions of leadership or advocacy in their community. Their status and prestige, and, thus, the ability to serve their community, are threatened by the possibility of being "found out."

Most "politically aware" Latins show a remarkable lack of understanding of gay-related issues. In a recent meeting of Hispanic women in a major US city, one participant expressed the opinion that "lesbianism is a sickness we get from American women and American culture." This is, obviously, another version of the myth about the free sexuality of American women so prevalent among Hispanics. But it is also an expression of the common belief that homosexuality is chosen behavior, acquired through the bad influence of others, like drug addiction. Socialist attitudes in this respect are extremely traditional, as attitudes of the Cuban revolution towards homosexuality clearly manifest. Thus, Hispanics who consider themselves radical and committed to civil rights remain extremely traditional when it comes to gay rights. These attitudes clearly add further stress to the lives of Latin women who have a homosexual orientation and who are invested in enhancing the lives of members of their communities.

They experience oppression in three ways: as women, as Hispanics and as lesbians. This last form of oppression is in fact experienced most powerfully from inside their own culture. Most Latin women who are lesbians have to remain "closeted" among their families, their colleagues and society at large. To be "out of the closet" only in an Anglo context deprives them of essential supports from their communities and families, and, in turn, increases their invisibility in the Hispanic culture, where only the openly "butch" types are recognized as lesbians. . . .

NOTES

1. J. B. Miller, *Toward a New Psychology of Women*, Boston, Beacon, 1976.

2. Calderón de la Barca, *El Alcalde de Zalamea*.

3. S. Brown, "Love Unites Them and Hunger Separates Them: Poor Women in the Dominican Republic," in Rayna Reiter (ed.), *Toward an Anthropology of Women*, New York, Monthly Review Press, 1975, p. 322.

4. O. M. Espín, "Hispanic Female Healers in Urban Centers in the United States," unpublished manuscript, 1983.

5. V. Abad, J. Ramos, and E. Boyce, "A Model for Delivery of Mental Health Services to Spanish-Speaking Minorities," *American Journal of Orthopsychiatry*, vol. 44, no. 4, 1974, pp. 584–95.

6. E. S. Le Vine and A. M. Padilla, *Crossing Cultures in Therapy: Pluralistic Counseling for the Hispanic*, Monterey, California, Brooks/Cole, 1980.

7. A. Aramoni, "Machismo," *Psychology Today*, vol. 5, no. 8, 1982, pp. 69–72.

8. O. M. Espín, "The 'Quinceañeras': A Latin American Expression of Women's Roles," unpublished paper presented at the national meeting of the Latin American Studies Association, Atlanta, 1975.

9. R. E. Cromwell and R. A. Ruiz, "The Myth of 'Macho' Dominance in Decision Making within Mexican and Chicano Families," *Hispanic Journal of Behavioral Sciences*, vol. 1, no. 4, 1979, p. 371.

10. H. Hidalgo and E. Hidalgo-Christensen, "The Puerto Rican Cultural Response to Female Homosexuality," in E. Acosta-Belén (ed.), *The Puerto Rican Woman*, New York, Praeger, 1979, p. 118.

A NEW POLITICS OF SEXUALITY

47

June Jordan

As a young worried mother, I remember turning to Dr. Benjamin Spock's *Common Sense Book of Baby and Child Care* just about as often as I'd pick up the telephone. He was God. I was ignorant but striving to be good: a good Mother. And so it was there, in that best-seller pocketbook of do's and don't's, that I came upon this doozie of a guideline: Do not wear miniskirts or other provocative clothing because that will upset your child, especially if your child happens to be a boy. If you give your offspring "cause" to think of you as a sexual being, he will, at the least, become disturbed; you will derail the equilibrium of his notions about your possible identity and meaning in the world.

It had never occurred to me that anyone, especially my son, might look upon me as an asexual being. I had never supposed that "asexual" was some kind of positive designation I should, so to speak, lust after. I was pretty surprised by Dr. Spock. However, I was also, by habit, a creature of obedience. For a couple of weeks I actually experimented with lusterless colors and dowdy tops and bottoms, self-consciously hoping thereby to prove myself as a lusterless and dowdy and, therefore, excellent female parent.

Years would have to pass before I could recognize the familiar, by then, absurdity of a man setting himself up as the expert on a subject that presupposed women as the primary objects for his patriarchal discourse—on motherhood, no less! Years passed before I came to perceive the perversity of dominant power assumed by men, and the perversity of self-determining power ceded to men by women.

A lot of years went by before I understood the dynamics of what anyone could summarize as the Politics of Sexuality.

I believe the Politics of Sexuality is the most ancient and probably the most profound arena for human conflict. Increasingly, it seems clear to me

This essay was adapted from the author's keynote address to the Bisexual, Gay, and Lesbian Student Association at Stanford University on April 29, 1991. It was published in *The Progressive*, July 1991.

that deeper and more pervasive than any other oppression, than any other bitterly contested human domain, is the oppression of sexuality, the exploitation of the human domain of sexuality for power.

When I say sexuality, I mean gender: I mean male subjugation of human beings because they are female. When I say sexuality, I mean heterosexual institutionalization of rights and privileges denied to homosexual men and women. When I say sexuality I mean gay or lesbian contempt for bisexual modes of human relationship.

The Politics of Sexuality therefore subsumes all of the different ways in which some of us seek to dictate to others of us what we should do, what we should desire, what we should dream about, and how we should behave ourselves, generally. From China to Iran, from Nigeria to Czechoslovakia, from Chile to California, the politics of sexuality—enforced by traditions of state-sanctioned violence plus religion and the law—reduces to male domination of women, heterosexist tyranny, and, among those of us who are in any case deemed despicable or deviant by the powerful, we find intolerance for those who choose a different, a more complicated—for example, an interracial or bisexual—mode of rebellion and freedom.

We must move out from the shadows of our collective subjugation—as people of color/as women/as gay/as lesbian/as bisexual human beings.

I can voice my ideas without hesitation or fear because I am speaking, finally, about myself. I am Black and I am female and I am a mother and I am bisexual and I am a nationalist and I am an antinationalist. And I mean to be fully and freely all that I am!

Conversely, I do not accept that any white or Black or Chinese man—I do not accept that, for instance, Dr. Spock—should presume to tell me, or any other woman, how to mother a child. He has no right. He is not a mother. My child is not his child. And, likewise, I do not accept that anyone—any woman or any man who is not inextricably part of the subject he or she dares to address—should attempt to tell any of us, the objects of her or his presumptuous discourse, what we should do or what we should not do.

Recently, I have come upon gratuitous and appalling pseudoliberal pronouncements on sexuality. Too often, these utterances fall out of the mouths of men and women who first disclaim any sentiment remotely related to homophobia, but who then proceed to issue outrageous opinions like the following:

- That it is blasphemous to compare the oppression of gay, lesbian, or bisexual people to the oppression, say, of black people, or of the Palestinians.

- That the bottom line about gay or lesbian or bisexual identity is that you can conceal it whenever necessary and, so, therefore, why don't you do just that? Why don't you keep your deviant sexuality in the closet and let the rest of us—we who suffer oppression for reasons of our ineradicable and always visible components of our personhood such as race or gender—get on with our more necessary, our more beleaguered struggle to survive?

Well, number one: I believe I have worked as hard as I could, and then harder than that, on behalf of equality and justice—for African-Americans, for the Palestinian people, and for people of color everywhere.

And no, I do not believe it is blasphemous to compare oppressions of sexuality to oppressions of race and ethnicity: Freedom is indivisible or it is nothing at all besides sloganeering and temporary, short-sighted, and short-lived advancement for a few. Freedom is indivisible, and either we are working for freedom or you are working for the sake of your self-interests and I am working for mine.

If you can finally go to the bathroom wherever you find one, if you can finally order a cup of coffee and drink it wherever coffee is available, but you cannot follow your heart—you cannot respect the response of your own honest body in the world—then how much of what kind of freedom does any one of us possess?

Or, conversely, if your heart and your honest body can be controlled by the state, or controlled by community taboo, are you not then, and in that case, no more than a slave ruled by outside force?

What tyranny could exceed a tyranny that dictates to the human heart, and that attempts to dictate the public career of an honest human body?

Freedom is indivisible; the Politics of Sexuality is not some optional "special-interest" concern for serious, progressive folk.

And, on another level, let me assure you: if every single gay or lesbian or bisexual man or woman active on the Left of American politics decided to stay home, there would be *no* Left left.

One of the things I want to propose is that we act on that reality: that we insistently demand reciprocal respect and concern from those who cheerfully depend upon our brains and our energies for their, and our, effective impact on the political landscape.

Last spring, at Berkeley, some students asked me to speak at a rally against racism. And I did. There were four or five hundred people massed on Sproul Plaza, standing together against that evil. And, on the next day, on that same plaza, there was a rally for bisexual and gay and lesbian rights, and students asked me to speak at that rally. And I did. There were fewer than seventy-five people stranded, pitiful, on that public space. And I said then what I say today:

That was disgraceful! There should have been just one rally. One rally: freedom is indivisible.

As for the second, nefarious pronouncement on sexuality that now enjoys mass-media currency: the idiot notion of keeping yourself in the closet—that is very much the same thing as the suggestion that black folks and Asian-Americans and Mexican-Americans should assimilate and become as "white" as possible—in our walk/talk/music/food/values—or else. Or else? Or else we should, deservedly, perish.

Sure enough, we have plenty of exposure to white everything so why would we opt to remain our African/Asian/Mexican selves? The answer is that suicide is absolute, and if you think you will survive by hiding who you really are, you are sadly misled: there is no such thing as partial or intermittent suicide. You can only survive if you—who you really are—do survive.

Likewise, we who are not men and we who are not heterosexist—we, sure enough, have plenty of exposure to male-dominated/heterosexist this and that.

But a struggle to survive cannot lead to suicide: suicide is the opposite of survival. And so we must not conceal/assimilate/integrate into the would-be dominant culture and political system that despises us. Our survival requires that we alter our environment so that we can live and so that we can hold each other's hands and so that we can kiss each other on the streets, and in the daylight of our existence, without terror and without violent and sometimes fatal reactions from the busybodies of America.

Finally, I need to speak on bisexuality. I do believe that the analogy is interracial or multiracial identity. I do believe that the analogy for bisexuality is a multicultural, multi-ethnic, multiracial world view. Bisexuality follows from such a perspective and leads to it, as well.

Just as there are many men and women in the United States whose parents have given them more than one racial, more than one ethnic identity and cultural heritage to honor; and just as these men and women must deny no given part of themselves except at the risk of self-deception and the insanities that must issue from that; and just as these men and women embody the principle of equality among races and ethnic communities; and just as these men and women falter and anguish and choose and then falter again and then anguish and then choose yet again how they will honor the irreducible complexity of their God-given human being—even so, there are many men and women, especially young men and women, who seek to embrace the complexity of their total, always-changing social and political circumstance.

They seek to embrace our increasing global complexity on the basis of the heart and on the basis of an honest human body. Not according to ideology. Not according to group pressure. Not according to anybody's concept of "correct."

This is a New Politics of Sexuality. And even as I despair of identity politics—because identity is given and principles of justice/equality/freedom cut across given gender and given racial definitions of being, and because I will call you my brother, I will call you my sister, on the basis of what you *do* for justice, what you *do* for equality, what you *do* for freedom and *not* on the basis of you who are, even so I look with admiration and respect upon the new, bisexual politics of sexuality.

This emerging movement politicizes the so-called middle ground: Bisexuality invalidates either/or formulation, either/or analysis. Bisexuality means I am free and I am as likely to want and to love a woman as I am likely to want and to love a man, and what about that? Isn't that what freedom implies?

If you are free, you are not predictable and you are not controllable. To my mind, that is the keenly positive, politicizing significance of bisexual affirmation:

To insist upon complexity, to insist upon the validity of all of the components of social/sexual complexity, to insist upon the equal validity of all of the components of social/sexual complexity.

This seems to me a unifying, 1990s mandate for revolutionary Americans planning to make it into the twenty-first century on the basis of the heart, on the basis of an honest human body, consecrated to every struggle for justice, every struggle for equality, every struggle for freedom.

Violence and Social Control

"THE MIND THAT BURNS IN EACH BODY": *Women, Rape, and Racial Violence*

48

Jacquelyn Dowd Hall

HOSTILITY FOCUSED ON HUMAN FLESH

FLORIDA TO BURN NEGRO AT STAKE: SEX CRIMINAL SEIZED FROM JAIL, WILL BE MUTILATED, SET AFIRE IN EXTRA-LEGAL VENGEANCE FOR DEED
—Dothan (Alabama) *Eagle,* October 26, 1934

After taking the nigger to the woods . . . they cut off his penis. He was made to eat it. Then they cut off his testicles and made him eat them and say he liked it.
—Member of a lynch mob, 1934[1]

Lynching, like rape, has not yet been given its history. Perhaps it has been too easily relegated to the shadows where "poor white" stereotypes dwell. Perhaps the image of absolute victimization it evokes has been too difficult to reconcile with what we know about black resilience and resistance. Yet the impact of lynching, both as practice and as symbol, can hardly be underestimated. Between 1882 and 1946 almost 5,000 people died by lynching. The lynching of Emmett Till in 1955 for whistling at a white woman, the killing of three civil rights workers in Mississippi in the 1960s, and the hanging of a black youth in Alabama in 1981 all illustrate the persistence of this tradition of ritual violence in the service of racial control, a tradition intimately bound up with the politics of sexuality.

Vigilantism originated on the eighteenth-century frontier where it filled a vacuum in law enforcement. Rather than passing with the frontier, however, lynching was incorporated into the distinctive legal system of southern slave society.[2] In the nineteenth century, the industrializing North moved toward a modern criminal justice system in which police, courts, and prisons administered an impersonal, bureaucratic rule of law designed to uphold property rights and discipline unruly workers. The South, in contrast, maintained order through a system of deference and customary authority in which all whites had informal police power over all blacks, slave owners meted out plantation justice undisturbed by any generalized rule of law, and the state encouraged vigilantism as part of its overall reluctance to maintain a strong system of formal authority that would have undermined the planter's prerogatives. The purpose of one system was class control, of the other, control over a slave population. And each tradition continued into the period after the Civil War. In the North, factory-like penitentiaries warehoused displaced members of the industrial proletariat. The South maintained higher rates of personal violence than any other region in the country and lynching crossed over the line from informal law enforcement into outright political terrorism.

White supremacy, of course, did not rest on force alone. Routine institutional arrangements denied to the freedmen and women the opportunity to own land, the right to vote, access to education, and participation in the administration of the law. Lynching reached its height during the battles of Reconstruction and the Populist revolt; once a new system of disfranchisement, debt peonage, and segregation was firmly in place, mob violence gradually declined. Yet until World War I, the average number of lynchings never fell below two or three a week. Through the twenties and thirties, mob violence reinforced white dominance by providing planters with a quasi-official way of enforcing labor contracts and crop lien laws and local officials with a means of extracting deference, regardless of the letter of the law. Individuals may have lynched for their own twisted reasons, but the practice continued only with tacit official consent.[3]

Most importantly, lynching served as a tool of psychological intimidation aimed at blacks as a group. Unlike official authority, the lynch mob was unlimited in its capriciousness. With care and vigilance, an individual might avoid situations that landed him in the hands of the law. But a lynch mob could strike anywhere, any time. Once the brush fire of rumor began, a manhunt was organized, and the local paper began putting out special editions announcing a lynching in progress, there could be few effective reprieves. If the intended victim could not be found, an innocent bystander might serve as well.

It was not simply the threat of death that gave lynching its repressive power. Even as outbreaks of mob violence declined in frequency, they were increasingly accompanied by torture and sexual mutilation. Descriptions of the first phase of Hitler's death sweep are chillingly applicable to lynching: "Killing was ad hoc, inventive, and in its dependence on imagination, peculiarly expressive . . . this was murder uncanny in its anonymous intimacy, a hostility so personally focused on human flesh that the abstract fact of death was not enough."[4]

At the same time, the expansion of communications and the development of photography in the late nineteenth and early twentieth centuries gave reporting a vividness it had never had before. The lurid evocation of human suffering implicated white readers in each act of aggression and drove home to blacks the consequences of powerlessness. Like whipping under slavery, lynching was an instrument of coercion intended to impress not only the immediate victim but all who saw or heard about the event. And the mass media spread the imagery of rope and faggot far beyond the community in which each lynching took place. Writing about his youth in the rural South in the 1920s, Richard Wright describes the terrible climate of fear:

> The things that influenced my conduct as a Negro did not have to happen to me directly; I needed but to hear of them to feel their full effects in the deepest layers of my consciousness. Indeed, the white brutality that I had not seen was a more effective control of my behavior than that which I knew. The actual experience would have let me see the realistic outlines of what was really happening, but as long as it remained something terrible and yet remote, something whose horror and blood might descend upon me at any moment, I was compelled to give my entire imagination over to it.[5]

A penis cut off and stuffed in a victim's mouth. A crowd of thousands watching a black man scream in pain. Such incidents did not have to occur very often, or be witnessed directly, to be burned indelibly into the mind.

NEVER AGAINST HER WILL

> White men have said over and over—and we have believed it because it was repeated so often—that not only was there no such thing as a chaste Negro woman—but that a Negro woman could not be assaulted, that it was never against her will.
>
> —Jessie Daniel Ames (1936)

Schooled in the struggle against sexual rather than racial violence, contemporary feminists may nevertheless find familiar this account of lynching's political function, for analogies between rape and lynching have often surfaced in the literature of the anti-rape movement. To carry such analogies too far would be to fall into the error of radical feminist writing that misconstrues the realities of racism in the effort to illuminate sexual subordination.[6] It is the suggestion of this essay, however, that there is a significant resonance between these two forms of violence. We are only beginning to understand the web of connections among racism, attitudes toward women, and sexual ideologies. The purpose of looking more closely at the dynamics of repressive violence is not to reduce sexual assault and mob murder to static equivalents but to illuminate some of the strands of that tangled web.

The association between lynching and rape emerges most clearly in their parallel use in racial subordination. As Diane K. Lewis has pointed out, in a patriarchal society, black men, as men, constituted a potential challenge to the established order.[7] Laws were formulated primarily to exclude black men from adult male prerogatives in the public sphere, and lynching meshed with these legal mechanisms of exclusion. Black women represented a more ambiguous threat. They too were denied access to the politico-jural domain, but since they shared this exclusion with women in general, its maintenance engendered less anxiety and required less force. Lynching served primarily to dramatize hierarchies among men. In contrast, the violence directed at black women illustrates the double jeopardy of race and sex. The records of the Freedmen's Bureau and the oral histories collected by the Federal Writers' Project testify to the sexual atrocities endured by black women as whites sought to reassert their command over the newly freed slaves. Black women were sometimes executed by lynch mobs, but more routinely they served as targets of sexual assault.

Like vigilantism, the sexual exploitation of black women had been institutionalized under slavery. Whether seized through outright force or voluntarily granted within the master-slave relation, the sexual access of white men to black women was a cornerstone of patriarchal power in the South. It was used as a punishment or demanded in exchange for leniency. Like other forms of deference and conspicuous consumption, it buttressed planter hegemony. And it served the practical economic purpose of replenishing the slave labor supply.

After the Civil War, the informal sexual arrangements of slavery shaded into the use of rape as a political weapon, and the special vulnerability of black women helped shape the ex-slaves' struggle for the prerequisites of freedom. Strong family bonds had survived the adversities of slavery; after freedom, the black family served as a bulwark against a racist society. Indeed, the sharecropping system that replaced slavery as the South's chief mode of production grew in part from the desire of blacks to withdraw from gang labor and gain control over their own work, family lives, and bodily integrity. The sharecropping family enabled women to escape white male supervision, devote their productive and reproductive powers to their own families, and protect themselves from sexual assault.[8]

Most studies of racial violence have paid little attention to the particular suffering of women.[9] Even rape has been seen less as an aspect of sexual oppression than as a transaction between white and black men. Certainly Claude Lévi-Strauss's insight that men use women as verbs with which to communicate with one another (rape being a means of communicating defeat to the men of a conquered tribe) helps explain the extreme viciousness of sexual violence in the post-emancipation era.[10] Rape *was* in part a reaction to the effort of the freedmen to assume the role of patriarch, able to provide for and protect his family. Nevertheless, as writers like Susan Griffin and Susan Brownmiller and others have made clear, rape is first and foremost a crime against women.[11] Rape sent a message to black men, but more centrally, it expressed male sexual attitudes in a culture both racist and patriarchal. . . .

In the United States, the fear and fascination of female sexuality was projected onto black women; the passionless lady arose in symbiosis with the primitively sexual slave. House slaves often served as substitute mothers; at a black woman's breast white men experienced absolute dependence on a being who was both a source of wish-fulfilling joy and of grief-producing disappointment. In adulthood, such men could find in this black woman a ready object for the mixture of rage and desire that so often underlies male heterosexuality. The black woman, already in chains, was sexually available, unable to make claims for support or concern; by dominating her, men could replay the infant's dream of unlimited access to the mother.[12] The economic and political challenge posed by the black patriarch might be met with death by lynching, but when the black woman seized the opportunity to turn her maternal and sexual resources to the benefit of her own family, sexual violence met her assertion of will. Thus rape reasserted white dominance and control in the private arena as lynching reasserted hierarchical arrangements in the public transactions of men.

LYNCHING'S DOUBLE MESSAGE

> The crowds from here that went over to see [Lola Cannidy, the alleged rape victim in the Claude Neal lynching of 1934] said he was so large he could not assault her until he took his knife and cut her, and also had either cut or bit one of her breast [sic] off.
>
> —Letter to Mrs. W. P. Cornell, October 29, 1934, Association of Southern Women for the Prevention of Lynching Papers

> . . . more than rape itself, the fear of rape permeates our lives. . . . and the best defense against this is not to be, to deny being in the body, as a self, to . . . avert your gaze, make yourself, as a presence in the world, less felt.
>
> —Susan Griffin, *Rape: The Power of Consciousness* (1979)

In the 1920s and 1930s, the industrial revolution spread through the South, bringing a demand for more orderly forms of law enforcement. Men in authority, anxious to create a favorable business climate, began to withdraw their tacit approval of extralegal violence. Yet lynching continued, particularly in rural areas, and even as white moderates criticized lynching in the abstract, they continued to justify outbreaks of mob violence for the one special crime of sexual assault. For most white Americans, the association between lynching and rape called to mind not twin forms of white violence against black men and women, but a very different image: the black rapist, "a monstrous beast, crazed with lust";[13] the white victim—young, blond, virginal; her manly Anglo-Saxon avengers. Despite the pull of modernity, the emotional logic of lynching remained: only swift, sure violence, unhampered by legalities, could protect white women from sexual assault.

The "protection of white womanhood" was a pervasive fixture of racist ideology. In 1839, for example, a well-known historian offered this commonly accepted rationale for lynching: black men find "something strangely alluring and seductive . . . in the appearance of the white woman; they are aroused and stimulated by its foreignness to their experience of sexual pleasures, and it moves them to gratify their lust at any cost and in spite of every obstacle." In 1937, echoing an attitude that characterized most local newspapers, the Jackson, Mississippi, *Daily News* published what it felt was the *coup de grace* to anti-lynching critics: "What would you do if your wife, daughter, or one of your loved ones was ravished? You'd probably be right there with the mob." Two years later, 65 percent of the white respondents in an anthropological survey believed that lynching was justified in cases of sexual assault.[14] Despite its tenacity, however, the myth of the black rapist was never founded on objective reality. Less than a quarter of lynch victims were even accused of rape or attempted rape. Down to the present, almost every study has underlined the fact that rape is overwhelmingly an intraracial crime, and the victims are more often black than white.[15]

A major strategy of anti-lynching reformers, beginning with Ida B. Wells in the 1880s and continuing with Walter White of the NAACP and Jessie Daniel Ames of the Association of Southern Women for the Prevention of Lynching, was to use such facts to undermine the rationalizations for mob violence. But the emotional circuit between interracial rape and lynching lay beyond the reach of factual refutation. A black man did not literally have to attempt sexual assault for whites to perceive some transgression of caste mores as a sexual threat. White women were the forbidden fruit, the untouchable property, the ultimate symbol of white male power. To break the racial rules was to conjure up an image of black over white, of a world turned upside down.

Again, women were a means of communication and, on one level, the rhetoric of protection, like the rape of black women, reflected a power struggle among men. But impulses toward women as well as toward blacks were played out in the drama of racial violence. The fear of rape was more than a hypocritical excuse for lynching; rather, the two phenomena were intimately intertwined. The "southern rape complex" functioned as a means of both sexual and racial suppression.[16]

For whites, the archetypal lynching for rape can be seen as a dramatization of cultural themes, a story they told themselves about the social arrangements and psychological strivings that lay beneath the surface of everyday life. The story such rituals told about the place of white women in southern society was subtle, contradictory, and demeaning. The frail victim, leaning on the arms of her male relatives, might be brought to the scene of the crime, there to identify her assailant and witness his execution. This was a moment of humiliation. A woman who had just been raped, or who had been apprehended in a clandestine interracial affair, or whose male relatives were pretending that she had been raped, stood on display before the whole community. Here was the quintessential Woman as Victim: polluted, "ruined for life," the object of fantasy and secret contempt. Humiliation, however, mingled with heightened worth as she played for a moment the role of the Fair Maiden violated and avenged. For this privilege—if the alleged assault had in fact taken place—she might pay with suffering in the extreme. In any case, she would pay with a lifetime of subjugation to the men gathered in her behalf.

Only a small percentage of lynchings, then, revolved around charges of sexual assault; but those that did received by far the most attention and publicity—indeed, they gripped the white imagination far out of proportion to their statistical significance. Rape and rumors of rape became the folk pornography of the Bible Belt. As stories spread the rapist became not just a black man but a ravenous brute, the victim a beautiful young virgin. The experience of the woman was described in minute and progressively embellished detail, a public fantasy that implied a group participation in the rape as cathartic as the subsequent lynching. White men might see in "lynch law" their ideal selves: patriarchs, avengers, righteous protectors. But, being men themselves, and sometimes even rapists, they must also have seen themselves in the lynch mob's prey.

The lynch mob in pursuit of the black rapist represented the trade-off implicit in the code of chivalry: for the right of the southern lady to protection presupposed her obligation to obey. The connotations of wealth and family background attached to the position of the lady in the antebellum South faded in the twentieth century, but the power of "ladyhood" as a value construct remained. The term denoted chastity, frailty, graciousness. "A lady," noted one social-psychologist, "is always in a state of becoming: one acts like a lady, one attempts to be a lady, but one never *is* a lady." Internalized by the individual, this ideal regulated behavior and restricted interaction with the world.[17] If a woman passed the tests of ladyhood, she could tap into the reservoir of protectiveness and shelter known as southern chivalry. Women who abandoned secure, if circumscribed, social roles forfeited the claim to personal security. Together the practice of ladyhood and the etiquette of chivalry controlled white women's behavior even as they guarded caste lines. . . .

THE DECLINE OF CHIVALRY

> As male supremacy becomes ideologically untenable, incapable of justifying itself as protection, men assert their domination more directly, in fantasies and occasionally in acts of raw violence.
>
> —Christopher Lasch, *Marxist Perspectives* (1978)

In the 1970s, for the second time in the nation's history, rape again attracted widespread public attention. The obsession with interracial rape, which peaked at the turn of the nineteenth century but lingered from the close of the Civil War into the 1930s, became a magnet for racial and sexual oppression. Today the issue of rape has crystallized important feminist concerns.

Rape emerged as a feminist issue as women developed an independent politics that made sexuality and personal life a central arena of struggle. First in consciousness-raising groups, where autobiography became a politicizing technique, then in public "speakouts," women broke what in retrospect seems a remarkable silence about a pervasive aspect of female experience. From that beginning flowed both an analysis that held rape to be a political act by which men affirm their power over women and strategies for change that ranged from the feminist self-help methods of rape crisis centers to institutional reform of the criminal justice and medical care systems. After 1976, the movement broadened to include wife-battering, sexual harassment, and, following the lead of Robin Morgan's claim that "pornography is the theory, rape the practice," media images of women.[18]

By the time Susan Brownmiller's *Against Our Will: Men, Women and Rape* gained national attention in 1975, she could speak to and for a feminist constituency already sensitized to the issue by years of practical, action-oriented work. Her book can be faulted for supporting a notion of universal patriarchy and timeless sexual victimization; it leaves no room for understanding the reasons for women's collaboration, their own sources of power (both self-generated and derived), the class and racial differences in their experience of discrimination and sexual danger. But it was an important milestone, pointing the way for research into a subject that has consistently been trivialized and ignored. Many grass-roots activists would demur from Brownmiller's assertion that all men are potential rapists, but they share her understanding of the continuum between sexism and sexual assault.[19]

The demand for control over one's own body—control over whether, when, and with whom one has children, control over how one's sexuality is expressed—is central to the feminist project because, as Rosalind Petchesky persuasively argues, it is essential to "a sense of being a person, with personal and bodily integrity," able to engage in conscious activity and to participate in social life.[20] It is the right to bodily integrity and self-determination that rape, and the fear of rape, so thoroughly undermines. Rape's devastating effect on individuals derives not so much from the sexual nature of the crime (and anti-rape activists have been concerned to revise the idea that rape is a "fate worse than death" whose victims, if no longer "ruined for life," are at least so traumatized that they must rely for recovery on therapeutic help rather than on their own resources) as from the experience of helplessness and loss of control, the sense of one's self as an object of rape. And women who may never be raped share, by chronic attrition, in the same helplessness, "otherness," lack of control. The struggle against rape, like the anti-lynching movement, addresses not only external dangers but also internal consequences: the bodily muting, the self-censorship that limits one's capacity to "walk freely in the world."[21]

The focus on rape, then, emerged from the internal dynamics of feminist thought and practice. But it was also a response to an objective increase in the crime. From 1969 to 1974, the number of rapes rose 49 percent, a greater increase than for any other violent crime. Undoubtedly rape statistics reflect general demographic and criminal trends, as well as a greater willingness of victims to report sexual attacks (although observers agree that rape is still the most underreported of crimes).[22] But there can be no doubt that rape is a serious threat and that it plays a prominent role in women's subordination. Using recent high-quality survey data, Allan Griswold Johnson has estimated that, at a minimum, 20 to 30 percent of girls now twelve years old will suffer a violent attack sometime in their lives. A woman is as likely to be raped as she is to experience a divorce or to be diagnosed as having cancer.[23]

In a recent anthology on women and pornography, Tracey A. Gardner has drawn a parallel between the wave of lynching that followed Reconstruction and the increase in rapes in an era of anti-feminist backlash.[24] Certainly, as women enter the workforce, postpone marriage, live alone or as single heads of households, they become easier targets for sexual assault. But observations like Gardner's go further, linking the intensification of sexual violence directly to the feminist challenge. Such arguments come dangerously close to blaming the victim for the crime. But they may also contain a core of truth. Sociological research on rape has only recently begun, and we do not have studies explaining the function and frequency of the crime under various historical conditions; until that work is done we cannot with certainty assess the current situation. Yet it seems clear that just as lynching ebbed and flowed with new modes of racial control, rape—both as act and idea—cannot be divorced from changes in the sexual terrain.

In 1940, Jessie Ames released to the press a statement that, for the first time in her career, the South could claim a "lynchless year," and in 1942, convinced that lynching was no longer widely condoned in the name of white womanhood, she allowed the Association of Southern Women for the Prevention of Lynching to pass quietly from the scene. The women's efforts, the larger, black-led anti-lynching campaign, black migration from the rural South, the spread of industry—these and other developments contributed to the decline of vigilante justice. Blacks continued to be victimized by covert violence and routinized court procedures that amounted to "legal lynchings." But after World War II, public lynchings, announced in the papers, openly accomplished, and tacitly condoned, no longer haunted the land, and the black rapist ceased to be a fixture of political campaigns and newspaper prose.

This change in the rhetoric and form of racial violence reflected new attitudes toward women as well as toward blacks. By the 1940s few southern leaders were willing, as Jessie Ames put it, to "lay themselves open to ridicule" by defending lynching on the grounds of gallantry, in part because gallantry itself had lost conviction.[25] The same process of economic development and national integration that encouraged the South to adopt northern norms of authority and control undermined the chivalric ideal. Industrial capitalism on the one hand and women's assertion of independence on the other weakened paternalism and with it the conventions of protective deference.[26] This is not to say that the link between racism and sexism was broken; relations between white women and black men continued to be severely sanctioned, and black men, to the present, have drawn disproportionate punishment for sexual assault. The figures speak for themselves: of the 455 men executed for rape since 1930, 405 were black, and almost all the complainants were white.[27] Nevertheless, "the protection of white womanhood" rang more hollow in the postwar New South and the fear of interracial rape became a subdued theme in the nation at large rather than an openly articulated regional obsession.

The social feminist mainstream, of which Jessie Ames and the anti-lynching association were a part, thus chipped away at a politics of gallantry that locked white ladies in the home under the guise of protecting them from the world. But because such reformers held to the genteel trappings of their role even as they asserted their autonomous citizenship, they offered reassurance that women's influence could be expanded without mortal danger to male prerogatives and power. Contemporary feminists have eschewed some of the comforting assumptions of their nineteenth-century predecessors: women's passionlessness, their limitation to social housekeeping, their exclusive responsibility for childrearing and housekeeping. They have couched their revolt in explicit ideology and unladylike behavior. Meanwhile, as Barbara Ehrenreich has argued, Madison Avenue has perverted the feminist message into the threatening image of the sexually and economically liberated woman. The result is a shift toward the rapaciousness that has always mixed unstably with sentimental exaltation and concern. Rape has emerged more clearly into the sexual domain, a crime against women most often committed by men of their own race rather than a right of the powerful over women of a subordinate group or a blow by black men against white women's possessors.[28]

It should be emphasized, however, that the connection between feminism and the upsurge of rape lies not so much in women's gains but in their assertion of rights within a context of economic vulnerability and relative powerlessness. In a perceptive article published in 1901, Jane Addams traced lynching in part to "the feeling of the former slave owner to his former slave, whom he is now bidden to regard as his fellow citizen."[29] Blacks in the post-Reconstruction era were able to express will and individuality, to wrest from their former masters certain concessions and build for themselves supporting institutions. Yet they lacked the resources to protect themselves from economic exploitation and mob violence. Similarly, contemporary feminist efforts have not yet succeeded in overcoming women's isolation, their economic and emotional dependence on men, their cultural training toward submission. There are few restraints against sexual aggression, since up to 90 percent of rapes go unreported, 50 percent of assailants who are reported are never caught, and seven out of ten prosecutions end in acquittal.[30] Provoked by the commercialization of sex, cut loose from traditional community restraints, and "bidden to regard as his fellow citizen" a female being whose subordination has deep roots in the psyches of both sexes, men turn with impunity to the use of sexuality as a means of asserting dominance and control. Such fear and rage are condoned when channeled into right-wing attacks on women's claim to a share in public power and control over their bodies. Inevitably they also find expression in less acceptable behavior. Rape, like lynching, flourishes in an atmosphere in which official policies toward members of a subordinate group give individuals tacit permission to hurt and maim.

In 1972 Anne Braden, a southern white woman and long-time activist in civil rights struggles, expressed her fear that the new anti-rape movement might find itself "objectively on the side of the most reactionary social forces" unless it heeded a lesson from history. In a pamphlet entitled *Open Letter to Southern White Women*—much circulated in regional women's liberation circles at the time—she urged anti-rape activists to remember the long pattern of racist manipulation of rape fears. She called on white women, "for their own liberation, to refuse any longer to be used, to act in the tradition of Jessie Daniel Ames and the white women who fought in an earlier period to end lynching," and she went on to discuss her own politicization through left-led protests against the prosecution of black men on false rape charges. Four years later, she joined the chorus of black feminist criticism of *Against Our Will*, seeing Brownmiller's book as a realization of her worst fears.[31]

Since this confrontation between the Old Left and the New, between a white woman who placed herself in a southern tradition of feminist anti-racism and a radical feminist from the North, a black women's movement has emerged, bringing its own perspectives to bear. White activists at the earliest "speakouts" had acknowledged "the racist image of black men as rapists," pointed out the large number of black women among assault victims, and debated the contradictions involved in looking for solutions to a race and class-biased court system. But not until black women had developed their own autonomous organizations and strategies were true alliances possible across racial lines.

A striking example of this development is the Washington, D.C., Rape Crisis Center. One of the first and largest such groups in the country, the center has evolved from a primarily white self-help project to an aggressive interracial organization with a multifaceted program of support services, advocacy, and community education. In a city with an 80 percent black population and more than four times as many women as men, the center has recruited black leadership by channeling its resources into staff salaries and steering clear of the pitfalls of middle-class voluntarism on the one hand and professionalism on the other. It has challenged the perception of the anti-rape movement as a "white woman's thing" by stressing not only rape's devastating effect on women but also its impact on social relations in the black community. Just as racism undermined working-class unity and lynching sometimes pitted poor whites against blacks, sexual aggression now divides the black community against itself. In a society that defines manhood in terms of power and possessions, black men are denied the resources to fulfill their expected roles. Inevitably, they turn to domination of women, the one means of manhood within their control. From consciousness-raising groups for convicted rapists to an intensive educational campaign funded by the city's public school system and aimed at both boys and girls from elementary through high school, the center has tried to alter the cultural plan for both sexes that makes men potential rapists and women potential victims.[32]

As the anti-rape movement broadens to include Third World women, analogies between lynching and rape and the models of women like Ida B. Wells and Jessie Daniel Ames may become increasingly useful. Neither lynching nor rape is the "aberrant behavior of a lunatic fringe."[33] Rather, both grow out of everyday modes of interaction. The view of women as objects to be possessed, conquered, or defiled fueled racial hostility; conversely, racism has continued to distort and confuse the struggle against sexual violence. Black men receive harsher punishment for raping white women, black rape victims are especially demeaned and ignored, and, until recently, the different historical experience of black and white women has hindered them from making common cause. Taking a cue from the women's anti-lynching campaign of the 1930s as well as from the innovative tactics of black feminists, the anti-rape movement must not limit itself to training women to avoid rape or depending on imprisonment as a deterrent, but must aim its attention at changing the behavior and attitudes of men. Mindful of the historical connection between rape and lynching, it must make clear its stand against *all* uses of violence in oppression.

NOTES

My title comes from Adrienne Rich, "Disloyal to Civilization: Feminism, Racism, Gynephobia," in Rich, *On Lies, Secrets and Silences: Selected Prose, 1966–1978* (New York: W. W. Norton, 1979), p. 299. Parts of this essay are taken from Jacquelyn Dowd Hall, *Revolt Against Chivalry: Jessie Daniel Ames and the Women's Campaign Against Chivalry* (New York: Columbia University Press, 1979), and full documentation can be found in that work. See also Hall, " 'A Truly Subversive Affair': Women Against Lynching in the Twentieth-Century South," in *Women of America: A History*, eds. Carol

Berkin and Mary Beth Norton (Boston: Houghton Mifflin, 1979). Thanks to Rosemarie Hester, Walter Dellinger, Loretta Ross, Nkenge Toure, Janet Colm, Kathleen Dowdy, and Nell Painter for their help and encouragement.

1. Quoted in Howard Kester, *The Lynching of Claude Neal* (New York: National Association for the Advancement of Colored People, 1934).

2. Michael Stephen Hindus, *Prison and Plantation: Crime, Justice, and Authority in Massachusetts and South Carolina, 1767–1878* (Chapel Hill: University of North Carolina Press, 1980), pp. xix, 31, 124, 253.

3. For recent overviews of lynching, see Robert L. Zangrando, *The NAACP Crusade Against Lynching, 1909–1950* (Philadelphia: Temple University Press, 1980); McGovern, *Anatomy of a Lynching;* and Hall, *Revolt Against Chivalry.*

4. Terrence Des Pres, "The Struggle of Memory," *The Nation,* 10 April 1982, p. 433.

5. Quoted in William H. Chafe, *Women and Equality: Changing Patterns in American Culture* (New York: Oxford University Press, 1977), p. 60.

6. Margaret A. Simons, "Racism and Feminism: A Schism in the Sisterhood," *Feminist Studies* 5 (Summer 1979): 384–401.

7. Diane K. Lewis, "A Response to Inequality: Black Women, Racism, and Sexism," *Signs* 3 (Winter 1977): 341–42.

8. Jacqueline Jones, *Freed Women?: Black Women, Work, and the Family During the Civil War and Reconstruction,* Working Paper No. 61, Wellesley College, 1980; Roger L. Ransom and Richard Sutch, *One Kind of Freedom: The Economic Consequences of Emancipation* (New York: Cambridge University Press, 1977), pp. 87–103.

9. Gerda Lerner, *Black Women in White America: A Documentary History* (New York: Random House, 1972), is an early and important exception.

10. Robin Morgan, "Theory and Practice: Pornography and Rape," *Take Back the Night: Women on Pornography,* ed. Laura Lederer (New York: William Morrow, 1980), p. 140.

11. Susan Griffin, "Rape: The All-American Crime," *Ramparts* (September 1971): 26–35; Susan Brownmiller, *Against Our Will: Men, Women, and Rape* (New York: Simon and Schuster, 1975). See also Kate Millet, *Sexual Politics* (Garden City, N.Y.: Doubleday & Co., 1970).

12. Dorothy Dinnerstein, *The Mermaid and the Minotaur: Sexual Arrangements and Human Malaise* (New York: Harper and Row, 1977). See also Phyllis Marynick Palmer, "White Women/Black Women: The Dualism of Female Identity and Experience," unpublished paper presented at the American Studies Association, September 1979, pp. 15–17. Similarly, British Victorian eroticism was structured by class relations in which upper-class men were nursed by lower-class country women. . . .

13. A statement made in 1901 by George T. Winston, president of the University of North Carolina, typifies these persistent images: "The southern woman with her helpless little children in a solitary farm house no longer sleeps secure. . . . The black brute is lurking in the dark, a monstrous beast, crazed with lust. His ferocity is almost demoniacal. A mad bull or a tiger could scarcely be more brutal" (quoted in Charles Herbert Stember, *Sexual Racism: The Emotional Barrier to an Integrated Society* [New York: Elsevier, 1976], p. 23).

14. Philip Alexander Bruce, *The Plantation Negro as a Freeman* (New York: Putnam's, 1889), pp. 83–84; Jackson *Daily News*, 27 May 1937; Hortense Powdermaker, *After Freedom: A Cultural Study in the Deep South* (1939; New York: Atheneum, 1969), pp. 54–55, 389.

15. For a contradictory view, see, for example, S. Nelson and M. Amir, "The Hitchhike Victim of Rape: A Research Report," in *Victimology: A New Focus. Vol. 5: Exploiters and Exploited*, eds. M. Agopian, D. Chappell, and G. Geis, and I. Drapkin and E. Viano (1975), p. 47; and "Black Offender and White Victim: A Study of Forcible Rape in Oakland, California," in *Forcible Rape: The Crime, the Victim, and the Offender* (New York: Columbia University Press, 1977).

16. Winthrop Jordan, *White over Black: American Attitudes Toward the Negro, 1550–1812* (Baltimore: Penguin Books, 1969); W. J. Cash, *The Mind of the South* (New York: Knopf, 1941), p. 117.

17. This reading of lynching as a "cultural text" is modeled on Clifford Geertz, "Deep Play: Notes on the Balinese Cockfight," in *The Interpretation of Cultures: Selected Essays by Clifford Geertz* (New York: Basic Books, 1973), pp. 412–53. For "ladyhood," see Greer Litton Fox, " 'Nice Girl': Social Control of Women Through a Value Construct," *Signs* 2 (Summer 1977): 805–17.

18. Noreen Connell and Cassandra Wilsen, eds., *Rape: The First Sourcebook for Women* (New York: New American Library, 1974); Morgan, "Theory and Practice."

19. Interview with Janet Colm, director of the Chapel Hill-Carrboro (North Carolina) Rape Crisis Center, April 1981. Two of the best recent analyses of rape are Ann Wolbert Burgess and Lynda Lytle Holmstrom, *Rape: Crisis and Recovery* (Bowie, Md.: Robert J. Brady Co., 1979) and Lorenne M. G. Clark and Debra J. Lewis, *Rape: The Price of Coercive Sexuality* (Toronto: Canadian Women's Educational Press, 1977).

20. Rosalind Pollack Petchesky, "Reproductive Freedom: Beyond 'A Woman's Right to Choose,' " *Signs* 5 (Summer 1980): 661–85.

21. Adrienne Rich, "Taking Women Students Seriously," in *Lies, Secrets and Silences*, p. 242.

22. Vivian Berger, "Man's Trial, Woman's Tribulation: Rape Cases in the Courtroom," *Columbia Law Review* 1 (1977): 3–12. Thanks to Walter Dellinger for this reference.

23. Allan Griswold Johnson, "On the Prevalence of Rape in the United States," *Signs* 6 (Fall 1980): 136–46.

24. Tracey A. Gardner, "Racism in Pornography and the Women's Movement," in *Take Back the Night*, p. 111.

25. Jessie Daniel Ames, "Editorial Treatment of Lynching," *Public Opinion Quarterly* 2 (January 1938): 77–84.

26. For a statement of this theme, see Christopher Lasch, "The Flight from Feeling: Sociopsychology of Sexual Conflict," *Marxist Perspectives* 1 (Spring 1978): 74–95.

27. Berger, "Man's Trial, Woman's Tribulation," p. 4. For a recent study indicating that the harsher treatment accorded black men convicted of raping white women is not limited to the South and has persisted to the present, see Gary D. LaFree, "The Effect of Sexual Stratification by Race on Official Reactions to Rape," *American Sociological Review* 45 (October 1980): 842–54. Thanks to Darnell Hawkins for this reference.

28. Barbara Ehrenreich, "The Women's Movement: Feminist and Antifeminist," *Radical America* 15 (Spring 1981): 93–101; Lasch, "Flight from Feeling." Because violence against women is so inadequately documented, it is impossible to make accurate racial comparisons in the incidence of the crime. Studies conducted by Menachen Amir in the late 1950s indicated that rape was primarily intraracial, with 77 percent of rapes involving black victims and black defendants and 18 percent involving whites. More recent investigations claim a somewhat higher percentage of interracial assaults. Statistics on reported rapes show that black women are more vulnerable to assault than white women. However, since black women are more likely than white women to report assaults, and since acquaintance rape, most likely to involve higher status white men, is the most underreported of crimes, the vulnerability of white women is undoubtedly much greater than statistics indicate (Berger, "Man's Trial, Woman's Tribulation," p. 3, n. 16; LaFree, "Effect of Sexual Stratification," p. 845, n. 3; Johnson, "On the Prevalence of Rape," p. 145).

29. Quoted in Aptheker, *Lynching and Rape*, pp. 10–11.

30. Berger, "Woman's Trial, Man's Tribulation," p. 6; Johnson, "On the Prevalence of Rape," p. 138.

31. Anne Braden, "A Second Open Letter to Southern White Women," *Generations: Women in the South*, a special issue of *Southern Exposure* 4 (Winter 1977), edited by Susan Angell, Jacquelyn Dowd Hall, and Candace Waid.

32. Interview with Loretta Ross and Nkenge Toure, Washington, D.C., May 12, 1981. See also Rape Crisis Center of Washington, D.C., *How to Start a Rape Crisis Center* (1972, 1977).

33. Johnson, "On the Prevalence of Rape," p. 137.

MORE POWER THAN WE WANT:

49

Masculine Sexuality and Violence

Bruce Kokopeli and George Lakey

Masculine sexuality involves the oppression of women, competition among men, and homophobia (fear of homosexuality). Patriarchy, the systematic domination of women by men through unequal opportunities, rewards, punishments, and the internalization of unequal expectations through sex role differentiation, is the institution which organizes these behaviors. Patriarchy is men having more power, both personally and politically, than women of the same rank. This imbalance of power is the core of patriarchy, but definitely not the extent of it.

Sex inequality cannot be routinely enforced through open violence or even blatant discriminatory agreements—patriarchy also needs its values accepted in the minds of people. If as many young *women* wanted to be physicians as men, and as many young men *wanted* to be nurses as women, the medical schools and the hospitals would be hard put to maintain the masculine domination of health care; open struggle and the naked exercise of power would be necessary. Little girls, therefore, are encouraged to think "nurse" and boys to think "doctor."

Patriarchy assigns a list of human characteristics according to gender: women should be nurturant, gentle, in touch with their feelings, etc.; men should be productive, competitive, super-rational, etc. Occupations are valued according to these gender-linked characteristics, so social work, teaching, housework, and nursing are of lower status than business executive, judge, or professional football player.

When men do enter "feminine" professions they disproportionately rise to the top and become chefs, principals of schools, directors of ballet, and teachers of social work. A man is somewhat excused from his sex role deviation if he at least dominates within the deviation. Domination, after all, is what patriarchy is all about.

From: *Off Their Backs . . . and on Our Own Two Feet* (Philadelphia: New Society Publishers, 1983), pp. 17–24. Reprinted by permission.

Access to powerful positions by women (i.e., those positions formerly limited to men) is contingent on the women adopting some masculine characteristics, such as competitiveness. They feel pressure to give up qualities assigned to females (such as gentleness) because those qualities are considered inherently weak by patriarchal culture. The existence, therefore, of a woman like Indira Gandhi in the position of a dictator in no way undermines the basic sexist structure which allocates power to those with masculine characteristics.

Patriarchy also shapes men's sexuality so it expresses the theme of domination. Notice the masculine preoccupation with size. The size of a man's body has a lot to say about his clout or his vulnerability, as any junior high boy can tell you. Many of these schoolyard fights are settled by who is bigger than whom, and we experience in our adult lives the echoes of intimidation and deference produced by our habitual "sizing up" of the situation.

Penis size is part of this masculine preoccupation, this time directed toward women. Men want to have large penises because size equals power, the ability to make a woman "really feel it." The imagery of violence is close to the surface here, since women find penis size irrelevant to sexual genital pleasure. "Fucking" is a highly ambiguous word, meaning both intercourse and exploitation/assault.

It is this confusion that we need to untangle and understand. Patriarchy tells men that their need for love and respect can only be met by being masculine, powerful, and ultimately violent. As men come to accept this, their sexuality begins to reflect it. Violence and sexuality combine to support masculinity as a character ideal. To love a woman is to have power over her and to treat her violently if need be. The Beatles' song "Happiness Is A Warm Gun" is but one example of how sexuality gets confused with violence and power. We know one man who was discussing another man who seemed to be highly fertile—he had made several women pregnant. "That guy," he said, "doesn't shoot any blanks."

Rape is the end logic of masculine sexuality. Rape is not so much a sexual act as an act of violence expressed in a sexual way. The rapist's mind-set—that violence and sexuality *can* go together—is actually a product of patriarchal conditioning, for most of us men understand the same, however abhorrent rape may be to us personally.

In war, rape is astonishingly prevalent even among men who "back home" would not do it. In the following description by a marine sergeant who witnessed a gang rape in Vietnam, notice that nearly all the nine-man squad participated:

They were supposed to go after what they called a Viet Cong whore. They went into her village and instead of capturing her, they raped her— every man raped her. As a matter of fact, one man said to me later that it was the first time he had ever made love to a woman with his boots on. The man who led the platoon, or the squad, was actually a private. The squad leader was a sergeant but he was a useless person and he let the private take over his squad. Later he said he took no part in the raid. It was against his morals. So instead of telling his squad not to do it, because they wouldn't listen to him anyway, the sergeant went into another side of the village and just sat and stared bleakly at the ground, feeling sorry for himself. But at any rate, they raped the girl, and then, the last man to make love to her, shot her in the head. [Vietnam Veterans Against the War, statement by Michael McClusker in *The Winter Soldier Investigation: An Inquiry Into American War Crimes.*]

Psychologist James Prescott adds to this account:

What is it in the American psyche that permits the use of the word "love" to describe rape? And where the act of love is completed with a bullet in the head! [*Bulletin of the Atomic Scientists,* November 1975, p. 17.]

MASCULINITY AGAINST MEN: THE MILITARIZATION OF EVERYDAY LIFE

Patriarchy benefits men by giving us a class of people (women) to dominate and exploit. Patriarchy also oppresses men, by setting us at odds with each other and shrinking our life space.

The pressure to win starts early and never stops. Working class gangs fight over turf; rich people's sons are pushed to compete on the sports field. British military officers, it is said, learned to win on the playing fields of Eton.

Competition is conflict held within a framework of rules. When the stakes are really high, the rules may not be obeyed; fighting breaks out. We men mostly relate through competition, but we know what is waiting in the wings. John Wayne is not a cultural hero by accident.

Men compete with each other for status as masculine males. Because masculinity equals power, this means we are competing for power. The ultimate proof of power/masculinity is violence. A man may fail to "measure up" to the macho stereotype in important ways, but if he can fight successfully with the person who challenges him on his deviance, he is still all right. The television policeman Baretta is strange in some ways: he is gentle with women and he cried when a man he loved was killed. However, he has what are probably the largest biceps in television and he proves weekly that he can beat up the toughs who come his way.

The close relationship between violence and masculinity does not need much demonstration. War used to be justified partly because it promoted "manly virtue" in a nation. Those millions of people in the woods hunting deer, in the National Rifle Association, and cheering on the bloodiest hockey teams are overwhelmingly men.

The world situation is so much defined by patriarchy that what we see in the wars of today is competition between various patriarchal ruling classes and governments breaking into open conflict. Violence is the accepted masculine form of conflict resolution. Women at this time are not powerful enough in the world situation for us to see mass overt violence being waged on them. But the violence is in fact there; it is hidden through its legitimization by the state and by culture.

In everyday middle-class life, open violence between men is of course rare. The defining characteristics of masculinity, however, are only a few steps removed from violence. Wealth, productivity, or rank in the firm or institution translate into power—the capacity (whether or not exercised) to dominate. The holders of power in even polite institutions seem to know that violence is at their fingertips, judging from the reactions of college presidents to student protest in the 1960s. We know of one urban "pacifist" man, the head of a theological seminary, who was barely talked out of calling the police to deal with a nonviolent student sit-in at "his" seminary!

Patriarchy teaches us at very deep levels that we can never be safe with other men (or perhaps with anyone!), for the guard must be kept up lest our vulnerability be exposed and we be taken advantage of. At a recent Quaker conference in Philadelphia, a discussion group considered the value of personal sharing and openness in the Quaker Meeting. In almost every case the women advocated more sharing and the men opposed it. Dividing by gender on that issue was predictable; men are conditioned by our life experience of masculinity to distrust settings where personal exposure will happen, especially if men are present. Most men find emotional intimacy possible only with women; many with only one woman; some men cannot be emotionally intimate with anyone.

Patriarchy creates a character ideal—we call it masculinity—and measures everyone against it. Many men as well as women fail the test and even men who are passing the test today are carrying a heavy load of anxiety about tomorrow. Because masculinity is a form of domination, no one can really rest secure. The striving goes on forever unless you are actually willing to give up and find a more secure basis for identity.

MASCULINITY AGAINST GAY MEN: PATRIARCHY FIGHTS A REAR GUARD ACTION

Homophobia is the measure of masculinity. The degree to which a man is thought to have gay feelings is the degree of his unmanliness. Because patriarchy presents sexuality as men over women (part of the general dominance theme), men are conditioned to have only that in mind as a model of sexual expression. Sex with another man must mean being dominated, which is very scary. A non-patriarchal model of sexual expression as the mutuality of equals doesn't seem possible; the transfer of the heterosexual model to same-sex relations can at best be "queer," at worst, "perverted."

In the recent book *Blue Collar Aristocrats*, by E. E. LeMasters, a working class tavern is described in which the topic of homosexuality sometimes comes up. Gayness is never defended. In fact, the worst thing you can call a man is homosexual. A man so attacked must either fight or leave the bar.

Notice the importance of violence in defending yourself against the charge of being a "pansy." Referring to your income or academic degrees or size of your car is no defense against such a charge. Only fighting will re-establish your respect as a masculine male. Because "gay" appears to mean "powerless," one needs to go to the masculine source of power—violence—for adequate defense.

Last year, the Argentinian government decided to persecute gays on a systematic basis. The Ministry of Social Welfare offered the rationale for this policy in an article in its journal, which also attacked lesbians, concluding that they should be put in jail or killed:

> As children they played with dolls. As they grew up, violent sports horrified them. As was to be expected, with the passage of time and the custom of listening to foreign mulattos on the radio, they became conscientious objectors. [*El Caudillo*, February 1975, excerpted in *Peace News*, July 11, 1975, p. 5]

The Danish government, by contrast with Argentina, has liberal policies on gay people. There is no government persecution and all government jobs are open to gays—except in the military and the diplomatic service! Two places where the nation-state is most keen to assert power are places where gays are excluded as a matter of policy.

We need not go abroad to see the connections between violence and homophobia. In the documentary film *Men's Lives*, a high school boy is interviewed on what it is like to be a dancer. While the interview is conducted, we see him working out, with a very demanding set of acrobatic exercises. The boy mentions that other boys think he must be gay. "Why is that?" the interviewer asks. "Dancers are free and loose," he replies; they are not big like football players; and "you're not trying to kill anybody."

Different kinds of homosexual behavior bring out different amounts of hostility, curiously enough. That fact gives us further clues to violence and female oppression. In prisons, for example, men can be respected if they fuck other men, but not if they are themselves fucked. (We use the word "fucked" intentionally for its ambiguity.) Often prison rapes are done by men who identify as heterosexual; one hole substitutes for another in this scene, for sex is in either case an expression of domination for the masculine mystique.

But for a man to be entered sexually, or to use effeminate gestures and actions, is to invite attack in prison and hostility outside. Effeminate gay men are at the bottom of the totem pole because they are *most like women*, which is nothing less than treachery to the Masculine Cause. Even many gay men shudder at drag queens and vigilantly guard against certain mannerisms because they, too, have internalized the masculinist dread of effeminacy.

John Braxton's report of prison life as a draft resister is revealing on this score. The other inmates knew immediately that John was a conscientious objector because he did not act tough. They also assumed he was gay, for the same reason. (If you are not masculine, you must be a pacifist and gay, for masculinity is a package which includes both violence and heterosexuality.)

A ticket of admission to masculinity, then, is sex with women, and bisexuals can at least get that ticket even if they deviate through having gay feelings as well. This may be why bisexuality is not feared as much as exclusive gayness among men. Exclusively gay men let down the Masculine Cause in a very important way—those gays do not participate in the control of women through sexuality. Control through sexuality matters because it is flexible; it usually is mixed with love and dependency so that it becomes quite subtle. (Women often testify to years of confusion and only the faintest uneasiness at their submissive role in traditional heterosexual relationships and the role sex plays in that.)

Now we better understand why women are in general so much more supportive of gay men than non-gay men are. Part of it of course is that heterosexual men are often paralyzed by fear. Never very trusting, such men find gayness one more reason to keep up the defenses. But heterosexual women are drawn to active support for the struggles of gay men because there is a common enemy—patriarchy and its definition of sexuality as domination. Both heterosexual women and gay men have experienced first hand the violence of sexism; we all have experienced its less open forms such as put-downs and discrimination, and we all fear its open forms such as rape and assault.

Patriarchy, which links characteristics (gentleness, aggressiveness, etc.) to gender, shapes sexuality as well, in such a way as to maintain male power. The Masculine Cause draws strength from homophobia and resorts habitually to violence in its battles on the field of sexual politics. It provides psychological support for the military state and is in turn stimulated by it.

THE POLICE AND THE BLACK MALE   50

Elijah Anderson

The police, in the Village-Northton [neighborhood] as elsewhere, represent society's formal, legitimate means of social control. Their role includes protecting law-abiding citizens from those who are not law-abiding by preventing crime and by apprehending likely criminals. Precisely how the police fulfill the public's expectations is strongly related to how they view the neighborhood and the people who live there. On the streets, color-coding often works to confuse race, age, class, gender, incivility, and criminality, and it expresses itself most concretely in the person of the anonymous black male. In doing their job, the police often become willing parties to this general color-coding of the public environment, and related distinctions, particularly those of skin color and gender, come to convey definite meanings. Although such coding may make the work of the police more manageable, it may also fit well with their own presuppositions regarding race and class relations, thus shaping officers' perceptions of crime "in the city." Moreover, the anonymous black male is usually an ambiguous figure who arouses the utmost caution and is generally considered dangerous until he proves he is not. . . .

To be white is to be seen by the police—at least superficially—as an ally, eligible for consideration and for much more deferential treatment than that

---

Ed. note: This article is based on the author's field research in two city neighborhoods he calls Village-Northton.

---

accorded to blacks in general. This attitude may be grounded in the back-grounds of the police themselves. Many have grown up in . . . "ethnic" neigh-borhoods. They may serve what they perceive as their own class and neighborhood interests, which often translates as keeping blacks "in their place"—away from neighborhoods that are socially defined as "white." In trying to do their job, the police appear to engage in an informal policy of monitoring young black men as a means of controlling crime, and often they seem to go beyond the bounds of duty. . . .

On the streets late at night, the average young black man is suspicious of others he encounters, and he is particularly wary of the police. If he is dressed in the uniform of the "gangster," such as a black leather jacket, sneakers, and a "gangster cap," if he is carrying a radio or a suspicious bag (which may be confiscated), or if he is moving too fast or too slow, the police may stop him. As part of the routine, they search him and make him sit in the police car while they run a check to see whether there is a "detainer" on him. If there is nothing, he is allowed to go on his way. After this ordeal the youth is often left afraid, sometimes shaking, and uncertain about the area he had previously taken for granted. He is upset in part because he is painfully aware of how close he has come to being in "big trouble." He knows of other youths who have gotten into a "world of trouble" simply by being on the streets at the wrong time or when the police were pursuing a criminal. In these circumstances, particularly at night, it is relatively easy for one black man to be mistaken for another. Over the years, while walking through the neighborhood I have on occasion been stopped and questioned by police chasing a mugger, but after explaining myself I was released.

Many youths, however, have reason to fear such mistaken identity or harassment, since they might be jailed, if only for a short time, and would have to post bail money and pay legal fees to extricate themselves from the mess. . . . When law-abiding blacks are ensnared by the criminal justice sys-tem, the scenario may proceed as follows. A young man is arbitrarily stopped by the police and questioned. If he cannot effectively negotiate with the officer(s), he may be accused of a crime and arrested. To resolve this situation he needs financial resources, which for him are in short supply. If he does not have money for any attorney, which often happens, he is left to a public defender who may be more interested in going along with the court system than in fighting for a poor black person. Without legal support, he may well wind up "doing time" even if he is innocent of the charges brought against him. The next time he is stopped for questioning he will have a record, which will make detention all the more likely.

Because the young black man is aware of many cases when an "innocent" black person was wrongly accused and detained, he develops an "attitude" toward the police. The street word for police is "the man," signifying a certain machismo, power, and authority. He becomes concerned when he notices "the man" in the community or when the police focus on him because he is outside his own neighborhood. The youth knows, or soon finds out, that he exists in a legally precarious state. Hence he is motivated to avoid the police, and his public life becomes severely circumscribed. . . .

Such scrutiny and harassment by local police make black youths see them as a problem to get beyond, to deal with, and their attempts affect their overall behavior. To avoid encounters with the man, some streetwise young men camouflage themselves, giving up the urban uniform and emblems that identify them as "legitimate" objects of police attention. They may adopt a more conventional presentation of self, wearing chinos, sweat suits, and generally more conservative dress. Some youths have been known to "ditch" a favorite jacket if they see others wearing one like it, because wearing it increases their chances of being mistaken for someone else who may have committed a crime.

But such strategies do not always work over the long run and must be constantly modified. For instance, because so many young ghetto blacks have begun to wear Fila and Adidas sweat suits as status symbols, such dress has become incorporated into the public image generally associated with young black males. These athletic suits, particularly the more expensive and colorful ones, along with high-priced sneakers, have become the leisure dress of successful drug dealers, and other youths will often mimic their wardrobe to "go for bad" in the quest for local esteem. Hence what was once a "square" mark of distinction approximating the conventions of the wider culture has been adopted by a neighborhood group devalued by that same culture. As we saw earlier, the young black male enjoys a certain power over fashion: whatever the collective peer group embraces can become "hip" in a manner the wider society may not desire. . . . These same styles then attract the attention of the agents of social control.

THE IDENTIFICATION CARD

Law-abiding black people, particularly those of the middle class, set out to approximate middle-class whites in styles of self-presentation in public, including dress and bearing. Such middle-class emblems, often viewed as "square," are not usually embraced by young working-class blacks. Instead, their connections with and claims on the institutions of the wider society seem to be symbolized by the identification card. The common identification card associates its holder with a firm, a corporation, a school, a union, or some other institution of substance and influence. Such a card, particularly from a prominent establishment, puts the police and others on notice that the youth is "somebody," thus creating an important distinction between a black man who can claim a connection with the wider society and one who is summarily judged as "deviant." Although blacks who are established in the middle class might take such cards for granted, many lower-class blacks, who continue to find it necessary to campaign for civil rights denied them because of skin color, believe that carrying an identification card brings them better treatment than is meted out to their less fortunate brothers and sisters. For them this link to the wider society, though often tenuous, is psychically and socially important. . . .

"DOWNTOWN" POLICE AND LOCAL POLICE

In attempting to manage the police—and by implication to manage themselves—some black youths have developed a working connection of the police in certain public areas of the Village-Northton. Those who spend a good amount of their time on these corners, and thus observing the police, have come to distinguish between the "downtown" police and the "regular" local police.

The local police are the ones who spend time in the area; normally they drive around in patrol cars, often one officer to a car. These officers usually make a kind of working peace with the young men on the streets; for example, they know the names of some of them and may even befriend a young boy. Thus they offer an image of the police department different from that displayed by the "downtown" police. The downtown police are distant, impersonal, and often actively looking for "trouble." They are known to swoop down arbitrarily on gatherings of black youths standing on a street corner; they might punch them around, call them names, and administer other kinds of abuse, apparently for sport. A young Northton man gave the following narrative about his experiences with the police.

> And I happen to live in a violent part. There's a real difference between the vio-lence level in the Village and the violence level in Northton. In the nighttime it's more dangerous over there.
>
> It's so bad now, they got downtown cops over there now. They doin' a good job bringin' the highway patrol over there. Regular cops don't like that. You can tell that. They even try to emphasize to us the certain category. High-way patrol come up, he leave, they say somethin' about it. "We can do our job over here." We call [downtown police] Nazis. They about six feet eight, seven feet. We walkin', they jump out. "You run, and we'll blow your nigger brains out." I hate bein' called a nigger. I want to say somethin' but get myself in trouble.

When a cop do somethin', nothing happen to 'em. They come from down-town. From what I heard some of 'em don't even wear their real badge num-bers. So you have to put up with that. Just keep your mouth shut when they stop you, that's all. Forget about questions, get against the wall, just obey 'em. "Put all that out right there"—might get rough with you now. They snatch you by the shirt, throw you against the wall, pat you hard, and grab you by the arms, and say, "Get outta here." They call you nigger this and little black this, and things like that. I take that. Some of the fellas get mad. It's a whole differ-ent world. . . .

You call a cop, they don't come. My boy got shot, we had to take him to the hospital ourselves. A cop said, "You know who did it?" We said no. He said, "Well, I hope he dies if y'all don't say nothin'." What he say that for? My boy said, "I hope your mother die," he told the cop right to his face. And I was grabbin' another cop, and he made a complaint about that. There were a lot of witnesses. Even the nurse behind the counter said the cop had no busi-ness saying nothin' like that. He said it loud, "I hope he dies." Nothin' like that should be comin' from a cop.

Such behavior by formal agents of social control may reduce the crime rate, but it raises questions about social justice and civil rights. Many of the old-time liberal white residents of the Village view the police with some ambivalence. They want their streets and homes defended, but many are convinced that the police manhandle "kids" and mete out an arbitrary form of "justice." These feelings make many of them reluctant to call the police when they are needed, and they may even be less than completely cooperative after a crime has been committed. They know that far too often the police simply "go out and pick up some poor black kid." Yet they do cooperate, if ambiva-lently, with these agents of social control. . . .

. . . Stories about police prejudice against blacks are often traded at Village get-togethers. Cynicism about the effectiveness of the police mixed with community suspicion of their behavior toward blacks keeps middle-class Villagers from embracing the notion that they must rely heavily on the formal means of social control to maintain even the minimum freedom of movement they enjoy on the streets.

Many residents of the Village, especially those who see themselves as the "old guard" or "old-timers," who were around during the good old days when antiwar and antiracist protest was a major concern, sigh and turn their heads when they see the criminal justice system operating in the ways described here. They express hope that "things will work out," that tensions will ease, that crime will decrease and police behavior will improve. Yet as incivility and crime become increasing problems in the neighborhood, whites become less tolerant of anonymous blacks and more inclined to embrace the police as their heroes. . . .

Gentrifiers and the local old-timers who join them, and some traditional residents continue to fear, care more for their own safety and well-being than for the rights of young blacks accused of wrong-doing. Yet reliance on the police, even by an increasing number of former liberals, may be traced to a general feeling of oppression at the hands of street criminals, whom many believe are most often black. As these feelings intensify and as more yuppies and students inhabit the area and press the local government for services, especially police protection, the police may be required to "ride herd" more stringently on the youthful black population. Thus young black males are often singled out as the "bad" element in an otherwise healthy diversity, and the tensions between the lower-class black ghetto and the middle and upper-class white community increase rather than diminish.

KOREAN AMERICANS VS. AFRICAN AMERICANS: *Conflict and Construction*

51

Sumi K. Cho

The violence and destruction that followed the Rodney King verdict again exploded the myth of a shared consensus around American justice and democracy. The blind injustice of the Simi Valley "not guilty" verdict produced a rainbow coalition of people—old, young, of all colors—who had few or no reservations about looting stores owned primarily by Koreans and Latinos. It soon became clear that the nation's professional, academic, political, and business elite were ill-equipped to deal with the complexity of issues before

Author's Note: I would like to thank Gil Gott, Pedro Noguera, Susan Lee, Ronnie Stevenson, Ann Park, Evageline Ordaz, Christopher Burroughs, and Bob Gooding-Williams for their helpful comments on this essay.

From Sumi Cho, "Korean Americans vs. African Americans: Conflict and Construction," in *Reading Rodney King/Reading Urban Uprising*, ed. Robert Gooding-Williams (New York: Routledge, 1993), pp. 196–211. Reprinted by permission of the author.

them, captured in the simple but straightforward plea by Rodney King: "Can we all get along?"

King's question hints at the real question confronting U.S. society: Who is the "we" that must get along? For too long, the political and academic tradition has defined U.S. race relations in terms of a Black/white binary opposition. For example, a CNN–*Time Magazine* poll taken immediately after the verdict surveyed "Americans" on their opinions regarding the verdict and the violence that followed. Yet the poll only sought the views of African Americans and whites regarding the future of race relations. The Black/white framing of race issues must give way to a fuller, more differentiated under-standing of a multiracial, multiethnic society divided along the lines of race, class, gender, and other axes in order to explicate effectively the Los Angeles explosion and to contribute to the long-term empowerment of those who, for a short time, exercised the power of their own agency.

Dominating the current debate within the Black/white racial paradigm are, on the one hand, the human-capital theorists[1] who assert that the degradation of "family values" caused the fires in L.A. Dan Quayle's now infamous Murphy Brown speech, for example, pointed the finger at a television sitcom for contrib-uting to the social and moral decay which lie behind the problems in South Central L.A. by portraying the single motherhood of a white woman from the professional class. On the other hand, there are structuralists[2] proposing that the U.S. adopt a "Marshall Plan" for its cities to address the institutional lockout of people of color from economic development.

While the structural explanations provide more illuminating insights than human-capital theories, they cannot fully explain the myriad of events in Los Angeles, particularly those that affected the Korean-American community. The portrayal of Asian Americans as the paragons of socioeconomic success contributed to the targeting of Korean Americans as a scapegoat by those above and below Koreans on the socioeconomic ladder during the L.A. riots. The King verdict and the failure of the U.S. economy to provide jobs and a decent standard of living for all of its peoples, the ostensible root causes of the rioting, were not the fault of Korean (or Latino) shopowners. If, as some have suggested, the Willie Horton imagery[3] is the myth and Rodney King's beating the practice, then likewise, the casting of Asian Americans as a model minority is the myth, and the looting and burning of Koreatown and Korean-owned stores is the practice.

Manipulation of Korean Americans into a "model minority" contributed to their "triple scapegoating" following the King verdict. The first layer of attack came from those who targeted Korean-owned stores for looting and arson. The second layer consisted of those in positions of power who were responsible for the sacrifice of Koreatown, Pico Union, and South Central Los Angeles to ensure the safety of wealthier, whiter communities. The final scapegoating came at the hands of the media, eager to sensationalize the events by excluding Korean perspectives from coverage and stereotyping the immigrant community. These three forces combined to blame the Korean-American community for the nation's most daunting economic and sociopolitical problems.

Rodney King's spontaneous reaction upon seeing the varied groups of color pitted against each other in L.A. reflected a deeper understanding of the commonalities among those groups than many intellectuals and politicians have shown: "I love people of color," he declared, although almost all the mainstream media edited out this statement. Further, his plea can be read as a challenge to the academic community and intellectual activists of racial and ethnic politics: "We're all stuck here for awhile. . . . Let's try to work it out."

Race-relations theorists must accept this challenge and go beyond the standard structuralist critique of institutional racism to incorporate the most difficult issues presented by the aftermath of the not-guilty verdicts in the Rodney King case. Specifically, intellectual activists must devise a new approach to understanding interethnic conflict between subordinated groups and work toward a proactive theory of social change. Such a theory would examine the structural conditions that influence patterns of conflict or cooperation between groups of color, racial ideologies and how they influence group prejudices, as well as the roles of human agency, community education, and political leadership and accountability. Such a theory should strive to explain and resolve interethnic conflicts to unite subordinated groups.

KOREAN AMERICANS IN L.A.: MODEL MINORITY MYTH AND PRAXIS

The View from Below: The Politics of Resentment

The deteriorated socioeconomic conditions of neglected inner cities have led scholars to compare ghettos like South Central Los Angeles (which includes the Southeast and South Central city-planning areas) to South African "bantustans" that serve solely as "holding space for blacks and browns no longer of use to the larger economy."[4] According to 1990 census data, African Americans constitute 24.5 percent of South Central's residents, and Latinos account for 46 percent (with a rapidly growing Latino population). Only 3.6 percent of the total land mass is zoned for industry.[5] South Central residents lack self-determination and political-economic power. In this context, Korean Americans who open stores in the neighborhood are resented by long-deprived residents and are seen as "outsiders" exerting unfair control and power in the community. The interaction between the two racial groups is structured strictly by market relationships: one is the consumer, the other is the owner. This market structuring of group relations has influenced the "Korean/Black conflict" and contributed to the course of events following the not-guilty verdicts.

Koreans first were scapegoated by rioters of all colors who looted stores and later set them afire. Reports of the early activity following the not-guilty verdict in the Rodney King beating centered around a crowd that had gathered in front of Tom's Liquor Store at Florence and Normandie in the late afternoon of 29 April 1992. One of the first targets of the group was the "swap meets" because they were Korean-owned. The swap meets were essentially indoor flea markets that operated every day during the week and offered discounted prices on consumer items such as electronics and clothing. Whether these stores provided a service to an underserved community by offering low-priced goods where megaretailers refused to tread or whether they shamelessly promoted consumerism to an economically disadvantaged population was not at issue. The most oft-stated reason for the targeting of these stores and Korean stores in general was a familiar refrain: Korean owners were rude to African-American and Latino customers. One Latino interviewed on television was asked why people were looting Koreans. "Because we hate 'em. Everybody hates them," he responded.

Much has been said about the rudeness of Korean owners. Some of the major media outlets that covered the tensions between African Americans and Korean Americans attempted to reduce the conflict to "cultural differences" such as not smiling enough, not looking into another person's eyes, not placing change in a person's hands. Although Koreans wanted very badly to believe in this reductionism, one making an honest assessment must conclude that far too many Korean shopowners had accepted widespread stereotypes about African Americans as lazy, complaining criminals.

The dominant U.S. racial hierarchy and its concomitant stereotypes are transferred worldwide to every country that the United States has occupied militarily. Korean women who married American GIs and returned to the United States after the Korean War quickly discovered the social significance of marrying a white versus an African-American GI. American racial hierarchies were telegraphed back home. When Koreans immigrate to the U.S., internationalized stereotypes are reinforced by negative depictions of African Americans in U.S. films, television shows, and other popular forms of cultural production. This stereotype, combined with the high crime rate inherent in businesses such as liquor or convenience stores (regardless of who owns them), produced the prejudiced, paranoid, bunker mentality of Soon Ja Du who shot Latasha Harlins, a 15-year-old African-American girl, in the back of the neck during a dispute over a bottle of orange juice. Since 1 January 1990, at least 25 Korean merchants have been killed by non-Korean gunmen.[6]

On the other hand, many African Americans also internalize stereotypes of Korean Americans. Asian Americans walk a fine line between being seen as model minorities and callous unfair competitors. The result is a split image of success and greed that goes together with callousness and racial superiority. Distinctions between Asian ethnicities are often blurred as are distinctions between Asians and Asian Americans. "Sins of the neighbor" are passed on to Koreans in the United States. For example, when then prime minister Yasuhiro Nakasone of Japan made his blundering remarks in 1986 that "the level of intelligence in the United States is lowered by the large number of Blacks, Puerto Ricans, and Mexicans who live there,"[7] this comment was often applied generally to represent the views of all Asians, including Koreans, although Korea has a long history of colonization at the hands of the Japanese.

In rationalizing the violence that followed the verdict, African-American leaders often repeated the myth that Korean immigrants unfairly compete with aspiring entrepreneurs from the Black community because Korean Americans receive preferential treatment over African Americans for bank and government loans. "Until banks make loans available to blacks as much as other people, there's going to be resentment," observed John Murray of Cal-Pac, a Black liquor-store-owners' association.[8] In reality, however, banks and government lenders uniformly reject loan applications for businesses located in poor, predominantly minority neighborhoods such as South Central Los Angeles, regardless of the applicant's color. Korean immigrants rarely receive traditional financing. Those who open liquor stores and small businesses often come over with some capital and/or borrow from family and friends. At times, groups of Koreans will act as their own financial institutions through informal rotating credit associations known as "kyes" (pronounced "keh").

Thus, the ability to open stores largely depends upon a class variable, as opposed to a racial one. In fact, Korean merchants at times purchased their liquor and grocery stores from African-American owners. Prior to the Watts riot in 1965, most of the liquor-store owners in South Central Los Angeles were Jewish. After the riots, many Jewish owners wanted out. The easing of government-backed loans and the low selling price of the stores in the mid-1960s opened the way for African-American entrepreneurs. The labor-intensive, high-risk nature of work combined with the deregulation of liquor pricing in 1978 and subsequent price wars that drove many small stores out of business led to a sell-off of liquor stores by African Americans in the late 1970s and early 1980s. The sell-off coincided with a large increase in Korean immigration to the U.S. during the same time period. African-American merchants often made a solid profit selling to Koreans.[9] Korean immigrants, many of whom held college degrees, could not find jobs commensurate with their training and education in the new country due to structural discrimination, occupational downgrading, and general antiforeign sentiment. A disproportionate number of Korean families turned to small businesses as a living. According to U.S. Census Bureau statistics, the rate of self-employment among Koreans is higher than that of any other ethnic group.[10] This self-employment, or "self-exploitation" as some have termed the practice, extends not only to the "head of the household," but often to the entire family, including children. Given the high educational background of this group, high self-employment rates in mom-and-pop grocery stores can hardly be viewed as "success."[11]

Nevertheless, the politics of resentment painted Koreans as callous and greedy invaders who got easy bank loans. As this depiction ran unchecked, it became increasingly easy to consider violence against such a contemptible group. The popular rap artist, Ice Cube, warned Korean shopowners in his song entitled "Black Korea" (on his *Death Certificate* album) to "pay respect to the black fist, or we'll burn your store right down to a crisp." The scene was set for disaster and required simply a spark to ignite a highly flammable situation. When the scorching injustice of the verdict was announced, Korean-owned businesses were scapegoated as the primary target for centuries of racial injustice against African Americans—injustice that predated the bulk of Korean immigration which occurred only after the lifting of discriminatory immigration barriers in 1965.

The View from Above: "Racist Love" and the Model Minority

The police nonresponse to the initial outbreak of violence represented a conscious sacrificing of South Central Los Angeles and Koreatown, largely inhabited by African Americans, Chicanos, Latinos, and Korean Americans, to ensure the safety of affluent white communities. The LAPD's intentional neglect in failing to respond to the initial outbreak of rioting set up a foreseeable conflict between African Americans, Chicanos, Latinos, and Korean Americans with destructive and deadly consequences. When it became evident that no police would come to protect Koreatown, Koreans took up arms in self-defense against other minority groups. Rather than focus on whether these actions by the Korean community were right or wrong, one should question the allocation of police-protection resources and the absolute desertion of Koreatown and South Central L.A. by the police. When Deputy Chief Matthew Hunt, who commands officers in South Los Angeles, asked Chief of Police Daryl Gates for greater preparation prior to the announcement of the verdicts, the chief refused, claiming smugly that such preparations were unnecessary.

It was a badly kept secret that angry mobs bent upon destruction and violence would be descending on Koreatown. When Korean Americans called the LAPD and local and state officials for assistance and protection, there was no response. It was very clear that Koreatown was on its own. In the absence of police protection, the community harnessed its resources through Radio Korea (KBLA) in Los Angeles to coordinate efforts to defend stores from attack. People from as far away as San Bernadino and Orange County came to help. Radio Korea reported on the movement of the crowds and instructed volunteers where to go. In short, the radio station and individual volunteers served as the police force for Koreatown. It is bitterly ironic that some Koreans defending stores were later arrested by police for weapon-permit violations.

In stark contrast, when looters began to work over major shopping malls such as the Fox Hills Mall in Culver City, they were quickly stopped by the police, with "merchants and residents praising [police] efforts."[12] Likewise, the police made sure that the downtown business interests were secured. The west-side edition of the *Los Angeles Times* even boasted that the police forces in the predominantly white communities of Santa Monica, Beverly Hills, and West Hollywood "emerged remarkably unscathed by the riots."[13] Other areas, especially communities of color such as South Central, Pico Union, and Koreatown, effectively became a "no-person's-land." Enter at your own risk. There was no protection for these areas. Although residents paid taxes for police protection, the LAPD made a conscious choice to stay out of the neighborhoods of color that were most at risk during the postverdict rioting.

Although Korean Americans donated substantial amounts of money to the election campaigns of state and local officials, no political assistance was forthcoming in their time of need. "We've made significant contributions to the city council, as well as county, state and federal politicians, and they failed us," said Bong Hwan Kim, executive director of the L.A. Korean Youth Center.[14] Attempts to contact Governor Wilson for help from the National Guard were in vain. The National Guard did not arrive on the scene until there was little left to do. Wilson's explanation was that the National Guard was late because they had run out of ammunition.[15] During the initial outbreak of violence, it is now known Chief Gates was attending an expensive fundraiser on the west side while these communities burned. In response, he later stated at his first press conference following the verdict that police were needed to protect the firefighters. Unfortunately for Gates, the fire department was unwilling to go along with this excuse. In a stinging report to the Fire Commission, L.A. Fire Chief Donald Manning complained that the LAPD dismissed their requests for protection during the outbreak of violence as "not a top priority," thereby delaying fire-department responses to the fires.[16] Moreover, Gates was stumped by an *L.A. Times* reporter who pressed him about the hours prior to the arsons, when some of the most sensationalized violence occurred outside of Tom's Liquor Store.

Many Koreans mistakenly believed that they would be taken care of if they worked hard, did not complain, and contributed handsomely to powerful politicians such as Governor Wilson and local officials. One hard lesson to be learned from the aftermath of the King verdict and the Korean-American experience is that a model minority is expediently forgotten and dismissed if white dominance or security is threatened. This point was also demonstrated through the Asian-American admissions scandals at the nation's premier institutions of higher education when Asian-American students began to outperform white students on SATs and GPAs. The rules of the game of "meritocracy" had to be changed to put a ceiling on Asian-American admissions, thereby producing the largest unstated affirmative-action policy benefiting whites.

The easy dismissal of Korean-American pleas for help is rooted in the relationship between the dominant power brokers and the model-minority group. The model-minority "thesis," popularized during the late 1960s and the mid-1980s during the Reagan administration, represents a political backlash to civil rights struggles. It is critical to understand that the model-minority thesis is no compliment for Asian Americans. It has been historically constructed to discipline activists in the Black Power, Black Panther, American Indian, United Farm Workers, and other radical social movements demanding institutional change.[17] Because it has this genesis, the embrace of Asian Americans as a model minority is an embrace of "racist love."[18] The basis of that love has a racist origin: to provide a public rationale for the ongoing subordination of non-Asian people of color. Because the embrace or love is not genuine, one cannot reasonably expect the architects truly to care about the health or well-being of the model minority. . . .

NOTES

1. The human-capital model is a variant of neoclassical economic theory applied to labor-market economics. The model attempts to explain a dependent variable such as income inequality as derivative of independent variables such as educational attainment, work experience, total number of hours worked per year, marital status, and English proficiency, among others. It explains Asian Americans' advancement by foregrounding their investments in human capital such as educational attainment. It would also explain the lack of mobility of certain groups such as Southeast Asians or African Americans who had not yet invested appropriately in human capital.

2. Structuralists emphasize the basic characteristics or "structures" of a social system. For example, internal colonial theory places the U.S. historical and contemporary actualities of race in the context of international colonialism that connects the Third

World abroad with the Third World within the U.S. This framework emphasizes similarities in the patterns of racial domination and exploitation experienced by people of color and distinguishes their incorporation into the U.S. from the processes of immigration and assimilation by which European Americans are incorporated into a nation. See Robert Blauner, "Colonized and Immigrant Minorities," *Racial Oppression in America* (New York: Harper and Row, 1972), 51–82.

3. President Bush's 1988 television campaign centered around an ad attacking Michael Dukakis for being "soft on crime." To support this claim, the TV ads played upon white fears by focusing on close-up "mug" shots of Willie Horton, an African-American convict who committed subsequent crimes while on an early-release program implemented by Dukakis, then governor of Massachusetts.

4. Cynthia Hamilton, *Apartheid in an American City: The Case of the Black Community in Los Angeles* (Van Nuys, CA: Labor Community Strategy Center, 1991), 1.

5. Ibid. Internal colonial theory is useful in viewing South Central as a Third World colony, within a First World nation, that lacks political and economic self-determination, similar to the relationship between Third World nations abroad and First World ones.

6. Eui-Young Yu, "We Saw Our Dreams Burn for No Reason," *San Francisco Examiner*, 24 May 1992, editorial page.

7. Robert C. Toh, "Blacks Pressing Japanese to Halt Slurs, Prejudice," *Los Angeles Times*, 13 December 1990.

8. Susan Moffat, "Shopkeepers Fight Back," *Los Angeles Times*, 15 May 1992.

9. Ibid.

10. At 17 percent, self-employment rates of foreign-born Korean men top those of white men (10 percent) and other people of color. U.S. Bureau of the Census, 1980 Census of Population, vol. 2, Subject Reports, *Asian and Pacific Islander Population in the United States: 1980*, table 45A. A 1990 survey conducted by California State University sociologist Eui-Young Yu found that nearly 40 percent of Korean families own a business.

11. The long hours worked by more members of the family under dangerous conditions, combined with the high educational levels of those owning businesses, calls into question the definition of success applied to Korean Americans. Some have suggested that a blue-collar union job offers better working conditions and a more reliable income than engaging in small businesses.

12. Nancy Hill-Holtzman and Mathis Chazanov, "Police Credited for Heading Off Spread of Riots," *Los Angeles Times*, 7 May 1992.

13. Ibid.

14. Steven Chin, "Innocence Lost: L.A.'s Koreans Fight to be Heard," *San Francisco Examiner*, 9 May 1992.

15. California National Guard Brigadier General Daniel Brennan stated that bullets and grenades were not loaded for transport to L.A. because there were no lights on the parade ground where the ammunition is stored. Daniel Weintraub, "National Guard Official Cites Series of Delays," *Los Angeles Times*, 7 May 1992.

16. Rich Connell and Richard Simon, "Top LAPD Officer, Fire Chief Cite Flaws in Police Response," *Los Angeles Times*, 8 May 1992.

17. *At a time when it is being proposed that hundreds of billions be spent to uplift Negroes and other minorities,* the nation's 300,000 Chinese are moving ahead on their own . . . with no help from anyone else. Still being taught in Chinatown is the old idea that people should depend on their own efforts . . . not a welfare check . . . in order to reach America's 'promised land' " (emphasis added). "Success Story of One Minority Group in U.S.," *U.S. News and World Report*, 26 December 1966, 73. Similarly, scholar Thomas Sowell argues in *Race and Economics* (New York: McKay and Co., 1975) that historic exclusion from U.S. political institutions paradoxically benefits Jews and Asian Americans. "[T]hose American ethnic groups that have succeeded best politically have not usually been the same as those who succeeded best economically. . . . [T]hose minorities that have pinned their greatest hopes on political action—the Irish and the Negroes, for example—have made some of the slower economic advances."

18. Frank Chin and Jeffrey Paul Chan, "Racist Love," in *Seen through Shuck* ed. Richard Kostelanetz (New York: Ballantine Books, 1972), 65–79.

FRATERNITIES AND RAPE ON CAMPUS

52

Patricia Yancey Martin and Robert A. Hummer

Rapes are perpetrated on dates, at parties, in chance encounters, and in specially planned circumstances. That group structure and processes, rather than individual values or characteristics, are the impetus for many rape

Author's note: We gratefully thank Meena Harris and Diane Mennella for assisting with data collection. The senior author thanks the graduate students in her fall 1988 graduate research methods seminar for help with developing the initial conceptual framework. Judith Lorber and two anonymous *Gender & Society* referees made numerous suggestions for improving our article and we thank them also.

From: *Gender & Society* 3 (December 1989): 457–473. Reprinted by permission of Sage Publications, Inc.

episodes was documented by Blanchard (1959) 30 years ago (also see Geis 1971), yet sociologists have failed to pursue this theme (for an exception, see Chancer 1987). A recent review of research (Muehlenhard and Linton 1987) on sexual violence, or rape, devotes only a few pages to the situational context of rape events, and these are conceptualized as potential risk factors for individuals rather than qualities of rape-prone social contexts.

Many rapes, far more than come to the public's attention, occur in fraternity houses on college and university campuses, yet little research has analyzed fraternities at American colleges and universities as rape-prone contexts (cf. Ehrhart and Sandler 1985). Most of the research on fraternities reports on samples of individual fraternity men. One group of studies compares the values, attitudes, perceptions, family socioeconomic status, psychological traits (aggressiveness, dependence), and so on, of fraternity and nonfraternity men (Bohrnstedt 1969; Fox, Hodge, and Ward 1987; Kanin 1967; Lemire 1979; Miller 1973). A second group attempts to identify the effects of fraternity membership over time on the values, attitudes, beliefs, or moral precepts of members (Hughes and Winston 1987; Marlowe and Auvenshine 1982; Miller 1973; Wilder, Hoyt, Doren, Hauck, and Zettle 1978; Wilder, Hoyt, Surbeck, Wilder, and Carney 1986). With minor exceptions, little research addresses the group and organizational context of fraternities or the social construction of fraternity life (for exceptions, see Letchworth 1969; Longino and Kart 1973; Smith 1964).

Gary Tash, writing as an alumnus and trial attorney in his fraternity's magazine, claims that over 90 percent of all gang rapes on college campuses involve fraternity men (1988, p. 2). Tash provides no evidence to substantiate this claim, but students of violence against women have been concerned with fraternity men's frequently reported involvement in rape episodes (Adams and Abarbanel 1988). Ehrhart and Sandler (1985) identify over 50 cases of gang rapes on campus perpetrated by fraternity men, and their analysis points to many of the conditions that we discuss here. Their analysis is unique in focusing on conditions in fraternities that make gang rapes of women by fraternity men both feasible and probable. They identify excessive alcohol use, isolation from external monitoring, treatment of women as prey, use of pornography, approval of violence, and excessive concern with competition as precipitating conditions to gang rape (also see Merton 1985; Roark 1987).

The study reported here confirmed and complemented these findings by focusing on both conditions and processes. We examined dynamics associated with the social construction of fraternity life, with a focus on processes that foster the use of coercion, including rape, in fraternity men's relations with women. Our examination of men's social fraternities on college and university campuses as groups and organizations led us to conclude that fraternities are a physical and sociocultural context that encourages the sexual coercion of women. We make no claims that all fraternities are "bad" or that all fraternity men are rapists. Our observations indicated, however, that rape is especially probable in fraternities because of the kinds of organizations they are, the kinds of members they have, the practices their members engage in, and a virtual absence of university or community oversight. Analyses that lay blame for rapes by fraternity men on "peer pressure" are, we feel, overly simplistic (cf. Burkhart 1989; Walsh 1989). We suggest, rather, that fraternities create a sociocultural context in which the use of coercion in sexual relations with women is normative and in which the mechanisms to keep this pattern of behavior in check are minimal at best and absent at worst. We conclude that unless fraternities change in fundamental ways, little improvement can be expected.

METHODOLOGY

Our goal was to analyze the group and organizational practices and conditions that create in fraternities an abusive social context for women. We developed a conceptual framework from an initial case study of an alleged gang rape at Florida State University that involved four fraternity men and an 18-year-old coed. The group rape took place on the third floor of a fraternity house and ended with the "dumping" of the woman in the hallway of a neighboring fraternity house. According to newspaper accounts, the victim's blood-alcohol concentration, when she was discovered, was .349 percent, more than three times the legal limit for automobile driving and an almost lethal amount. One law enforcement officer reported that sexual intercourse occurred during the time the victim was unconscious: "She was in a life-threatening situation" (*Tallahassee Democrat*, 1988b). When the victim was found, she was comatose and had suffered multiple scratches and abrasions. Crude words and a fraternity symbol had been written on her thighs (*Tampa Tribune*, 1988). When law enforcement officials tried to investigate the case, fraternity members refused to cooperate. This led, eventually, to a five-year ban of the fraternity from campus by the university and by the fraternity's national organization.

In trying to understand how such an event could have occurred, and how a group of over 150 members (exact figures are unknown because the fraternity refused to provide a membership roster) could hold rank, deny knowledge of the event, and allegedly lie to a grand jury, we analyzed newspaper articles about the case and conducted open-ended interviews with a variety of respondents about the case and about fraternities, rapes, alcohol use, gender relations, and sexual activities on campus. Our data included over 100 newspaper articles on the initial gang rape case; open-ended interviews with Greek (social fraternity and sorority) and non-Greek (independent) students (N = 20); university administrators (N = 8, five men, three women); and alumni advisers to Greek organizations (N = 6). Open-ended interviews were held also with judges, public and private defense attorneys, victim advocates, and state prosecutors regarding the processing of sexual assault cases. Data were analyzed using the grounded theory method (Glaser 1978; Martin and Turner 1986). In the following analysis, concepts generated from the data analysis are integrated with the literature on men's social fraternities, sexual coercion, and related issues.

FRATERNITIES AND THE SOCIAL CONSTRUCTION OF MEN AND MASCULINITY

Our research indicated that fraternities are vitally concerned—more than with anything else—with masculinity (cf. Kanin 1967). They work hard to create a macho image and context and try to avoid any suggestion of "wimpishness," effeminacy, and homosexuality. Valued members display, or are willing to go along with, a narrow conception of masculinity that stresses competition, athleticism, dominance, winning, conflict, wealth, material possessions, willingness to drink alcohol, and sexual prowess vis-à-vis women.

Valued Qualities of Members

When fraternity members talked about the kind of pledges they prefer, a litany of stereotypical and narrowly masculine attributes and behaviors was recited and feminine or woman-associated qualities and behaviors were expressly denounced (cf. Merton 1985). Fraternities seek men who are "athletic," "big guys," good in intramural competition, "who can talk college sports." Males "who are willing to drink alcohol," "who drink socially," or "who can hold their liquor" are sought. Alcohol and activities associated with the recreational use of alcohol are cornerstones of fraternity social life. Nondrinkers are viewed with skepticism and rarely selected for membership.[1]

Fraternities try to avoid "geeks," nerds, and men said to give the fraternity a "wimpy" or "gay" reputation. Art, music, and humanities majors, majors in traditional women's fields (nursing, home economics, social work, education), men with long hair, and those whose appearance or dress violate current norms are rejected. Clean-cut, handsome men who dress well (are clean, neat, conforming, fashionable) are preferred. One sorority woman commented that "the top ranking fraternities have the best looking guys."

One fraternity man, a senior, said his fraternity recruited "some big guys, very athletic" over a two-year period to help overcome its image of wimpiness. His fraternity had won the interfraternity competition for highest grade-point average several years running but was looked down on as "wimpy, dancy, even gay." With their bigger, more athletic recruits, "our reputation improved; we're a much more recognized fraternity now." Thus a fraternity's reputation and status depends on members' possession of stereotypically masculine qualities. Good grades, campus leadership, and community service are "nice" but masculinity dominance—for example, in athletic events, physical size of members, athleticism of members—counts most.

Certain social skills are valued. Men are sought who "have good person-alities," are friendly, and "have the ability to relate to girls" (cf. Longino and Kart 1973). One fraternity man, a junior, said: "We watch a guy [a potential pledge] talk to women . . . we want guys who can relate to girls." Assessing a pledge's ability to talk to women is, in part, a preoccupation with homosexu-ality and a conscious avoidance of men who seem to have effeminate manners or qualities. If a member is suspected of being gay, he is ostracized and informally drummed out of the fraternity. A fraternity with a reputation as wimpy or tolerant of gays is ridiculed and shunned by other fraternities. Militant heterosexuality is frequently used by men as a strategy to keep each other in line (Kimmel 1987).

Financial affluence or wealth, a male-associated value in American culture, is highly valued by fraternities. In accounting for why the fraternity involved in the gang rape that precipitated our research project had been recognized recently as "the best fraternity chapter in the United States," a university official said: "They were good-looking, a big fraternity, had lots of BMWs [expensive, German-made automobiles]." After the rape, newspaper stories described the fraternity members' affluence, noting the high number of members who owned expensive cars (*St. Petersburg Times,* 1988).

The Status and Norms of Pledgeship

A pledge (sometimes called an associate member) is a new recruit who occupies a trial membership status for a specific period of time. The pledge period (typically ranging from 10 to 15 weeks) gives fraternity brothers an opportu-nity to assess and socialize new recruits. Pledges evaluate the fraternity also and decide if they want to become brothers. The socialization experience is structured partly through assignment of a Big Brother to each pledge. Big Brothers are expected to teach pledges how to become a brother and to support them as they progress through the trial membership period. Some pledges are repelled by the pledging experience, which can entail physical abuse; harsh discipline; and demands to be subordinate, follow orders, and engage in demeaning routines and activities, similar to those used by the military to "make men out of boys" during boot camp.

Characteristics of the pledge experience are rationalized by fraternity members as necessary to help pledges unite into a group, rely on each other, and join together against outsiders. The process is highly masculinist in execution as well as conception. A willingness to submit to authority, follow orders, and do as one is told is viewed as a sign of loyalty, togetherness, and unity. Fraternity pledges who find the pledge process offensive often drop out. Some do this by openly quitting, which can subject them to ridicule by brothers and other pledges, or they may deliberately fail to make the grades necessary for initiation or transfer schools and decline to reaffiliate with the fraternity on the new campus. One fraternity pledge who quit the fraternity he had pledged described an experience during pledgeship as follows:

> This one guy was always picking on me. No matter what I did, I was wrong. One night after dinner, he and two other guys called me and two other pledges into the chapter room. He said, "Here, X, hold this 25 pound bag of ice at arms' length 'til I tell you to stop." I did it even though my arms and hands were killing me. When I asked if I could stop, he grabbed me around the throat and lifted me off the floor. I thought he would choke me to death. He cussed me and called me all kinds of names. He took one of my fingers and twisted it until it nearly broke. . . . I stayed in the fraternity for a few more days, but then I decided to quit. I hated it. Those guys are sick. They like seeing you suffer.

Fraternities' emphasis on toughness, withstanding pain and humiliation, obedience to superiors, and using physical force to obtain compliance contributes to an interpersonal style that de-emphasizes caring and sensitivity but fosters intragroup trust and loyalty. If the least macho or most critical pledges drop out, those who remain may be more receptive to, and influenced by, masculinist values and practices that encourage the use of force in sexual relations with women and the covering up of such behavior (cf. Kanin 1967).

Norms and Dynamics of Brotherhood

Brother is the status occupied by fraternity men to indicate their relations to each other and their membership in a particular fraternity organization or group. Brother is a male-specific status; only males can become brothers, although women can become "Little Sisters," a form of pseudomembership. "Becoming a brother" is a rite of passage that follows the consistent and often lengthy display by pledges of appropriately masculine qualities and behaviors. Brothers have a quasi-familial relationship with each other, are normatively said to share bonds of closeness and support, and are sharply set off from nonmembers. Brotherhood is a loosely defined term used to represent the bonds that develop among fraternity members and the obligations and expectations incumbent upon them (cf. Marlowe and Auvenshine [1982] on fraternities' failure to encourage "moral development" in freshman pledges).

Some of our respondents talked about brotherhood in almost reverential terms, viewing it as the most valuable benefit of fraternity membership. One senior, a business-school major who had been affiliated with a fairly high-status fraternity throughout four years on campus, said:

> Brotherhood spurs friendship for life, which I consider its best aspect, although I didn't see it that way when I joined. Brotherhood bonds and unites. It instills values of caring about one another, caring about community, caring about ourselves. The values and bonds [of brotherhood] continually develop over the four years [in college] while normal friendships come and go.

Despite this idealization, most aspects of fraternity practice and conception are more mundane. Brotherhood often plays itself out as an overriding concern with masculinity and, by extension, femininity. As a consequence, fraternities comprise collectivities of highly masculinized men with attitudinal qualities and behavioral norms that predispose them to sexual coercion of women (cf. Kanin 1967; Merton 1985; Rapaport and Burkhart 1984). The norms of masculinity are complemented by conceptions of women and femininity that are equally distorted and stereotyped and that may enhance the probability of women's exploitation (cf. Ehrhart and Sandler 1985; Sanday 1981, 1986).

Practices of Brotherhood

Practices associated with fraternity brotherhood that contribute to the sexual coercion of women include a preoccupation with loyalty, group protection and secrecy, use of alcohol as a weapon, involvement in violence and physical force, and an emphasis on competition and superiority.

Loyalty, group protection, and secrecy. Loyalty is a fraternity preoccupation. Members are reminded constantly to be loyal to the fraternity and to their brothers. Among other ways, loyalty is played out in the practices of group protection and secrecy. The fraternity must be shielded from criticism. Members are admonished to avoid getting the fraternity in trouble and to bring all problems "to the chapter" (local branch of a national social fraternity) rather than to outsiders. Fraternities try to protect themselves from close scrutiny and criticism by the Interfraternity Council (a quasi-governing body composed of representatives from all social fraternities on campus), their fraternity's national office, university officials, law enforcement, the media, and the public. Protection of the fraternity often takes precedence over what is procedurally, ethically, or legally correct. Numerous examples were related to us of fraternity brothers' lying to outsiders to "protect the fraternity."

Group protection was observed in the alleged gang rape case with which we began our study. Except for one brother, a rapist who turned state's evidence, the entire remaining fraternity membership was accused by university and criminal justice officials of lying to protect the fraternity. Members consistently failed to cooperate even though the alleged crimes were felonies, involved only four men (two of whom were not even members of the local chapter), and the victim of the crime nearly died. According to a grand jury's findings, fraternity officers repeatedly broke appointments with law enforcement officials, refused to provide police with a list of members, and refused to cooperate with police and prosecutors investigating the case (*Florida Flambeau*, 1988).

Secrecy is a priority value and practice in fraternities, partly because full-fledged membership is premised on it (for confirmation, see Ehrhart and Sandler 1985; Longino and Kart 1973; Roark 1987). Secrecy is also a boundary-maintaining mechanism, demarcating in-group from out-group, us from them. Secret rituals, handshakes, and mottoes are revealed to pledge brothers as they are initiated into full brotherhood. Since only brothers are supposed to know a fraternity's secrets, such knowledge affirms membership in the fraternity and separates a brother from others. Extending secrecy tactics from protection of private knowledge to protection of the fraternity from criticism is a predictable development. Our interviews indicated that individual members knew the difference between right and wrong, but fraternity norms that emphasize loyalty, group protection, and secrecy often overrode standards of ethical correctness.

Alcohol as weapon. Alcohol use by fraternity men is normative. They use it on weekdays to relax after class and on weekends to "get drunk," "get crazy," and "get laid." The use of alcohol to obtain sex from women is pervasive—in other words, it is used as a weapon against sexual reluctance. According to several fraternity men whom we interviewed, alcohol is the major tool used to gain sexual mastery over women (cf. Adams and Abarbanel 1988; Ehrhart and Sandler 1985). One fraternity man, a 21-year-old senior, described alcohol use to gain sex as follows: "There are girls that you know will fuck, then some you have to put some effort into it. . . . You have to buy them drinks or find out if she's drunk enough. . . ."

A similar strategy is used collectively. A fraternity man said that at parties with Little Sisters: "We provide them with 'hunch punch' and things get wild.

We get them drunk and most of the guys end up with one." " 'Hunch punch,' " he said, "is a girls' drink made up of overproof alcohol and powdered Kool-Aid, no water or anything, just ice. It's very strong. Two cups will do a number on a female." He had plans in the next academic term to surreptitiously give hunch punch to women in a "prim and proper" sorority because "having sex with prim and proper sorority girls is definitely a goal." These women are a challenge because they "won't openly consume alcohol and won't get openly drunk as hell." Their sororities have "standards committees" that forbid heavy drinking and easy sex.

In the gang rape case, our sources said that many fraternity men on campus believed the victim had a drinking problem and was thus an "easy make." According to newspaper accounts, she had been drinking alcohol on the evening she was raped; the lead assailant is alleged to have given her a bottle of wine after she arrived at his fraternity house. Portions of the rape occurred in a shower, and the victim was reportedly so drunk that her assailants had difficulty holding her in a standing position (*Tallahassee Democrat*, 1988a). While raping her, her assailants repeatedly told her they were members of another fraternity under the apparent belief that she was too drunk to know the difference. Of course, if she was too drunk to know who they were, she was too drunk to consent to sex (cf. Allgeier 1986; Tash 1988).

One respondent told us that gang rapes are wrong and can get one expelled, but he seemed to see nothing wrong in sexual coercion one-on-one. He seemed unaware that the use of alcohol to obtain sex from a woman is grounds for a claim that a rape occurred (cf. Tash 1988). Few women on campus (who also may not know these grounds) report date rapes, however; so the odds of detection and punishment are slim for fraternity men who use alcohol for "seduction" purposes (cf. Byington and Keeter 1988; Merton 1985).

Violence and physical force. Fraternity men have a history of violence (Ehrhart and Sandler 1985; Roark 1987). Their record of hazing, fighting, property destruction, and rape has caused them problems with insurance companies (Bradford 1986; Pressley 1987). Two university officials told us that fraternities "are the third riskiest property to insure behind toxic waste dumps and amusement parks." Fraternities are increasingly defendants in legal actions brought by pledges subjected to hazing (Meyer 1986; Pressley 1987) and by women who were raped by one or more members. In a recent alleged gang rape incident at another Florida university, prosecutors failed to file charges but the victim filed a civil suit against the fraternity nevertheless (*Tallahassee Democrat*, 1989).

Competition and superiority. Interfraternity rivalry fosters in-group identification and out-group hostility. Fraternities stress pride of membership and superiority over other fraternities as major goals. Interfraternity rivalries take many forms, including competition for desirable pledges, size of pledge class, size of membership, size and appearance of fraternity house, superiority in intramural sports, highest grade-point averages, giving the best parties, gaining the best or most campus leadership roles, and, of great importance, attracting and displaying "good looking women." Rivalry is particularly intense over members, intramural sports, and women (cf. Messner 1989).

FRATERNITIES' COMMODIFICATION OF WOMEN

In claiming that women are treated by fraternities as commodities, we mean that fraternities knowingly, and intentionally, *use* women for their benefit. Fraternities use women as bait for new members, as servers of brother's needs, and as sexual prey.

Women as bait. Fashionably attractive women help a fraternity attract new members. As one fraternity man, a junior, said, "They are good bait." Beautiful, sociable women are believed to impress the right kind of pledges and give the impression that the fraternity can deliver this type of woman to its members. Photographs of shapely, attractive coeds are printed in fraternity brochures and videotapes that are distributed and shown to potential pledges. The women pictured are often dressed in bikinis, at the beach, and are pictured hugging the brothers of the fraternity. One university official says such recruitment materials give the message: "Hey, they're here for you, you can have whatever you want," and, "we have the best looking women. Join us and you can have them too." Another commented: "Something's wrong when males join an all-male organization as the best place to meet women. It's so illogical."

Fraternities compete in promising access to beautiful women. One fraternity man, a senior, commented that "the attraction of girls [i.e., a fraternity's success in attracting women] is a big status symbol for fraternities." One university official commented that the use of women as a recruiting tool is so well entrenched that fraternities that might be willing to forgo it say they cannot afford to unless other fraternities do so as well. One fraternity man said, "Look, if we don't have Little Sisters, the fraternities that do will get all the good pledges." Another said, "We won't have as good a rush [the period during which new members are assessed and selected] if we don't have these women around."

In displaying good-looking, attractive, skimpily dressed, nubile women to potential members, fraternities implicitly, and sometimes explicitly, promise sexual access to women. One fraternity man commented that "part of what being in a fraternity is all about is the sex" and explained how his fraternity uses Little Sisters to recruit new members:

We'll tell the sweetheart [the fraternity's term for Little Sister], "You're gorgeous; you can get him." We'll tell her to fake a scam and she'll go hang all over him during a rush party, kiss him, and he thinks he's done wonderful and wants to join. The girls think it's great too. It's flattering for them.

Women as servers. The use of women as servers is exemplified in the Little Sister program. Little Sisters are undergraduate women who are rushed and selected in a manner parallel to the recruitment of fraternity men. They are affiliated with the fraternity in a formal but unofficial way and are able, indeed required, to wear the fraternity's Greek letters. Little Sisters are not full-fledged fraternity members, however; and fraternity national offices and most universities do not register or regulate them. Each fraternity has an officer called Little Sister Chairman who oversees their organization and activities. The Little Sisters elect officers among themselves, pay monthly dues to the fraternity, and have well-defined roles. Their dues are used to pay for the fraternity's social events, and Little Sisters are expected to attend and hostess fraternity parties and hang around the house to make it a "nice place to be." One fraternity man, a senior, described Little Sisters this way: "They are very social girls, willing to join in, be affiliated with the group, devoted to the fraternity." Another member, a sophomore, said: "Their sole purpose is social—attend parties, attract new members, and 'take care' of the guys."

Our observations and interviews suggested that women selected by fraternities as Little Sisters are physically attractive, possess good social skills, and are willing to devote time and energy to the fraternity and its members. One undergraduate woman gave the following job description for Little Sisters to a campus newspaper:

It's not just making appearances at all the parties but entails many more responsibilities. You're going to be expected to go to all the intramural games to cheer the brothers on, support and encourage the pledges, and just be around to bring some extra life to the house. [As a Little Sister] you have to agree to take on a new responsibility other than studying to maintain your grades and managing to keep your checkbook from bouncing. You have to make time to be a part of the fraternity and support the brothers in all they do. (*The Tomahawk*, 1988)

The title of Little Sister reflects women's subordinate status; fraternity men in a parallel role are called Big Brothers. Big Brothers assist a sorority primarily with the physical work of sorority rushes, which, compared to fraternity rushes, are more formal, structured, and intensive. Sorority rushes take place in the daytime and fraternity rushes at night so fraternity men are free to help. According to one fraternity member, Little Sister status is a benefit to women because it gives them a social outlet and "the protection of the brothers." The gender-stereotypic conceptions and obligations of these Little Sister and Big Brother statuses indicate that fraternities and sororities promote a gender hierarchy on campus that fosters subordination and dependence in women, thus encouraging sexual exploitation and the belief that it is acceptable.

Women as sexual prey. Little Sisters are a sexual utility. Many Little Sisters do not belong to sororities and lack peer support for refraining from unwanted sexual relations. One fraternity man (whose fraternity has 65 members and 85 Little Sisters) told us they had recruited "wholesale" in the prior year to "get lots of new women." The structural access to women that the Little Sister program provides and the absence of normative supports for refusing fraternity members' sexual advances may make women in this program particularly susceptible to coerced sexual encounters with fraternity men.

Access to women for sexual gratification is a presumed benefit of fraternity membership, promised in recruitment materials and strategies and through brothers' conversations with new recruits. One fraternity man said: "We always tell the guys that you get sex all the time, there's always new girls. . . . After I became a Greek, I found out I could be with females at will." A university official told us that, based on his observations, "no one [i.e., fraternity men] on this campus wants to have 'relationships.' They just want to have fun [i.e., sex]." Fraternity men plan and execute strategies aimed at obtaining sexual gratification, and this occurs at both individual and collective levels.

Individual strategies include getting a woman drunk and spending a great deal of money on her. As for collective strategies, most of our undergraduate interviewees agreed that fraternity parties often culminate in sex and that this outcome is planned. One fraternity man said fraternity parties often involve sex and nudity and can "turn into orgies." Orgies may be planned in advance, such as the Bowery Ball party held by one fraternity. A former fraternity member said of this party:

> The entire idea behind this is sex. Both men and women come to the party wearing little or nothing. There are pornographic pinups on the walls and usually porno movies playing on the TV. The music carries sexual overtones. . . . They just get schnockered [drunk] and, in most cases, they also get laid.

When asked about the women who come to such a party, he said: "Some Little Sisters just won't go. . . . The girls who do are looking for a good time, girls who don't know what it is, things like that."

Other respondents denied that fraternity parties are orgies but said that sex is always talked about among the brothers and they all know "who each other is doing it with." One member said that most of the time, guys have sex with their girlfriends "but with socials, girlfriends aren't allowed to come and it's their [members'] big chance [to have sex with other women]." The use of alcohol to help them get women into bed is a routine strategy at fraternity parties.

CONCLUSIONS

In general, our research indicated that the organization and membership of fraternities contribute heavily to coercive and often violent sex. Fraternity houses are occupied by same-sex (all men) and same-age (late teens, early twenties) peers whose maturity and judgment is often less than ideal. Yet fraternity houses are private dwellings that are mostly off-limits to, and away from scrutiny of, university and community representatives, with the result that fraternity house events seldom come to the attention of outsiders. Practices associated with the social construction of fraternity brotherhood emphasize a macho conception of men and masculinity, a narrow, stereotyped conception of women and femininity, and the treatment of women as commodities. Other practices contributing to coercive sexual relations and the cover-up of rapes include excessive alcohol use, competitiveness, and normative support for deviance and secrecy (cf. Bogal-Allbritten and Allbritten 1985; Kanin 1967).

Some fraternity practices exacerbate others. Brotherhood norms require "sticking together" regardless of right or wrong; thus rape episodes are unlikely to be stopped or reported to outsiders, even when witnesses disapprove. The ability to use alcohol without scrutiny by authorities and alcohol's frequent association with violence, including sexual coercion, facilitates rape in fraternity houses. Fraternity norms that emphasize the value of maleness and masculinity over femaleness and femininity and that elevate the status of men and lower the status of women in members' eyes undermine perceptions and treatment of women as persons who deserve consideration and care (cf. Ehrhart and Sandler 1985; Merton 1985).

Androgynous men and men with a broad range of interests and attributes are lost to fraternities through their recruitment practices. Masculinity of a narrow and stereotypical type helps create attitudes, norms, and practices that predispose fraternity men to coerce women sexually, both individually and collectively (Allgeier 1986; Hood 1989; Sanday 1981, 1986). Male athletes on campus may be similarly disposed for the same reasons (Kirshenbaum 1989; Telander and Sullivan 1989).

Research into the social contexts in which rape crimes occur and the social constructions associated with these contexts illuminate rape dynamics on campus. Blanchard (1959) found that group rapes almost always have a leader who pushes others into the crime. He also found that the leader's latent homosexuality, desire to show off to his peers, or fear of failing to prove himself a man are frequently an impetus. Fraternity norms and practices contribute to the approval and use of sexual coercion as an accepted tactic in relations with women. Alcohol-induced compliance is normative, whereas, presumably, use of a knife, gun, or threat of bodily harm would not be because the woman who "drinks too much" is viewed as "causing her own rape" (cf. Ehrhart and Sandler 1985).

Our research led us to conclude that fraternity norms and practices influence members to view the sexual coercion of women, which is a felony crime, as sport, a contest, or a game (cf. Sato 1988). This sport is played not between men and women but between men and men. Women are the pawns or prey in the interfraternity rivalry game; they prove that a fraternity is successful or prestigious. The use of women in this way encourages fraternity men to see women as objects and sexual coercion as sport. Today's societal norms support young women's right to engage in sex at their discretion, and coercion is unnecessary in a mutually desired encounter. However, nubile young women say they prefer to be "in a relationship" to have sex while young men say they prefer to "get laid" without a commitment (Muehlenhard and Linton 1987). These differences may reflect, in part, American puritanism and men's fears of sexual intimacy or perhaps intimacy of any kind. In a fraternity context, getting sex without giving emotionally demonstrates "cool" masculinity. More important, it poses no threat to the bonding and loyalty of the fraternity brotherhood (cf. Farr 1988). Drinking large quantities of alcohol before having sex suggests that "scoring" rather than intrinsic sexual pleasure is a primary concern of fraternity men.

Unless fraternities' composition, goals, structures, and practices change in fundamental ways, women on campus will continue to be sexual prey for fraternity men. As all-male enclaves dedicated to opposing faculty and administration and to cementing in-group ties, fraternity members eschew any hint of homosexuality. Their version of masculinity transforms women, and men with womanly characteristics, into the out-group. "Womanly men" are ostracized; feminine women are used to demonstrate members' masculinity. Encouraging renewed emphasis on their founding values (Longino and Kart 1973), service orientation and activities (Lemire 1979), or members' moral development (Marlowe and Auvenshine 1982) will have little effect on fraternities' treatment of women. A case for or against fraternities cannot be made by studying individual members. The fraternity qua group and organization is at issue. Located on campus along with many vulnerable women, embedded in a sexist society, and caught up in masculinist goals, practices, and values, fraternities' violation of women—including forcible rape—should come as no surprise.

NOTE

1. Recent bans by some universities on open-keg parties at fraternity houses have resulted in heavy drinking before coming to a party and an increase in drunkenness among those who attend. This may aggravate, rather than improve, the treatment of women by fraternity men at parties.

REFERENCES

Allgeier, Elizabeth. 1986. "Coercive Versus Consensual Sexual Interactions." G. Stanley Hall Lecture to American Psychological Association Annual Meeting, Washington, DC, August.

Adams, Aileen and Gail Abarbanel. 1988. *Sexual Assault on Campus: What Colleges Can Do.* Santa Monica, CA: Rape Treatment Center.

Blanchard, W. H. 1959. "The Group Process in Gang Rape." *Journal of Social Psychology* 49:259–66.

Bogal-Allbritten, Rosemarie B. and William L. Allbritten. 1985. "The Hidden Victims: Courtship Violence Among College Students." *Journal of College Student Personnel* 43:201–4.

Bohrnstedt, George W. 1969. "Conservatism, Authoritarianism and Religiosity of Fraternity Pledges." *Journal of College Student Personnel* 27:36–43.

Bradford, Michael. 1986. "Tight Market Dries Up Nightlife at University." *Business Insurance* (March 2): 2, 6.

Burkhart, Barry. 1989. Comments in Seminar on Acquaintance/Date Rape Prevention: A National Video Teleconference, February 2.

Burkhart, Barry R. and Annette L. Stanton. 1985. "Sexual Aggression in Acquaintance Relationships." Pp. 43–65 in *Violence in Intimate Relationships*, edited by G. Russell. Englewood Cliffs, NJ: Spectrum.

Byington, Diane B. and Karen W. Keeter. 1988. "Assessing Needs of Sexual Assault Victims on a University Campus." Pp. 23–31 in *Student Services: Responding to Issues and Challenges.* Chapel Hill: University of North Carolina Press.

Chancer, Lynn S. 1987. "New Bedford, Massachusetts, March 6, 1983-March 22, 1984: The 'Before and After' of a Group Rape." *Gender & Society* 1:239–60.

Ehrhart, Julie K. and Bernice R. Sandler. 1985. *Campus Gang Rape: Party Games?* Washington, DC: Association of American Colleges.

Farr, K. A. 1988. "Dominance Bonding Through the Good Old Boys Sociability Network." *Sex Roles* 18:259–77.

Florida Flambeau. 1988. "Pike Members Indicted in Rape." (May 19):1, 5.

Fox, Elaine, Charles Hodge, and Walter Ward. 1987. "A Comparison of Attitudes Held by Black and White Fraternity Members." *Journal of Negro Education* 56:521–34.

Geis, Gilbert. 1971. "Group Sexual Assaults." *Medical Aspects of Human Sexuality* 5:101–13.

Glaser, Barney G. 1978. *Theoretical Sensitivity: Advances in the Methodology of Grounded Theory.* Mill Valley, CA: Sociology Press.

Hood, Jane. 1989. "Why Our Society Is Rape-Prone." *New York Times*, May 16.

Hughes, Michael J. and Roger B. Winston, Jr. 1987. "Effects of Fraternity Membership on Interpersonal Values." *Journal of College Student Personnel* 45:405–11.

Kanin, Eugene J. 1967. "Reference Groups and Sex Conduct Norm Violations." *The Sociological Quarterly* 8:495–504.

Kimmel, Michael, ed. 1987. *Changing Men: New Directions in Research on Men and Masculinity.* Newbury Park, CA: Sage.

Kirshenbaum, Jerry. 1989. "Special Report, An American Disgrace: A Violent and Unprecedented Lawlessness Has Arisen Among College Athletes in all Parts of the Country." *Sports Illustrated* (February 27): 16–19.

Lemire, David. 1979. "One Investigation of the Stereotypes Associated with Fraternities and Sororities." *Journal of College Student Personnel* 37:54–57.

Letchworth, G. E. 1969. "Fraternities Now and in the Future." *Journal of College Student Personnel* 10:118–22.

Longino, Charles F., Jr., and Cary S. Kart. 1973. "The College Fraternity: An Assessment of Theory and Research." *Journal of College Student Personnel* 31:118–25.

Marlowe, Anne F. and Dwight C. Auvenshine. 1982. "Greek Membership: Its Impact on the Moral Development of College Freshmen." *Journal of College Student Personnel* 40:53–57.

Martin, Patricia Yancey and Barry A. Turner. 1986. "Grounded Theory and Organizational Research." *Journal of Applied Behavioral Science* 22:141–57.

Merton, Andrew. 1985. "On Competition and Class: Return to Brotherhood." *Ms.* (September): 60–65, 121–22.

Messner, Michael. 1989. "Masculinities and Athletic Careers." *Gender & Society* 3:71–88.

Meyer, T. J. 1986. "Fight Against Hazing Rituals Rages on Campuses." *Chronicle of Higher Education* (March 12):34–36.

Miller, Leonard D. 1973. "Distinctive Characteristics of Fraternity Members." *Journal of College Student Personnel* 31:126–28.

Muehlenhard, Charlene L. and Melaney A. Linton. 1987. "Date Rape and Sexual Aggression in Dating Situations: Incidence and Risk Factors." *Journal of Counseling Psychology* 34:186–96.

Pressley, Sue Anne. 1987. "Fraternity Hell Night Still Endures." *Washington Post* (August 11):B1.

Rapaport, Karen and Barry R. Burkhart. 1984. "Personality and Attitudinal Characteristics of Sexually Coercive College Males." *Journal of Abnormal Psychology* 93:216–21.

Roark, Mary L. 1987. "Preventing Violence on College Campuses." *Journal of Counseling and Development* 65:367–70.

Sanday, Peggy Reeves. 1981. "The Socio-Cultural Context of Rape: A Cross-Cultural Study." *Journal of Social Issues* 37:5–27.

———. 1986. "Rape and the Silencing of the Feminine." Pp. 84–101 in *Rape*, edited by S. Tomaselli and R. Porter. Oxford: Basil Blackwell.

St. Petersburg Times. 1988. "A Greek Tragedy." (May 29):1F, 6F.

Sato, Ikuya. 1988. "Play Theory of Delinquency: Toward a General Theory of 'Action.'" *Symbolic Interaction* 11:191–212.

Smith, T. 1964. "Emergence and Maintenance of Fraternal Solidarity." *Pacific Sociological Review* 7:29–37.

Tallahassee Democrat. 1988a. "FSU Fraternity Brothers Charged" (April 27):1A, 12A.

———. 1988b. "FSU Interviewing Students About Alleged Rape" (April 24):1D.

———. 1989. "Woman Sues Stetson in Alleged Rape" (March 19):3B.

Tampa Tribune. 1988. "Fraternity Brothers Charged in Sexual Assault of FSU Coed." (April 27):6B.

Tash, Gary B. 1988. "Date Rape." *The Emerald of Sigma Pi Fraternity* 75(4):1–2.

Telander, Rick and Robert Sullivan. 1989. "Special Report, You Reap What You Sow." *Sports Illustrated* (February 27):20–34.

The Tomahawk. 1988. "A Look Back at Rush, A Mixture of Hard Work and Fun" (April/May):3D.

Walsh, Claire. 1989. Comments in Seminar on Acquaintance/Date Rape Prevention: A National Video Teleconference, February 2.

Wilder, David H., Arlyne E. Hoyt, Dennis M. Doren, William E. Hauck, and Robert D. Zettle. 1978. "The Impact of Fraternity and Sorority Membership on Values and Attitudes." *Journal of College Student Personnel* 36:445–49.

Wilder, David H., Arlyne E. Hoyt, Beth Shuster Surbeck, Janet C. Wilder, and Patricia Imperatrice Carney. 1986. "Greek Affiliation and Attitude Change in College Students." *Journal of College Student Personnel* 44:510–19.

V

Social Change and the Politics of Empowerment

Power is typically equated with domination and control over people or things. Social institutions depend on this version of power to reproduce hierarchies of race, class, and gender. Exploration of the experiences of African Americans, Latinos, women, Native Americans, Asian Americans, and the poor reveals much about how dominant groups exercise power, but centering on the experiences of historically marginalized groups also reveals much about resistance to oppression.

In different ways and under different conditions, people engage in individual acts of resistance and organized political activism to challenge race, class, and gender oppression, but dominant ideologies and the social institutions they defend try to obscure the individual and collective political activism of everyday citizens. By making the political activism of historically marginalized groups invisible, social institutions suppress the strength of these groups, making them more easily exploited. The articles in "Political Activism: Making a Difference" address this forced invisibility. Each article explores a different dimension of empowerment in the context of everyday life, demonstrating that people can and do work for social change. Although academic credentials, positions of authority, and economic resources do much to help individuals challenge hierarchies of race, class, and gender, these articles suggest that individual and collective actions of ordinary persons form the bedrock for any lasting social change.

For Chang Jok Lee, the elderly immigrant Chinese-American mother of eight and grandmother of eleven interviewed by Nancy Diao in "From Homemaker to Housing Advocate," political activism is not a theoretical issue but part of her everyday life. Her story demonstrates how individuals with ostensibly few resources for resistance can exercise political power. In "Sharing the Shop Floor," Stan Gray provides a firsthand account of the daily issues he confronted as a union organizer. He discusses the importance of holistic analyses of race, class, and gender in political activism and the necessity of collective action. "Growing Numbers, Growing Force," Kathleen Kautzer's account of how older women organized on a national level to improve their lives, parallels Stan Gray's rendition of the difficulties of organizing for change. Once again, we see how individuals deemed the least likely to organize on their own behalf not only were effective but also addressed the interlocking nature of race, class, and gender in their activism. Roberta Praeger, a welfare mother, gives a compelling account of her own successful struggle to become an activist working for "A World Worth Living In." Praeger's narrative demonstrates the importance of community and the value of collective struggle in empowering oneself and in working to empower others. Praeger found that she had to resist oppressive beliefs and actions from those closest to her daily life. Oppression accompanied by love is often difficult to see, let alone resist. Praeger engaged in a silent internal struggle to reject dominant ideologies telling her that she was lesser. She offers us an inside view of resisting internalized oppression, where racist, sexist, homophobic, or other dominant ideologies encourage the victims of oppression to collude in their own domination.

Because existing institutions are structured around race, class, and gender oppression, gaining power within them as they currently exist is unlikely to improve significantly the lives of African Americans, Native Americans, women, Asian Americans, the poor, gays and lesbians, Latinos, and other historically marginalized groups. Replacing one type of dominant group with another of a different color or gender may improve the lives of some, but it does not address the fundamental inequalities that pervade existing social institutions.

Investigating forms of power used by historically marginalized groups offers one way of rethinking social change and reconceptualizing the politics of empowerment. As Black feminist theorist bell hooks (1984: 93) suggests, "Sexism has never rendered women powerless. It has either suppressed their strength or exploited it. Recognition of that strength, that power, is a step women together can take toward liberation." African Americans, Native Americans, women, Latinos, gays and lesbians, and the poor and working class have never been powerless; the question is how to identify and use those forms of power that often go unrecognized.

Reenvisioning and exercising power to bring about social change requires a sense of purpose and a vision that encourages us to look beyond what already exists. We must learn to question what is possible, at the same time that we challenge existing social conditions. For example, what type of housing would result if Chang Jok Lee were central in her community's decision-making process? If men and women truly learned to "share the shop floor," how might economic security be better provided for all? What would distinguish a "world worth living in" from the one we now have?

The four essays in "Envisioning Change" provide new visions about what is possible *and* what is necessary. Each offers inclusive thinking, reconstructed knowledge, and a distinctive perspective on the kinds of questions we should be considering in working toward Praeger's "world worth living in." One theme that pervades the concluding essays is rethinking the nature of difference and confronting the enormous consequences of racism in this society. "We have *all* been programmed to respond to the human differences between us with fear and loathing," contends Audre Lorde in "Age, Race, Class, and Sex," "and to handle that difference in one of three ways: ignore it, and if that is not possible, copy it if we think it is dominant, or destroy it if we think it is subordinate." Race, class, and gender have created distinctive histories among us and have encouraged us to think about differences in ways that provide few models for relating across differences as equals. As Cornel West argues in "Race Matters," we have been paralyzed by seeing race only as a problem of and for Black people, not as a problem endemic in the society. As he says, "we need to begin with a frank acknowledgment of the basic humanness and Americanness of each of us."

These essays envision how we might address the question of difference. Rayna Green's "Culture and Gender in Indian America" suggests that Native-American culture offers alternative ways of conceptualizing difference that would enrich Western social institutions. "In Indian cultures I go back to my families and there is no division," observes Green. "There are distinctions, certainly, about the way people are treated, and the authority they have, but I want to reclaim the power to move through categories, so that I do not have to stay fixed in any one place." Her account of the utility of reclaiming all parts of her past provides a model for allowing diverse groups to speak for themselves. It allows us to imagine alternative ways of being that are not based on hierarchical distinctions among race, class, and gender groups. For example, one such distinction—the vast differences in wealth that characterize the class system—is viewed quite differently in Native American culture. According to Green, in Native American traditions, real wealth lies in "our own hearts, and not in something that is a commodity beyond it." The richest person is the one who "gives the most away, not the person who keeps the most for themselves."

Through the image of the barred room in "Coalition Politics: Turning the Century," Bernice Reagon explores the creation of nurturing spaces and communities where we can recover from the damaging effects of race, class, and gender oppression. "Nationalism is crucial to a people if you are going to ever impact as a group in your interest," suggests Reagon. African American communities, for example, offer a space where Blacks can retreat from the difficulties of oppression, learn ways of supporting one another, and identify the positive qualities of difference based on group identity, but Reagon also points to the limited value of these spaces. Nationalism and the celebration of difference that can accompany it become limiting if a group does not look beyond its own room and learn to build coalitions with other groups. Cornel West is similarly critical of nationalist movements, such as Afrocentrism. Such movements are important, he argues, because they put the achievements and sufferings of oppressed people at the center of our thinking, but in silencing discussion and analysis of how race, homophobia, class, and gender matter, nationalist movements "reinforce narrow discussion of race" and do not provide the framework from which to build the common good.

Building a more inclusive and equitable society requires coalition building. This is hard work that creates tensions and discomforts. Green illustrates this problem in her discussion of the tensions created at a women's studies conference when White feminists misunderstood the actions of Native American people because of their lack of knowledge of Native American culture. This tension can be further aggravated when groups of different power within hierarchies of race, class, and gender try to build coalitions. Although the White feminists in Green's example had good intentions, they also wanted to retain their power to define the conference. They wanted to include Native American women, but only on terms acceptable to them. As a result, effective communication was limited by differences in power that allowed one group to define the terms of the discourse.

Audre Lorde offers another view of how coalitions can be built across differences generated by different histories and varying amounts of power. Lorde cautions us to reject the so-called mythical norm, one key indicator of group power. Usually defined as White, thin, male, young, heterosexual,

Christian, and financially secure, within this norm reside the trappings of power. Coalition building in such a context requires not just trying to relate across differences but also identifying the differences in power that accrue to groups based on their proximity to the mythical norm. Truly effective coalition building involves rejecting the entire norm, not just that part of it that oppresses one's own group. African Americans of both genders must resist sexism; antiracist policies must inform feminist political activism; and poverty cannot be eliminated without challenging its race, class, and age-specific dimensions.

"Change means growth, and growth can be painful," observes Audre Lorde. "But we sharpen self-definition by exposing the self in work and struggle together with those whom we define as different from ourselves, although sharing the same goals." Building coalitions across differences and empowering historically marginalized groups requires seeing the connections among groups currently differentiated by inequalities of race, class, and gender. Doing so fosters much-needed social change and eventually empowers us all; not doing so leaves us vulnerable to the hatred, violence, and fear that can destroy us. We can only have the tools to imagine and build a multicultural democracy if we see how race, class, and gender shape our common history and our common destiny and if, as Cornel West suggests, we base this new spirit and vision on empathy and compassion.

Reference

hooks, bell. 1984. *From Margin to Center*. Boston: South End Press.

Political Activism: Making a Difference

GROWING NUMBERS, GROWING FORCE: *Older Women Organize*

Kathleen Kautzer

53

We have played the game according to rules you set down. Now the rules have changed and we're called out, but we have one half to one third of our lives still to go, and we will not be shelved.

—Tish Sommers, "Epilogue"[1]

With these words, Tish Sommers, a self-described "free-lance agitator," aptly summarized the spirit and message of the Older Women's movement that she labored to initiate and build over the past fifteen years. (Sommers died in the fall of 1985.)

Sommers described her generation of American women as "playing by the rules" because her cohorts, by and large, accepted traditional female roles in both the family and the labor market. This generation had very

[1]This essay is based on ongoing work for my dissertation on the Older Women's League (OWL). It could not have been written without the collaboration and support of Tish Sommers and Alice Quinlan.

From: Rochelle Lefkowitz and Ann Withorn (eds.), *For Crying Out Loud: Women and Poverty in the United States* (New York: Pilgrim Press, 1986), pp. 89–98. Reprinted by permission of the Pilgrim Press, Cleveland, Ohio.

limited exposure to feminism because their youth and adolescence occurred between 1930 and 1950, when the first wave of the feminist movement had subsided, and the second wave had not yet coalesced. During World War II many had enjoyed brief stints of employment as "Rosie, the Riveter" recruited to replace male workers who entered military service. However, at the end of the war, many women were persuaded, with varying degrees of reluctance and resistance, to relinquish their lucrative and challenging jobs to returning soldiers, or as was often the situation, to continue to work at less well-paid "women's jobs." In order to achieve this transition, American producers and public officials subjected this generation to an intense propaganda campaign extolling the homemaker role as the most suitable and rewarding occupation available to women.

Now that Sommers' generation has reached late adulthood many of her cohorts are discovering that "the rules have changed and we've been called out." In other words, contrary to their expectations, older women are encountering penalties rather than rewards for their lifelong service in female roles because of changing demographic and social trends and discriminatory social policies. Rising divorce rates and increasing gaps in the longevity rates of males and females have left today's generation of older women more likely than earlier generations to survive over an extended period without spousal companionship or support. After being socialized to view their husband as their provider, many older women find themselves lacking the resources needed to insure their own economic survival. As a result of the combined effects of age and sex discrimination in the labor market, older women are frequently unable to find employment or earn adequate wages. Equally important, older women find their retirement security jeopardized by both Social Security and pension programs, which penalize women for their lower lifetime earnings and/or their status as dependent spouses. Lastly, the passive, self-effacing, and subservient behavior women acquire during their lengthy playing of female roles inhibits their ability to assert their rights and protest the omnipresent discrimination they encounter.

Fortunately, Sommers' assertion that her generation of American women "will not be shelved" is being echoed by many of her cohorts, more than 13,000 of whom have become members of the Older Women's League (OWL), the first and only advocacy organization in the United States for mid-life and older women. Founded in 1981 by Tish Sommers and Laurie Shields, OWL is dedicated to the goal of "bridging the gap between the women's movement and aging activism." . . .

EMERGENCE AND GROWTH OF
THE OLDER WOMEN'S MOVEMENT

With the rallying slogan of "Don't Agonize, Organize," Tish Sommers attained wide recognition among both feminist and aging activists for her keen political instincts, her witty and captivating speaking style, her energetic and persistent leadership, and her path-breaking insights regarding the problems and potentialities of older women.

The origins of Sommers' interest in social justice and political activism can be traced to her experience when, as a young, foreign, exchange dance student residing with a Jewish family during the 1930s, she witnessed Hitler's rise to power. Following this early encounter with the horrors of racism and militarism in Nazi Germany, Sommers pursued a lifelong career as a volunteer community organizer supporting poverty programs and the Civil Rights Movement. Sommers' organizing talents reached their most complete expression when her own encounters with discriminatory treatment sparked her interest in launching a collective protest movement of older women.

When Sommers became divorced at age fifty-seven she experienced many of the problems of other displaced homemakers: she had difficulty getting credit in her own name and was unable to purchase medical insurance because of medical history. She also found that her background as a former homemaker and community organizer commanded little respect from employers and society as a whole. As an experienced political activist, Sommers quickly recognized herself to be the victim of structural inequities that negatively affected many women of her generation. Shortly thereafter she concluded that her peers were clearly "ripe" for political mobilization based on their overwhelmingly positive response to her public appeals for collective protest by and for older women.

Sommers' formal career as an advocate for older women began in 1973 when she convinced the National Organization for Women (NOW) to form a Task Force on Older Women. As coordinator of this task force, Sommers attracted a dedicated group of older women activists. Although Sommers and her colleagues maintained an ongoing dialogue with organizations representing women and the elderly, they eventually recognized the need to build their own advocacy organization focused exclusively on the unique problems of older women that remained largely ignored or misunderstood, even by feminist and aging activists.

Sommers' first attempt at organization building focused on displaced homemakers. In 1974 she formed a partnership with Laurie Shields, a former advertising executive and recently widowed displaced homemaker, and together they founded the Alliance for Displaced Homemakers (ADH). At present the ADH has been supplanted by the Displaced Homemaker Network (DHN), with headquarters in Washington, which serves as an information clearinghouse and national advocacy organization for displaced homemakers.

The women who became participants in the displaced-homemaker movement represented a wide range of backgrounds including (1) middle- and upper-income women who entered the ranks of the "new poor" upon loss of spousal support; (2) welfare mothers who lost Aid to Families with Dependent Children benefits when their eldest child turned eighteen; (3) homemakers with graduate degrees whose credentials were considered "outdated" by employers; and (4) women of color who remained "underemployed" as a result of their dual burdens as a family wage earner and care giver.

In spite of its limited resources and small membership base, ADH compiled an impressive list of legislative victories: by 1980, thirty states had passed displaced-homemaker legislation, and the U.S. Congress had officially recognized displaced homemakers as a "disadvantaged group" eligible for priority services under the 1978 Comprehensive Employment and Training Act (CETA) Reauthorization Act. . . .

Shields and Sommers always viewed the displaced-homemaker movement as a first step toward their long-term goal of building a mass membership organization for older women. In 1978 they laid the groundwork for realizing this goal by establishing the Older Women's League Education Fund (OWLEF), to engage in public education and consciousness raising regarding the problems shared by today's generation of older women.

Documenting the problems of older women proved to be a challenging task for OWLEF because of the lack of prior research. Feminist researchers had focused primarily on issues of concern to younger women (child care, reproductive rights, hiring policies) and had barely tapped issues of concern to women during the latter stages of the life cycle (retirement policies, health insurance, tending disabled family members, nursing home care, age-sex discrimination). Gerontological research was even more disappointing. For example, in the field of retirement research, inclusion of female subjects was "almost unheard of" before 1975. Many gerontologists justified their preoccupation with male retirees by arguing that retirement was primarily a "male" problem because of the role discontinuity men experience when forced to abandon their lifelong role as family wage earner. Even when women were included as research subjects, gerontologists usually failed to explain or highlight the high rates of unemployment, poverty, and institutionalization encountered by older women.[2] Lastly, government statistics are organized into categorical schemes that disguise the disadvantaged status of older women. For example, the Department of Labor does not cross-classify their employment statistics by age and sex, thereby making it impossible to pinpoint how older women fare relative to other employee groups. Similarly, homemakers are not recognized as an occupational category, thereby making it impossible to identify the number of older women who are full-time homemakers or displaced homemakers.

Shields and Sommers recognized that as long as the problems of older women remain undocumented and unpublicized, they would remain invisible. After all, the very existence of displaced homemakers had escaped public attention and concern until the displaced-homemaker's movement emerged and provided them with an official label and evidence of their disadvantaged status. Consequently, OWLEF prepared a series of policy papers outlining inequities in divorce settlements, Social Security, pension plans, the labor market, and health insurance. These papers—the OWLEF Gray Papers—received favorable reviews from policymakers in the field of the aging. In turn, this heightened awareness and receptivity to OWL issues on the part of professionals concerned with aging greatly enhanced efforts to solicit funds and find prestigious sponsors for OWLEF activities from organizations that serve the elderly.

In 1980 Sommers served as convenor of the Mini-Conference on Older Women (funded by the U.S. Administration on Aging) and invited delegates to remain an extra day at the close of the conference to participate in the launching of the Older Women's League. Three hundred of the four hundred conference delegates responded to this appeal, and the Older Women's League (OWL) was born.

To date OWL has attracted more than 13,000 members and has established ninety-five local chapters. The composition of OWL's membership is not, however, truly representative of the older female population. According to a recent membership survey, only 4 percent of OWL members are ethnic minorities and only 8 percent are full-time homemakers. A majority of OWL members have college degrees, are currently employed, and are between the ages of fifty and sixty-five. In the interest of building a more representative organization, OWL leaders are currently experimenting with a variety of strategies for increasing the number of members who have low incomes and/or belong to minority groups.

Shields and Sommers take pride in OWL's expanding capacity to engage in organizing and advocacy activities. For example, OWL's staff currently includes a government-relations specialist and several field organizers. Grants from private and public foundations enable OWL to provide grants and leadership training to local chapters and to publish an extensive range of consciousness-raising and educational material. OWL also receives pro-bono services from a prestigious media consulting firm, which designs public-service ads that cleverly dramatize OWL's mission. (One popular ad features a picture of a weeping Statue of Liberty with a caption that reads: "When it comes to older women, this country takes a lot of liberties.")

In discussing OWL's political objectives, OWL leaders consistently emphasize that genuine and enduring equity for older women can be accomplished only by eliminating structural inequities that are responsible for the multiple forms of discrimination they presently encounter. Reforms viewed necessary to achieve this long-range goal include the following:

- Social Security coverage for homemakers to eliminate the penalties imposed on them as a result of their "dependent" status
- full employment and expanded employment training programs to ensure employment to displaced homemakers and other older women who are subject to widespread discrimination in the current labor market
- comparable pay legislation to establish more equitable pay scales in traditionally female occupations

It should be noted that none of these reforms can be characterized as exclusively older women's issues; instead they would eliminate inequities encountered by women of all age groups. A variety of feminist organizations consequently include such proposals on their reform agendas. Needless to say, however, OWL leaders recognize that reforms of this magnitude can be achieved only as a result of an intensive and protracted struggle, led by a broad-based coalition of organizations representing women, the elderly, and their allies. . . .

Sommers' own words best describe the guiding principles of the older women's movement:

> One way we have worked in OWL is to take our experience, some of it bitter, and turn it into good energy to make this a better society for ourselves and those who will follow us. "We build a new road to aging and the road to aging builds us." Facing death and planning for it, squeezing the sweet juice out of adversity, is part of what OWL is all about.[3]

NOTES

1. Tish Sommers, "Epilogue" in *Displaced Homemakers: Organizing for a New Life*, Laurie Shields (New York: McGraw-Hill, 1981). This book is a firsthand account of the genesis and evolution of the displaced-homemakers' movement.

2. Maximilian Szinovacz, ed., *Women's Retirement* (Beverly Hills, Calif.; Sage Publications, 1982). This anthology documents and criticizes the sexist biases of many

gerontologists and presents summaries of a variety of recent research studies focused exclusively on female retirees. Another excellent feminist critique of gerontological research is Diane Beeson, "Women in Studies of Aging: A Critique and Suggestion," *Social Problems* 23 (1975): 52–59. Since the Older Women's League Education Fund (OWLEF) was founded in 1978, gerontologists have expressed an increasing interest and concern with older women's issues. Owing to a recent influx of female scholars in the field, there has also been a dramatic expansion of research focused exclusively on older women. This research is summarized in *The Mature Women in America, A Selected Annotated Bibliography 1979–1982* (available from the National Council on the Aging, Inc., Washington, DC 20004).

3. Letter to Older Women's League membership "On Death and Dying" by Tish Sommers, May 1985. (Copies may be obtained from OWL, 1325 G. Street, N.W., Washington, D.C. 20005.)

SHARING THE SHOP FLOOR   54

Stan Gray

On an October evening in 1983, a group of women factory workers from Westinghouse came to the United Steelworkers hall in Hamilton, Ontario, to tell their story to a labor federation forum on affirmative action. The women told of decades of maltreatment by Westinghouse—they had been confined to job ghettoes with inferior conditions and pay, and later, when their "Switchgear" plant was shut down, they had fought to be transferred to the other Westinghouse plants in the city. They had to battle management and the resistance of some, though not all, of their brothers in the shops. They won the first round, but when the recession hit, many were laid off regardless of seniority and left with little or no income in their senior years.

By the night of the forum I had worked at Westinghouse for ten years and had gone through the various battles for equality in the workplace. As I listened to the women, I thought of how much their coming into our plant had changed me, my fellow workers, and my brother unionists.

From: *Canadian Dimension* 18 (June 1984). Reprinted by permission.

The women were there to tell their own story because the male staff officials of their union, United Electrical Workers, had prevented the women's committee of the local labor council from presenting their brief. The union claimed it was inaccurate, the problems weren't that bad, and it didn't give union officials the credit for leading the fight for women's rights. The Westinghouse women gave their story and then the union delivered a brief of its own, presenting a historical discussion of male-female relations in the context of the global class struggle, without mentioning Westinghouse or Hamilton or any women that it represented.[1]

This kind of thing happens in other cities and in other unions. The unanimous convention resolutions in support of affirmative action tend to mask a male resistance within the unions and on the shop floor. Too many men pay lip service to women's rights but leave the real fighting to the women. They don't openly confront the chauvinism of their brothers on the shop floor and in the labor movement. Yet an open fight by men against sexism is an important part of the fight for sexual equality. It is also important because sexism is harmful for working men, in spite of whatever benefits they gain in the short term; it runs counter to their interests and undermines the quality of their trade unionism.

I was one of those unionists who for years sat on the fence in this area until sharp events at work pushed me off. I then had to try to deal with these issues in practice. The following account of the debates and struggles on the shop floor at Westinghouse concentrates on the men rather than on the women's battle; it focuses on the men's issues and tries to bring out concretely the interests of workingmen in the fight against sexism.

MY EDUCATION BEGINS

My education in the problems of the Westinghouse women began in November 1978, when I was recalled to work following a bitter and unsuccessful five-month strike. The union represented eighteen hundred workers in three plants that produced turbines, motors, transformers, and switchgear equipment. When I was recalled to work it wasn't to my old Beach Road plant—where I had been a union steward and safety rep—but to an all-female department in the Switchgear plant and to a drastic drop in my labor grade. The plant was mostly segregated; in other words, jobs (and many departments) were either male or female. There were separate seniority lists and job descriptions. The dual-wage, dual-seniority system was enshrined in the collective agreement signed and enforced by both company and union.

At Switchgear I heard the complaints of the women, who worked the worst jobs in terms of monotony, speed, and work discipline but received lower pay, were denied chances for promotion, and were frequently laid off. They complained too of the union, accusing the male leadership of sanctioning and policing their inferior treatment. In cahoots with management, it swept the women's complaints under the carpet. From the first day it was obvious to me that the company enforced harsher standards for the women. They worked harder and faster, got less break time, and were allowed less leeway than the men. When I was later transferred to the all-male machine shop, the change was from night to day.

Meanwhile the men's club that ran the union made its views known to me early and clearly. The staff rep told me that he himself would never work with women. He boasted that he and his friends in the leadership drank in the one remaining all-male bar in the city. The local president was upset when he heard that I was seriously listening to the complaints of the women workers. He told me that he always just listened to their unfounded bitching, said "yes, yes, yes," and then completely ignored what he had been told. I ought to do the same, was his advice. Although I had just been elected to the executive in a rank-and-file rebellion against the old guard, he assumed that a common male bond would override our differences. When I persisted in taking the women's complaints seriously, the leadership started to ridicule me, calling me "the Ambassador" and saying they were now happy that I was saving them the distasteful task of listening to the women's bitching.

Then in 1979 the boom fell at Switchgear: the company announced it would close the plant. For the women, this was a serious threat. In the new contract the seniority and wage lists had been integrated, thanks to a new Ontario Human Rights Code. But would the women be able to exercise their seniority and bump or transfer to jobs in the other Hamilton plants, or would they find themselves out in the street after years at Switchgear?

DIVIDE AND CONQUER

By this time I had been recalled to my old department at the Beach Road plant, thanks to shop-floor pressure by the guys. There was a lot of worry in the plants about the prospect of large-scale transfers of women from Switchgear. A few women who had already been transferred had met with harassment and open hostility from the men. Some of us tried to raise the matter in the stewards' council, but the leadership was in no mood to discuss and confront sexism openly. The union bully boys went after us, threatening, shouting, breast beating, and blaming the women for the problems.

Since the union structures weren't going to touch the problem, we were left to our own resources in the shop. I worked in the Transformer Division, which the management was determined to keep all male. As a steward I insisted that the Switchgear women had every right to jobs in our department, at least to training and a trial period as stipulated by seniority. Since this was a legal and contractual right, management developed a strategy of Divide and Rule: present the women as a threat to men's jobs; create splits and get the hourly men to do the bosses' dirty work for them. Management had a secondary objective here, which was to break our shop-floor union organization. Since the trauma of the strike and post-strike repression, a number of stewards and safety reps had patiently rebuilt the union in the plant, block by block—fighting every grievance, hazard, and injustice with a variety of tactics and constructing some shop-floor unity. We did so in the teeth of opposition from both company and union, whose officials were overly anxious to get along peacefully with each other. A war of the sexes would be a weapon in management's counteroffensive against us.

For months before the anticipated transfers, foremen and their assorted rumor mongers stirred up the pot with the specter of the Invasion of the Women. Two hundred Switchgear women would come and throw all Beach Road breadwinners out in the street; no one's job would be safe. Day after day, week after week, we were fed the tales: for example, that fourteen women with thirty years' seniority were coming to the department in eight days and no male would be protected. Better start thinking now about unemployment insurance.

In the department next to mine a few transfers of women were met with a vicious response from the men. Each side, including the militant steward, ended up ratting on the other to the boss. The men were furious and went all over the plant to warn others against allowing any "cunts" or "bitches" into their departments.

Meanwhile I had been fighting for the women to be called into new jobs opening up in the iron-stacking area of my department. The union's business agent had insisted that women couldn't physically handle those and other jobs. But I won the point with the company. The major influx of women would start here.

For weeks before their arrival, the department was hyper-alive, everyone keyed to the Invasion of the Women. I was approached by one of the guys, who said that a number of them had discussed the problem and wanted me, as their steward, to tell management the men didn't want the women in here and would fight to keep them out.

The moment was a personal watershed for me. As I listened to him, I knew that half measures would no longer do. I would now have to take the bull by the horns.

Over the years I had been dealing with male chauvinism in a limited fashion. As a health and safety rep, I had to battle constantly with men who would knowingly do dangerous work because it was "manly" to do so and because it affirmed their masculine superiority. The bosses certainly knew how to use guys like that to get jobs done quickly. With a mixture of sarcasm, force, and reason, I would argue, "It's stupidity not manliness to hurt yourself. Use your brains, don't be a hero and cripple yourself; you're harming all of us and helping the company by breaking the safety rules we fought so hard to establish, rules that protect all of us."

From this I was familiar with how irrational, self-destructive, and anti-collective the male ego could be. I also felt I had learned a great deal from the women's movement, including a never-ending struggle with my own sexism. Off and on I would have debates with my male co-workers about women's liberation. But all this only went so far. Now with the approaching invasion and the Great Fear gripping the department, I had to deal with an angry male sexism in high gear. I got off the fence.

I told this guy, "No. These women from Switchgear are our sisters, and we have fought for them to come into our department. They are our fellow workers with seniority rights, and we want them to work here rather than get laid off. If we deny them their seniority rights, it hurts us, for once that goes down the drain, none of us has any protection. It is our enemies, the bosses, who are trying to do them out of jobs here. There's enough work for everyone; even if there weren't, seniority has to rule. For us as well as for them. The guys should train the women when they come and make them feel welcome."

And with that reply, the battle was on. For the next few weeks the debate raged hot and heavy, touching on many basic questions, drawing in workers from all over the plant. Many men made the accusation that the women would be the bosses' fifth column and break our unity. They would side with the foremen, squeal on us, outproduce us, and thereby force speed-ups. The women were our enemy, or at least agents of the enemy, and would be used by *them* against *us*. Many of them pointed to the experience of the next department over, where, since the influx of a few women, the situation had been steadily worsening.

The reply was that if we treated the women as sisters and friends they'd side with us not the boss. Some of us had worked in Switchgear and knew it was the *men* there who got favored treatment. What's more, our own shop-floor unity left a lot to be desired and many of our male co-workers engaged in squealing and kowtowing to the boss. Some of us argued sarcastically that women could never equal some of our men in this area.

We argued that we had common class interests with our sisters against the company, particularly in protecting the seniority principle.

It was easy to tease guys with the contradictions that male double standards led them to. Although they were afraid the women would over-produce, at the same time they insisted that women wouldn't be physically strong enough to do our "man's work." Either they could or they couldn't was the answer to that one, and if they could, they deserved the jobs. It would be up to us to initiate them into the department norms. Many of the guys said that the women would never be able to do certain of the heavy and rotten jobs. As steward and safety rep I always jumped on that one: we shouldn't do those jobs either. Hadn't we been fighting to make them safer and easier for ourselves? Well, they answered, the women would still not be able to do all the jobs. Right, I would say, but how many guys here have we protected from doing certain jobs because of back or heart problems, or age, or simply personal distaste? If the women can't do certain jobs, we treat them the same way as men who can't. We don't victimize people who can't do everything the company wants them to. We protect them: as our brothers, and as our sisters.

By pointing out the irrationalities of the sexist double standards, we were pushing the guys to apply their class principles—universal standards of equal treatment. Treat the women just as we treat men regarding work tasks, seniority, illness, and so on.

COUNTERING SEXISM

Male sexist culture strives to degrade women to nothing but pieces of flesh, physical bodies, mindless animals . . . something less than fully human, which the men can then be superior to. Name-calling becomes a means of putting women in a different category from *us*, to justify different and inferior treatment.

Part of the fight to identify the women as co-workers was therefore the battle against calling them "cunts" or "bitches." It was important to set the public standard whereby the women were labeled as part of us, not *them*. I wouldn't be silent with anyone using these sexist labels and pushed the point very aggressively. Eventually everyone referred to "the women."

After a while most of the men in the department came to agree that having the women in and giving them a chance was the right thing to do by any standard of fairness, unionism, or solidarity, and was required by the basic human decency that separates *us* from *them*. But then the focus shifted to other areas. Many men came back with traditional arguments against women in the work force. They belong at home with the kids, they're robbing male bread-winners of family income and so forth. But others disagreed: most of the guys' wives worked outside the home or had done so in the past; after all a family needed at least two wages these days. Some men answered that in bad times a family should have only one breadwinner so all would have an income. Fine, we told them, let's be really fair and square: you go home and clean the house and leave your wife at work. Alright, they countered, they could tolerate women working who supported a family, but not single women. And so I picked out four single men in our department and proposed they be immediately sacked.

Fairness and equality seemed to triumph here too. The guys understood that everyone who had a job at Westinghouse deserved equal protection. But then, some men found another objection. As one, Peter, put it, "I have no respect for any women who could come in to work here in these rotten conditions." The comeback was sharp: "What the hell are *you* putting up with this shit for? Why didn't you refuse to do that dirty job last month? Don't *you* deserve to be treated with respect?"[2]

As the Invasion Date approached I got worried. Reason and appeals to class solidarity had had a certain impact. Most of the guys were agreeing, grudgingly, to give the women a chance. But the campaign had been too short; fear and hostility were surfacing more and more. I was worried that there would be some ugly incident the first day or two that would set a pattern.

Much of the male hostility had been kept in check because I, as the union steward, had fought so aggressively on the issue. I decided to take this one step further and use some intimidation to enforce the basics of public behavior. In a tactic I later realized was a double-edged sword, I puffed myself up, assumed a cocky posture, and went for the jugular. I loudly challenged the masculinity of any worker who was opposed to the women. What kind of man is afraid of women? I asked. Only sissies and wimps are threatened by equality. A *real man* has nothing to be afraid of; he wants strong women. Any man worth his salt doesn't need the crutch of superiority over his sisters; he fears no female. A real man lives like an equal, doesn't step on women, doesn't degrade his sisters, doesn't have to rule the roost at home in order to affirm his manhood. Real men fight the boss, stand up with self-respect and dignity, rather than scapegoat our sisters.

I was sarcastic and cutting with my buddies: "This anti-woman crap of yours is a symbol of weakness. Stand up like a real man and behave and work as equals. The liberation of the women is the best thing that ever came along. . . . It's in *our* interests." To someone who boasted of how he made his wife cook his meals and clean his floors, I'd ask if she wiped his ass too? To the porno addicts I'd say, "You like that pervert shit? What's wrong with the real thing? Can you only get it up with those fantasies and cartoon women? Afraid of a real woman?" I'd outdo some of the worst guys in verbal intimidation and physical feats. Then I'd lecture them on women's equality and on welcoming our sisters the next week. I zeroed in on one or two of the sick types and physically threatened them if they pulled off anything with the women.

All of this worked, as I had hoped. It established an atmosphere of intimidation; no one was going to get smart with the women. Everyone would stand back for a while, some would cooperate, some would be neutral, and those I saw as "psycho-sexists" would keep out.

The tactic was effective because it spoke directly to a basic issue. But it was also effective because it took a leaf from the book of the psycho-sexists themselves.

At Westinghouse as elsewhere, some of the men were less chauvinistic and more sensible than others, but they often kept quiet in a group. They allowed the group pattern to be set by the most sexist bullies, whose style of woman baiting everyone at least gave in to. The psycho-sexists achieved this result because they challenged, directly or by implication, the masculinity of any male who didn't act the same way. All the men, whatever their real inclinations, are intimidated into acting or talking in a manner degrading to women. I had done the same thing, but in reverse. I had challenged the masculinity of any worker who would oppose the women. I had scared them off.

THE DAY THE WOMEN ARRIVED

The department crackled with tension the morning The Women arrived. There were only two of them to start with. The company was evidently scared by the volatile situation it had worked so hard to create. They backed off a direct confrontation by assigning my helper George and me to work with the women.

The two women were on their guard: Betty and Laura, in their late thirties, were expecting trouble. They were pleasantly shocked when I said matter-of-factly that we would train them on the job. They were overjoyed when I explained that the men had wanted them in our department and had fought the bosses to bring them here.

It was an unforgettable day. Men from all corners of the plant crept near the iron-stacking area to spy on us. I explained the work and we set about our tasks. We outproduced the standard rate by just a hair so that the company couldn't say the women weren't able to meet the normal requirements of the job.

My strategy was to get over the hump of the first few days. I knew that once the guys got used to the women being there, they'd begin to treat them as people, not as "women" and their hysteria would go away. It was essential to avoid incidents. Thus I forced the guys to interact with them. Calling over one of the male opponents, I introduced him as Bruce the Slinger who knew all the jobs and was an expert in lifts and would be happy to help them if asked and could always be called on to give a hand. This put him on the spot. Finally he flashed a big smile, and said, "Sure, just ask and I'd be pleased to show you anything, and to begin with, here's what to watch out for. . . ."

The morning went by. There were no incidents. From then on it was easy. More guys began to talk to the two women. They started to see them as Betty with four kids who lived on the mountain and knew wiring and was always cheerful; or Laura, who was a friend of John's uncle and was cranky early in the morning, who could easily operate the crane but had trouble with the impact gun, and who liked to heat up meat pies for lunch. After all, these men lived and worked with women all of their lives outside the plant—mothers, sisters, wives, in-laws, friends, daughters, and girlfriends. Having women at work was no big deal once they got over the trauma of the invasion of this male preserve. Just like helping your sister-in-law hang some wallpaper.

As the news spread, more and more women applied to transfer to our department. They were integrated with minimum fuss. The same thing happened in several adjoining departments. Quickly, men and women began to see each other as people and co-workers, not as enemies. Rather than man vs. woman it was John, Mary, Sue, Peter, Alice, George, and Laura. That Christmas we had a big party at someone's home—men and women of the department, drinking and dancing. The photos and various raucous tales of that night provided the basis for department storytelling for the next three months.

Was this, then, peace between the sexes? The integration of men and women as co-workers in the plant? Class solidarity triumphing over sex antagonism? Not quite. Although they were now together, it was not peace. The result was more complicated, for now the war between the sexes was being extended from the community into the workplace.

WORKPLACE CULTURE

As our struggle showed, sexism coexists and often is at war with class consciousness and with the trade union solidarity that develops among factory men. Our campaign was successful to the extent that it was able to sharply polarize and push the contradictions between these two tendencies in each individual. With most of the men, their sense of class solidarity triumphed over male chauvinism.

Many of the men had resisted the female invasion of the workplace because for them it was the last sanctum of male culture. It was somewhere they could get away from the world of women, away from responsibility and children and the civilized society's cultural restraints. In the plant they could revel in the rough and tumble of a masculine world of physical harshness, of constant swearing and rough behavior, of half-serious fighting and competition with each other and more serious fighting with the boss. It was eight hours full of filth and dirt and grease and grime and sweat—manual labor and a manly atmosphere. They could be vulgar and obscene, talk about football and car repairs, and let their hair down. Boys could be boys.

The male workplace culture functions as a form of rebellion against the discipline of their society. Outside the workplace, women are the guardians of the community. They raise the kids and enforce some degree of family and collective responsibility. They frequently have to force this upon men, who would rather go drinking or play baseball while the women mind the kids, wash the family's clothes, attend to problems with the neighbors and in-laws, and so on. Like rebellious teenage sons escaping mother's control, male wage earners enter the factory gates, where in their male culture they feel free of the restraints of these repressive standards.

Even if all factory men don't share these attitudes, a large proportion do, to a greater or lesser degree.

The manly factory culture becomes an outlet for accumulated anger and frustration. But this is a vicious circle because the tedious work and the subordination to the bosses is in large part the very cause of the male worker's dissatisfaction. He is bitter against a world that has kept him down, exploited his labor power, bent him to meet the needs of production and profit, cheated him of a better life, and made the daily grind so harsh. Working men are treated like dirt everywhere: at work they are at the bottom of the heap and under the thumb of the boss; outside they are scorned by polite society. But, the men can say, we are better than them all in certain ways; we're doing men's work; it's physically tough; women can't do it; neither can the bankers and politicians. Tough work gives a sense of masculine superiority that compensates for being stepped on and ridiculed. All that was threatened by the Women's Invasion.

However, this male workplace culture is not one-sided, for it contains a fundamentally positive sense of class value. The workingmen contrast themselves to other classes and take pride in having a concrete grasp of the physical world around them. The big shots can talk fancy and manipulate words, flout their elegance and manners. But we control the nuts and bolts of production, have our hands on the machines and gears and valves, the wires and lathes and pumps, the furnaces and spindles and batteries. We're the masters of the real and the concrete; we manipulate the steel and the lead, the wood, oil, and aluminum. What we know is genuine, the real and specific world of daily life.

Workers are the wheels that make a society go round, the creators of social value and wealth. There would be no fancy society, no civilized conditions if it were not for our labor.

The male workers are contemptuous of the mild-mannered parasites and soft-spoken vultures who live off our daily sweat: the managers and directors, the judges and entertainers, the lawyers, the coupon clippers, the administrators, the insurance brokers, the legislators . . . all those who profit from the shop floor, who build careers for themselves with the wealth we create. All that social overhead depends upon our mechanical skills, our concrete knowledge, our calloused hands, our technical ingenuity, our strained muscles and backs.

The Dignity of Labor, but society treats us like a pack of dumb animals, mere bodies with no minds or culture. We're physical labor power; the intelligence belongs to the management class. Workers are sneeringly regarded as society's bodies, the middle class as society's mind. One is inferior; the other is superior and fully human. The workers are less than human, close to animals, society's beasts of burden.

[handwritten margin note: Body is denigrated here too.]

The male workplace culture tends to worship this self-identity of vulgar physicalness. It is as if the men enjoy wallowing in a masculine filth. They brag of being the wild men of the factory. Say it loud: I'm a brute and I'm proud.

Sexism thus undermines and subverts the proud tradition of the dignity of labor. It turns a class consciousness upside down by accepting and then glorifying the middle-class view of manual labor and physical activity as inferior, animalistic, and crude. When workers identify with the savages that the bosses see them as, they develop contempt for themselves. It is self-contempt to accept the scornful labels, the negative definitions, the insulting dehumanized treatment, the cartoon stereotypes of class chauvinism: the super-masculine menials, the industrial sweathogs.

Remember Peter, who couldn't respect a woman who would come to work in this hellhole. It was obviously a place where he felt he had lost his own self-respect. My reply to him was that he shouldn't put up with that rotten treatment, *that the men also deserved better*. We should be treated with dignity. Respect yourself—fight back like a man, not a macho fool who glorifies that which degrades him.

Everything gets turned inside out. It is seen as manly to be treated as less than a man, as just a physical, instinctual creature. But this is precisely how sexist society treats women: as mindless bodies, pieces of flesh . . . "biology is destiny." You would think that male factory workers and the women's movement would be natural allies, that they'd speak the same language. They share a common experience of being used as objects, dehumanized by those on top.

Men in the factory are treated not as persons, but as bodies, replaceable numbers, commodities, faceless factors of production. The struggles of workingmen and of women revolve around similar things. The right to choice on abortion, for example, revolves around the right for women to control their own bodies. Is this not what the fight for health and safety on the shop floor is all about? To have some control over our bodies, not to let the bastards do what they want with our lives and limbs, to wreck us in their search for higher profits.

But male chauvinism turns many workingmen away from their natural allies, away from a rational and collective solution to their problems, diverting them from class unity with their sisters into oppressors and degraders of their sisters. Robbed of their real manhood—their humanity as men—they get a false sense of manhood by lording over women.

PLAYING THE FOREMAN AT HOME

Many men compensate for their wage-labor status in the workplace by becoming the boss at home. Treated terribly in the factory, he plays foreman after work and rules with authority over his wife and kids. He thus gains at home that independence he loses on the shop floor. He becomes a part-time boss himself with women as his servants. This becomes key to his identity and sense of self-esteem. Working-class patriarchs, rulers of the roost.

This sense of authority has an economic underpinning. The male worker's role as primary breadwinner gives him power over the family and status in society. It also makes him the beneficiary of the woman's unpaid labor in the household.

A wage laborer not only lacks independence, he also lacks property, having nothing but his labor power to sell. Sexism gives him the sense of property, as owner of the family. His wife or girlfriend is his sexual property. As Elvis sang, "You are my only possession, you are my everything." This domination and ownership of a woman are basic to how he sees himself.

These things are powerful pressures toward individualism, a trait of the business class: foreman of the family, man of property, possessiveness. They elevate the wage earner above the category of the downtrodden common laborer, and in doing so divert him from the collective struggle with his brothers and sisters to change their conditions. Capitalism is based on competitiveness and encourages everyone to be better than the next guy, to rise up on the backs of your neighbors. Similarly the male chauvinist seeks superiority over others, of both sexes. Men tend to be competitive, always putting one another down, constantly playing one-upmanship. Men even express appreciation and affection for each other through good-natured mutual insults.

Sexist culture thus undermines the working-class traditions of equality and solidarity and provides a recruiting ground for labor's adversaries. Over the years at Westinghouse I had noticed that a high proportion of workers who became foremen were extreme chauvinists—sexual braggarts, degraders of women, aggressive, individualistic, ambitious, ever willing to push other workers around. Male competition is counterproductive in the shop or union, where we ought to cooperate as equals and seek common solutions. The masculine ego makes for bad comradeship, bad brotherhood. It also makes it difficult for chauvinistic men to look at and deal objectively with many situations because their fragile egos are always on the line. They have to keep up a facade of superiority and are unable to handle criticism, no matter how constructive. Their chauvinistic crutches make them subjective, irrational, unreliable, and often self-destructive, as with men who want to work or drive dangerously.

Workingmen pay a high price for the limited material benefits they get from sexist structures. It is the bosses who make the big bucks and enjoy the real power from the inferior treatment of women.

THE NEXT ROUND AND A PEEK INTO THE WOMEN'S WORLD

Battles continued about the women getting a crack at the more skilled and high-paying assembly jobs up the floor. Next we won the fight against the company, which was trying to promote junior men. This time women were there to fight for themselves, and there were male stewards from other departments who backed them up. The shop floor was less hostile, many of the men being sympathetic or neutral.

But despite the general cooperation, most men still maintained that the women were inferior workers. The foremen did their best to foster sex divisions by spreading stories of all the mistakes the women supposedly made. They would reserve the worst jobs for the men, telling them the women couldn't do them. The men would thus feel superior while resenting the women's so-called privileges and the women would feel grateful for not having to do these jobs. The supervisors forged a common cause with some of the guys against the women. They fed their male egos and persuaded them to break safety rules, outproduce, and rat on other workers. The male bond often proved stronger than the union bond, and our collective strength suffered as a result.

As for myself, I was learning and changing a lot as a result of my experiences. I would often meet with the women at the lunch table to plan strategy. These sessions affected me in many ways. They were good talks, peaceful and constructive, with no fighting and argument, no competition, all of us talking sensibly about a common problem and figuring out how to handle it as a group. It was a relaxed and peaceful half hour, even when we had serious differences.

This was in marked contrast to the men's lunch tables, which were usually boisterous and raucous during those months. There was a lot of yelling and shouting, mutual insults, fists pounding, and throwing things at one another. When you ate at the women's table, you sat down to rest and relax. When you ate at the men's table, you sat down to fight.

I had read and heard a lot from my feminist friends about this so-called woman's world of warmth, cooperation, and friendship, as contrasted to men's norm of aggression, violence, and competition. Although I had always advocated women's liberation and respected the women's movement, I paid only lip service, if that, to this distinction, and was in fact more often scornful of this "women's world." Over the years I had become a more aggressive male, which I saw as distinct from being a chauvinist or sexist male. In the world of constant struggle, I thought, you had to be aggressive or go under. We'd have peace and love in the socialist future, some distant day.

As a unionist it became very clear to me that the women almost automatically acted like a collective. And in those months of going back and forth between the men's and women's tables, I took a long and serious look at this women's world. It was an unnerving but pleasant experience to sit down among friends, without competition and put-downs, not to have to watch out for flying objects, not to be on the alert for nerve-shattering noises, to be in a non-threatening atmosphere. There was obviously something genuine there and it seemed to offer a better way. It also became obvious to me that the gap between the sexes was enormous and that men and women were far from speaking a common language.

NEW STRUGGLES AND THE RECESSION

In the months ahead there were new struggles. There was the fight to form a women's committee in the union in order to bring women's demands to the fore, to combat sexism among the male workers, and to give women a forum for developing their own outlook, strategy, and leadership. We launched that fight in the fall of 1981, with Mary, a militant woman in our rank-and-file group, in the lead. The old guard, led by the union's national president, fought us tooth and nail. The battle extended over a number of months and tumultuous membership meetings, and we eventually lost as the leadership railroaded through its chauvinistic policy. No women's committee was formed. In time, however, the leadership came to support women's rights formally, even though they did little to advance the cause in practice.

In the spring of 1982 the recession finally caught up with us at Westinghouse and there were continuous layoffs in every division. Our bargaining power shrank, everyone was afraid for his or her own job, and the contract became little more than a piece of paper as the company moved aggressively to roll back the clock on our hard-won traditions on seniority rules, health and safety regulations, and so on. Bitterness and frustration were everywhere.

The company went after the women. Their seniority rights were blatantly ignored as they were transferred to "chip and grind" duties—the least skilled, the heaviest, dirtiest, and most unpleasant jobs. The progress the men had made also seemed to vanish. From the first day of the layoff announcements, many rallied to the call of "Get the women out first." Those most hostile to women came back out in the open and campaigned full blast. They found many sympathetic responses on the shop floor: protect the breadwinners and, what's more, no women should be allowed to bump a male since they're not physically capable of doing the jobs anyway. It was the war of the sexes all over again, but far worse now because the situation allowed little leeway. There was some baiting of the women, and the plant became a tension-ridden, hateful place for all workers.

The bosses managed to seize back many of the powers the shop floor had wrenched away from them over the years, and even to create newer and deeper divisions within the workforce. But the recession was not all-powerful. We still managed to win all our battles on health and safety, and the shop floor continued to elect our militant shop stewards.

Some of the women gave in to the inevitable and were laid off despite their seniority. But others fought back and fought well. Some of them were even able to gain the sympathy of the male workers who had at first stood aside or resisted them. In some cases, the men joined in and helped the women retain their jobs.

Obviously, things had changed a great deal amongst the men since the first women began to come into the division. *When the chips were down, many men took their stand with their sisters against the company—despite the recession. . . .*

Workingmen share basic common interests with our sisters. When more of us recognize this, define and speak about these interests in our own way, and act in common with women, then we will be able to start moving the mountains that stand in our way.

NOTES

1. United Electrical Workers (UEW) is a union whose militant rhetoric is rarely matched by its actual behavior. For example, it has passed resolutions at its national conventions favoring the formation of women's committees at the national and local levels, but what is on paper often does not match daily reality. Some leaders have a habit of advocating a position that suits the political needs of the moment rather than consistent principles. These limitations are not peculiar to UEW, which is much like the rest of the labor movement, despite its sometimes radical rhetoric. Like most of the labor movement, it has its good and its bad locals, its good and its bad leaders. Like a lot of other unions it has moved toward a better position on "women's" issues, although with a lot of sharp contradictions and see-saws in behavior along the way, given its authoritarian style.

2. The names of the plant workers in this article are not their real ones.

FROM HOMEMAKER TO HOUSING ADVOCATE: *An Interview with* Mrs. *Chang Jok Lee*

55

Nancy Diao

I first met Mrs. Lee in 1976 at a rally in front of the International Hotel and then again in 1985 in the midst of a financial crisis in the San Francisco Housing Authority. The agency was more than $9 million in debt, and its executive director Carl Williams had been asked to resign by Mayor Diane Feinstein. During this time, the Ping Yuen Residents Improvement Association (PYRIA) remained the best organized and most effective tenant association in the city.[1] Much of its strength was due to the consistent participation of Mrs. Lee. She had been the backbone of a monumental effort to protect

From: Asian Women United of California (eds)., *Making Waves: An Anthology of Writings By and About Asian American Women* (Boston: Beacon Press, 1989), pp. 377–386. Copyright © 1989 by Asian Women United. Reprinted by permission of Beacon Press.

the rights of low-income tenants in San Francisco's Chinatown. This is an unusual role for an immigrant woman whose Chinese tradition frowns upon women activists.

What struck me was Mrs. Lee's dedication to working for social change, an unusual choice for a woman her age. Instead of playing mah-jongg with her contemporaries, she prefers to attend community meetings, testify at city hearings, and help fellow tenants settle disputes. Mrs. Lee is in her late fifties, but looks much younger. With glasses and short black hair, permed and fashionably kept, she is always well groomed and impeccably dressed. For a Chinese woman, she is rather big-framed, but looks sturdy and confident. She speaks her mind freely, from telling stories about her favorite granddaughter to tales about growth pains with the tenant association or gossip in the Chinese community. Though she speaks a combination of Chinese dialects, with a mixture of some English words, she looks you straight in the eye when she talks. You can't help but notice her sincerity and passion.

GROWING UP IN JAPAN

In a 1985 interview, Mrs. Lee told of how poverty and discrimination have plagued her since her childhood in Japan. Born in 1927, in Kobe, she was the third of eight children, and the second girl. Her family suffered the hard life of Chinese immigrants in Japan, and she remembers growing up poor, segregated from the Japanese.

> My family was very poor when I was born. We didn't even have money to buy soy sauce. When I was two years old, my father got a job as a chef in the Egyptian Embassy, so our entire family lived in the servants' quarters of the embassy. My mother helped out with the housecleaning and ironing.
>
> We didn't have much contact with the Japanese except when we went shopping. In Kobe, the Chinese operated pastry, coffee, tailoring, and other shops and had two Chinese schools. I went to the Mandarin school until the sixth grade, but we didn't have enough money for me to continue; my sister went only to night school.

The heavy responsibilities she assumed as a girl helped to groom her for her later leadership role in the Chinatown tenants' group. When her family returned to the Zhongshan district in southeastern China during the Sino-Japanese War, and while her father and older brother remained in Japan, Mrs. Lee had to bear the bulk of caring for the family though she was only eleven years old. This responsibility continued even after the family reunited in Japan one year later. More aggressive and verbal than her older, frail sister, Mrs. Lee represented the family at air raid exercises and in food ration lines. After the sixth grade, she worked in a Taiwanese-owned shoe factory and then in a candy factory to help with the family income.

Along with other Chinese in Japan, she and her family were subjected to many forms of discrimination because Japan and China were on opposite sides of a war.

> Some pharmacies would use slogans like "Can even kill the Nanking Bloodsuckers" as advertisements for the effectiveness of pesticides.[2] In many Japanese shops we would have to wait until the Japanese customers were served first. We were also discriminated against in employment and were only able to get lower class jobs regardless of our education, skills, and abilities. This situation forced many of us to start our own small businesses such as cafe/restaurants, tailor shops, and painting stores.

When the United States began bombing Kobe in 1944, life became even harsher for Mrs. Lee's family.

> Whole families died in air raid shelters, smothered by smoke. When the planes came, everyone in my family went into caves or shelters; only my older brother and I stayed behind to watch our house. One time our house was fire-bombed, and I tried to put out the fire by stomping, but in vain. My brother went looking for me all over the place, but the fire and smoke had spread so fast that he couldn't see anything. Fortunately I had escaped, and he did thereafter. We lost our house, and our family split up. I was sent to live with a family friend who came from the Fukien province.

ROMANCE LEADS TO AMERICA

While living with the Fukienese family and working for them to earn her keep, Mrs. Lee met her future husband, George, through her first boyfriend. George was a Chinese American GI who was stationed in Yokohama after the surrender of the Japanese government at the end of World War II. When asked how she met George, Mrs. Lee giggled. Her face lit up and she blushed. Then her eyes softened with a watery glow. Compared to her first boyfriend, who treated her like a "good little workhorse," George was considerate and romantic, though they didn't talk much in those days.

> George always treated me with kindness and respect, very different from my first boyfriend. He always saved me a seat on the bus and gave me little gifts, whereas my boyfriend never showed any appreciation. [For instance], when my boyfriend's family's house was bombed, and he lost all of his belongings, I stayed up all night to knit him a sweater. He never even said "thank you."

Mrs. Lee and George married in 1946, and their first son was born one year later. When the son was just six months old, George returned to the United States while Mrs. Lee remained in Japan with her parents till her husband came back to get her. They arrived in San Francisco in 1950, and in two years moved into one of the Ping Yuen public housing apartments in Chinatown. She remembers her life being full, but also one of poverty.

> We were so poor that most of the time we didn't even have a penny in the house, but I wasn't scared or worried. We raised eight children, four boys and four girls, and from them I learned some English.

When the children were small, Mrs. Lee spent all of her time raising them; but as they grew older, she found more time to think about her own needs and interests. She began to become more active in the community, beginning first with just singing and socializing, and then onto more serious work.

> When the children were all grown up, I started learning Mandarin and singing songs at the Asian Community Center.[3] Since I went to a Mandarin school in Japan, I wanted to keep it up. While I was learning Mandarin, I had the opportunity to read a lot of newspapers and books, and went to May Day celebrations with George. Ever since George got disabled from a car accident in 1972, he has had a lot of free time to get involved in community issues.

CONFRONTATION WITH HOUSING ISSUES

The Asian Community Center was a commercial tenant of the International Hotel block, which soon became the focal point of the early conflict of interests between low-income tenants and land developers.[4] Mrs. Lee's association with the center eventually led to her involvement with community housing issues.

> In 1977, when my youngest daughters, Sylvia, Patricia, and Teresa, were twenty-one, nineteen, and ten, I became involved in the International Hotel struggle. I would take my youngest daughter, Teresa, to meetings and classes with me. Because I knew some of the tenants who lived in the International Hotel, I got upset when I saw leaflets about the possibility of them being evicted; I did not want to see them homeless. The young people at the Asian Community Center encouraged me to go to meetings on the third floor of the I-Hotel. It took me a while to get used to meetings and rallies, but eventually I even spoke with bullhorns at demonstrations.

> On August 3, 1977, the night of the eviction, George, Teresa, Patricia, and I were there. It was a warm night; there were four hundred policemen on horses, in addition to the tactical squad. The I-Hotel was surrounded by thousands of people—Asian, white, black, young and old, including many from Reverend Jim Jones's church who came in busloads. It seemed that we all stood on the sidewalk for hours. Suddenly the horses charged. I screamed, and everywhere there was yelling, screaming, and crying. We wanted the horses to stop charging, but the tactical squad used their billy clubs to hold us back on the sidewalk. As the horses rushed and trampled, the human chain around the hotel broke. People fell down. Tears poured out of my eyes as I heard Hongisto (then chief of police) breaking down the door to the I-Hotel. We stayed in front of the I-Hotel until three in the morning—watching every tenant being either dragged out or carried out of the hotel; then we went to Portsmouth Square.

Even as Mrs. Lee's support of the I-Hotel continued, she transferred more energy toward improving living conditions in Ping Yuen. In 1977 all the pipes in Ping Yuen were rotting, but in spite of repeated calls to the San Francisco Housing Authority, nothing was being done to fix the problem. Eventually George and some members of the tenant association initiated a massive petition drive to get the plumbing repaired. At the end of 1977, the housing authority finally repaired all of the pipes and painted the exterior walls of half of the buildings. And George was elected president of the association.

A year later, when the housing authority proved unresponsive to meeting the tenants' demands for better security, Mrs. Lee participated in the Ping Yuen tenants' first rent strike. The action was instigated by the brutal rape and murder of tenant Judy Wong.[5] It was an intense period for Mrs. Lee.

> I remember passing out leaflets door to door, talking to the tenants, attending lots of meetings, and collecting rent for fifteen days of each month at the association office on Pacific Avenue. The strike lasted for four months, with numerous press conferences and tedious negotiations with the housing authority, at the end of which we got our security guards.

The second strike followed at the end of 1979, when housing authority groundskeepers and office workers struck for higher pay. The city-wide Public Housing Tenants Association (PHTA) wanted to strike in support, but only the Ping Yuen tenants actually did. When the city employees went back to work, the Public Housing Tenants Association withdrew their support. But the Ping Yuen group continued to strike for maintenance issues, such as fixing apartment interiors and elevators, repairing floors, and painting. It was a long and drawn-out fight, but the tenants' persistence brought them victory.

> We started with eighty households, but some tenants discontinued their strike support for fear of eviction. We held many meetings and visited people door to door. We also had membership drives and sponsored activities to keep the striking tenants together. Since I was the treasurer, I collected the rent, put it in escrow, and kept the books. After two years, we finally got our demands met.

At the end of the strike, most of the tenants chose to donate 50 percent of the escrow interest, about ten thousand dollars, to PYRIA for a color television in the community room and a banquet at Asia Garden. At the banquet the tenants surprised Mrs. Lee and George with two round-trip tickets to Japan to show their appreciation for the couple's efforts in the strike.

During her husband's term as president of the improvement association, from 1979 to 1981, Mrs. Lee worked on two major projects that brought additional benefits to the Ping Yuen tenants. In 1979, at the request of the tenant population, Mrs. Lee went door to door at least two hours a day to sign up enough tenants to pressure Cablevision to install cable television services. Second, Mrs. Lee took over the coordination of the vegetable garden and established new rules: each member had an opportunity to have a garden and the size of all the lots was made equal. She thereby abolished all favoritism in the distribution of garden plots.

REACTIONS TO ACTIVISM

Though Mrs. Lee can now act fearlessly, this was not so when she first became active in the community.

> At first I was scared, or rather, kind of embarrassed. I didn't speak English and was not used to speaking in front of people. But after a while, I got used to it. As long as I am fighting for a just cause, then I am not scared.

Since Mrs. Lee's own family has remained in Japan, and George is also alone in the United States, neither has had to face pressure and criticism from relatives, who traditionally might have frowned on women's activism. She and her husband have, however, had some run-ins with the more conservative element of the community.

> I didn't really get much reaction from getting involved in I-Hotel, but when I became active in the business of the association, I started getting a lot of harassment. The wall near our apartment was often spray painted with the word

"commies!" with a black arrow pointing to our apartment. Everytime we challenged the previous PYRIA administration's way of doing things, we were called "commies." There were also flyers and posters attacking us.

Neither have her relationships with other tenants always been smooth. Some have criticized her for "doing too much." Take, for example, the laundromat project.[6]

> One of the officers of the association says that I am stupid to sweep the floors of the laundryroom. But when the laundryroom is dirty, I just can't stand it. It took so much out of us to get this project done; I feel like it's my own. So, when people don't clean up after themselves and youths abuse the furniture and write on walls, it really hurts me. But what hurts me more is when other officers nag at me for "doing too much." If they do some and if everybody does something, then I wouldn't have to do so much. Sometimes I squeeze in the sweeping when the baby is taking a nap.

After a recent officers' meeting, Mrs. Lee went home crying. The stress brought her a few sleepless nights and some additional white hair. At times like these, she wonders about whether her efforts are worth all the headaches and talks about quitting, but she stays. She remains undaunted about making Ping Yuen a better place to live and confident about the tenants' overall good feelings towards her.

> Deep down, I know a lot of tenants really like me. They respect me and support George. The maintenance worker, Mr. Wong, complains about the youths not listening to him, but I don't have any problems with them. I just tell them to get out [of the laundryrooms] and they do. Most of the tenants listen to me, and whenever there is something bothering them, they always either ask me questions or ask me to help them.

Sometimes even her children scold her for "wasting her time." Yet other times they have helped out by protecting her at demonstrations or doing errands.

> Some of my children get down on me for doing so much volunteer work. They say that I am crazy for spending so much time on the association when I don't get paid. They don't really understand me. I am happier when I am active, though there is nothing material to gain. It keeps me young. I don't have much white hair or wrinkles [*she points to her head*], do I?
> Sylvia doesn't get down on me for doing so much. She just doesn't want me and George to be taken advantage of; she helps me out a lot. She is the one who taught me how to do books, how to do a membership drive, and keep a membership list. Her husband took off work a couple of times to take

care of their daughter whom I watch [five days a week], so that I could be freed up to go to the public hearings on the Orangeland Project at the City Planning Commission.[7]

Teresa . . . knows that I am happier when I am active. She doesn't complain when I am not home to cook dinner, and sometimes she even translates for me.

CONCLUSION: BALANCING LIFE'S DEMANDS

When asked if it has been hard to balance all the demands in her life—being a wife, mother of eight, grandmother of eleven now (eight when she was interviewed), and a housing activist—she laughed:

From these activities, I learned that there is nothing to fear. I feel alive when I come out to do things. But I do take a lot of abuse from people— gripes, complaints, blames, and a lot of headaches. Even George and I have differences sometimes, and he is very stubborn. But basically we are alike, so things don't get too bad at home. At least he doesn't bug me about housework or cooking; sometimes we just go out to eat. . . .

In the past a lot of the community leaders courted George and me. They always invited us to events and asked us to help. Now no one comes. I guess they realize that they can't just use us anymore. I try to keep up with the issues. Sometimes I get upset about association business, and I can't sleep at night. But most of the time, being active keeps me alive. I don't play mah-jongg or go to Reno, so I take care of my granddaughter, and I go to meetings.

On 10 July 1985, Mrs. Chang Jok Lee was honored for her dedication and hard work with the Ping Yuen Residents Improvement Association at the eighth anniversary celebration of the Chinatown Neighborhood Improvement Resources Center, which has spearheaded much of the effort to retain housing in San Francisco's Chinatown. In front of 550 people, she said in Chinese, "I don't really deserve this, but I know that if we all work together, anything can be done." Then, the fifty-eight-year-old grandmother smiled and curtsied.

NOTES

1. The Ping Yuen Residents Improvement Association (PYRIA), formed in 1966, serves as the official representative of 430 households of mostly monolingual Chinese public housing tenants in San Francisco's Chinatown. The Pings, as the four buildings are commonly called, constitute 15 percent of the low-income housing in Chinatown, have a waiting list with more than two thousand names, and a waiting period of five to ten years.

2. Nanking (Namking) is a city in the southern part of China. Nanking bloodsuckers are particularly poisonous and vicious worms from the area.

3. The Asian Community Center was a volunteer community service and advocacy organization which provided English, Mandarin, and singing lessons for community people. The center is now closed.

4. The International Hotel, or I-Hotel, was a San Francisco residential hotel in the last foothold of Manilatown and on the edge of Chinatown. In 1965 Four Seas Corporation, owned by real estate investors from Thailand, bought the hotel with plans to demolish the building and convert the space into mixed-use development, including offices. It became the rallying point for a concerted community-wide effort to stem the loss of low-income housing in the area.

5. Judy Wong was raped and murdered in North Ping Yuen in 1978. The elevators had not been functioning for six months, forcing Wong to use the stairs, where she was attacked.

6. In 1983 PYRIA received city funds to renovate the Ping Yuen laundromats.

7. The Orangeland Project was designed to include both low-income housing and mixed-use commercial development. But its construction would have involved the dislocation of 195 elderly and family tenants residing on the site and fourteen commercial and neighborhood shops. The Ping Yuen tenants supported the Orangeland tenants' efforts to keep their homes. With communitywide support, the Orangeland tenants were able to stay in their homes, and the original development project was relocated to another site.

A WORLD WORTH LIVING IN

56

Roberta Praeger

As an impoverished woman I live with the exhaustion, the frustration, the deprivation of poverty. As a survivor of incest I struggle to overcome the emotional burden. One thing has led to another in my life as the causes of

From: Rochelle Lefkowitz and Ann Withorn (eds.), *For Crying Out Loud: Women and Poverty in the U.S.* (New York: Pilgrim Press, 1986). Reprinted by permission of Pilgrim Press, Cleveland, Ohio.

poverty, of incest, of so many issues have become increasingly clear. My need to personalize has given way to a realization of social injustice and a commitment to struggle for social change.

LIVING ON WELFARE

I live alone with my four-year-old child, Jamie. This state (Massachusetts) allocates $328 a month to a family of two living on Aid to Families with Dependent Children (AFDC). This sum places us, along with other social service recipients, at an income 40 percent below the federal poverty line. In today's economy, out of this sum of money, we are expected to pay for rent, utilities, clothing for two, child-care expenses, food not covered by food stamps, and any other expenses we may incur.

My food stamps have been cut to the point where they barely buy food for half the month. I have difficulty keeping up with the utility bills, and my furniture is falling apart. Furniture breaks, and there is no money to replace it. Things that others take for granted, such as sheets and towels, become irreplaceable luxuries.

Chaos exists around everything, even the most important issues, like keeping a roof over one's head. How are people expected to pay rent for their families on the shameful amount of income provided by the Welfare Department? The answer, in many cases, is reflected in the living conditions of welfare recipients. Some of us live in apartments that should be considered uninhabitable. We live with roaches, mice, sometimes rats, and floors about to cave in. I live in subsidized housing. It's that or the street. My rent without the subsidy is $400 a month, $72 more than my entire monthly income. It took over a year of red tape between the time of my first application to the time of final acceptance into the program, all the while watching the amount of my rent climb higher and higher. What becomes of the more than 80 percent of AFDC recipients who are not subsidized because there isn't enough of this housing available?

Emergencies are dealt with in the best way possible. One cold winter day, Jamie broke his ankle in the day-care center. It was the day my food stamps were due to arrive. With no food in the house, I had to take him, on public transportation, to the hospital emergency room and then walk to the supermarket with my shopping cart in a foot of snow. This was not an unusual event in my life. All AFDC mothers get caught up in situations like this, because we are alone, because we have few resources and little money.

Ronald Reagan's war on the poor has exacerbated an already intolerable situation. Human service programs have been slashed to the bone. Regulations governing the fuel assistance program have been changed in ways that now make many of the impoverished ineligible. Energy assistance no longer pays my utility bills. For some these changes have meant going without needed fuel, thereby forcing people to endure freezing temperatures.

The food stamp situation has gone from bad to worse. The amount of money allocated for the program has been drastically reduced. My situation reflects that of most welfare recipients. Last year, my food stamps were cut back from $108 to $76 a month, barely enough to buy food for two weeks. Reagan doesn't even allow us to work to supplement our meager income. His reforms resulted in a law, the Omnibus Budget Reconciliation Act, that, in one fell swoop, instituted a number of repressive work-related changes. Its main impact came when it considerably lowered the amount of money a recipient can earn before the termination of benefits. Under this new law even a low-paying, part-time job can make a person no longer eligible for assistance.

The complexity of our lives reaches beyond economic issues. Monday through Friday, I work as an undergraduate student at the University of Massachusetts. On weekends, when in two-parent families, one parent can sometimes shift the responsibility to the other, I provide the entertainment for my child. All the household chores are my responsibility, for I have no one to share them with. When Jamie is sick I spend nights awake with him. When I am sick, I have no one to help me. I can't do things others take for granted, such as spend an evening out at the movies, because I don't have enough money to pay both the admission fee and for child care. Even if I did, I would be too exhausted to get out the front door. Often I wind up caught in a circle of isolation.

What kind of recognition do I and other welfare mothers get for all this hard work? One popular image of welfare recipients pictures us as lazy, irresponsible women, sitting at home, having babies, and living off the government. Much of society treats us like lepers, degrading and humiliating us at every turn, treating us as if we were getting something for nothing. One day I walked into a small grocery store wearing a button that read "Stop Reagan's War on the Poor." The proprietor of the store looked at my button and said to me, "You know, all those people on welfare are rich." Most welfare recipients would say anything rather than admit to being on welfare because of the image it creates. My brother-in-law had the audacity to say to me in conversation one day, "People are poor because they're lazy. They don't want to work."

And the Welfare Department shares this image of the recipient with the general public. From the first moment of contact with the department, the client is treated with rudeness, impatience, mistrust, and scorn. She is intimidated by constant redeterminations, reviews, and threats of being cut off. Her life is controlled by a system wracked with ineptness and callous indifference. Two years ago, unable to pay my electric bill, I applied for emergency assistance. It took the Welfare Department so long to pay the bill that the electric company turned the power off. We lived for two days without electricity before my constant badgering of the Welfare Department and the utility company produced results.

The department gives out information that is misleading and/or incomplete. The recipient is made to feel stupid, guilty, and worthless, a "problem" rather than a person. When I first applied I had to answer all sorts of questions about my personal life time and again. Many of my replies were met with disbelief. I sat there for hours at a time, nine months pregnant, waiting to be interviewed. And that was just the beginning of hours and hours of waiting, of filling in forms for the programs that keep us and our children alive.

THE PERSONAL IS POLITICAL

For most women in this situation, suffering is nothing new. Poverty is seldom an isolated issue. It's part of a whole picture. Other issues complicate our lives. For myself, as for many of us, suffering is complicated by memories, the results of trauma brought forward from childhood. Under frilly pink dresses and little blue sailor suits lay horror stories shared by many. The memories I bring with me from my childhood are not very pretty.

I was born in a Boston neighborhood in 1945. My father was a linoleum installer, my mother a homemaker. I want to say that my childhood was colored by the fact that I was an abused child. It's still difficult for me to talk about some things to this day.

My mother didn't give me a life of my own. When I was an infant she force-fed me. At age nine, she was still spoon-feeding me. Much of the time she didn't let me out of her sight. She must have seen school as a threat to her control, for she kept me home half the time. In me she saw not a separate person but an extension of herself. She felt free to do as she wished with my body. Her attempts to control my elimination process have had a lasting impact on my sexuality. The methods she used have been documented for their use in cases of mother-daughter incest. She had an obsessive-compulsive desire to control what went into me, to control elimination, to control everything about me. I had no control over anything. I couldn't get out from under what was happening to me psychologically. Powerlessness, frustration, and emotional insecurity breed a chain of abuse as men abuse women and children and women abuse children.

In the face of all the adverse, perverted attention received from my mother, I turned to my father for love and affection. We became close. In time, though, it became clear he knew as little about child rearing as my mother. His own deprived childhood had taught him only bitterness.

Throughout the years I knew him, he had gambled literally thousands of dollars away at the horse track. He left me alone outside the track gate at age five or six when the sign read "No children allowed." When I was twelve, he set fire to our house. He had run out of money for gambling purposes, and the house was insured against fire. When the insurance money arrived, my mother somehow managed to intercept it and bought new furniture. When my father found out what she had done, he went on a rampage. I had a knickknack shelf, charred from the fire. He threw it across the room, splinters of glass flying everywhere, and then he hit my mother. This was not an uncommon scene in my childhood.

By the time I reached eleven, my father was beginning to see me in a different light. At this point, the closeness that had developed between him and me still existed. And he proceeded to take advantage of it.

My mother belonged to a poker club, and once a week she would leave the house to go to these sessions. On these occasions my father would come over to me and remove both my clothes and his. He would then use my body to masturbate until he reached orgasm. He attempted to justify these actions by saying, "A man has to have sex and your mother won't." This occurred a number of times when I was between the ages of eleven and thirteen. I cried the last time he did this and he stopped molesting me sexually. In incestuous situations there doesn't have to be any threat; very often there isn't, because parents are in a position of trust, because parents are in a position of power, and because the child needs love. Children are in a developmental stage where they have no choice.

I went to public school and did very well. When I was sixteen, the school authorities told me that no matter how well I did, no matter how high my grades were, they could not keep me in school if I appeared only half the time. So I dropped out. I just spent the whole time sitting in front of the television until my mother died.

I didn't understand the extent of my mother's sickness in her treatment of me. Fear that if my mother knew what my father was doing she would kill him seemed realistic to me at the time. And so, I kept silent. I looked at my parents, as all children do, as authority figures. Longing for a way out of the situation, I felt trapped. In the face of all this I felt overwhelmed, afraid, and isolated. I retreated from reality into a world of fantasy. For only through my imagination could I find any peace of mind or semblance of happiness. The abuse I had taken all these years began to manifest itself in psychosomatic symptoms. Periodically, I started suffering intense abdominal pain. Once I fainted and fell on the bathroom floor. There were three visits to the hospital emergency room. My doctor had misdiagnosed the symptoms as appendicitis.

When I was seventeen, my father, after a major argument with my mother, moved out. Without money to pay the rent, with no job training or other resources, we rented out a room. Ann, the young woman renting the room, was appalled at the situation she discovered. She began to teach me some basic skills such as how to wash my own clothes. For the first time in my life I related positively to someone. My father moved back two months later. For two years this is how the situation remained, my mother, father, Ann, and I all living together.

In 1965 my mother was diagnosed as having lung cancer. Three months later she was dead. I felt nothing, no sorrow, no anger, no emotion. I had long ago learned how to bury my emotions deep down inside of me. I packed my belongings and moved out with Ann, in the midst of my father's ranting and raving.

Living in an apartment with Ann made things seem to improve on the surface. My life was quieter. I held a steady job for the first time. There was a little money to spend. I could come and go as I pleased. It felt good, and I guess, at that point I thought my life had really changed. It took me a long time to realize what an adult who has had this kind of childhood must still go through. I had nightmares constantly. There were times when I went into deep depressions. I didn't know what was going on, what was happening to me, or why.

After holding my first job for three years, I went through a series of jobs in different fields: sales, hairdressing, and, after a period of training, nursing. It seemed I was not functioning well in any area. I fell apart in any situation where demands were made of me. I had no tolerance for hierarchy. I discovered that I performed best when I was acting as charge nurse. Unfortunately, in my role as a Licensed Practical Nurse, I usually wound up low person on the totem pole, generally having to answer to someone else. In time, it became clear to me, that control was a major factor in a variety of situations that I encountered.

My self-confidence and self-esteem were abysmally low. Every time something went wrong my first thought was, "There must be something wrong with me." I didn't know then, what I know now. Children never blame their parents for the wrong that is done. They blame themselves. This feeling carries on into adulthood until it is difficult not to blame oneself for everything that does not go right.

I was depressed, suicidal, frightened out of my wits, and completely overwhelmed by life. After a series of failed relationships and a broken marriage, I wound up alone, with a small child, living on welfare. The emotional burden I carried became complicated by the misery and exhaustion of poverty.

Why have I survived? Why am I not dead? Because I'm a survivor. Because I have Jamie and I love him more than words can say. In him I see the future, not the past.

A MAJOR CHANGE

I survived because in the midst of all this something happened, something that was to turn out to be the major guiding force in my life. In 1972 the owner of my building sent out notices threatening eviction if the tenants did not pay a huge rent increase. Everyone in the building was aghast at the prospect and so formed a tenant union to discuss alternatives. Through my activities with the union, I learned of an organization that did community work throughout the city. Cambridge Tenants Organizing Committee (CTOC) was a multi-faceted organization involved in work around issues such as tenants' rights, welfare advocacy, antiracist work, and education concerning sexism in society. I began working with a group of people unlike any that I had been exposed to in the past. They treated me as a person capable of assuming responsibility and doing any job well. My work with CTOC included organizing tenant unions throughout the city, counseling unemployed workers, and attending countless demonstrations, marches, picketings, and hearings. As a group we organized and/or supported eviction blocking, we helped defend people against physical racist attacks, and we demonstrated at the state house for continuation of rent control. I wasn't paid for my work. What I earned was far more important than money. I learned respect for myself as a woman. I learned the joys and pitfalls of working collectively and began assimilating more information than I had at any other time of my life.

And I flourished. I went to meeting after meeting until they consumed almost all of my nonwork time. Over the years I joined other groups. One of my major commitments was to a group that presented political films. My politics became the center of my life.

Through counseling and therapy groups with therapists who shared my political perspective, the guilt I had shouldered all these years began to lessen. Within a period of three years, through groups and conventions, I listened to and/or spoke with more than three hundred other survivors of incest. I heard stories that would make your hair stand on end. New learning led to making connections. Although the true extent of incest is not known, owing to the fact that sexual abuse within families usually goes unreported, various statistics estimate that 100,000 to 250,000 children are sexually molested each year in the United States. Other studies show that one out of every three to four women in this country is a victim of sexual abuse as a child. Incest Resources, my primary resource for group counseling, believes these statistics heavily underestimate the extent of actual abuse. So do I. Although the guilt for the abuse lies with the abuser, the context of the problem reaches far beyond, beyond me or my parents, into society itself as power and inequality surround us.

Connections with other people have become part and parcel of my life as my life situation has led me into work surrounding the issue of poverty. Although I had known for years of the existence of the Coalition for Basic Human Needs (CBHN)—a progressive group composed almost totally of welfare recipients—my political work, reflecting my life situation, had led me in other directions. Now I found myself alone with a toddler, living on AFDC. When Jamie was two, I returned to school to acquire additional skills. My academic work led me into issues concerning poverty as I became involved in months of activity, along with students and faculty and a variety of progressive women's groups, constructing a conference on the issue of women and poverty. Members of CBHN were involved in work on the conference, and we connected.

Social change through welfare-rights struggles became a major focus in my life as I began working with CBHN. Collectively, we sponsored legislative bills that would improve our lives financially while, at the same time, we taught others how we actually live. Public education became intertwined with legislative work as I spoke at hearings on the reality of living on AFDC. Political support became intertwined with public education as I spoke on the work of CBHN to progressive groups and their constituencies, indicting this country's political system for the impoverishment of its people.

CBHN is composed of chapters representing various cities and towns across Massachusetts. Grass-roots organizing of welfare recipients takes place within local chapters, while the organization as a whole works on statewide issues. We mail out newsletters, hold press conferences, and initiate campaigns. We are currently involved in our most ambitious effort. Whereas in the past our work has been directed toward winning small goals such as a clothing allowance or a small increase in benefits, this time we have set about to bring welfare benefits up to the poverty level, a massive effort involving all of our past strategies and more. We have filed a bill with the state legislature. Public education has become concentrated in the campaign as we plan actions involving the work of welfare-rights groups and individuals throughout the state.

The courage we share as impoverished women has been mirrored throughout this effort. At a press conference to announce the campaign a number of us gave truth to the statement that we and our children go without the basic necessities of life. One woman spoke of sending her child to school without lunch because there was no food in the house. Others had no money for winter jackets or shoes for their children. Sharing the reality of living at 40 percent below the poverty level brings mutual support as well as frustration and anger. In mid-April hundreds of welfare recipients—women carrying infants, the disabled, and the homeless—came from all over Massachusetts to rally in front of the state house along with our supporters and testify to the legislature in behalf of our "Up to the Poverty Level" bill. Courage and determination rang out in statements such as this:

> Take the cost of implementing this program and weigh it, if it must be weighed at all; then weigh it against the anguish suffered by the six-month-old twins who starved to death in a Springfield housing project.
> The day that the state legislature has to scrape the gilt off the dome of the state house and sell if for revenue is the day that this state can answer to us that there is not the means to do this.

The complexity of the situation comes home to me again and again as I sit in the CBHN office answering the telephone and doing welfare advocacy work. I've spoken with women on AFDC who have been battered to within an inch of their lives, some who are, like myself, survivors of child abuse, and others who are reeling from the effects of racism as well as poverty.

In this society so many of us internalize oppression. We internalize the guilt that belongs to the system that creates the conditions people live in. When we realize this and turn our anger outward in an effort to change society, then we begin to create a world worth living in.

The work isn't easy. Many of us become overwhelmed as well as overextended. It takes courage and fortitude to survive. For we live in a society laden with myths and an inequality that leads to human suffering. In order to alleviate the suffering and provide the equality each and every one of us deserves, we must effect social change. If we are to effect social change, then we must recognize social injustice and destroy the myths it creates. Little did I realize, years ago, when I first began this work how far it reached beyond my own survival. For those of us involved in creating a new society are doing the most important work that exists.

Envisioning Change

AGE, RACE, CLASS, AND SEX: *Women Redefining Difference*

Audre Lorde

57

Much of Western European history conditions us to see human differences in simplistic opposition to each other: dominant/subordinate, good/bad, up/down, superior/inferior. In a society where the good is defined in terms of profit rather than in terms of human need, there must always be some group of people who, through systematized oppression, can be made to feel surplus, to occupy the place of the dehumanized inferior. Within this society, that group is made up of Black and Third World people, working-class people, older people, and women.

As a forty-nine-year-old Black lesbian feminist socialist mother of two, including one boy, and a member of an interracial couple, I usually find myself a part of some group defined as other, deviant, inferior, or just plain wrong. Traditionally, in american society, it is the members of oppressed, objectified groups who are expected to stretch out and bridge the gap between the actualities of our lives and the consciousness of our oppressor. For in order to survive, those of us for whom oppression is as american as apple pie have always had to be watchers, to become familiar with the language and manners of the oppressor, even sometimes adopting them for some illusion of protection.

Paper delivered at the Copeland Colloquium, Amherst College, April 1980.

From: Audre Lorde, *Sister Outsider* (Freedom, Calif.: Crossing Press, 1984), pp. 114–123. Reprinted by permission.

Whenever the need for some pretense of communication arises, those who profit from our oppression call upon us to share our knowledge with them. In other words, it is the responsibility of the oppressed to teach the oppressors their mistakes. I am responsible for educating teachers who dismiss my children's culture in school. Black and Third World people are expected to educate white people as to our humanity. Women are expected to educate men. Lesbians and gay men are expected to educate the heterosexual world. The oppressors maintain their position and evade responsibility for their own actions. There is a constant drain of energy which might be better used in redefining ourselves and devising realistic scenarios for altering the present and constructing the future.

Institutionalized rejection of difference is an absolute necessity in a profit economy which needs outsiders as surplus people. As members of such an economy, we have *all* been programmed to respond to the human differences between us with fear and loathing and to handle that difference in one of three ways: ignore it, and if that is not possible, copy it if we think it is dominant, or destroy it if we think it is subordinate. But we have no patterns for relating across our human differences as equals. As a result, those differences have been misnamed and misused in the service of separation and confusion.

Certainly there are very real differences between us of race, age, and sex. But it is not those differences between us that are separating us. It is rather our refusal to recognize those differences, and to examine the distortions which result from our misnaming them and their effects upon human behavior and expectation.

Racism, the belief in the inherent superiority of one race over all others and thereby the right to dominance. Sexism, the belief in the inherent superiority of one sex over the other and thereby the right to dominance. Ageism. Heterosexism. Elitism. Classism.

It is a lifetime pursuit for each one of us to extract these distortions from our living at the same time as we recognize, reclaim, and define those differences upon which they are imposed. For we have all been raised in a society where those distortions were endemic within our living. Too often, we pour the energy needed for recognizing and exploring difference into pretending those differences are insurmountable barriers, or that they do not exist at all. This results in a voluntary isolation, or false and treacherous connections. Either way, we do not develop tools for using human difference as a springboard for creative change within our lives. We speak not of human difference, but of human deviance.

Somewhere, on the edge of consciousness, there is what I call a *mythical norm,* which each one of us within our hearts knows "that is not me." In america, this norm is usually defined as white, thin, male, young, heterosexual, christian, and financially secure. It is with this mythical norm that the trappings of power reside within this society. Those of us who stand outside that power often identify one way in which we are different, and we assume that to be the primary cause of all oppression, forgetting other distortions around difference, some of which we ourselves may be practicing. By and large within the women's movement today, white women focus upon their oppression as women and ignore differences of race, sexual preference, class, and age. There is a pretense to a homogeneity of experience covered by the word *sisterhood* that does not in fact exist.

Unacknowledged class differences rob women of each others' energy and creative insight. Recently a women's magazine collective made the decision for one issue to print only prose, saying poetry was a less "rigorous" or "serious" art form. Yet even the form our creativity takes is often a class issue. Of all the art forms, poetry is the most economical. It is the one which is the most secret, which requires the least physical labor, the least material, and the one which can be done between shifts, in the hospital pantry, on the subway, and on scraps of surplus paper. Over the last few years, writing a novel on tight finances, I came to appreciate the enormous differences in the material demands between poetry and prose. As we reclaim our literature, poetry has been the major voice of poor, working class, and Colored women. A room of one's own may be a necessity for writing prose, but so are reams of paper, a typewriter, and plenty of time. The actual requirements to produce the visual arts also help determine, along class lines, whose art is whose. In this day of inflated prices for material, who are our sculptors, our painters, our photographers? When we speak of a broadly based women's culture, we need to be aware of the effect of class and economic differences on the supplies available for producing art.

As we move toward creating a society within which we can each flourish, ageism is another distortion of relationship which interferes without vision. By ignoring the past, we are encouraged to repeat its mistakes. The "generation gap" is an important social tool for any repressive society. If the younger members of a community view the older members as contemptible or suspect or excess, they will never be able to join hands and examine the living memories of the community, nor ask the all important question, "Why?" This gives rise to a historical amnesia that keeps us working to invent the wheel every time we have to go to the store for bread.

We find ourselves having to repeat and relearn the same old lessons over and over that our mothers did because we do not pass on what we have learned, or because we are unable to listen. For instance, how many times has this all been said before? For another, who would have believed that once again our daughters are allowing their bodies to be hampered and purgatoried by girdles and high heels and hobble skirts?

Ignoring the differences of race between women and the implications of those differences presents the most serious threat to the mobilization of women's joint power.

As white women ignore their built-in privilege of whiteness and define *woman* in terms of their own experience alone, then women of Color become "other," the outsider whose experience and tradition is too "alien" to comprehend. An example of this is the signal absence of the experience of women of Color as a resource for women's studies courses. The literature of women of Color is seldom included in women's literature courses and almost never in other literature courses, nor in women's studies as a whole. All too often, the excuse given is that the literatures of women of Color can only be taught by Colored women, or that they are too difficult to understand, or that classes cannot "get into" them because they come out of experiences that are "too different." I have heard this argument presented by white women of otherwise quite clear intelligence, women who seem to have no trouble at all teaching and reviewing work that comes out of the vastly different experiences of Shakespeare, Molière, Dostoyefsky, and Aristophanes. Surely there must be some other explanation.

This is a very complex question, but I believe one of the reasons white women have such difficulty reading Black women's work is because of their reluctance to see Black women as women and different from themselves. To examine Black women's literature effectively requires that we be seen as whole people in our actual complexities—as individuals, as women, as human—rather than as one of those problematic but familiar stereotypes provided in this society in place of genuine images of Black women. And I believe this holds true for the literatures of other women of Color who are not Black.

The literatures of all women of Color recreate the textures of our lives, and many white women are heavily invested in ignoring the real differences. For as long as any difference between us means one of us must be inferior, then the recognition of any difference must be fraught with guilt. To allow women of Color to step out of stereotypes is too guilt provoking, for it threatens the complacency of those women who view oppression only in terms of sex. *[margin note: or wrong?]*

Refusing to recognize difference makes it impossible to see the different problems and pitfalls facing us as women.

Thus, in a patriarchal power system where whiteskin privilege is a major prop, the entrapments used to neutralize Black women and white women are not the same. For example, it is easy for Black women to be used by the power structure against Black men, not because they are men, but because they are

Black. Therefore, for Black women, it is necessary at all times to separate the needs of the oppressor from our own legitimate conflicts within our communities. This same problem does not exist for white women. Black women and men have shared racist oppression and still share it, although in different ways. Out of that shared oppression we have developed joint defenses and joint vulnerabilities to each other that are not duplicated in the white community, with the exception of the relationship between Jewish women and Jewish men.

On the other hand, white women face the pitfall of being seduced into joining the oppressor under the pretense of sharing power. This possibility does not exist in the same way for women of Color. The tokenism that is sometimes extended to us is not an invitation to join power; our racial "otherness" is a visible reality that makes that quite clear. For white women there is a wider range of pretended choices and rewards for identifying with patriarchal power and its tools.

Today, with the defeat of ERA, the tightening economy, and increased conservatism, it is easier once again for white women to believe the dangerous fantasy that if you are good enough, pretty enough, sweet enough, quiet enough, teach the children to behave, hate the right people, and marry the right men, then you will be allowed to co-exist with patriarchy in relative peace, at least until a man needs your job or the neighborhood rapist happens along. And true, unless one lives and loves in the trenches it is difficult to remember that the war against dehumanization is ceaseless.

But Black women and our children know the fabric of our lives is stitched with violence and with hatred, that there is no rest. We do not deal with it only on the picket lines, or in dark midnight alleys, or in the places where we dare to verbalize our resistance. For us, increasingly, violence weaves through the daily tissues of our living—in the supermarket, in the classroom, in the elevator, in the clinic and the schoolyard, from the plumber, the baker, the saleswoman, the bus driver, the bank teller, the waitress who does not serve us.

Some problems we share as women, some we do not. You fear your children will grow up to join the patriarchy and testify against you, we fear our children will be dragged from a car and shot down in the street, and you will turn your backs upon the reasons they are dying.

The threat of difference has been no less blinding to people of Color. Those of us who are Black must see that the reality of our lives and our struggle does not make us immune to the errors of ignoring and misnaming difference. Within Black communities where racism is a living reality, differences among us often seem dangerous and suspect. The need for unity is often misnamed as a need for homogeneity, and a Black feminist vision mistaken for betrayal of our common interests as a people. Because of the continuous battle against racial erasure that Black women and Black men share, some Black women still refuse to recognize that we are also oppressed as women, and that sexual hostility against Black women is practiced not only by the white racist society, but implemented within our Black communities as well. It is a disease striking the heart of Black nationhood, and silence will not make it disappear. Exacerbated by racism and the pressures of powerlessness, violence against Black women and children often becomes a standard within our communities, one by which manliness can be measured. But these woman-hating acts are rarely discussed as crimes against Black women.

As a group, women of Color are the lowest paid wage earners in america. We are the primary targets of abortion and sterilization abuse, here and abroad. In certain parts of Africa, small girls are still being sewed shut between their legs to keep them docile and for men's pleasure. This is known as female circumcision, and it is not a cultural affair as the late Jomo Kenyatta insisted, it is a crime against Black women.

Black women's literature is full of the pain of frequent assault, not only by a racist patriarchy, but also by Black men. Yet the necessity for and history of shared battle have made us, Black women, particularly vulnerable to the false accusation that anti-sexist is anti-Black. Meanwhile, womanhating as a recourse of the powerless is sapping strength from Black communities, and our very lives. Rape is on the increase, reported and unreported, and rape is not aggressive sexuality, it is sexualized aggression. As Kalamu ya Salaam, a Black male writer points out, "As long as male domination exists, rape will exist. Only women revolting and men made conscious of their responsibility to fight sexism can collectively stop rape."*

Differences between ourselves as Black women are also being misnamed and used to separate us from one another. As a Black lesbian feminist comfortable with the many different ingredients of my identity, and a woman committed to racial and sexual freedom from oppression, I find I am constantly being encouraged to pluck out some one aspect of myself and present this as the meaningful whole, eclipsing or denying the other parts of self. But this is a destructive and fragmenting way to live. My fullest concentration of energy is available to me only when I integrate all the parts of who I am, openly, allowing power from particular sources of my living to flow back and forth freely through all my different selves, without the restrictions of externally imposed definition. Only then can I bring myself and my energies as a whole to the service of those struggles which I embrace as part of my living.

---

*From "Rape: A Radical Analysis, An African-American Perspective" by Kalamu ya Salaam in *Black Books Bulletin*, vol. 6, no. 4 (1980).

A fear of lesbians, or of being accused of being a lesbian, has led many Black women into testifying against themselves. It has led some of us into destructive alliances, and others into despair and isolation. In the white women's communities, heterosexism is sometimes a result of identifying with the white patriarchy, a rejection of that interdependence between women-identified women which allows the self to be, rather than to be used in the service of men. Sometimes it reflects a die-hard belief in the protective coloration of heterosexual relationships, sometimes a self-hate which all women have to fight against, taught us from birth.

Although elements of these attitudes exist for all women, there are particular resonances of heterosexism and homophobia among Black women. Despite the fact that woman-bonding has a long and honorable history in the African and African-american communities, and despite the knowledge and accomplishments of many strong and creative women-identified Black women in the political, social and cultural fields, heterosexual Black women often tend to ignore or discount the existence and work of Black lesbians. Part of this attitude has come from an understandable terror of Black male attack within the close confines of Black society, where the punishment for any female self-assertion is still to be accused of being a lesbian and therefore unworthy of the attention or support of the scarce Black male. But part of this need to misname and ignore Black lesbians comes from a very real fear that openly women-identified Black women who are no longer dependent upon men for their self-definition may well reorder our whole concept of social relationships.

Black women who once insisted that lesbianism was a white woman's problem now insist that Black lesbians are a threat to Black nationhood, are consorting with the enemy, are basically un-Black. These accusations, coming from the very women to whom we look for deep and real understanding, have served to keep many Black lesbians in hiding, caught between the racism of white women and the homophobia of their sisters. Often, their work has been ignored, trivialized, or misnamed, as with the work of Angelina Grimke, Alice Dunbar-Nelson, Lorraine Hansberry. Yet women-bonded women have always been some part of the power of Black communities, from our unmarried aunts to the amazons of Dahomey.

And it is certainly not Black lesbians who are assaulting women and raping children and grandmothers on the streets of our communities.

Across this country, as in Boston during the spring of 1979 following the unsolved murders of twelve Black women, Black lesbians are spearheading movements against violence against Black women.

What are the particular details within each of our lives that can be scrutinized and altered to help bring about change? How do we redefine difference for all women? It is not our differences which separate women, but our reluctance to recognize those differences and to deal effectively with the distortions which have resulted from the ignoring and misnaming of those differences.

As a tool of social control, women have been encouraged to recognize only one area of human difference as legitimate, those differences which exist between women and men. And we have learned to deal across those differences with the urgency of all oppressed subordinates. All of us have had to learn to live or work or coexist with men, from our fathers on. We have recognized and negotiated these differences, even when this recognition only continued the old dominant/subordinate mode of human relationship, where the oppressed must recognize the masters' difference in order to survive.

But our future survival is predicated upon our ability to relate within equality. As women, we must root out internalized patterns of oppression within ourselves if we are to move beyond the most superficial aspects of social change. Now we must recognize differences among women who are our equals, neither inferior nor superior, and devise ways to use each others' difference to enrich our visions and our joint struggles.

The future of our earth may depend upon the ability of all women to identify and develop new definitions of power and new patterns of relating across difference. The old definitions have not served us, nor the earth that supports us. The old patterns, no matter how cleverly rearranged to imitate progress, still condemn us to cosmetically altered repetitions of the same old exchanges, the same old guilt, hatred, recrimination, lamentation, and suspicion.

For we have, built into all of us, old blueprints of expectation and response, old structures of oppression, and these must be altered at the same time as we alter the living conditions which are a result of those structures. For the master's tools will never dismantle the master's house.

As Paulo Freire shows so well in *The Pedagogy of the Oppressed,** the true focus of revolutionary change is never merely the oppressive situations which we seek to escape, but that piece of the oppressor which is planted deep within each of us, and which knows only the oppressors' tactics, the oppressors' relationships.

Change means growth, and growth can be painful. But we sharpen self-definition by exposing the self in work and struggle together with those whom we define as different from ourselves, although sharing the same goals. For Black and white, old and young, lesbian and heterosexual women alike, this can mean new paths to our survival.

---

*Seabury Press, New York, 1970.

We have chosen each other
and the edge of each others battles
the war is the same
if we lose
someday women's blood will congeal
upon a dead planet
if we win
there is no telling
we seek beyond history
for a new and more possible meeting.*

COALITION POLITICS: *Turning the Century***

58

Bernice Johnson Reagon

I've never been this high before. I'm talking about the altitude. There is a lesson in bringing people together where they can't get enough oxygen, then having them try to figure out what they're going to do when they can't think properly. I'm serious about that. There probably are some people here who can breathe, because you were born in high altitudes and you have big lung cavities. But when you bring people in who have not had the environmental conditioning, you got one group of people who are in a strain—and the group of people who are feeling fine are trying to figure out why you're staggering around, and that's what this workshop is about this morning.

I wish there had been another way to graphically make me feel it because I belong to the group of people who are having a very difficult time being here. I feel as if I'm gonna keel over any minute and die. That is often what it feels

---

*From "Outlines," unpublished poem.

**Based upon a presentation at the West Coast Women's Music Festival 1981, Yosemite National Forest, California.

From: Barbara Smith (ed.), *Home Girls: A Black Feminist Anthology* (New York: Kitchen Table Press, 1983), pp. 356–368. Reprinted by permission.

like if you're *really* doing coalition work. Most of the time you feel threatened to the core and if you don't, you're not really doing no coalescing.

I'm Bernice Reagon. I was born in Georgia, and I'd like to talk about the fact that in about twenty years we'll turn up another century. I believe that we are positioned to have the opportunity to have something to do with what makes it into the next century. And the principles of coalition are directly related to that. You don't go into coalition because you just *like* it. The only reason you would consider trying to team up with somebody who could possibly kill you, is because that's the only way you can figure you can stay alive.

A hundred years ago in this country we were just beginning to heat up for the century we're in. And the name of the game in terms of the dominant energy was technology. We have lived through a period where there have been things like railroads and telephones, and radios, TV's and airplanes, and cars, and transistors, and computers. And what this has done to the concept of human society and human life is, to a large extent, what we in the latter part of this century have been trying to grapple with. With the coming of all that technology, there was finally the possibility of making sure no human being in the world would be unreached. You couldn't find a place where you could hide if somebody who had access to that technology wanted to get to you. Before the dawning of that age you had all these little cute villages and the wonderful homogeneous societies where everybody looked the same, did things the same, and believed the same things, and if they didn't, you could just kill them and nobody would even ask you about it.

We've pretty much come to the end of a time when you can have a space that is "yours only"—just for the people you want to be there. Even when we have our "women-only" festivals, there is no such thing. The fault is not necessarily with the organizers of the gathering. To a large extent it's because we have just finished with that kind of isolating. There is no hiding place. There is nowhere you can go and only be with people who are like you. It's over. Give it up.

Now every once in awhile there is a need for people to try to clean out corners and bar the doors and check everybody who comes in the door, and check what they carry in and say, "Humph, inside this place the only thing we are going to deal with is X or Y or Z." And so only the X's or Y's or Z's get to come in. That place can then become a nurturing place or a very destructive place. Most of the time when people do that, they do it because of the heat of trying to live in this society where being an X or Y or Z is very difficult, to say the least. The people running the society call the shots as if they're still living in one of those little villages, where they kill the ones they don't like or put them in the forest to die. (There are some societies where babies are born and if they are not wanted for some reason they are put over in a corner. They do that here too, you know, put them in garbage cans.) When somebody else is running a society like that, and you are the one who would be put out to die, it gets too hard to stay out in that society all the time. And that's when you find a place, and you try to bar the door and check all the people who come in. You come together to see what you can do about shouldering up all of your energies so that you and your kind can survive.

There is no chance that you can survive by staying *inside* the barred room. (Applause) That will not be tolerated. The door of the room will just be painted red and then when those who call the shots get ready to clean house, they have easy access to you.

But that space while it lasts should be a nurturing space where you sift out what people are saying about you and decide who you really are. And you take the time to try to construct within yourself and within your community who you would be if you were running society. In fact, in that little barred room where you check everybody at the door, you act out community. You pretend that your room is a world. It's almost like a play, and in some cases you actually grow food, you learn to have clean water, and all of that stuff, you just try to do it all. It's like, "If *I* was really running it, this is the way it would be."

Of course the problem with the experiment is that there ain't nobody in there but folk like you, which by implication means you wouldn't know what to do if you were running it with all of the other people who are out there in the world. Now that's nationalism. I mean it's nurturing, but it is also nationalism. At a certain stage nationalism is crucial to a people if you are going to ever impact as a group in your own interest. Nationalism at another point becomes reactionary because it is totally inadequate for surviving in the world with many peoples. (Applause)

Sometimes you get comfortable in your little barred room, and you decide you in fact are going to live there and carry out all of your stuff in there. And you gonna take care of everything that needs to be taken care of in the barred room. If you're white and in the barred room and if everybody's white, one of the first things you try to take care of is making sure that people don't think that the barred room is a racist barred room. So you begin to talk about racism and the first thing you do is say, "Well, maybe we better open the door and let some Black folks in the barred room." Then you think, "Well, how we gonna figure out whether they're X's or not?" Because there's nothing in the room but X's. (Laughter) You go down the checklist. You been working a while to sort out who you are, right? So you go down the checklist and say, "If we can find Black folk like that we'll let them in the room." You don't really want Black folks, you are just looking for yourself with a little color to it.

And there are those of us Black folk who are like that. So if you're lucky you can open the door and get one or two. Right? And everything's wonderful. But no matter what, there will be one or two of us who have not bothered to be like you and you know it. We come knocking on your door and say, "Well, you let them in, you let me in too." And we will break your door down trying to get in. (Laughter) As far as we can see we are also X's. Cause you didn't say, "THIS BARRED ROOM IS FOR WHITE X'S ONLY." You just said it was for X's. So everybody who thinks they're an X comes running to get into the room. And because you trying to take care of everything in this room, and you know you're not racist, you get pressed to let us all in.

The first thing that happens is that the room don't feel like the room anymore. (Laughter) And it ain't home no more. It is not a womb no more. And you can't feel comfortable no more. And what happens at that point has to do with trying to do too much in it. You don't do no coalition building in a womb. It's just like trying to get a baby used to taking a drink when they're in your womb. It just don't work too well. Inside the womb you generally are very soft and unshelled. You have no covering. And you have no ability to handle what happens if you start to let folks in who are not like you.

Coalition work is not work done in your home. Coalition work has to be done in the streets. And it is some of the most dangerous work you can do. And you shouldn't look for comfort. Some people will come to a coalition and they rate the success of the coalition on whether or not they feel good when they get there. They're not looking for a coalition; they're looking for a home! They're looking for a bottle with some milk in it and a nipple, which does not happen in a coalition. You don't get a lot of food in a coalition. You don't get fed a lot in a coalition. In a coalition you have to give, and it is different from your home. You can't stay there all the time. You go to the coalition for a few hours and then you go back and take your bottle wherever it is, and then you go back and coalesce some more.

It is very important not to confuse them—home and coalition. Now when it comes to women—the organized women's movement—this recent thrust—we all have had the opportunity to have some kind of relationship with it. The women's movement has perpetuated a myth that there is some common experience that comes just cause you're women. And they're throwing all these festivals and this music and these concerts happen. If you're the same kind of women like the folk in that little barred room, it works. But as soon as some other folk check the definition of "women" that's in the dictionary (which you didn't write, right?) they decide that they can come because they are women, but when they do, they don't see or hear nothing that is like them. Then they charge, "This ain't no women's thing!" (Applause) Then if you try to address that and bring them in, they start to play music that ain't even women's music!

(Laughter and hoots) And you try to figure out what happened to your wonderful barred room. It comes from taking a word like "women" and using it as a code. There is an in-house definition so that when you say "women only" most of the time that means you had better be able—if you come to this place—to handle lesbianism and a lot of folks running around with no clothes on. And I'm being too harsh this morning as I talk to you, but I don't want you to miss what I'm trying to say. Now if you come and you can't handle that, there's another term that's called "woman-identified." They say you might be a woman but you're not woman-identified, and we only want women who are "woman-identified." That's a good way to leave a lot of women out of your room.

So here you are and you grew up and you speak English and you know about this word "woman" and you know you one, and you walk into this "woman-only" space and you ain't there. (Laughter) Because "woman" in that space does not mean "woman" from your world. It's a code word and it traps, and the people that use the word are not prepared to deal with the fact that if you put it out, everybody that thinks they're a woman may one day want to seek refuge. And it ain't no refuge place! And it's not safe! It should be a coalition! It may have been that in its first year the Michigan National "Women-Only" festival was a refuge place. By the fourth year it was a place of coalition, and it's not safe anymore. (Applause) It ain't safe for nobody who comes. When you walk in there you in trouble—and everybody who comes is trying to get to their home there. At this festival [Yosemite] they said: whatever you drink, bring it with you—tea, honey, you know, whatever it is—and we will provide hot water. Now I understand that you got here and there was no hot water. Can't get nothing! That is the nature of coalition. (Laughter) You have to give it all. It is not to feed you; you have to feed it. And it's a monster. It never gets enough. It always wants more. So you better be sure you got your home someplace for you to go to so that you will not become a martyr to the coalition. Coalition *can* kill people; however, it is not by nature fatal. You do not have to die because you are committed to coalition. I'm not so old, and I don't know nothing else. But you do have to know how to pull back, and you do have to have an old-age perspective. You have to be beyond the womb stage.

None of this matters at all very much if you die tomorrow—that won't even be cute. It only matters if you make a commitment to be around for another fifty more years. There are some grey haired women I see running around occasionally, and we have to talk to those folks about how come they didn't commit suicide forty years ago. Don't take everything they say because some of the stuff they gave up to stay around ain't worth considering. But be sure you get on your agenda some old people and try to figure out what it will be like if you are a raging radical fifty years from today.

Think about yourself that way. What would you be like if you had white hair and had not given up your principles? It might be wise as you deal with coalition efforts to think about the possibilities of going for fifty years. It calls for some care. I'm not gonna be suicidal, if I can help it. Sometimes you don't even know you just took a step that could take your head off cause you can't know everything when you start to coalesce with these people who sorta look like you in just one aspect but really they belong to another group. That is really the nature of women. It does not matter at all that biologically we have being women in common. We have been organized to have our primary cultural signals come from some other factors than that we are women. We are not from our base acculturated to be women people, capable of crossing our first people boundaries—Black, White, Indian, etc.

Now if we are the same women from the same people in this barred room, we never notice it. That stuff stays wherever it is. It does not show up until somebody walks into the room who happens to be a woman but really is also somebody else. And then out comes who we really are. And at that point you are not a woman. You are Black or you are Chicana or you are Disabled or you are Racist or you are White. The fact that you are a woman is not important at all and it is not the governing factor to your existence at that moment. I am now talking about bigotry and everybody's got it. I am talking about turning the century with some principles intact. Today wherever women gather together it is not necessarily nurturing. It is coalition building. And if you feel the strain, you may be doing some good work. (Applause) So don't come to no women's festival looking for comfort unless you brought it in your little tent. (Laughter) And then if you bring it in your tent don't be inviting everybody in because everybody ain't your company, and then you won't be able to stand the festival. Am I confusing you? Yes, I am. If coalition is so bad, and so terrible, and so uncomfortable, why is it necessary? That's what you're asking. Because the barred rooms will not be allowed to exist. They will all be wiped out. That is the plan that we now have in front of us.

Now these little rooms were created by some of the most powerful movements we have seen in this country. I'm going to start with the Civil Rights movement because of course I think that that was the first one in the era we're in. Black folks started it, Black folks did it, so everything you've done politically rests on the efforts of my people—that's my arrogance! Yes, and it's the truth; it's my truth. You can take it or leave it, but that's the way I see it. So once we did what we did, then you've got women, you've got Chicanos, you've got the Native Americans, and you've got homosexuals, and you got all of these people who also got sick of somebody being on their neck. And maybe if they come together, they can do something about it. And I claim all of you as coming from something that made me who I am. You can't tell me that you ain't in the Civil Rights movement. You are in the Civil Rights movement that we created that just rolled up to your door. But it could not stay the same, because if it was gonna stay the same it wouldn't have done you no good. Some of you would not have caught yourself dead near no Black folks walking around talking about freeing themselves from racism and lynching. So by the time our movement got to you it had to sound like something you knew about. Like if I find out you're gay, you gonna lose your job.

There were people who came South to work in the movement who were not Black. Most of them were white when they came. Before it was over, that category broke up—you know, some of them were Jewish, not simply white, and some others even changed their names. Say if it was Mary when they came South, by the time they were finished it was Maria, right? It's called finding yourself. At some point, you cannot be fighting oppression and be oppressed yourself and not feel it. Within the Black movement there was also all of the evils of the society, so that anything that was happening to you in New York or the West Coast probably also happened to you in another way, within the movement. And as you became aware of that you tried to talk to these movement people about how you felt. And they say, "Well let's take that up next week. Because the most important thing now is that Black people are being oppressed and we must work with that." Watch these mono-issue people. They ain't gonna do you no good. I don't care who they are. And there are people who prioritize the cutting line of the struggle. And they say the cutting line is this issue, and more than anything we must move on this issue and that's automatically saying that whatever's bothering you will be put down if you bring it up. You have to watch these folks. Watch these groups that can only deal with one thing at a time. On the other hand, learn about space within coalition. You can't have everybody sitting up there talking about everything that concerns you at the same time or you won't get no place. . . .

It must become necessary for all of us to feel that this is our world. And that we are here to stay and that anything that is here is ours to take and to use in our image. And watch that "our"—make it as big as you can—it ain't got nothing to do with that barred room. The "our" must include everybody you have to include in order for you to survive. You must be sure you understand that you ain't gonna be able to have an "our" that don't include Bernice Johnson Reagon, cause I don't plan to go nowhere! That's why we have to have coalitions. Cause I ain't gonna let you live unless you let me live. Now there's danger in that, but there's also the possibility that we can both live—if you can stand it. . . .

CULTURE AND GENDER IN INDIAN AMERICA

59

Rayna Green

I don't have a theory or line of argument this morning. I want to go through a series of vignettes, all of which cast a different light, cast a different slant, and give a slightly different ear to each other.

My first story is about two friends of mine, both Sioux women who were up at the Capitol one day. Up on top of the Capitol there's a statue. It's a marvelous statue. If you get very close to that statue, you'll see that it's clearly a female figure, and it might look something like Miss Liberty standing out in New York Harbor. Nobody knows much about her. These two young Sioux women I know went up to the guard and they said, "What's the statue? What's it about?" He said, "Well, you know, that statue, a lot of people seem to think it's an Indian. Her name is Freedom. But Freedom isn't an Indian. She's a woman." And my two friends laughed to think that in this country freedom could be either. But the truth is, in this country freedom is both. And that's what I want to talk about today. And I want to talk about justice and liberty, and about America. I want to talk about home, family, and about women. I want to talk about changing our names, and taking hold of our names. And I want to talk about coming home.

Freedom is an Indian and a woman. She's also Black and a woman. She's also Jewish, Vietnamese, and Salvadoran, and a woman. She is all of those things. But her iconography is clear. She comes from the fifteenth century, when the first images of the New World went back to Europe. Freedom, in the early days of America, was pictured as this large, bare-breasted Indian woman. She was a queen. She was our kind of girl. She was pictured with her foot on the head of an alligator, her spear in her hands. Pineapples, corn, all these wonderful bounteous crops spilling out of her arms. Her warriors stood behind her. She was in control; she was the New World, the promise of everything that everyone wanted.

And they took her away from us. As the two centuries moved forward and things happened here that we now must pay for, dearly, she changed. They took away her flesh. They covered her breasts. She couldn't be naked, the symbol of innocence in the fifteenth, sixteenth century; the symbol of virtue.

From: *Sojourner: The Women's Forum* 15 (September 1989): 20–22. Reprinted by permission.

They had to cover her and make her less savage, less pagan. They took away her alligator. I mean, can you imagine taking a girl's alligator away? She must have been angry. They took away all of the fruits of her fields. And she became a Renaissance little wonder woman icon, like Minerva, draped in little tasteful white garments with her breasts covered. And they wouldn't even let her be like Minerva; they robbed her of her power. Thin and powerless, with a tiny little diadem on her head. She never needed a crown to know she was a queen. But now, she's changed. And that's who stands in the New York harbor. Someone who changed. And that's who stands on top of the Capitol. But we need to know her name. Her name is Freedom, and she is who I have described. And we need to go back to her to begin to look to the future.

That future is vague and muddy, though. That future is opaque. We all have difficulties now knowing who we are, and what our names are. We try to put on different names, give ourselves a kind of identity. We struggle through various ways of looking at an identity. Indian people have been forced to confront lots of different faces in the mirror, and those faces are confusing for all of us. At one level, Indians are totally insignificant. They exist in no number to matter—to the economy, to the judicial system, to anything else. There are fewer Indians in America than Vietnamese. And so why do Indians matter? I'll tell you why they matter: simply because of the history I've spoken of.

But Indians cannot simply be functions of the historic past. They can't simply be reminders of an America that once was, bad or good. They can't simply function as vague, ghostly reminders of poverty index levels, of hunger and homelessness in America; that's irrelevant. What is relevant is that there is a metaphor here that stands for all of us in some ways and doesn't encompass the experience of others in another. What is relevant is that as my friend Roberta Hill Whiteman says, "Indians know how to wait." And it is the waiting that will dignify us all. The waiting for freedom and justice to come in the form of an Indian woman, once again, to reclaim us all. In Roberta's words:

> *Look west long enough, the moon will grow*
> *inside you.*
> *Coyote hears her song, he'll*
> *teach you now.*
> *Mirrors follow trails of blood and lightning.*
> *Mother needs the strength of one like you.*
> *Let blood*
> *dry, but seize the lightning. Hold it like your*
> *mother*
> *rocks the trees. In your fear, watch the road,*
> *breathe deeply.*
> *Indians know how to wait.*

(from "Lines for Marking Time")

What are we waiting for? Are we waiting for a moment like that which happened at the National Women's Studies Association Conference in Minneapolis last year? The planning group invited a young Indian girl to dance for the opening event. Typically, and profoundly, she came with her uncle, and a group of young men who drummed for her. In Indian culture, an uncle is like your father. An uncle raises you. An aunt raises you. Your own father and mother are perhaps even less significant in some ways. Her uncle came with her because her father had just died. And her uncle, to honor her, and to honor the women who had come to see her, spoke for her.

In our world, people speak for you when you're honored. It's a gift to speak for someone. And when that man, who was honoring everyone there by his presence, rose to speak for her, he was booed. Because in an environment where we've gotten our signals crossed, we don't know the faces of other people, we don't know how they live, and we cannot speak to them directly. He gave them even a further gift, he explained to them why she was not wearing her jingle dress. (A jingle dress is a wonderful buckskin or cloth dress filled with little tin coins that make a marvelous noise when a young lady dances.) She was in her menstrual period, and a girl does not wear a jingle dress when she's menstruating, because the noise of those coins, you see, is a prayer, and it's a prayer for power. Music goes up, music calls down the spirits to look at you, and asks for power. Because a menstruating woman is already so powerful, to wear the jingle dress is to really risk a problem; to call down uncontrolled power, perhaps. He gave them the gift of telling them this. He was explaining something rather arcane, something that people don't just discuss in public. It was a women's event. He wanted to reach out. And they booed him for that, because they thought he was talking about pollution.

I'm not here, as I said, to accuse. This is not accusatory. That is not what Indian women and Indian men are about. This is about knowing our own names and knowing our faces. A gift was refused because no one knew it was a gift. We have got to come forward and know the gifts that different people give us. And that's why a meeting like this is essential—to look in the face of different gifts and to learn to honor each other, by accepting the terms on which those gifts are given.

Sometimes, because things get so confused in moments like that, we're forced to change our names. Sometimes we have to change our shape. Shape shifters are important for all of us; all of our worlds have shape shifters. Sometimes we have to shift shape because guns are aimed at us, and we must escape. Sometimes the old shape has become too uncomfortable. Sometimes, like the queen in the early days, our original form is taken—we're sent to the diet center, forced on cultural aerobics until we change. And I say it's time to change our shape because we want to, to change our names because we want to.

Sometimes when we're forced to change those shapes it's painful; sometimes it's a joy. In Indian cultures, there is a tradition of name-changing and shape-shifting, and it's an important tradition to look to for all of us because it enables us to be empowered; it enables us to be in control. We are all women, certainly; we are all men; we are all gay; we are all straight; we are all old; we are all young. And in Indian cultures I go back to my families and there is no division. There are distinctions, certainly, about the way people are treated, and the authority they have, but I want to reclaim the power to move through categories, so that I do not have to stay fixed in any one place. Jesse [Jackson] said a couple of years ago, in the presidential elections, "God ain't finished with me yet, she never will be finished." We can all move and grow.

Some of the categories become so restrictive, we have to be able to move out of them. One category that Indian people suffer from is that of "half-breed." It's staggering to think about that kind of marginality; to suggest that someone is a quart low of whatever it is that makes them real is to take their life and breath and squeeze it until it stops. We're all half-breeds, if you come right down to it. As my mom used to say, "Heinz 57." We locate our internal space, perhaps, in one place that gives us a name to call ourselves that we're proud of, but we are all out there on the margins (and there are no people on the margin like women, because we have to shift into so many shapes). But the category of "half-breed" is tragic, damaging in so many ways, we've got to give it up. It is like Freedom. We've got to put a name to it that enables us to stand up again.

I am a German Jew. I feel comfortable with that. I've lived my life as a German Jew. One of my grandmothers was a German Jew who became a profound Texan. I will not deny her. I will not look at my mother, with her blue eyes, her white skin, and deny her. My father is only one part of me. He must be claimed, too. And I gravitate toward that part of him, and his world, and my other grandmother, who gave me a name and space. But I will not deny any one part of that world to force Indian people, native people, any people to live on a margin, where they cannot define their own existence as a whole, whatever parts may be there, just to rob them of a future, and to force them to die somewhere in a past.

My German grandmother was a remarkable person. She is my mother, the primary character in my life. She shaped and formed me; she gave me stories, language, and songs. In many ways, she gave me more than my Indian grandmother, who was afraid because of all the things that had happened to her; afraid to sing, to breathe, to leave town. Her pain transferred over, so I took joy from the maternal grandmother, and I took the name from the other one. We take what we can from each one of our relatives and honor them.

My German grandmother was an extraordinary woman who loved to dance, loved to sing, loved to tell dirty stories. She was the mistress of them all. She wanted to be a dancer on the Palace stage. Dressed to kill, she'd play whorehouse songs on the piano, or sad ones that would make us cry and beg for more. On that summer porch, we believed she could have been anything, living in her ruby pleasures. Oh, she glittered then, dancing across that summer porch, dancing the stories that made me dream over her shattered breath. She is mine, and she is yours, too. Never walk away from all the faces you've known in your life, who gave you birth. That's the Indian way, certainly. But it's all of our way. If we only can have the courage to look back for them.

Some of the shape-shifting sometimes makes it important to walk away from being female. Being female is painful. Being male is painful, never more than lately, as we look into the faces of our young men on the streets of urban cities. We see death in their faces, and it pains us. If only they had the freedom we have, to walk into a female body, into a female metaphor, and say, "That's not my world. Those guns, that dope, those drugs are not my world. That world that lures me only because my name is man, that world that pains me because I have to live trapped in a metaphor, that will not work for me."

We have an option that makes us live, an option women have always had, an extraordinary option. Some women take it in different ways. Some women call themselves mothers, some sisters, some lovers. Some women call themselves men, and walk into Coyote's terrible dream, which enables us to move through the changes. We can take any of those options at any one time. But to take those options means we have to teach our baby boys to grow up and be all those things, too. We don't have to claim the men in us by being tough. We only have to look in our children's and our brothers' faces, and bring them up to live with us.

There's a gift that moves in Indian country, and that gift is an extraordinary gift. It is the gift of giving itself. All of us have it in our worlds. At dances, various ceremonial occasions, you'll see women walk over and put shawls on other people's backs—men's backs, too. In ceremonial occasions these shawls get piled so high, you never know how anybody stands up under them. (I want to eartag a shawl during a powwow season sometime like biologists do animals, see how it migrates across the room.) Those gifts are extraordinary.

I want to take the metaphor of that shawl, and I want to wrap it around your shoulders now, and say you are my sister, you are my mother, you are my friend, you are my brother, you are my husband, you are my uncle, you are my aunt, come into that dance circle. It is the gift that keeps on moving, because it brings us into the circle. The richest person in Indian country is the person who gives the most away, not the person who keeps the most for themselves. And this is a gift that America needs. We would not have homeless people on our streets, if we truly believed we lived in Indian country. If freedom was really the Indian woman we know she is, we would not have people living out on the margin, children selling themselves for one shot, for a pint of Thunderbird. We have got to wrap that shawl around the rest of us, and women can do that.

To talk about culture in this country, to talk about gender, is to talk about giving. And it is in the heart and face of our own cultures, and our own passions that we can look and see the gift that we have to wrap again, and keep moving, and keep giving. But in order to do that, we have to know what real wealth is. Real wealth lies in our own hearts, and not in something that is a commodity beyond it. Indian country knows that. The gift that moves will carry us to that place. I think of the warrior women who reputedly used to carry their husbands' and brothers' and friends' bodies off the field and take up the bow or the gun or the spear themselves. And I say, this is not militance, this is not warrior behavior (although at one level it is: it is what is required to survive). All of us in our communities now are carrying those boys' bodies off the field. What will we do when there are no more of them? Will we become the warriors? Are we willing to take that battle on? Perhaps the gift is not to keep thinking of it as a battle, but to think of it as a role for us all, in the survival of our people, in the raising of our children. What is a real warrior woman—in Indian terms, even? Certainly, it is not to take up a spear. If you have to, you do, to defend your life or that of your children. But I don't want to talk about defensiveness; I want to talk about survival. And there are keys to survival.

In 1642, a group of British got off a boat in Virginia somewhere and migrated to what we call North Carolina; and there they met a delegation of Cherokees, led by a man who had been a warrior. His name was Outacitty, which means man-killer. He was a great warrior, a red chief, sent by the Beloved Woman and the clan-mothers to make war. But this time they had asked him not to make war. The Beloved Woman of the Nation, Ghigau had asked him to become a white chief, a peace chief, and to go and make peace with these people. And so Outacitty rode up to meet them. The first thing he said to them was, "Where are your women?" These men had come to do serious business, and they had no women with them. Peace is a very serious business. No act of war, no act of peace in my country is made without the women there. And Outacitty was shocked: the British dared to come without their women. "Where are your women?" he said. And he went back and reported that there was a problem here. "We cannot do business with these men," he said. They were clearly missing half of the people needed to do business with.

And in 1987, a young woman named Wilma Mankiller was elected principal chief of the Cherokee nation, and I say to you she is him, come back, and she knows it. And the old people knew it. When we'd go out to campaign with her, the old ladies would say, "Good name, good name." And they didn't mean war. They didn't mean hunting; they didn't mean power. They meant she's back. She has returned; the Beloved has returned. You see, she is all of those converged together. Like all of us can be. Warrior, peacemaker, mother, father, Beloved woman, Beloved man, the white and the red merged together, to take our own story back.

And that's an interesting story. It's a story about family. Everybody in Indian country talks about the family as the center of their lives, just as in Black culture, in Irish-American communities. Because to talk about family is to talk about community, about survival, about the future.

The central figure in this next story about family is a woman called Buffalo Birdwoman who could not give up the old way. She was a great farmer. When the time came to make her change, they brought in the tractors, they brought in the freight wagons, and she said, "I don't want to do that. I'll stick to my digging stick. Because I grow corn better than your corn. I know how to grow corn, and I will not give up on corn songs. I will not forget the corn songs. The corn is my family, my mother, my grandmother, the corn gave me birth. And to give up my corn songs is to give up my family." Tradition is not a yoke around our neck, if we know its name. Tradition is not the chains that bind us, if we know how to use it. Tradition is not the deadly past wrapped around us like a coffin. Tradition, for the Indian family, for the Indian woman, is simply remembering who you are, and it is that story we must reclaim once more. Hear Linda Hogan, Chickasaw:

> calling myself home
>
> There were old women
> who lived on amber.
> Their dark hands
> laced the shells of turtles
> together, pebbles inside
> and they danced
> with rattles strong on their legs.
>
> There is a dry river
> between them and us.
> Its banks divide up our land.
> Its bed was the road
> I walked to return.

We are plodding creatures like the turtle
born of an old people.
We are nearly stone
turning slow as the earth.
Our mountains are underground
they are so old.

This land is the house
we have always lived in.
The women,
their bones are holding up the earth.

The red tail of a hawk
cuts open the sky
and the sun
brings their faces back
with the new grass.

Dust from yarrow
is in the air,
the yellow sun.
Insects are clicking again.

I came back to say good-bye
to the turtle
to those bones
to the shells locked together
on his back
gold atoms dancing underground.

The turtle in the stories of some Indian people is our mother. On her back the earth grew. We were born in the mud of her back. It doesn't matter which story you claim. Whether you think it was Corn or the Turtle Mother or the Spider Woman that gave you birth. Coyote came from all these. His tricking lies give us the ability to change our shapes. He is necessary to us. But the earth and where we were born is that woman's back and that woman's breast; she cuts it open to feed us and make us whole again.

Sun over the horizon, a sweating yellow force, our continuance. The uncountable distance that sweeps through our hands, the first prayers in the morning. It is this that I believe in. The galloping sun. In my whole life, a rider. It is that round earth—I call it Indian country, you call it the name you need to call it—the moon, the stars, and that sweating sun, that enables me to be the writer I am. If I choose not to climb on that galloping horse, that galloping sun, it is my own choice. But it is a choice I cannot make. I need to come home. I need my family. I need you, my brothers and sisters, my father, my uncle, and I need my aunties. The metaphor of family is simply one that works in Indian country because it brings that circle round. We all join in the dance that brings us to a place called America, where Freedom may be lots of things. And you can put the shape and face to her you wish.

The bottom line is very simple; it is morning once more. And that galloping sun races across the horizon. For me to come home to Indian country means I must climb on that sun and race across the horizon, with other people. I don't want to leave any of you behind. America has a way of leaving some of us out on the edge. I say it's time for the women of America, all of us—we're not separate, we belong with each other—to reclaim our families. To climb on that sun with me in that eternal morning. Once again, your whole life, a rider.

Ed. note: The poems quoted in this article are from *That's What She Said* edited by Rayna Green (Indiana University Press, 1984).

RACE MATTERS

60

Cornel West

Since the beginning of the nation, white Americans have suffered from a deeper inner uncertainty as to who they really are. One of the ways that has been used to simplify the answer has been to seize upon the presence of black Americans and use them as a marker, a symbol of limits, a metaphor for the "outsider." Many whites could look at the social position of blacks and feel that color formed an easy and reliable gauge for determining to what extent one was or was not American. Perhaps that is why one of the first epithets that many European immigrants learned when they got off the boat was the term "nigger"—it made them feel instantly American. But this is tricky magic. Despite his racial difference and social status, something indisputably American about Negroes not only raised doubts about the white man's value system but aroused the troubling suspicion that whatever else the true American is, he is also somehow black.

Ralph Ellison, "What America Would Be Like without Blacks" (1970)

What happened in Los Angeles in April of 1992 was neither a race riot nor a class rebellion. Rather, this monumental upheaval was a multiracial, trans-class, and largely male display of justified social rage. For all its ugly, xenopho-bic resentment, its air of adolescent carnival, and its downright barbaric behavior, it signified the sense of powerlessness in American society. Glib attempts to reduce its meaning to the pathologies of the black underclass, the criminal actions of hoodlums, or the political revolt of the oppressed urban masses miss the mark. Of those arrested, only 36 percent were black, more than a third had full-time jobs, and most claimed to shun political affiliation. What we witnessed in Los Angeles was the consequence of a lethal linkage of economic decline, cultural decay, and political lethargy in American life. Race was the visible catalyst, not the underlying cause.

The meaning of the earthshaking events in Los Angeles is difficult to grasp because most of us remain trapped in the narrow framework of the dominant liberal and conservative views of race in America, which with its worn-out vocabulary leaves us intellectually debilitated, morally disempowered, and personally depressed. The astonishing disappearance of the event from public dialogue is testimony to just how painful and distressing a serious engagement with race is. Our truncated public discussions of race suppress the best of who and what we are as a people because they fail to confront the complexity of the issue in a candid and critical manner. The predictable pitting of liberals against conservatives, Great Society Democrats against self-help Republicans, rein-forces intellectual parochialism and political paralysis.

The liberal notion that more government programs can solve racial problems is simplistic—precisely because it focuses *solely* on the economic dimension. And the conservative idea that what is needed is a change in the moral behavior of poor black urban dwellers (especially poor black men, who, they say, should stay married, support their children, and stop committing so much crime) highlights immoral actions while ignoring public responsibility for the immoral circumstances that haunt our fellow citizens.

The common denominator of these views of race is that each still sees black people as a "problem people," in the words of Dorothy I. Height, president of the National Council of Negro Women, rather than as fellow American citizens with problems. Her words echo the poignant "unasked question" of W. E. B. Du Bois, who, in *The Souls of Black Folk* (1903), wrote:

> They approach me in a half-hesitant sort of way, eye me curiously or compas-sionately, and then instead of saying directly, How does it feel to be a prob-lem? they say, I know an excellent colored man in my town. . . . Do not these Southern outrages make your blood boil? At these I smile, or am interested, or reduce the boiling to a simmer, as the occasion may require. To the real ques-tion, How does it feel to be a problem? I answer seldom a word.

Nearly a century later, we confine discussions about race in America to the "problems" black people pose for whites rather than consider what this way of viewing black people reveals about us as a nation.

This paralyzing framework encourages liberals to relieve their guilty consciences by supporting public funds directed at "the problems"; but at the same time, reluctant to exercise principled criticism of black people, liberals deny them the freedom to err. Similarly, conservatives blame the "problems" on black people themselves—and thereby render black social misery invisible or unworthy of public attention.

Hence, for liberals, black people are to be "included" and "integrated" into "our" society and culture, while for conservatives they are to be "well behaved" and "worthy of acceptance" by "our" way of life. Both fail to see that the presence and predicaments of black people are neither additions to nor defections from American life, but rather *constitutive elements of that life.*

To engage in a serious discussion of race in America, we must begin not with the problems of black people but with the flaws of American society— flaws rooted in historic inequalities and longstanding cultural stereotypes. How we set up the terms for discussing racial issues shapes our perception and response to these issues. As long as black people are viewed as a "them," the burden falls on blacks to do all the "cultural" and "moral" work necessary for healthy race relations. The implication is that only certain Americans can define what it means to be American—and the rest must simply "fit in."

The emergence of strong black-nationalist sentiments among blacks, especially among young people, is a revolt against this sense of having to "fit in." The variety of black-nationalist ideologies, from moderate views of Supreme Court Justice Clarence Thomas in his youth to those of Louis Farrakhan today, rest upon a fundamental truth: white America has been historically weak-willed in ensuring racial justice and has continued to resist fully accepting the humanity of blacks. As long as double standards and differential treatment abound—as long as the rap performer Ice-T is harshly condemned while former Los Angeles Police Chief Dayrl F. Gates's antiblack comments are received in polite silence, as long as Dr. Leonard Jeffries's anti-Semitic statements are met with vitriolic outrage while presidential candidate Patrick J. Buchanan's anti-Semitism receives a genteel response— black nationalisms will thrive.

Afrocentrism, a contemporary species of black nationalism, is a gallant yet misguided attempt to define an African identity in a white society perceived to be hostile. It is gallant because it puts black doings and sufferings, not white anxieties and fears, at the center of discussion. It is misguided because—out of fear of cultural hybridization and through silence on the issue of class, retrograde views on black women, gay men, and lesbians, and a reluctance to link race to the common good—it reinforces the narrow discussions about race.

To establish a new framework, we need to begin with a frank acknowledgement of the basic humanness and Americanness of each of us. And we must acknowledge that as a people—*E Pluribus Unum*—we are on a slippery slope toward economic strife, social turmoil, and cultural chaos. If we go down, we go down together. The Los Angeles upheaval forced us to see not only that we are not connected in ways we would like to be but also, in a more profound sense, that this failure to connect binds us even more tightly together. The paradox of race in America is that our common destiny is more pronounced and imperiled precisely when our divisions are deeper. The Civil War and its legacy speak loudly here. And our divisions are growing deeper. Today, eighty-six percent of white suburban Americans live in neighborhoods that are less than 1 percent black, meaning that the prospects for the country depend largely on how its cities fare in the hands of a suburban electorate. There is no escape from our interracial interdependence, yet enforced racial hierarchy dooms us as a nation to collective paranoia and hysteria—the unmaking of any democratic order.

The verdict in the Rodney King case which sparked the incidents in Los Angeles was perceived to be wrong by the vast majority of Americans. But whites have often failed to acknowledge the widespread mistreatment of black people, especially black men, by law enforcement agencies, which helped ignite the spark. The verdict was merely the occasion for deep-seated rage to come to the surface. This rage is fed by the "silent" depression ravaging the country—in which real weekly wages of all American workers since 1973 have declined nearly 20 percent, while at the same time wealth has been upwardly distributed.

The exodus of stable industrial jobs from urban centers to cheaper labor markets here and abroad, housing policies that have created "chocolate cities and vanilla suburbs" (to use the popular musical artist George Clinton's memorable phrase), white fear of black crime, and the urban influx of poor Spanish-speaking and Asian immigrants—all have helped erode the tax base of American cities just as the federal government has cut its supports and programs. The result is unemployment, hunger, homelessness, and sickness for millions.

And a pervasive spiritual impoverishment grows. The collapse of meaning in life—the eclipse of hope and absence of love of self and others, the breakdown of family and neighborhood bonds—leads to the social deracination and cultural denudement of urban dwellers, especially children. We have created rootless, dangling people with little link to the supportive networks—

family, friends, school—that sustain some sense of purpose in life. We have witnessed the collapse of the spiritual communities that in the past helped Americans face despair, disease, and death and that transmit through the generations dignity and decency, excellence and elegance.

The result is lives of what we might call "random nows," of fortuitous and fleeting moments preoccupied with "getting over"—with acquiring pleasure, property, and power by any means necessary. (This is not what Malcolm X meant by this famous phrase.) Post-modern culture is more and more a market culture dominated by gangster mentalities and self-destructive wantonness. This culture engulfs all of us—yet its impact on the disadvantaged is devastating, resulting in extreme violence in everyday life. Sexual violence against women and homicidal assaults by young black men on one another are only the most obvious signs of this empty quest for pleasure, property, and power.

Last, this rage is fueled by a political atmosphere in which images, not ideas, dominate, where politicians spend more time raising money than debating issues. The functions of parties have been displaced by public polls, and politicians behave less as thermostats that determine the climate of opinion than as thermometers registering the public mood. American politics has been rocked by an unleashing of greed among opportunistic public officials—who have followed the lead of their counterparts in the private sphere, where, as of 1989, 1 percent of the population owned 37 percent of the wealth and 10 percent of the population owned 86 percent of the wealth—leading to a profound cynicism and pessimism among the citizenry.

And given the way in which the Republican Party since 1968 has appealed to popular xenophobic images—playing the black, female, and homophobic cards to realign the electorate along race, sex, and sexual-orientation lines—it is no surprise that the notion that we are all part of one garment of destiny is discredited. Appeals to special interests rather than to public interests reinforce this polarization. The Los Angeles upheaval was an expression of utter fragmentation by a powerless citizenry that includes not just the poor but all of us.

What is to be done? How do we capture a new spirit and vision to meet the challenges of the post-industrial city, post-modern culture, and post-party politics?

First, we must admit that the most valuable sources for help, hope, and power consist of ourselves and our common history. As in the ages of Lincoln, Roosevelt, and King, we must look to new frameworks and languages to understand our multilayered crisis and overcome our deep malaise.

Second, we must focus our attention on the public square—the common good that undergirds our national and global destinies. The vitality of any public square ultimately depends on how much we *care* about the quality of our lives together. The neglect of our public infrastructure, for example—our water and sewage systems, bridges, tunnels, highways, subways, and streets—reflects not only our myopic economic policies, which impede productivity, but also the low priority we place on our common life.

The tragic plight of our children clearly reveals our deep disregard for public well-being. About one out of every five children in this country lives in poverty, including one out of every two black children and two out of every five Hispanic children. Most of our children—neglected by overburdened parents and bombarded by the market values of profit-hungry corporations—are ill-equipped to live lives of spiritual and cultural quality. Faced with these facts, how do we expect ever to constitute a vibrant society?

One essential step is some form of large-scale public intervention to ensure access to basic social goods—housing, food, health care, education, child care, and jobs. We must invigorate the common good with a mixture of government, business, and labor that does not follow any existing blueprint. After a period in which the private sphere has been sacralized and the public square gutted, the temptation is to make a fetish of the public square. We need to resist such dogmatic swings.

Last, the major challenge is to meet the need to generate new leadership. The paucity of courageous leaders—so apparent in the response to the events in Los Angeles—requires that we look beyond the same elites and voices that recycle the older frameworks. We need leaders—neither saints nor sparkling television personalities—who can situate themselves within a larger historical narrative of this country and our world, who can grasp the complex dynamics of our peoplehood and imagine a future grounded in the best of our past, yet who are attuned to the frightening obstacles that now perplex us. Our ideals of freedom, democracy, and equality must be invoked to invigorate all of us, especially the landless, propertyless, and luckless. Only a visionary leadership that can motivate "the better angels of our nature," as Lincoln said, and activate possibilities for a freer, more efficient, and stable America—only that leadership deserves cultivation and support.

This new leadership must be grounded in grass-roots organizing that highlights democratic accountability. Whoever *our* leaders will be as we approach the twenty-first century, their challenge will be to help Americans determine whether a genuine multiracial democracy can be created and sustained in an era of global economy and a moment of xenophobic frenzy.

Let us hope and pray that the vast intelligence, imagination, humor, and courage of Americans will not fail us. Either we learn a new language of empathy and compassion, or the fire this time will consume us all.